# 5th International Workshop on Heat/Mass Transfer Advances for Energy Conservation and Pollution Control (IWHT2019)

AA007450

Journal of Physics: Conference Series Volume 1369

Novosibirsk, Russia
13 - 16 August 2019

ISBN: 978-1-7138-1126-8
ISSN: 1742-6588

**Printed from e-media with permission by:**

Curran Associates, Inc.
57 Morehouse Lane
Red Hook, NY 12571

CURRAN ASSOCIATES INC. proceedings.com

**Some format issues inherent in the e-media version may also appear in this print version.**



Institute of Physics
Dirac House, Temple Back
Bristol BS1 6BE UK

Phone: 44 1 17 929 7481
Fax: 44 1 17 920 0979

techtracking@iop.org

**Additional copies of this publication are available from:**

Curran Associates, Inc.
57 Morehouse Lane
Red Hook, NY 12571 USA
Phone: 845-758-0400
Fax: 845-758-2633
Email: curran@proceedings.com
Web: www.proceedings.com

# TABLE OF CONTENTS

**THE 5-TH INTERNATIONAL WORKSHOP ON HEAT/MASS TRANSFER ADVANCES FOR ENERGY CONSERVATION AND POLLUTION CONTROL (IWHT2019)**

The 5-th International Workshop on Heat-Mass Transfer Advances for Energy Conservation and Pollution Control IWHT2019 is regularly held since 2011. The objective of thisWorkshop is to organize an international forum to exchange latest technological information, as well as dissemination of high-quality research and development on heat/mass transfer problems in the area pertinent to energy conservation and pollution control. This volume includes papers reflecting the results of latest research on the problems and achievements inthe heat and mass transfer inrelation to energyconservation and the development ofmodern methods of pollution control.

Today, Energy and Environment are the two major most urgent topics of the world. Both energy shortage and climate change are the most important challenges of the world, in general. Although there are various unique technologies to meet these challenges, it is widely encountered that heat and mass transfer are the key processes determining equipment performance and process efficiency. The results of various studies presented in this collection illustrate the direct relationship between experimental and computational solutions of heat and mass transfer problems, and the successful development of highly efficient environmentally friendly equipment, used in advanced energy systems for thermal energy storage, as well as in chemical and cryogenic industries.

The 5$^{th}$ IWHT Workshop was held in 2019. It covers the following topics:

- Advanced energy systems;
- CFD and numerical heat/mass transfer;
- Energy conservation and storage techniques;
- Environmental heat/mass transfer;
- Heat/mass transfer at high temperature;
- Heat/mass transfer enhancement techniques;
- Advanced materials, including micro-nanocoatings, for energy and chemical technologies;
- Heat/mass transfer in compact heat exchangers;
- Heat/mass transfer in pollutant emission control systems;
- Heat/mass transfer in renewable and clean energy systems;
- Multiphase heat/mass transfer;
- Micro/nanoscale heat/mass transfer;
- Heat/mass transfer in porous media;
- Radiative heat transfer;
- Other related areas.

Photo of the Workshop participants at the opening ceremony.

The number of IWHT-2019 participants reached 225 researchers (among them, 93 scientists were from Russia, and 132 from other countries) with a good proportion of young researchers. There were 8 plenary, 28 keynote, and 29 invited lectures, as well as 107 oral and 86 poster presentations. The plenary lectures have been presented by Professors Zhixiong (James) Guo, Sergey Alekseenko, Wojciech Lipiński, Bengt Sundén, Ruzhu Wang, Hendrik Kuhlmann, Qiuwang Wang, and Leonid Vassiliev. This issue contains 70 papers and is of particular value and interest to researchers, scientists, engineers, andpractitioners, who are working in heat/mass transfer field of renewable and sustainable energy systems,energy conservation and pollution reduction.

The Organizing Committee of the Workshop is thankful to the editorial members and reviewers for their contribution when preparing this volume. We would like to express our sincere gratitude to all authors for their research contributions, and also to organizers of the Workshop. The Organizing Committee gratefully acknowledges the Russian Foundation for Basic Research for financial support (grant # 19-08-20002).

W.Q. Tao
Chairman of International Scientific and Advisory Committee
of the IWHT-2019

A.N. Pavlenko, Q. W. Wang
Chairmen of International Organizing Committee
of the IWHT-2019

# Peer review statement

All papers published in this volume of *Journal of Physics: Conference Series* have been peer reviewed through processes administered by the proceedings Editors. Reviews were conducted by expert referees to the professional and scientific standards expected of a proceedings journal published by IOP Publishing.

5th International Workshop on Heat/Mass Transfer Advances
for Energy Conservation and Pollution Control
August 13-16, 2019, Novosibirsk, Russia

# Management of thermal and mechanic flow characteristics in the output channels of a turbocharger centrifugal compressor

**L V Plotnikov[1], B P Zhilkin[1] and Yu M Brodov[1]**

[1]Ural Federal University named after the first President of Russia B.N.Yeltsin, Mira str. 19, Ekaterinburg 620002, Russia

E-mail: plotnikovlv@mail.ru

Abstract. It is known that the thermal and mechanical characteristics of the air flow in the output channel of a turbocharger compressor largely determine the effectiveness of the gas exchange processes quality of a piston engine. The studies were carried out on an experimental installation containing a turbocharger, output channels of different configurations, a measuring base, and a data collection system. It was found that stabilization of the flow in the compressor output channel leads to a significant increase in heat transfer intensity (up to 25 %) compared to the baseline pipeline while simultaneously reducing the turbulence number by up to 30 %. A more significant increase in heat transfer intensity (up to 30 %) was observed in the output channel of the compressor with grooves compared to the base channel while simultaneously increasing the turbulence number by up to 12 %. The proposed configuration of the output channels of the compressor can be used to intensify heat transfer for the natural cooling of the air during the intake process. The configuration with a leveling grid can be used to stabilize the gas-dynamic flow parameters in order to reduce the hydraulic resistance of the intake system of a turbocharged engine.

## 1. Introduction

One of the main trends in the development of internal combustion engines (ICE) is to increase power and efficiency while maintaining mass and dimensional parameters. An effective solution to this problem is to install a turbocharger (TC) on the engine. It is known that the thermal and mechanical characteristics of the gas flows in the TC compressor output channel largely determine the quality of the gas exchange processes, the efficiency of the turbocharger and the piston engine [1-3]. Analysis of the results of modern research in the field of gas flows heat exchange in turbocharging systems indicates the relevance and great interest of specialists to this subject. Below, the works are noted, in which the influence of the configuration of inlet and outlet channels of a centrifugal compressor on its efficiency was studied based on mathematical modeling of gas-dynamics and heat transfer (in stationary and non-stationary conditions) [4, 5]. Deng et al. [6] and Leufvén et al. [7] developed mathematical models of TC (taking into account the mutual influence of the compressor and the turbine on each other) in order to simulate gas dynamics and heat transfer, as well as assess the efficiency of the turbocharger. There are also experimental works on this topic. The influence of different configurations of the compressor input devices on its performance, as well as on the noise level and safety margin was evaluated in the article [8]. Hirano et al. [9] and Gancedo et al. [10] investigated the gas dynamics and heat transfer flows in the input and output channels for different types of TC centrifugal compressors in order to increase their efficiency. Another article is devoted to the experimental study of the flow structure in the TC compressor wheel and the output channel using the PIV method [11]. Besides, there are some works on the comprehensive improvement of the configurations of intake and exhaust systems, TC operating modes and features of piston ICEs aimed at increasing their efficiency [12-14].

At the same time, there are virtually no studies related to the development of methods for controlling the thermo-mechanics of the flows in the compressor output channel in order to increase the efficiency of the intake system for ICEs with TC. The purpose of this study is to obtain new data on the effect of the configuration of the centrifugal compressor output channel on the non-stationary thermal and mechanical characteristics of gas flows.

## 2. Experimental setup and measurement base

The studies were conducted on an experimental setup (Fig. 1), which contained the following main elements: a turbocharger (TKR6), a bearing lubrication system, a TC rotor speed control system, an output channel, a measuring channel, and an automated data collection and processing system.

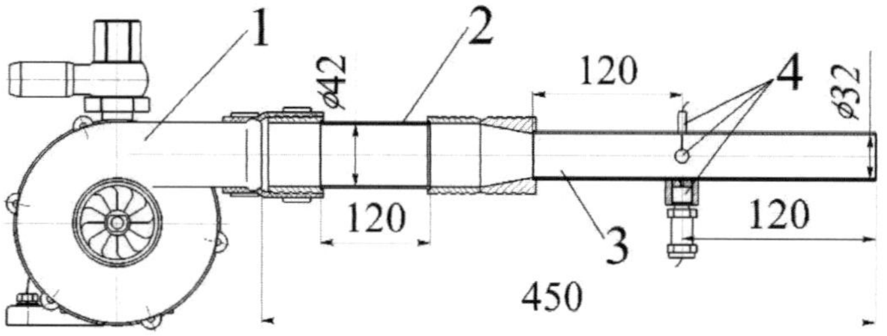


**Figure 1.** The main elements of the experimental setup: 1 – centrifugal compressor of the turbocharger; 2 – compressor output channel; 3 – measuring channel; 4 – installation sites for the thermo-anemometer sensor and pressure sensor.

The turbocharger consisted of a centrifugal compressor and a single-stage turbine. The TC rotor was rotated by supplying compressed air from an external source to the turbine blades. The variation range of the TC rotor rotational speed $n_{TC}$ ranged from 10000 rpm to 60000 rpm. The gas temperature in the studied system was about 40-45 °C.

Measurements of the instantaneous values of the velocity and pressure of the air flow, as well as the local heat transfer coefficient and the TC rotor rotation frequency, were carried out in the course of the experiments. A digital contactless tachometer was used to measure the rotational speed of the TC rotor. A pressure sensor was used to measure the instantaneous values of the static pressure $p_x$ in the flow after the compressor. The speed of the pressure sensor was less than 1 ms. A constant-temperature thermo-anemometer was used to determine the instantaneous values of the air flow rate $w_x$ and the local heat transfer coefficient $\alpha_x$. In our case, the sensitive element of the thermo-anemometer sensors was a nichrome filament with a diameter of 5 μm and a length of 5 mm. In this study, the determination of the local heat transfer coefficient in the gas flow is based on the idea of a heat transfer hydrodynamic analogy (Reynolds analogy). The analogy is based on the assumption of the unity of the processes of momentum and heat transfer in a turbulent flow and establishes a quantitative relationship between heat transfer and hydraulic resistance (Stanton criterion). The time constant of the thermo-anemometer was about 2 ms. The maximum systematic error in measuring the velocity wx was 5.4%, and the local heat transfer coefficient αx was 10.0%. A detailed description of the method for determining the local heat transfer coefficient and the calculation of experimental errors for this study are given in [15].

Straight, smooth pipe with a length of 120 mm and an internal diameter of 42 mm was used as the basic configuration of the compressor output channel. The leveling grid, according to the honeycomb principle, was installed in the pipe under consideration in order to stabilize the flow (Fig. 2, *a*). For another channel configuration, grooves were made on the inner surface of the pipe in order to intensify heat exchange (Fig. 2, *b*). At the same time, the length and inner diameter of the compressor output channel remained unchanged.

The efficiency of the studied configurations was estimated by a number of parameters: flow turbulence number, heat transfer intensity, average pressure, flow characteristics.

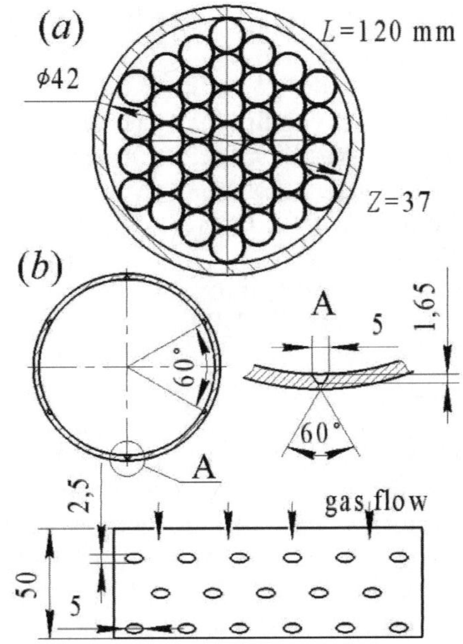


**Figure 2.** Configurations of the turbocharger compressor output channels: a – channel with leveling grid (honeycomb); b – channel with grooves.

## 3. Analysis of experimental results

The instantaneous values of flow rates and local heat transfer coefficients in time for three different configurations of the compressor output channel at the rotor TC rotation frequency equal to 10000 rpm are shown in Fig. 3

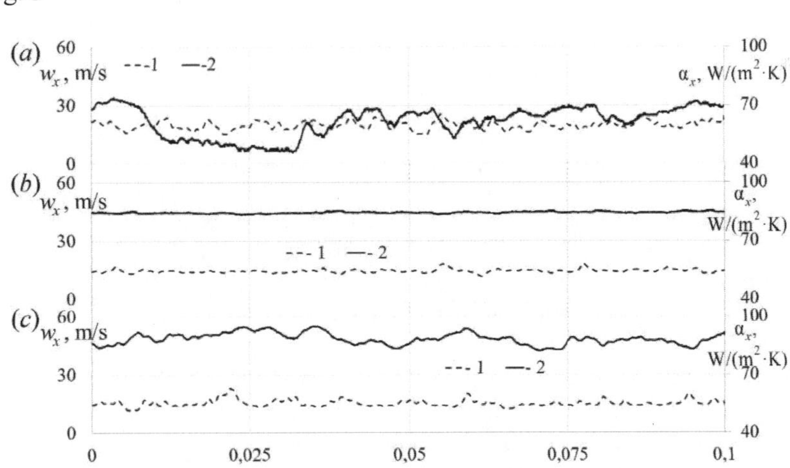


**Figure 3.** Dependences of air flow rate $w_x$ (1) and local heat transfer coefficient $\alpha_x$ (2) on time in the compressor output channel of different configurations ($n_{TC}$ = 10 000 rpm): a – base channel; b – channel with leveling grid; c – channel with grooves.

From Fig. 3, *a* it can be seen that significant amplitudes of air flow velocity pulsations are observed in the basic output channel of the compressor (standard deviation is 2.03). These pulsations are the result of the operation of the centrifugal compressor blades [16].

As expected, the installation of the leveling grid in the compressor output channel leads to a significant smoothing of the amplitudes of the velocity pulsations and the local heat transfer coefficient (Fig. 3, *b*). The standard deviation from the average velocity is 0.858. It can also be noted that there is a noticeable increase in the local heat transfer coefficient in the output channel with a leveling grid by about 25 %. The presence of grooves in the compressor output channel causes a noticeable increase in the local heat transfer coefficient by almost 30 % (compared to the base output channel); at the same time a slight smoothing of the amplitudes of the flow rate pulsations is observed (Fig. 3, *c*). The standard deviation from the average speed is 1.77. The results obtained in this study are in good agreement with those of other authors [17, 18]. It is known that various grooves and holes on the surface of pipelines lead to a significant intensification of heat exchange with a slight increase in the hydraulic resistance of the system.

The influence of the configuration of the compressor output channel on the turbulence number Tu can be seen in Fig. 4. The figure shows that the greatest differences in Tu are observed low and medium TC rotor speeds (from 10 000 to 40 000 rpm). For example, the turbulence number is reduced by almost 30 % when installing the leveling grid in the output channel (at $n_{TC}$ = 10 000 rpm) compared to the base channel, and this difference does not exceed 15% at $n_{TC}$ = 40 000 rpm. According to the authors, flow stabilization is associated with the equalization of the velocity field in honeycomb, i.e. the flow is stabilized after the impact of the compressor blades on it.

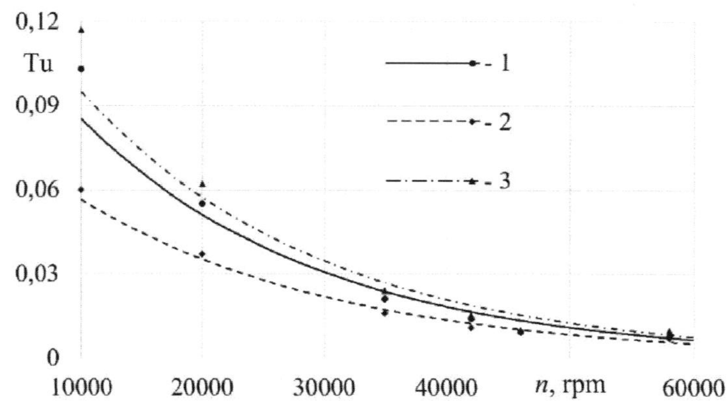


**Figure 4.** Dependences of the turbulence number Tu on the turbocharger rotor speed $n_{TC}$ in the compressor output channel of different configurations: 1 – base channel; 2 – channel with leveling grid (honeycomb); 3 – channel with grooves.

The presence of grooves in the compressor output channel leads to an increase in the turbulence number by 9-12% compared to the base channel. According to the authors, this is due to the formation of vortices behind the grooves and the general turbulization of the flow.

It should be noted that the air flow rate through the output channel of the compressor remained virtually unchanged (within the experimental error – 5.5%) with all channel configurations at a certain fixed rotational speed of the rotor TC.

The effect of different gas-dynamic conditions in the output channels of different configurations on the local heat transfer coefficient is shown in Fig. 5.

The figure shows that the installation of a leveling grid and the presence of grooves in the compressor output channel lead to an intensification of heat exchange in comparison with the base channel. Intensification is observed at all TC rotor rotation frequencies. At the same time, the installation of a leveling grid in the output channel causes an increase in the local heat transfer coefficient by 13-25 %, and the presence of grooves – by 15-30 %. The greatest differences (20-30%) are observed at low rotor TC rotation frequencies $n_{TC}$ (no more than 20000 rpm). It should be noted that the physical mechanism of heat transfer intensification for different configurations of the compressor output channel is different. In the case of the output channel with grooves, it consists in the formation of very significant secondary flow rates generated by the grooves [17, 18]. In turn, the leveling grid in the compressor output channel stabilizes the flow and contributes to the formation of a stable boundary layer with a corresponding intensification of heat transfer between the flow core and the channel walls.

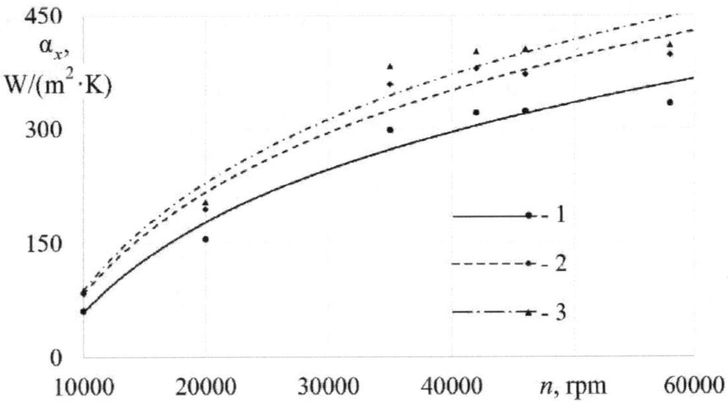


**Figure 5.** Dependences of the local heat transfer coefficient $\alpha_x$ on the turbocharger rotor speed $n_{TC}$ in the compressor output channel of different configurations: 1 – base channel; 2 – channel with leveling grid (honeycomb); 3 – channel with grooves.

The obtained results can have a positive impact on the operation of turbocharged internal combustion engines since the intensification of heat exchange will provide natural cooling of the air in the intake system of the engine without increasing the hydraulic resistance.

## 4. Conclusion

Based on the study, the following main conclusions can be drawn.

1. It has been established that installation of a leveling grid in the output channel of the TK compressor results in:

- a decrease in the amplitude of the velocity pulsations by 2-3 times and a decrease in the turbulence number up to 30% as compared to the base channel;
- intensification of heat transfer in the channel by 13-25% compared to the base channel.

2. It is shown that the presence of grooves in the output channel of the TC compressor leads to:

- an increase in the turbulence number by 9-12 % compared to the base channel;
- growth of the local heat transfer coefficient up to 30 % compared to the base channel.

3. The obtained data expands the knowledge base on the influence of the configuration of the output channel of a centrifugal compressor on the thermal and mechanical characteristics of gas flows. These data can be used to design promising intake systems for turbocharged piston engines.

**Acknowledgments**
The work has been supported by the Russian Science Foundation (grant No. 18-79-10003).

**References**

[1] Heywood J B 1988 *Internal combustion engine fundamentals* (New York: McGraw-Hill) p 458

[2] Romagnoli A, Manivannan A, Rajoo S, Chiong M S, Feneley A, Pesiridis A, Martinez-Botas R F 2017 *Renewable Sustainable En. Rev.* **79** 1442–1460
[3] Plotnikov L V 2017 *IOP Conf. Series: J. Phys* **899** 042008
[4] Zhang M, Zheng X 2018 *Appl. Therm. Eng.* **131** 933–946
[5] De Bellis V, Bontempo R 2018 *Energy* **142** 507–517
[6] Deng Q, Burke R D, Zhang Q, Pohorelsky L 2017 *J. Eng. Gas Turb. Power* **139(6)** 062603
[7] Leufvén O, Eriksson L 2016 *Int. J. Engine Res.* **17(2)** 153–168
[8] Galindo J, Tiseira A, Navarro R, Tarí D, Meano C M 2017 *Appl. Therm. Eng.* **110** 875–882
[9] Hirano T, Ogawa T, Yasui R, Tsujita H 2017 *J. Therm. Science* **26(1)** 11–17
[10] Gancedo M, Gutmark E, Guillou E 2016 *Exp. Fluids* **57(2)** 16
[11] Torregrosa A J, Broatch A, Pastor J V, García-Tíscar J, Sharma R K, Cheung R 2018 *Exp. Therm. Fluid Science* **99** 420–432
[12] Hou H, Wang L, Wang R, Yang Y 2017 *J. Therm. Science* **26(2)** 97–106
[13] Wang T J 2018 *J. Mech. Science Techn.* **32(7)** 3465–3472
[14] Bozza F, De Bellis V, Teodosio L 2017 *Int. J. Engine Res.* **18(8)** 810–823
[15] Plotnikov L V, Zhilkin B P 2017 *Int. J. Heat Mass Transfer* **115** 1182–1191
[16] Plotnikov L, Grigor'ev N, Kochev N 2019 *EPJ Web Conf.* **196** 00007
[17] Isaev S A, Leontiev A I, Kornev N V, Hassel E, Chudnovskii Y P 2015 *High Temp.* **53(3)** 375–386
[18] Isaev S A, Schelchkov A V, Leontiev A I, Baranov P A, Gulcova M E 2016 *Int. J. Heat Mass Transfer* **94** 426–448

5th International Workshop on Heat/Mass Transfer Advances
for Energy Conservation and Pollution Control
August 13-16, 2019, Novosibirsk, Russia

# Spatial Temperature Distribution of an Open Square Capillary Tube Wall during Passage of a Taylor Liquid Plug

**E Ya Gatapova[1], S Khandekar[2]**

[1]Kutateladze Institute of Thermophysics SB RAS, Novosibirsk, Russia

[2]Department of Mechanical Engineering, Indian Institute of Technology Kanpur, Kanpur, India

gatapova@itp.nsc.ru , samkhan@iitk.ac.in

Abstract. Taylor plug/bubble flow is a predominant two-phase flow pattern occurring in several microfluidic devices such as in microchannel heat sinks, fuel cells, pulsating heat pipes, lab-on-chips, microreactors, etc. Such flows are characterized by an alternating sequence of liquid plug and gas bubbles, whose ensuing local transport characteristics (heat and mass transfer) are complex and still not fully understood. Presence of gas-liquid interfaces, thin-film dynamics, moving contact lines, wettability, contact angle hysteresis, are some of the physical phenomena involved in Taylor bubble flow systems that manifests into its complex transport characteristics. Understanding the local interfacial transport mechanism is of vital importance to develop a comprehensive model for micro-scale devices utilizing Taylor bubble flow. In this background, we probe locally into the local thermo-hydrodynamics of 'unit-cell' of Taylor plug flow, which is essentially an isolated liquid plug moving inside a mini square capillary copper tube (side = 2 mm), surrounded by gas from both its sides, and try discerning local evaporation dynamics, with the help of IR thermography. This study focuses on the interplay of conjugate conduction, local thin film hydrodynamics formed behind the moving liquid plug, and its subsequent evaporation. We obtain local temperature distribution in the meniscus region, as the plug moves through the capillary tube. As the cold plug passes through the tube, local temperatures drop drastically, the transient reaches the heater-tube interface a bit delayed and there the drop in temperature is modest. The study clearly reveals that there is a strong thermo-hydrodynamic coupling between the tube wall and the moving Taylor liquid plug.

## 1. Introduction

Miniaturization of two-phase flow and passive heat transfer systems, such as loop heat pipes or pulsating heat pipes, leads to complex transport mechanisms in these surface tension dominated systems [1, 2]. Figure 1 shows a photograph of a typical Taylor flow pattern inside a 1 mm square glass capillary tube, which is one of the prominent flow patterns in such micro/mini two-phase systems. Inset in the figure highlights a 'unit-cell' of such a Taylor flow, which is essentially an isolated liquid plug surrounded by gas/vapor on either of its sides; a thin liquid film is always present around the gas/vapor phase. Fundamental understanding of such flows is difficult due to the presence of deformable two-phase interfaces and is the motivation of this paper, which focusses on the thermal transport behavior of the 'unit-cell' of the Taylor plug flows [3, 4].

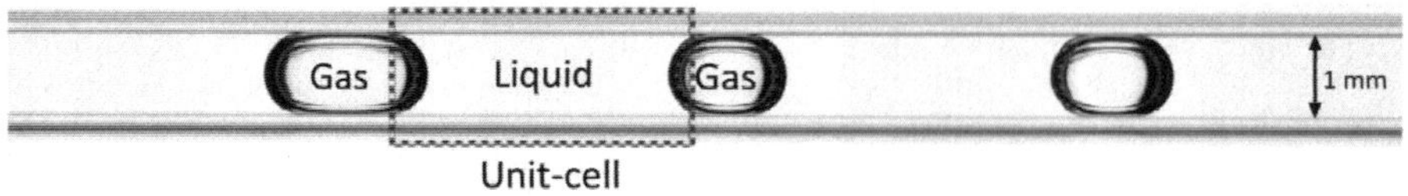

Fig. 1. Typical Taylor plug flow pattern, prominent in micro-channel two-phase flow.

Evaporation from the thin film region formed around the gas/vapor region can lead to significant heat transfer enhancement [5]. Such a fundamental understanding has implications in improving the design and efficiency of many engineering devices, such as pulsating/loop heat pipes [3, 4]. In the present experiments, one small Taylor liquid plug (of ethanol) is made to move inside a copper tube (side = 2 mm), which is initially maintained at a fixed temperature. The temperature transience of the setup is recorded by IR thermography, as the cold liquid ethanol plug passes through it and gets evaporated.

## 2. Experimental setup and protocol

The experimental setup is shown in Figure 2 (a). It consists of a high-speed video camera for flow visualization (in case of usage of a glass tube - although here, we present results with only a copper tube), which is synchronized with an IR-camera Titanium - 570M (115 Hz for 640×512 pxl). The main test cell is a copper tube 50 mm length, with a square cross-section and inner size of 2 mm x 2 mm. The tube is centrally heated by two 20 mm long copper heaters, both from the upper and the bottom sides, respectively. The working fluid is liquid ethanol (99%), a small plug of which (length ~ 8 mm to 15 mm) is made to pass through the pre-heated copper tube, by a syringe pump (Cole-Parmer). Copper heaters (Peltier elements) and the tube were coated with black graphite (emissivity ε ~ 0.96, calibrated) for facilitating IR thermography. The transient temperature field was measured and recorded as the liquid plug of ethanol passed through the preheated copper tube. The heater temperature range was kept from 26°C to 81°C, by adjusting the power supply to the Peltier elements.

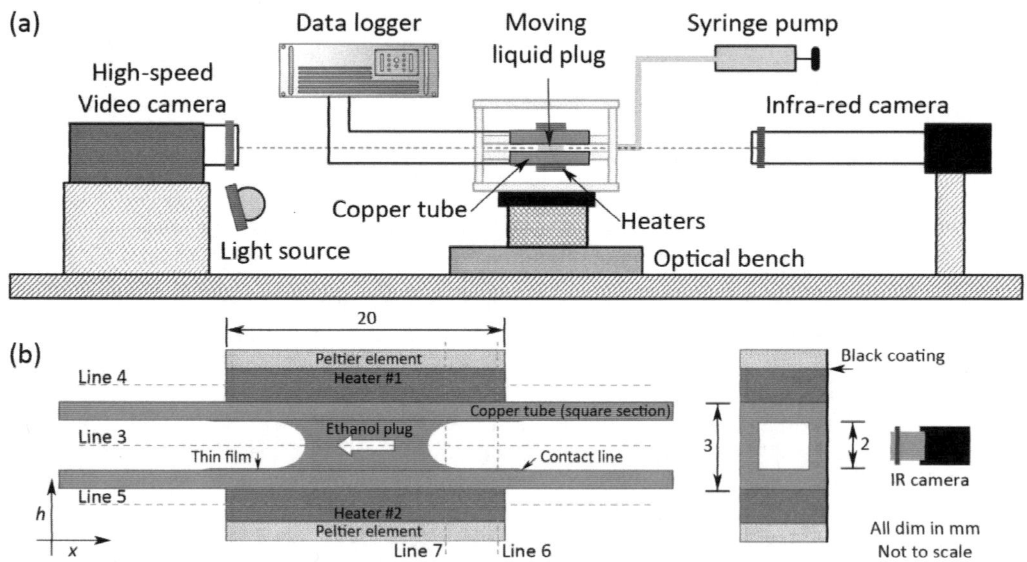

Fig. 2. (a) Details of the experimental setup to capture the temperature transience of a heated square copper capillary tube when a small liquid plug is made to flow in it. (b) Schematic showing the closer details of the tube and heater arrangement with identification of the lines along which IR temperature profiles are reported, i.e., Line 3 – temperature along the center of the tube, Line 4 – temperature along the upper heater, Line 5 – temperature along the bottom heater, Line 6 – temperature across the tube near the right entrance, Line 7 – temperature across the tube near its center.

Journal of Physics: Conference Series **1369** (2019) 012003 doi:10.1088/1742-6596/1369/1/012003

## 3. Results and discussion

Figures 3 and 4 show the data on temperature fields of the tube wall and the heater surface, which is in direct contact with the tube. The thermograph data is given here is for different times, where $t = 0$ sec corresponds to the field, when the tube is only wetted by ethanol, but still empty, without the plug mass inside it. The heater temperature is 75.5 °C at this stage (refer Fig. 4 (b) also later). As time progresses, the ethanol plug enters the heated copper tube and approaches the heater zone. We specify here the liner temperature profiles (lines) retrieved at different locations on the setup, as indicated earlier in Figure 2(b). Figure 3(a) depicts the corresponding recorded IR thermographs at three different times, as the plug passes through the heated copper tube, along with the calibrated temperature scale alongside it. Figure 3 (b) shows the temperature profiles across the tube wall, along Line 6, which is near the right entrance, as was located in Figure 2 (b). One can see a significant decrease of the temperature across the middle part of the tube wall, of the order of $\Delta T = 10$°C, when the liquid plug is injected in it and is 'seen' to pass through the tube at $t = 7.375$ sec. At the same time, the longitudinal temperature profile of the tube wall, near its contact with the heater surface is decreasing, by ~$\Delta T = 3$°C. Subsequently, the liquid plug gets evaporated during its leftward traverse, and the temperature levels of the tube wall get raised to the original higher levels before the plug entered inside it.

Longitudinal temperature profiles in the central section of the tube wall (Line 3) also highlights the significant decrease, by ~ $\Delta T = 5$°C in its middle part, and by ~ $\Delta T = 10$°C in the locations where the liquid ethanol plug connects the two opposite walls of the tube, i.e. the spreading thin film region of the meniscus (Figure 4 (a)). The strong decrease in temperature of the tube wall when it comes in contact with liquid plug is followed by a gradual increase back to the original levels, as the plug moves forward and eventually gets fully evaporated. The heater wall temperature, which is specified approximately 1 mm lower from the boundary of the tube (Line 5), decreases much more gradually, as is evident in Figure 4 (b). It was clearly discernible that the local heat transfer is a strong function of plug geometry, its speed, evaporation rate and conjugate interaction with the tube wall material.

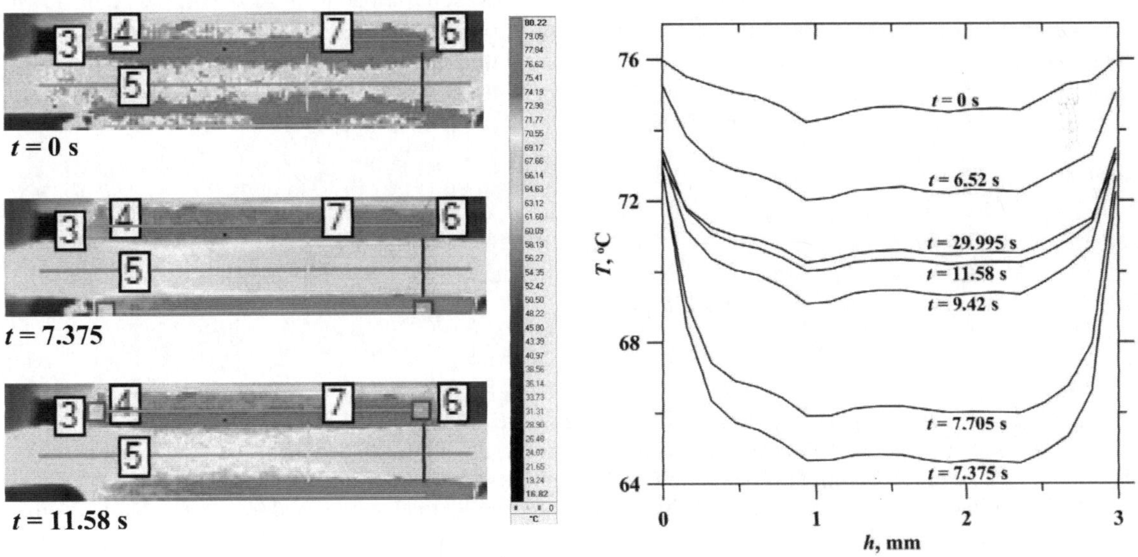


(a) (b)

Fig. 3. (a) IR thermographs of the setup at different times with the calibrated temperature scale, and, (b) temporal evolution of the temperature profile for Line 6.

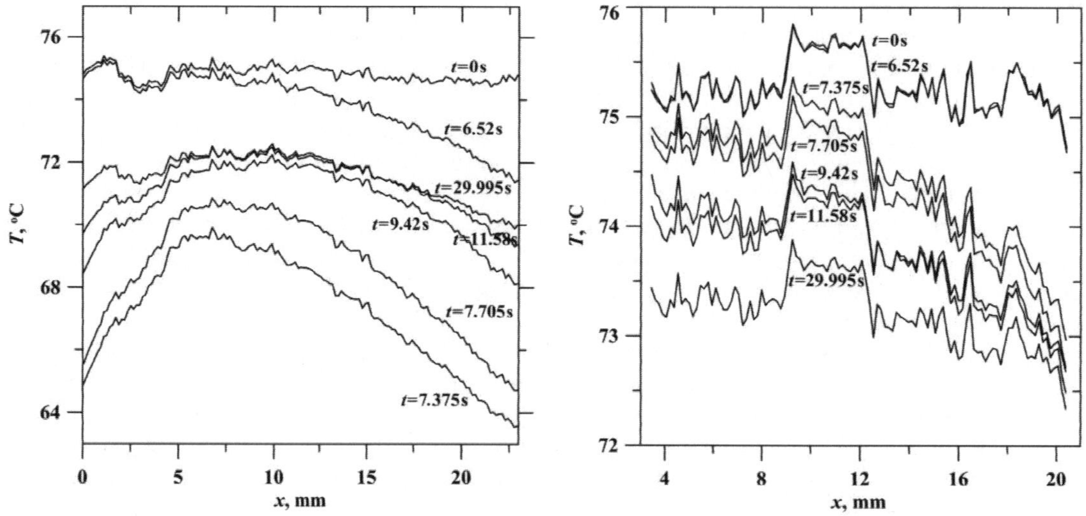


**(a)** **(b)**

Fig. 4. Evolution of the temperature profiles for (a) Line 3 (along the central part of the tube) and, (b) Line 5 (along the bottom heater) – refer Figure 2 (b) for reference line locations.

In summary, an experimental study of the interaction of liquid plug moving in the copper rectangular tube has been carried out using an IR-scanner. The local dynamics of the temperature field during the formation and evaporation of a liquid plug has been investigated. There is a significant decrease in the temperature of the transverse temperature in the region which is fully occupied by the liquid plug, as well as the longitudinal temperature profile along the middle of the channel wall. The boundary between the copper tube and the adjoining heater also demonstrates a significant temperature decreases (~10°C) after the liquid ethanol plug at room temperature enters the heated tube, which is initially maintained at ~75.5°C tube. Results clearly indicate the strong local enhancement in heat transfer due to evaporation of liquid plug, especially in the thin film region. It is clear that the system manifests a strong transient conjugate heat transfer coupling between the cold liquid plug and the hot tube wall, which should form the basis of estimating local heat transfer coefficient, in systems involving Taylor flows, such as pulsating heat pipes.

## Acknowledgements

This work was financially supported by the Ministry of Education and Science of the Russian Federation (Agreement 14.613.21.0067, Project Identifier RFMEFI61317X0067) and Department of Science and Technology, Government of India (Project Identifier Code HPCSLED-2016), under the BRICS Multi-regional Cooperation Framework.

## References

[1] Srinivasan V., Marty-Jourjon V., Khandekar S., L.efèvre F., Bonjour J., Evaporation of an isolated liquid plug moving inside a capillary tube, Int. J. Heat Mass Transf., Vol. 89, pp. 176-185, 2015.

[2] Chauris N., Ayel V., Bertin Y., Romestant C., Evaporation of a liquid film deposited on a capillary heated tube: Experimental analysis by infrared thermography of its thermal footprint, Int. J. Heat Mass Transf., Vol. 86, pp. 492-507, 2015.

[3] Rao M., Lefèvre F., Khandekar S., Bonjour J., Heat and mass transfer mechanisms of a self-sustained thermally driven oscillating liquid–vapour meniscus, Int. J. Heat Mass Transf., Vol. 86, pp. 519-530, 2015.

[4] Jasvanth V.S., Ambirajan A., Arakeri J.H., Experimental study on evaporation of pentane from a heated capillary slot, Int. J. Heat Mass Transf., Vol. 95, pp. 466-476, 2016.

[5] Gatapova E. Ya., Kabov O. A., Ajaev V. S., Evaporation and interface dynamics in microregion on heated substrate of non-uniform wettability, Int. J. Heat Mass Transf., Vol. 142, 118355, 2019.

# ONE-STEP SYNTHESIS OF BINARY MN–CE OXIDES BY A REDOX-PRECIPITATION AS HIGHLY EFFICIENT THEMALCATALYSTS FOR THE RESIDUAL METHANE OXIDIZING

**Liu Zhong, Qingyan Fang*, Xin Li, Cheng Zhang, and Gang Chen**

Coal Combustion, School of Energy and Power Engineering, Huazhong University of Science and Technology, Wuhan 430074, China

E-mail: qyfang@hust.edu.cn

Abstract. A series of Mn-Ce catalysts with different ratio of Mn-to-Ce were prepared by a redox-precipitation method, and their catalytic performance in lean methane combustion was investigated. A plethora of structural (XRD), textural (BET, FSEM), and spectroscopic (XPS) methods have been applied for characterization. The results show that the catalytic performance of the investigated samples is strongly related to the ratios of Mn-to-Ce. Mn-Ce catalyst with an atomic Mn/Ce ratio of 9/1 provided with a larger specific surface area, pore volumes, abundant manganese species of higher valence ($Mn^{4+}$), Cerium species of lower valence($Ce^{3+}$), and lattice oxygen.

## 1. Introduction

Methane was classified as one of pollution, due to the warming potential of methane is over 20 times than that of $CO_2$ (100 years)[1]. Mine Gas, about 8% of total methane emissions, is one of the most important methane emission [2]. Catalytic combustion is a promising way to remove trace amounts of methane (0.1-1%) in VAM, due to lower cost and safety, compared with other methods [2]. And the key to this technology is the catalyst`s development of ignition at low temperature and stable at high temperature [3].

Noble and non-noble metal catalysts are the main kinds of catalysts. Noble metal-based catalysts exhibit higher efficiency in low-temperature windows, however, the higher price has turned scientist`s attention toward inexpensive and thermally stable materials, such as Metal Oxide Catalyst. Metal Oxide catalysts could be more suitable materials, comparing with the noble-metal-containing catalysts.

Zhang et al.[4] reported that Ni-Mn-O solid solution is beneficial for lean methane catalytic combustion at low temperature, due to abundant highly dispersed higher oxidation state of manganese ($Mn^{4+}$). Hu et al.[5] reported that the presence of ceria has a strong influence on the crystal structure, surface phase composition and reducibility of catalyst, which is beneficial for catalytic performance.

The above results present excellent redox ability of Mn and Ce metal. The Mn-Ce catalyst has been investigated extensively in terms of $NH_3$-SCR[6], CWAO[7], etc, especially in low temperature. However, there are only a few reports about lean methane catalytic combustion by Mn-Ce catalyst [1,8–11], due to the particle size, the surface texture of catalyst has the effect on the performance of the catalyst.

Many methods for preparing Mn-Ce catalysts, such as co-precipitation, sol-gel, and hydrothermal methods, have been reported [12]. Co-precipitation has been applied extensively in the industry

Journal of Physics: Conference Series **1369** (2019) 012004 doi:10.1088/1742-6596/1369/1/012004

compared with other methods, due to features of the simple process and low cost. However, uneven pH value results in agglomeration. The redox-precipitation is an improved method of co-precipitation, which could reduce the influence of swing pH value, improve the distribution of Mn-Ce oxides and produce $\alpha$-$MnO_2$[6]. The performance of $\alpha$-$MnO_2$ is better than that of other manganese oxides[13].

In this paper, a series of Mn-Ce catalysts with different ratio of Mn-to-Ce were prepared by a redox-precipitation, and their catalytic performance in lean methane combustion was investigated. The physicochemical properties of the catalyst were characterized, furtherly. The ratio optimization of the catalyst was performed, which is an important parameter of the methane combustion process.

## 2. Experimental procedure

### *2.1. Catalyst Synthesis*

A series of Mn-Ce catalysts with different ratio of Mn-to-Ce were prepared by a redox-precipitation, following the similar procedures as described elsewhere[14,15]. Briefly, consisting in the titration at constant pH value (4.5±0.3, by addition of 0.4M KOH) of the $Mn(NO_3)_2$ (AR 50wt.% in H2O) nitrates precursor with a solution of $(NH_4)_2Ce(NO_3)_6$ (AR 99.0%) and $KMnO_4$ (GR). After titration, the precipitation was washed by Ultra-pure water and ethanol three times, after that, the solid was dried at 100℃ (12 h) and further calcined (6 h) in the air at 500℃. According to the Mn molar content in the catalyst (30%, 50%, 70%, 90%, 100%), the catalyst prepared in this way is denoted as Mnx (x=30, 50, 70, 90, 100).

### *2.2. Characterization*

The specific surface area and pore volume of the catalysts were calculated, using the Brunauer-Emmett-Teller (BET) and Barrett-Joyner-Halend (BJH) equation based on data obtained by an $N_2$ adsorption-desorption experiment on a physical adsorption station (ASAP2020, Micromeritics). The XRD pattern was recorded on X'pert$^3$ Powder X-ray diffractometer (PANalytical, Inc.) operated at 60 kV and 60 mA with a scan angle of 5°-85°. A Nova NanoSEM 450 microscope was also used for image observation of the samples at the accelerating voltage of 10 kV. The XPS spectra were conducted with XPS equipment (AXIS-ULTRADLD-600W Kratos). Transmittance spectra were recorded using a UV-Vis spectrophotometer (Shimadzu SolidSpec-3700) at room temperature.

### *2.3. Catalytic performance measurements*

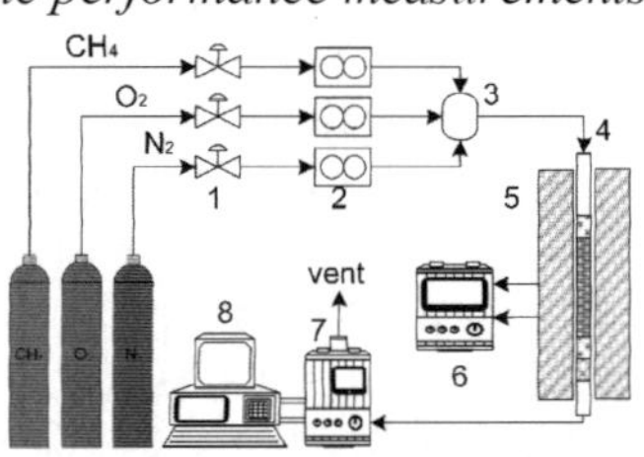


**Figure 1.** experimental apparatus for catalyst performance test: 1.Pressure reducing value; 2. Mass flowmeter; 3. Mix tanker; 4. Reaction tube; 5. Tube furnace; 6. Temperature controller; 7. Gas chromatography; 8. Microcomputer workstation

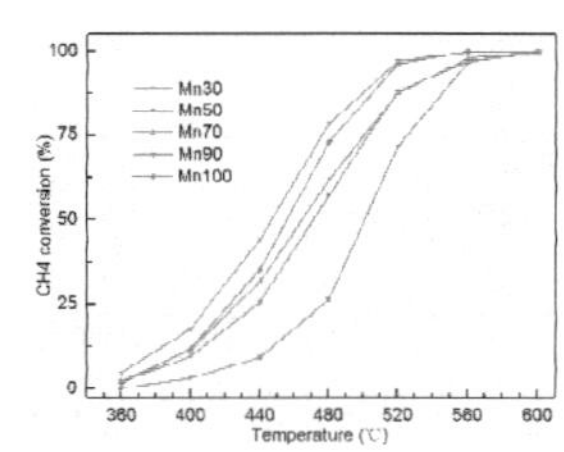


**Figure 2.** $CH_4$ conversion of a series of catalysts. Reaction conditions: CH4 1vol%, O2 10vol%/N2 balance, total flow rate 100ml/min, WSHV=30000ml/(h.g).

The performance of a series of catalysts in the oxidation of methane was measured in a fixed-bed reactor(id=6mm), as shown in Figure 1. The reaction gases typically consisted of 1% $CH_4$, 10% $O_2$ and the balance $N_2$ to simulate the flue gas. The concentrations of $CH_4$, CO, and $CO_2$ at reactor inlet and outlet were measured by GC9700II (FULI Instruments) with two detectors of FID and TCD ( no CO and $H_2$ were detected during the experiments, and methane was totally oxidized to $CO_2$). The reaction gas concentrations were accessed to the reactor after 30 mins later of the temperature of the tube furnace becoming the elevated value, In order to make sure the accuracy of the measurement.

Conversions of CH4 was calculated as follows:

$$X_{CH_4} = \frac{C_{in} - C_{out}}{C_{in}} \times 100\%$$

Where $C_{in}$ and $C_{out}$ are the $CH_4$ concentration in the feed gas and products (vol%), respectively;

## 3. Results and discussion

### *3.1. Catalytic performance*

The performance of Mnx is presented in Fig. 2. The results indicated the performance of catalysts were excellence over 560℃, and all conversion is over than 90%. Mostly, the temperature to reach a methane conversion of 50% (T50) is regarded as a criterion of the catalytic performance for methane combustion, especially for $MnO_x$ [1]. The Mn90 catalyst showed optimal performance, and 50% of methane conversion occurs at ~446℃. The performance of catalysts is improved with the content of Mn increasing from 30 to 90%, while the performance decreases with the Mn content increasing furtherly.

Compared with Mn90 catalyst, the Mn100 catalyst exhibited the same conversion efficiency at ~456 ℃, 10℃ higher than the Mn90 catalyst, indicating that Mn90 had higher catalytic activity.

### *3.2. BET results*

The specific surface areas and pore volumes of Mn-Ce catalysts are listed in Table 1. The specific surface of catalysts tended to increase first and then decrease, indicating giving an optimum parameter of specific surface areas and pore volumes with an optimal Mn content. And the specific surface areas and pore volumes of Mn90 were larger than those of Mn100, which is consistent with the literature [15]. The specific surface areas and pore volumes of Mn50, Mn70, and Mn90 are larger than those of Mn30 and Mn100 catalysts. There are more active sites over the catalyst, larger specific surface areas, and pore volumes, beneficial for the performance of catalyst improving.

**Table 1.** BET results of Mnx catalysts

| Sample | Bet Surface Area($m^2/g$) | Pore Volume($*10^{-2}cm^3/g$) | Average Pore(nm) |
|---|---|---|---|
| Mn30 | 100 | 39.16 | 10.7177 |
| Mn50 | 118 | 78.59 | 21.6493 |
| Mn70 | 117 | 76.29 | 23.6467 |
| Mn90 | 111 | 62.38 | 21.8270 |
| Mn100 | 39 | 25.30 | 25.7554 |

### *3.3. XRD results*

The XRD patterns of Mnx catalysts are shown in Fig. 3. it is clearly observed that the crystal peaks of $CeO_2$ (JCPDS 34-0394), $Mn_5O_8$ (JCPDS 39-1218) and $MnO_2$ (JCPDS 24-0735) when the ratio of Mn-to-Ce is lower than 7/3. There are lower intensity and broader half-peak width of peaks, with the ratio increasing from 3/7 to 7/3, which indicates stronger interaction of Mn and Ce [8]. However, the new crystal peaks of $KMn_8O_{16}$ (JCPDS 29-1020) showed the characteristic peaks (12.75 and 18.06, especially) and the peaks of $MnO_2$ disappear. There is no report about $KMn_8O_{16}$ in Mn-Ce catalyst for lean methane combustion[1,8–11], and just one report about the Mn-Ce catalyst made by a redox-precipitation for benzyl alcohol oxidation[16]. According to the literature, the performance of methane catalytic combustion decrease in the order of $KMn_8O_{16} \approx MnO_2 > Mn_5O_8$, which is consistent with our experience result.

Compared with Mn100, the intensity of Mn90 catalyst`s peaks is lower and broader, which indicated MnOx noncrystalline or Mn-Ce solid-solution catalyst. This means that the catalyst with a given number Cerium can improve the electron transfer between different Mn ion, and enhance the performance of the catalyst[6].

### *3.4. SEM and EDS results*

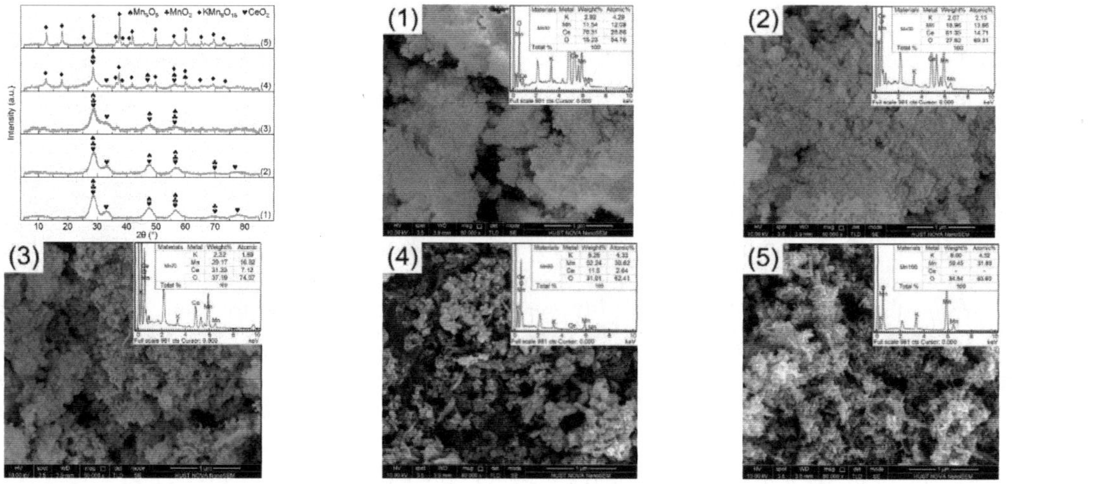


**Figure 3.** XRD pattern, SEM images and EDS result of Mnx catalysts: (1) Mn30; (2) Mn50; (3) Mn70; (4) Mn90; (5) Mn100

SEM images of Mnx catalysts are shown in figure 3. The texture and morphology of Mnx catalysts are markedly different. There are serious sintering and agglomeration when the ratio of Mn-to-Ce are 3/7 and 5/5. The size of catalyst particles would be smaller with Mn content increasing. There are many nanorods and nanosphere, when the ratio is 9/1 or 10/0, especially. As is known, the performance of the catalyst is better with smaller average particle size, consistent with the literature [6].

To evaluate the effect of different ratio on the dispersion of Mn and Ce on the catalyst surface, the surface Mn/Ce atomic ratios were calculated based on the EDS data, showed in figure 3. The ratio of Mn-to-Ce on the catalyst surface is lower than the theoretical value when the ratio of Mn-to-Ce is no more than 7/3. The Mn element on the surface is lower than the theoretical value, due to the interaction of different elements. The all ionic radii of $Mn^{2+}$(0.083nm), $Mn^{3+}$(0.066nm) and $Mn^{4+}$(0.056nm) are smaller than that of $Ce^{3+}$ (0.103nm) and $Ce^{4+}$(0.099nm), due to partial Mn cations doping into Ce lattice, forming Mn-Ce solid solution[1]. However, when the ratio of Mn-to-Ce is 9/1, the ratio of Mn-to-Ce is higher than the theoretical value, which may be due to the $KMn_8O_{16}$ crystal. The incorporation of Ce cations dopes into the $KMn_8O_{16}$ lattice in Mn90 sample, due to that the $MnO_6$ octahedra has a 2 × 2 tunnel structure with the size of 0.46 × 0.46nm[16] and the tunnel radius of $MnO_6$ octahedra is bigger than ionic radii of $Ce^{3+}$ and $Ce^{4+}$.

### *3.5. XPS Analysis*

**Table 2.** surface atomic composition of Mnx catalysts from XPS spectra

| Sample | Mn2p3/2 position(eV) | | $\frac{Mn^{4+}}{Mn^{3+}}$ | $\frac{Ce^{3+}}{Ce^{3+}+Ce^{4+}}$ | O1s position(eV) | | | $\frac{O_\alpha}{O_\alpha+O_\beta}$ |
|---|---|---|---|---|---|---|---|---|
| | $Mn^{4+}$ | $Mn^{3+}$ | | | Oα | Oβ | Oω | |
| Mn30 | 642.9 | 641.9 | 0.72 | 0.11 | 529.5 | 531.4 | 533.0 | 0.64 |
| Mn50 | 642.5 | 641.6 | 0.94 | 0.17 | 529.4 | 531.4 | 533.0 | 0.70 |
| Mn70 | 642.4 | 641.6 | 1.95 | 0.19 | 529.5 | 531.4 | 533.0 | 0.83 |
| Mn90 | 642.6 | 641.6 | 3.56 | 0.28 | 529.7 | 531.6 | 533.0 | 0.85 |
| Mn100 | 642.6 | 641.6 | 7.67 | - | 529.7 | 531.6 | 533.0 | 0.79 |

XPS spectra of Mnx was shown in Figure 4 and detail data were listed in Table 2. Figure 4(a) shows the Mn2p XPS spectra of Mnx catalysts. Based on the literature and previous XRD analysis, the value state of Mn in the catalyst made by the redox-precipitation method is between 3+ and 4+, one mixed valence[13]. Mostly, the peaks approximately 642.4-642.9 eV and 641.6-641.9 eV[13] are attributed to the presence of $Mn^{4+}$ and $Mn^{3+}$ species, respectively (the $Mn^{2+}$ peak was ignored). It`s clear that the $Mn^{4+}$ content in catalyst increases gradually with Mn content increasing, which is consistent with the XRD results. The ratio of $Mn^{4+}$-to-$Mn^{3+}$ in Mn100 is about 7.67, which is similar to that in $KMn_8O_{16}$, indicating the existence of $KMn_8O_{16}$. The combustion rate of lean methane catalytic combustion is positively related to the ratio of $Mn^{4+}$-to-$Mn^{3+}$ in Mn-Ce catalyst[17].

Figure 4(b) shows the Ce3d XPS spectra of Mnx catalysts. The peaks named as u```, u``, u, v```, v`` and v arise from the $Ce^{4+}$ contribution, while $Ce^{3+}$ is characterized by two other peaks [12]. The proportion of $Ce^{3+}$ in the catalyst increase from 0.11 to 0.28, when the ratio of Mn-to-Ce increases from 3/7 to 9/1, consistent with the changing trend of the catalyst performance. The oxygen vacancy and the active undersaturated bond would increase with the increase in the concentration of $Ce^{3+}$, promoting the oxygen adsorption and transfer processes, and improving the redox-ability of the catalyst [18].

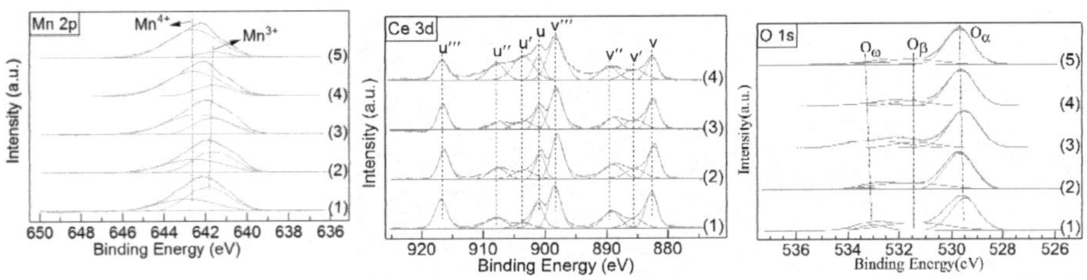


**Figure 4.** The XPS spectra of Mnx catalysts: (1) Mn30; (2) Mn50; (3) Mn70; (4) Mn90; (5) Mn100

Figure 4(c) shows the O1s XPS spectra of Mnx catalysts. There is a double-peak structure at a range of 529.4-529.9eV. O1s spectra were divided into three types peaks: the peak at BE of 529.6-529.9 eV was ascribed to the lattice oxygen (denoted as $O_\alpha$), the peak at BE of 531.4-531.6 eV was ascribed to chemisorbed oxygen (denoted as $O_\beta$), and the peak at a higher BE (above 533.0 eV) corresponded to adsorbed water (denoted as $O_\omega$)[6]. The process of organic compounds catalytic combustion occurs mostly via the Mars and van Krevelen mechanism [17], and the lattice oxygen of manganese oxide was considered to be the primary active species for activating the C-H bonds above 400°C, so the performance of catalyst increases, as the number of the lattice oxygen increases. The proportion of $O_\alpha$ in active oxygen tended to rise, then fall, with Mn content increasing, which is consistent with the change tendency of catalyst performance.

Thus, there are three parameters determining the performance of methane catalytic combustion: (1) higher value state of manganese species, (2) lower value state of Cerium species, and (3) higher content of lattice oxygen on the surface of the catalyst.

## 4. Results

The results obtained for Mnx catalysts, synthesized by a redox-precipitation, confirmed their performance for total combustion of $CH_4$ at low concentration (1 vol %). The results of this investigation lead us to draw the following conclusions:

- The performance of Mn90 catalyst is best (T50=~446℃) at 3000 ml/(h.g);
- The appropriate ratio of Mn-to-Ce is beneficial for the bigger specific surface areas and pore volumes.
- The appropriate ratio of Mn-to-Ce is beneficial for more $KMn_8O_{16}$, which improves the Mn content, oxygen vacancy, and lattice oxygen concentration.
- The appropriate ratio of Mn-to-Ce is beneficial for fine size for crystalline grain.

## References

[1] Fiuk MM, Adamski A. Activity of MnOx–CeO2 catalysts in combustion of low concentrated methane. *CATAL TODAY* 2015;**257**:131–5.

[2] Setiawan A, Kennedy EM, Stockenhuber M. Development of Combustion Technology for Methane Emitted from Coal-Mine Ventilation Air Systems. *ENERGY TECHNOL-GER* 2017;**5**:521–38.

[3] Li X. Progress of development and utilization of coalbed methane in the world. *Coal Process Compr Util* 2006:41–5.

[4] Zhang Y G, Qin Z F, Wang G F, Zhu H Q, Dong M, Li S N, Wu Z W, Li Z K, Wu Z H, Zhang J, Hu T D, Fan W B, Wang J G. Catalytic performance of MnOx–NiO composite oxide in lean methane combustion at low temperature. *APPL CATAL B-ENVIRON* 2013;**129**:172–81.

[5] Hu Z, Qiu S, You Y, Guo Y, Guo Y, Wang L, Zhan W, Lu G. Hydrothermal synthesis of NiCeOx nanosheets and its application to the total oxidation of propane. *APPL CATAL B-ENVIRON* 2018;**225**:110–20.

[6] Sheng Z, Hu Y, Xue J, Wang X, Liao W. A novel co-precipitation method for preparation of Mn–Ce/TiO2 composites for NOx reduction with NH3 at low temperature. *ENVIRON TECHNOL* 2012;**33**:2421–8.

[7] Cybulski A. Catalytic Wet Air Oxidation: Are Monolithic Catalysts and Reactors Feasible? *Ind Eng Chem Res* 2007;**46**:4007–33.

[8] Li S, Song P, Zhang J, He X, Xie Y, Zhang Y, Wang R, Li Z, Zhu H. Morphological effect of CeO2-MnOx catalyst on their catalytic performance in lean methane combustion. *J Fuel Chem Technol* 2018;**46**:615–24.

[9] ChangChun L, Jun-Jie Y, Zheng J, YanXin T, ZhengPing H, XuWen H. Methane catalytic combustion on Ce (1-x) MnxO (2-a) mixed oxide catalysts. *CHINESE J INORG CHEM* 2007;**02**:217–24.

[10] Palmqvist AEC, Johansson EM, Järås SG, Muhammed M. Total oxidation of methane over doped nanophase cerium oxides.*CATAL LETT* 1998;**56**:69–75.

[11] Zhang H, Yang W, Li D, Wang X. Influence of preparation method on the performance of Mn–Ce–O catalysts. *React Kinet Catal Lett* 2009;**97**:263–8.

[12] Zhang P, Lu H, Zhou Y, Zhang L, Wu Z, Yang S, Shi H, Zhu Q, Chen Y, Dai S. Mesoporous MnCeOx solid solutions for low temperature and selective oxidation of hydrocarbons. *NAT COMMUN* 2015;**6**:8446.

[13] Wang X, Liu Y, Zhang Y, Zhang T, Chang H, Zhang Y, Jiang L. Structural requirements of manganese oxides for methane oxidation: XAS spectroscopy and transition-state studies. *APPL CATAL B-ENVIRON* 2018;**229**:52–62.

[14] Arena F, Trunfio G, Negro J, Fazio B, Spadaro L. Basic Evidence of the Molecular Dispersion of MnCeOx Catalysts Synthesized via a Novel "Redox-Precipitation" Route. *Chem Mater* 2007;**19**:2269–76.

[15] Arena F, Gumina B, Lombardo AF, Espro C, Patti A, Spadaro L, Spiccia L. Nanostructured MnOx catalysts in the liquid phase selective oxidation of benzyl alcohol with oxygen: Part I. Effects of Ce and Fe addition on structure and reactivity. *APPL CATAL B-ENVIRON* 2015;**162**:260–7.

[16] Dharmarathna S, King'ondu CK, Pedrick W, Pahalagedara L, Suib SL. Direct sonochemical synthesis of manganese octahedral molecular sieve (OMS-2) nanomaterials using cosolvent systems, their characterization, and catalytic applications. *CHEM MATER* 2012;**24**:705–712.

[17] Han Y, Chen L, Ramesh K, Widjaja E, Chilukoti S, Kesumawinata Surjami I, Chen J. Kinetic and spectroscopic study of methane combustion over α-Mn2O3 nanocrystal catalysts. *J CATAL* 2008;**253**:261–8.

Kumar M, Bhatt V, Kim J, Kim J-S, Kim B-S, Lee C-Y, Singh B, Yun J-H. Role of Ce3+ valence state and surface oxygen vacancies on enhanced electrochemical performance of single step solvothermally synthesized CeO2 nanoparticles. *Electrochimica Acta* 2018;**284**:709–20.

5th International Workshop on Heat/Mass Transfer Advances
for Energy Conservation and Pollution Control
August 13-16, 2019, Novosibirsk, Russia

# Modeling Ice Melting with Selective Absorption of Radiation

**S D Sleptsov[1], N A Savvinova[2], M A Grishin[1]**

[1] Kutateladze Institute of Thermophysics, SB RAS, Novosibirsk, 1 acad. Lavrentiev ave., 630090, Russia

[2] Institute of Physics and Technologies, Ammosov North-East Federal University, Yakutsk, 20 Belinsky str, 677000, Russia

sleptsov@itp.nsc.ru

**Abstract.** Numerical simulation of the melting of an ice layer on a vertical substrate heated by long-wave radiation in a single-phase formulation of the Stefan problem is carried out. A simple model for taking into consider a thin film of water on the irradiated surface is proposed. Ice is presented as selectively absorbing material with two absorption bands. Calculations shows satisfactory agreement with the experiment.

## 1. Introduction

Simulation of ice melting is needed both to understand various natural processes and ensuring the safety of population, buildings, and engineering equipment in northern latitudes. Krass and Merzlikin [1] performed a detailed analysis of the modern state of snowpack and ice thermophysics and presented formulations of a new class of problems, where snow and ice are considered as light-scattering media that possess volume absorption and reflection. Modeling of ice heating and subsequent melting are formulated as Stefan problems in a semitransparent medium. Numerical and experimental investigations of heat transfer in semitransparent media with due allowance for thermal radiation in the case of two-phase and three-phase Stefan problems were widely described. Mathematical modeling of radiative-conductive heat transfer in the single-phase formulation of the Stefan problem was performed in [2, 3], but there are very few experimental studies that could be used verification of computational model.

Seki et al [4] performed computational and experimental investigations aimed at modeling radiation-induced melting of ice under the condition of incident long-wave and short-wave radiation fluxes from two types of lamps: halogen lamp and a lamp with a Nichrome filament. The ice layer is located on a vertical nontransparent substrate in a climatic chamber at a constant temperature of 0 °C. In the mathematical model of the process, the authors neglected the presence of a thawed water film on the surface and calculations was carried out in the single-phase Stefan's problem. They compared the rates of melting and heating of the non-radiated ice side and obtained satisfactory agreement between the experimental and computational results. The effect short-wave radiation on formation of highly rough surfaces in ice was shown. In computations, Seki et al [4] used fitting parameters and direct integration by the Bouguer law to take radiation into account.

The present paper is aimed at verification of our formulation of the problem and the method of the numerical solution of the radiative part of the problem, which takes into consider selective volume absorption of radiation by the ice layer [3]. Verification of the computational model is carried out by comparison with the experimental data presented in [4].

## 2. Formulation of the Problem

Figure 1 shows the geometrical scheme of problem, where a layer of clear and non-scattering ice of thickness $L_0$ is adhered onto a substrate located vertically in a climatic chamber with a constant temperature $T_\infty$. The right surface of the ice layer is irradiated by a lamp with a filament temperature of 800 K. The radiation of such a lamp approximately models the Sun radiation in a cloudy day. The ice boundaries are gray and diffusely absorb, reflect and transmit of radiation in such a way that $A_i + R_i + D_i = 1$, where $A_i$, $R_i$, and $D_i$ are the absorption, reflection, and transmission coefficients of the ice surface, respectively, $i = 1, 2$. The left boundary of the substrate is maintained at a constant temperature $T_{sub} = 256.15$ K, which is the initial temperature $T(x, 0)$, and ambient air is kept $T_\infty = 273.15$ K.

The solution of the problem includes two stages. At the first stage, radiative-conductive heat transfer is considered until the right boundary of ice $T(L_0,t)$ reaches the phase transition temperature $T_f$. At the second stage, the Stefan problem with a fixed temperature of the right boundary $T(L(t), t) = T_f$ is considered, and it is assumed that the irradiated surface is covered by a thin water film flowing down under the action of the gravitational forces. The presence of thin water films was noted in [4], but it was neglected in calculations. The positions of the interphase boundaries $L(t)$ is determined from solution of the boundary value problem.

The equation of energy conservation for ice is written as:

$$c_p\rho\frac{\partial T(x,t)}{\partial t}=\frac{\partial}{\partial x}\left(\lambda\frac{\partial T(x,t)}{\partial x}-E_\nu(x,t)\right),\ 0<x<L(t). \qquad (1)$$

Here $c_p$ is the specific heat at constant pressure, $\rho$ is the density, $\lambda$ is the thermal conductivity, and $E_\nu(x,t)=E_\nu^+(x,t)-E_\nu^-(x,t)$ is the flux density of the resultant radiation flux.

It is assumed that the surface of the substrate on the left boundary is maintained at constant temperature $T_{sub}$ and there is no heat flow from the substrate to the ice [1]

$$\frac{\partial T}{\partial x}=0 \text{ at } x=0. \qquad (2)$$

At the first stage, the surface of the right boundary is exposed to radiation from a long-wavelength radiation source, and heat removal due to convection is taken into account:

$$\lambda\frac{\partial T}{\partial x}-h(T_\infty-T)-|E_{res,2}|=0, \text{ at } x=L_0, \qquad (3)$$

Where $|E_{res,2}|=A_2\left(E_\nu^+(x,t)+E^*\right)-\varepsilon_2\sigma_0T^4(x,t)$. Equations (1)–(3) are added with the initial condition: $T(x,0)=T_{sub}$.

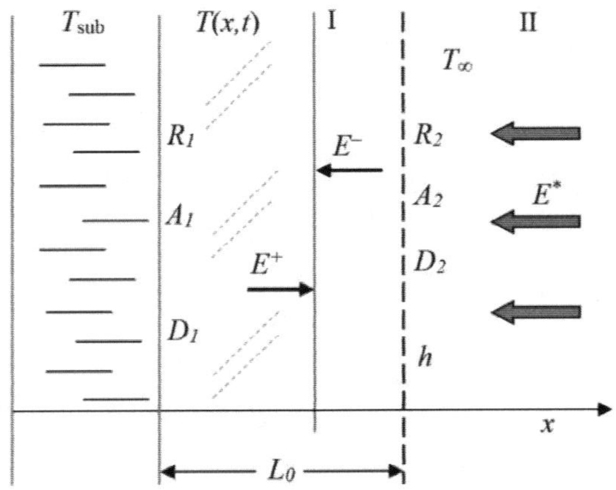


Fig. 1. Geometrical scheme of the problem

At the stage of ice melting, the temperature of the right boundary surface $x = L(t)$ is fixed: $T(x,t) = T_f$. The boundary condition (3) transforms to the Stefan condition with allowance for the thin water film formed on the surface. We assume that the water film is isothermal, and the temperature difference over its thickness is negligible:

$$\lambda \frac{\partial T}{\partial x} + h\left(T_{fil} - T_\infty\right) - \left|E_{res,fil}\right| = \rho\gamma \frac{\partial L}{\partial t} \tag{4}$$

where $\left|E_{res,fil}\right|$ has the form

$$\left|E_{res,fil}\right| = A_2\left(E_\nu^+(x,t) + E^*\right) - \varepsilon_2\sigma_0\left(T^4(x,t) - T_{fil}^4\right), \; x = L(t). \tag{5}$$

Here $T_f = 273.15$ K is the melting point of ice, $T_{fil} = 277.15$ K is the water film temperature and $\gamma$ is latent heat of ice melting. Condition (4) takes into account the heat removal from the external surface of the thin film of water. The volumetric thermal radiation of the film and the right side of ice is included into condition (5).

The assumption of the presence of a thin film of water on the ice surface does not contradict the single-phase approximation of the Stefan problem, since radiation does not absorb radiation in the film itself and it acts only as an additional boundary condition on the interfacial surface with constant values. The thermal problem is solved only in the thickness of ice on a vertical substrate.

The radiation flux densities in dimensionless form $\Phi_\nu^\pm = E_\nu^\pm / \left(4\sigma_0 T_f^4\right)$, $\Phi_\nu = \sum_j \left(\Phi_j^+ - \Phi_j^-\right)$ and involved into (1 – 5) are determined by solving the equation of radiation transfer in a plane layer of a radiating and selectively absorbing medium with a known distribution of temperature over the layer ($j$ is the number of the spectral band) [5, 6].

Wide range possibilities in terms of solution simplicity and efficiency of obtaining results are provided by a modified mean flux method [5, 7]. In the framework of this method, the integral-differential equation of radiation transfer is reduced to the system of two nonlinear differential equations for a plane layer of a semitransparent selectively absorbing medium.

The differential dimensionless analogue of the radiation transfer equation for the hemispherical fluxes $\Phi_\nu^\pm$ is presented in the following form [6, 7]:

$$\left.\begin{aligned} &\frac{d}{d\tau_j}\left(\Phi_j^+(\tau,\eta)-\Phi_j^-(\tau,\eta)\right)+\left(m_j^+(\tau)\Phi_j^+(\tau,\eta)-m_j^-(\tau)\Phi_j^-(\tau,\eta)\right)=n^2\Phi_0 \\ &\frac{d}{d\tau_j}\left(m_j^+(\tau)l_j^+(\tau)\Phi_j^+(\tau,\eta)-m_j^-(\tau)l_j^-(\tau)\Phi_j^-(\tau,\eta)\right)+\left(\Phi_j^+(\tau,\eta)-\Phi_j^-(\tau,\eta)\right)=0 \end{aligned}\right\} \quad (6)$$

The boundary conditions for system (6) in the dimensionless variables are written as

$$\left.\begin{aligned} &\tau_{j,\mathrm{I}}=0: \Phi_{j,\mathrm{I}}^+ = A_1 n^2\frac{\theta^4}{4}+D_1\frac{\theta^4}{4}+\left(1-\frac{1-R_1}{n^2}\right)\Phi_{j,\mathrm{I}}^- \\ &\tau_{j,\mathrm{I}}=\alpha_j L(t): \Phi_{j,\mathrm{I}}^- = A_2 n^2\frac{\theta^4}{4}+D_2F^*+\left[1-\frac{1-R_2}{n^2}-A_2\left(\frac{1+n^2}{n^2}\right)\right]\Phi_{j,\mathrm{I}}^+ \\ &\tau_{j,\mathrm{II}}=\alpha_j L(t)+\infty: \Phi_{j,\mathrm{II}}^- = F^*, \end{aligned}\right\} \quad (7)$$

Here $\Phi_0 = n^2 B_\mathrm{v}/\left(4\sigma_0 T_f^4\right)$ is the dimensionless density of the equilibrium radiation flux, $B_\mathrm{v}$ is the Planck
Function of blackbody radiation, $n$ is the refractive index of ice, and $\tau_j=\alpha_j L(t)$ is the spectral optical thickness of the layer at the time $t$. The values of the coefficients $m^\pm$ and $l^\pm$ are determined from the recurrent relation derived by means of formal solution of the radiation transfer equation; $j$ is the number of the spectral band [7]. Layers I and II refer to ice and the ambient air, respectively (Fig. 1).

The boundary value problem (1)–(5) is solved by a finite-difference method. The nonlinear system of implicit difference equations is solved by the sweep and iteration method. Solving of the radiative problem involves iterations, and the boundary-value problem (6)–(7) is solved at each iteration by the matrix factorization method. Rapid convergence of such a solution method allows obtaining results with a high degree of accuracy.

## 3. Analysis of Results

The results of numerical modeling presented below were obtained for a vertically located clear, non-scattering ice layer with the following physical parameters: initial ice thickness $L_0$ = 0.045 m, temperature of the left boundary of the substrate and initial temperature of the substrate and ice layer $T_{sub}$ = 256.15 K, the temperature of the atmosphere inside the chamber is maintained at a $T_\infty = 273.15$ K equal to the melting point of ice $T_f$, and constant density of the incident radiation flux $E^*$=1162.22 Wt/m$^2$. The thermophysical properties of ice are the thermal conductivity $\lambda = 1.9$ Wt/(m·K), thermal diffusivity $a$ = 9.3·10$^{-7}$ m$^2$/s, and latent heat of the phase transition $\gamma$ = 335 kJ/kg. The optical parameters are as follows: refractive index of ice $n$=1.31, reflection coefficients $R_1$ = 0.5 and $R_2$ = 0.063, and emissivity of the boundaries $\varepsilon_1$=1−$R_1$ and $\varepsilon_2$ = 0.97. The averaged spectral parameters for ice are listed in the table. In solving the problem, the heat transfer coefficient was varied: $h$ =14.35 Wt/m$^2$ at the stage of ice heating and $h$ =8 Wt/m$^2$ at the stage of the phase transition.

Spectral dependences of the absorption coefficient of ice

| $j$ | $\nu_j$, $10^{14}$ Hz | $\lambda_j$, μm | $\alpha_j$, m$^{-1}$ |
|---|---|---|---|
| 1 | 9,09– 2,02 | 0,33 – 1,2 | 0,001 |
| 2 | 2,02 – 1,18 | 1,2 – ∞ | 1 |

Figure 2 shows the temperature field of the ice layer during its heating (curves between 1 and 2) and melting (curves between 2 and 3). The character of the curves at the first stage is determined by

Journal of Physics: Conference Series **1369** (2019) 012005 doi:10.1088/1742-6596/1369/1/012005

the high absorption coefficient on the irradiated surface. At the stage of heating, the curves have a monotonic and nonlinear character and a comparatively high gradient in the bulk of ice. This fact is related to the small optical thickness and high absorption coefficient of the right boundary. The curves at the second stage show a linear behavior. This fact is associated with the use of the assumption about the emergence of a thin film of water on the irradiated surface.

Figure 3 shows the density field of the resultant radiation fluxes. The curves have a linear character. The values at the left boundary are higher than those on the right boundary due to the relatively high value of reflection coefficient. At the phase transition stage (curves between 2 and 3), the curves tend to one point on the left boundary. This is due to the additional consideration of a thin film of water on a right side of ice and a decrease in the volume of the medium.

In fig. 4 and 5 melting rate and an increase in temperature of the left boundary of the ice layer and their comparison with experimental data [4] are shown, respectively. At the beginning and middle of the melting rate (Fig. 4), the calculations better coincide with the experiment, but diverge at the end. According to predictions, the ice layer melts more than the experiment shows (discrepancy is 3 cm). The increase of temperature at the left boundary of ice (Fig. 5), according to calculations, also shows a slight excess at the end of the time interval (by 0.5 °C). Thus, taking into account the appearance of a thin film of water on the irradiated surface shows satisfactory agreement with the experiment.

## 4. Conclusion

A numerical simulation of the melting of ice caused by the radiation of a long-wave lamp, approximately simulating the condition of solar radiation on a cloudy day are carried out. In considering the radiation part, the selective absorption of radiation by ice and the appearance of a thin film of water on the irradiated ice surface with a temperature above $T_f$ were taken into account. A comparison of the calculation of the rate of ice melting and the temperature increase of the left boundary with the experiment shows satisfactory agreement, and thus allows us to consider the verification of the single-phase Stefan problem for a semitransparent medium as realized.

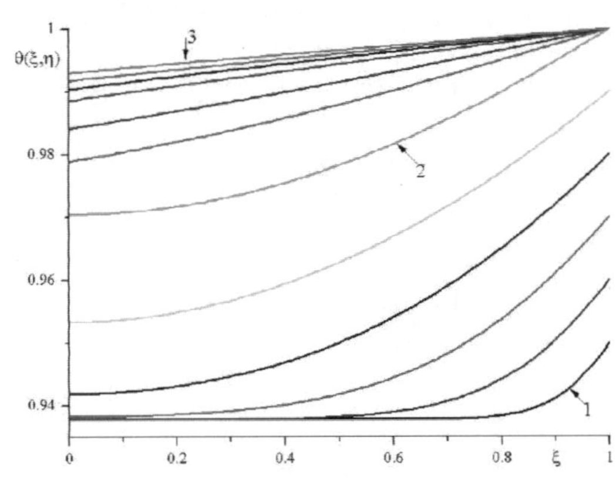


Figure 2. Temperature field of the ice layer: curves between 1 and 2 is heating stage, curves between 2 and 3 is melting stage

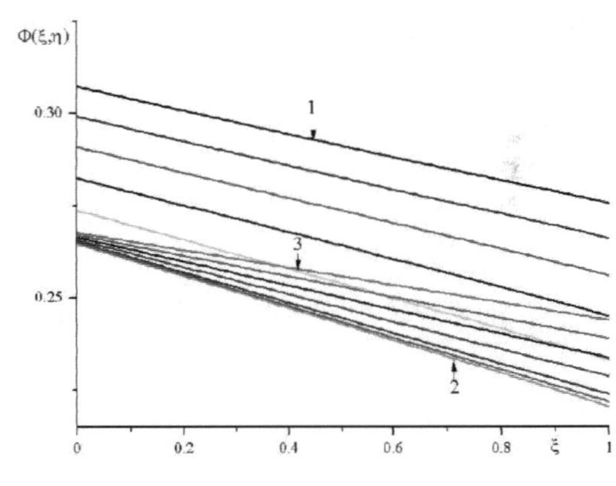


Figure 3. Resultant radiation field in the ice layer: curves between 1 and 2 is heating stage, curves between 2 and 3 is melting stage

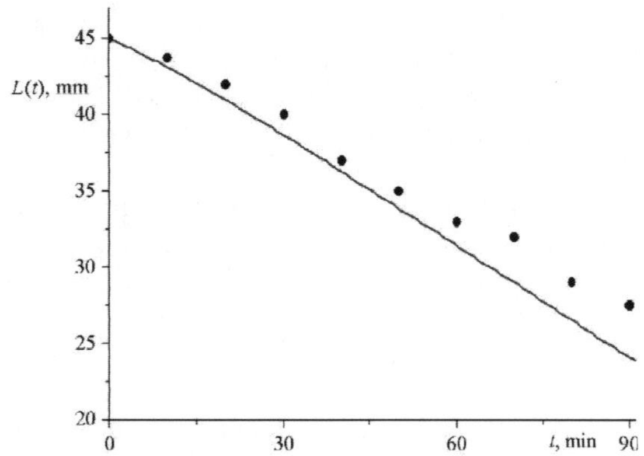


Figure 4. Rate of ice melting (the solid line is calculations and • the experimental data [4], respectively).

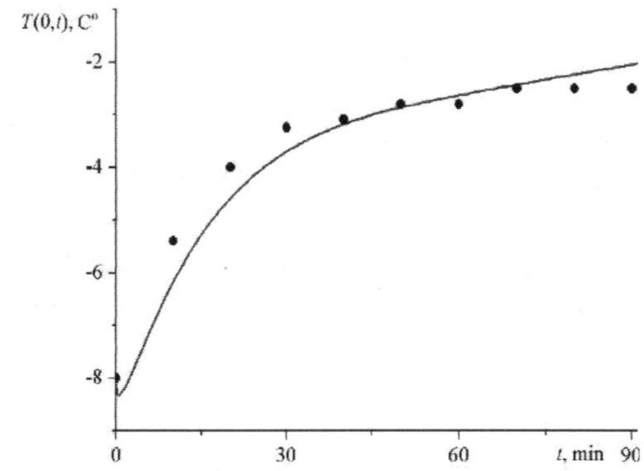


Figure 5. Temperature increase at the left boundary of ice (the solid line is calculations and • the experimental data [4], respectively).

**Acknowledgments**

The study of single-phase Stefan problem was carried out under state contract with IT SB RAS (AAAA-A17-117022850029-9), the study of radiative-conductive heat transfer in ice was financially supported by RFBR (Project No. 18-48-140011 r_a ).

## References

[1] Krass M S and Merzlikin V G, 1990 *Radiation Thermophysics of Snow and Ice* (Leningrad: Gidrometeoizdat), p. 262

[2] Le Dez V, Yousefian F, Vaillon R, Lemonnier D and Lallemand M 1996 *J. Phys. III France* **6** 373–390

[3] Rubtsov N A, Savvinova N A and Sleptsov S D 2015 *J. Engng Thermophysics* **2** 123–138

[4] Seki N, Sugawara M and Fukusaki S 1979 *Wärme- und Stoffübertragung* **2** 137–144

[5] Rubtsov N A, Timofeev A M and Savvinova N A 2003 *Combined Heat Exchange in Semi-transparent Media* (Novosibirsk: SB RAS Publ.) p 198

[6] Sleptsov S D, Rubtsov N A and Savvinova N A 2016 *Thermophys. Aeromech.* **1** 109–117

[7] Rubtsov N A *Radiative Heat Transfer In Continuous Media* 1984 (Novosibirsk: Nauka) p 277

5th International Workshop on Heat/Mass Transfer Advances
for Energy Conservation and Pollution Control
August 13-16, 2019, Novosibirsk, Russia

# Flow and heat transfer characters in the integral internal cooling channel of a turbine blade

**Tao Guo[1], Huiren Zhu[2], and Zhixiang Zhou[1]**

[1] School of Power and Energy, Northwestern Polytechnical University, Xi'an, P.R.China, guotao@nwpu.edu.cn
[2] Hunan Aviation Power Plant Research Institute, Changsha, Hunan, P.R.China

**Abstract.** A scaled model of the integral internal cooling channel of a turbine blade was established for the experimental investigation. The model was made by Perspex for its transparency so that the thermochromic liquid crystal measurement could be used easily. The cooling channel is composed of 3 legs of ribbed channels. Rib with angle 45°was settled in both pressure side and suction side, which forms a cross-rib structure. Three legs were connected by two 180° turns. The main inlet and two addition inlets were arranged in the root of the model channel. Two discharge outlets were located in the tip. A row of outlet holes was distributed along the trailing edge. The channel area gradually decreases along the blade height direction, while bending and torsion occur because of the blade airfoil shape. The aspect shapes and connection methods of the channel were kept in accordance with the real blade. The inlet Reynolds number is from 10000 to 32000. Five outlet discharge ratios were alternated in this investigation. The detailed heat transfer distributions, both in the pressure side and suction side, were measured by transient liquid crystal technic. The characters of heat transfer distribution and pressure drop along the channel were displayed. The effects of flow discharge ratio on heat transfer and pressure coefficient were also described.

## 1. Introduction

In order to increase the specific power and thermal efficiency, the inlet temperature of gas turbine is getting higher and higher. The reasonable measure must be taken to ensure the safety of turbine. Although both the gas temperature and allowable temperature of turbine metal are increasing gradually, the gas temperature increases faster. Effective cooling method and reasonable cooling design are inevitable technical approaches.

The ribbed channel is widely used for internal cooling in turbine blades. In view of the basic flow structure in the simplified ribbed channel, J C Han et al. [1] investigated the heat transfer performance in the channel in the static state, and obtained the relationship between the heat transfer performance of the channel with structural parameters, such as the aspect ratio of the channel, the rib arrangement, and aerodynamic parameters, such as Reynolds number. Casarsa [2] and Graham et al. [3] reported more detailed flow distribution obtained through experiments and clarified the flow structure. The results revealed more intuitively the mechanism of heat transfer enhancement in the ribbed channel.

Usually, the real internal cooling channel of the turbine blade is a multi-leg channel which is combined with straight passage through U-turn. Due to the U-turn connection, the flow structure is more complex, and the heat transfer performance may differ from the straight channel, which was used in heat transfer investigation. Some research studied U-shape (two legs) and S-shape (three legs) multi-leg ribbed channels, and the flow and heat transfer characteristics of each cooling unit were analyzed. S.Y. Son et al. [4] performed the experimental investigation using PIV and transient liquid crystal measurement technology. The flow structure and temperature distribution around the bending area in the smooth and ribbed U-shape channel were measured. The results show that the heat transfer enhancement is closely related to the shape, strength, position and rotation direction of the secondary flow. The mechanism of second flow producing by the combined effect of the rib and bending of the channel, and further the influence of the second flow to heat transfer characteristics were elaborated. Smith M. A. et al. [5] studied the heat transfer characteristics of the serpentine ribbed channel with different aspect ratios, and found that the heat transfer enhancement was more significant under the high Reynolds number condition. Under the combined action of the vortex formed by turning and the secondary flow generated by rib, the heat transfer in the middle chord passage is greatly enhanced. Murata A. et al. [6] investigated the effect of rib orientation on the flow pattern in a channel with a 180° turn rectangular channel. The results show that the heat transfer distribution and performance in the bending area is directly influenced by the flow field caused by a sharp turn. In order to obtain higher heat transfer performance, the combined effect of rib orientation and bending effect should be considered together. Schuler M. et al. [7] studied the heat transfer in a U-shape trailing edge channel with lateral bleeding. Compared with the channel without lateral bleeding, the heat transfer of the side wall and the region near the bleeding holes is augmented while air flow out through lateral holes. As some fluids flowed out of the outlet hole, the cooling flow rate in the passage was reduced, and the heat transfer coefficient of the downstream passage was decreased.

Schabacker J. et al. [8] and Murata A. et al. [9] performed the flow character investigation in U-shape channel. The flow pattern affected by U-turn and heat transfer augment mechanism is presented.

The investigations mentioned above are performed in a simplified rectangular channel. In the actual blade, leading edge and trailing edge channel is usually not rectangle. The cooling channel may screw because of the blade airfoil. To consider the effect of these factors and the influence of the cooling channels connection, the heat transfer was measured by transient liquid crystal technique in a complete internal channel. The heat transfer distribution and pressure coefficient were characterized. The results may provide a useful reference for the design and improvement of turbine blade cooling structure.

**Nomenclature**

| | |
|---|---|
| c | Specific heat of model metal |
| Cp | Pressure coefficient |
| D | The equivalent diameter of the channel |
| h | Heat transfer coefficient |
| P* | Total pressure |
| Re | Reynolds number |
| T | Temperature |
| t | Time |
| U | Velocity |
| x | Wall thickness |

*Greek symbols*

| | |
|---|---|
| $\rho$ | density |
| $\lambda$ | thermal conductivity |
| $\mu$ | viscosity |

*Subscript*

| | |
|---|---|
| in | inlet of passage |
| out | outlet of passage |

## 2. Experiment approach and facility

The structure of the internal channel is shown in Figure 1(a). The channel is a 3-leg serpentine channel which consists of 3 ribbed passage and 2 U-shape turning. The channel is bent and screwed because of the blade airfoil shape. In the leading edge and trailing edge, the section of the passage is almost triangle. The section of the passage in the mid-chord region is approximately rectangle. The channel has 3 inlets. The most of fluid flow into the channel in the root of the leading edge. The inlet located in root near the mid-chord and trailing is for supply some coolant for trailing edge passage. At the top of the leading edge passage and trailing edge passage, there is a bleeding hole separately. A row of bleeding holes is located in the trailing edge.

The experiment model is scaled 6 times. The shape of the experiment model is the same as the generic channel. The bending and screwing characters are also kept. Transparent plexiglass was used as the material of the test model for acquiring the image of the liquid crystal. The model is split into two parts along the blend bending direction for liquid spraying. The model can be assembled by bolts and ensure the airtightness. In order to ensure the uniformity of air intake, there is an inlet cavity at the front of inlets. The flow rate of the two outlet holes at the tip of the blade was measured by a flow meter. The 45° inclined ribs are equipped in the wall of the pressure side and suction side of the channel. The ribs in the opposite wall are arranged vertically.

Figure 2 is a schematic diagram of the experimental system. The air in the experimental system is provided by the compressor and the tank. The air flow is measured and regulated by flow meter and valve. The regulated air flows into the model through the heater and solenoid valve group. The air flow out from two top outlets is measured and regulated by a flow meter and valve for discharge ratio. The thermocouples are arranged in channel of the model. The temperature of the air flow is measured under the computer control. The color history of the liquid crystal on the inner surface of the channel is recorded by the camera.

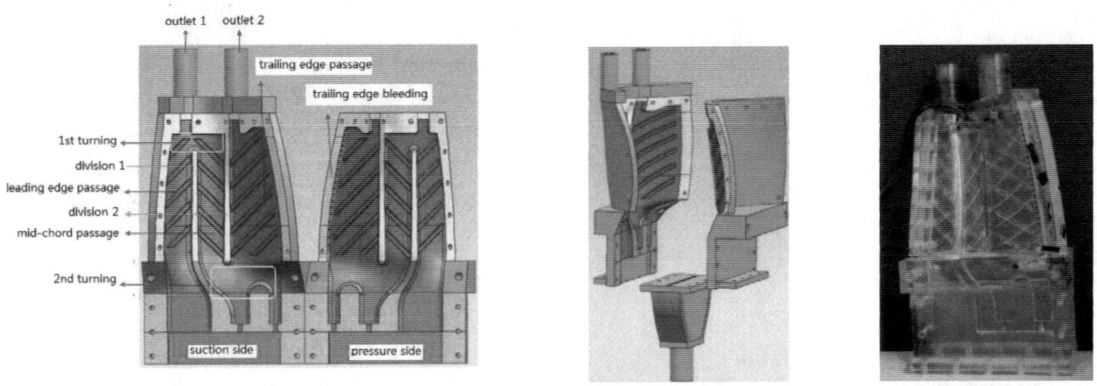


(a) Internal structure (b) Assemble sketch (c) Photo

**Figure 1.** Experiment model

Before the test, the power of the heater was shut down. The solenoid valve 1 was opened. The solenoid valve 2 was closed. The flow rate at the inlet of the model was regulated by valve in front of the model. The outlet discharge ratio was regulated by the valve at the back of the top bleeding hole. After the experimental condition are achieved, the solenoid valve 1 was closed and the solenoid valve 2 was opened. The air flow was diverted away from the model while the heater increased the air temperature to the preset value. At the beginning of the test the solenoid valve group switched the situation, the hot air flowed into the model at the desired flow rate, temperature. At the same time, the temperature of air and color are being captured.

Before of the test, the model has uniform initial temperature distribution. While the temperature of the air contacting the model rises suddenly, heat convection takes place between the air and the wall. The temperature of the wall increases and the heat conduct into the wall. The temperature change history relates to the heat transfer coefficient, the initial temperature, the thermal physical properties and the air temperature history. Therefore, under the condition that the thermal properties of the experimental model are known, as long as the initial temperature of the experimental model, the initial temperature of the model, temperature rise process of the airflow and the change history of the wall temperature are recorded, the heat transfer coefficient of the wall surface can be calculated according to the thermal conductivity

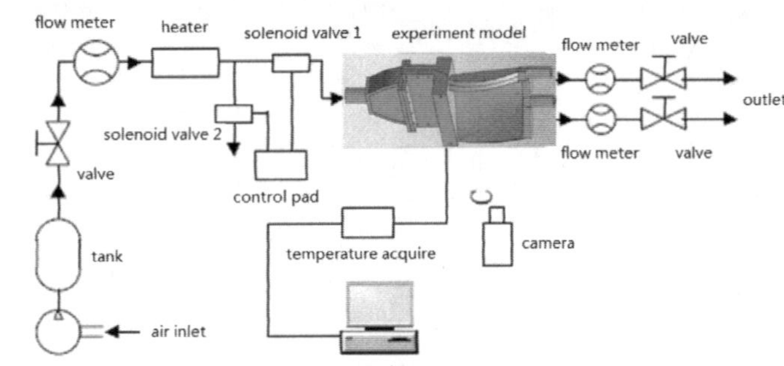


**Figure 2.** Experimental system

theory. In order to simplify the analysis and data processing, the experiment is usually completed in a relatively short time. At this condition, the heat conduction inside the model from one side has not reached another surface, that is, the experimental model is in a semi-infinite heat conduction state. The analysis solution can be used to calculate the heat transfer coefficient.

## 3. Results and discussions

### *3.1. Heat transfer distribution*

The inlet Reynolds number is from 10000 to 32000. The heat transfer distribution at different Re and discharge ratio are similar. Figure 4 shows the flow pattern and heat transfer distribution at the condition of Re=17000, outlet discharge ratio is 28:29:43.

#### *3.1.1. Pressure side*

Figure 3(a) and (b) show the flow pattern near the pressure side and heat transfer distribution of the pressure side wall. In this side, although the shape of the inlet channel was changed before the test region, the inlet velocity is nearly uniform. The heat transfer distribution is uniform approximately. The inclined ribs are directed to the top of the leading edge. As a result in simplified channel, the second flow generated by inclined ribs interference with the side wall, the heat transfer enhancement effect is best at sharp angle region back of the rib. The high heat transfer region directs to downstream along the ribs and becomes weakened. Therefore, the heat transfer at the leading side is higher than that in the division side.

There are two higher heat transfer regions at the turning of the leading edge passage. One higher heat transfer region is formed at sharp region downstream the rib. This region is close to the sharp angle at the back of the rib and the division of channel. The shape of the region deviates to the second passage caused by the turning of the passage. The other is at the outer of this high heat transfer region. In the Figure 3(a), this region is caused by bleeding at the top of the leading edge passage. The bleeding makes more fluid flow to the wall after the rib. The experimental result shown in Figure 3(b) indicates that these two regions can be distinguished but are very close. Because of the centrifugal effect caused by turning of passage, the airflow impinges the blade tip and forms another high heat transfer region at the top of the passage. After that, the fluid continues to flow along the outer side of the bend. At the top of the second division, the high heat transfer region is formed caused by the joint action of bending and rib. In addition to the strong overall heat transfer in region 2, part of the fluid flows along the leading edge of the rib and forms a strong heat transfer region which effectively improves the heat transfer in the backflow region.

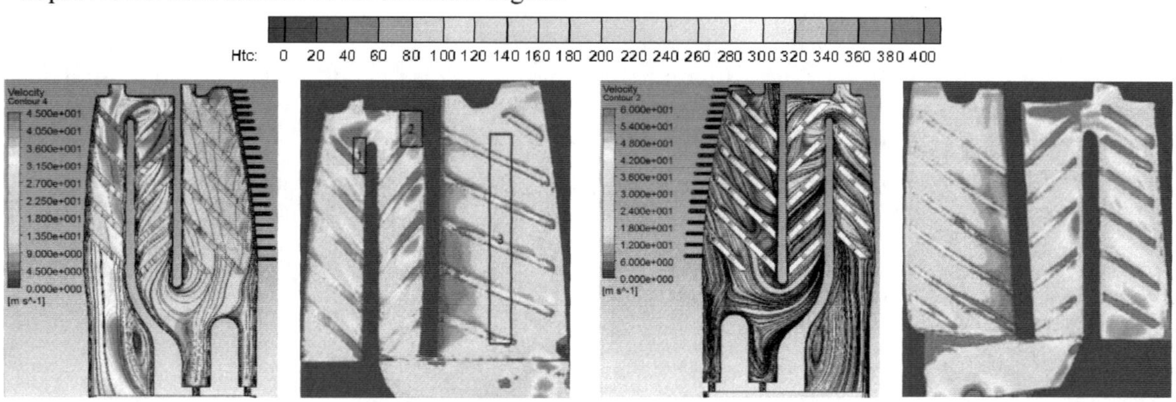


(a) Flow pattern (pressure side, CFD) (b) Heat transfer distibution (pressure side, experiment) (c) Flow pattern (suction side, CFD) (d) Heat transfer distibution (suction side, experiment)

**Figure 3.** Flow pattern and heat transfer distribution (Re=17000)

Because of the rib disturbing and the centrifugal effect of turning, the heat transfer in the second passage is strongest near the trailing edge side. The heat transfer is weakened along the passage. The heat transfer effect is stronger at the inlet of the second passage. This effect is almost throughout the whole passage. The overall heat transfer is stronger than that in the first passage.

Near the turning at the root, the bending effect and air supplement effect act on this region together. As the Figure 3(a) shows, the flow area is expanded. The velocity here is slow down and the centrifugal effect is small. On the other hand, the supplement effect makes the airflow from the second channel closer to the inner of the bend. This helps to reduce the low heat transfer region caused by the backflow.

The heat transfer in the trailing edge passage is the weakest among the three passages. Affect by the turning and bleeding at trailing edge, the air flow into the passage deviate to the trailing edge. Part of airflow bleeds out from the hole in trailing edge. The backflow region near the division is relatively broader. Therefore, the broader low heat transfer region is produced. The bleeding at the top of the third passage makes part of airflow to the top. This is a positive effect to reduce the low heat transfer region and improve the heat transfer performance to some extent. There is a higher heat transfer region 3 which located in the middle of the passage. This region is generated by higher velocity flow caused by air supplement of the middle inlet.

*3.1.2. Suction side*

Figure 3 (c) and (d) show the flow pattern near the suction surface and heat transfer distribution of the suction side wall. Because the cross rib structure is used in the internal cooling channel, and the direction of the rib on the suction side is opposite to that on the pressure side. Compared with the pressure side, in the leading edge channel, the heat transfer is better near the leading edge. At the entrance of the passage, because of the shape change and screw of the channel, unlike the pressure side, the flow velocity at the entrance of the suction side is uniform. A stronger heat transfer region caused by higher inlet velocity can be found. The orientation of the rib at the tip of the blade makes the airflow easier to deflect and enter the turning zone. After the last rib in the first passage, the flow crosses the rib is weak. A lower heat transfer is formed. In the downstream half of the bend, the cross-rib flow enhances the heat transfer in the tip area.

The region which is back of the first rib in the second passage is the backflow region of turning. This region has not obviously heat transfer enhancement caused by ribs. After the third rib, the separated flow caused by turning attaches to the division again and generates stronger heat transfer.

Although the rib orientation is opposite in the suction side and pressure side, the heat transfer distribution is similar. The heat transfer near the division is weaker than that near the tailing edge side. The bleeding at top of the passage improves the heat transfer near the tip region.

*3.2. Effect of discharge ratio*

The heat transfer distributions in the case of Re=21000 and various discharge ratio are shown in Figure 4. The results at other Re is similar to this. Although the heat transfer distribution is different in the suction and pressure side, the trend of heat transfer with discharge ratio is similar. All three bleeding outlets are located after the first passage. Therefore the change of discharge ratio seldom affects the first passage. Generally speaking, the flow rate of outlet 1 increase may lead to the change of flow pattern in first turning and reduce the flow rate enter into the second passage. The bleeding through outlet 1 affects these two

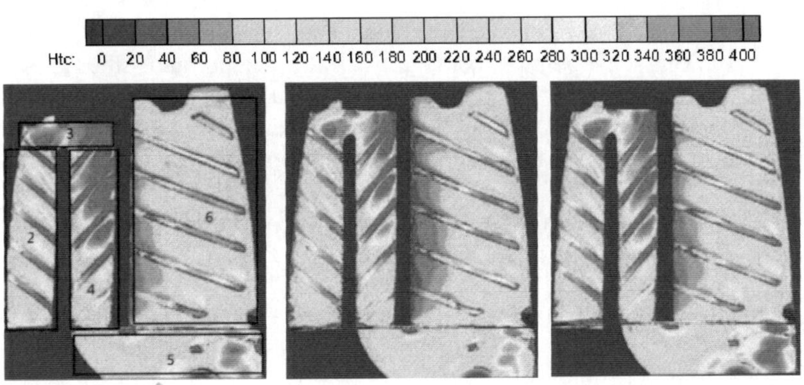


(a) 10:36:54 (b) 28:00:72 (c) 40:24:36

**Figure 4.** Heat transfer in different discharge ratio (Re=21000) regions significantly. In the case of the discharge ratio is 10:36:54, the flow rate in second passage is maximum, the heat transfer is strongest. In another case, the heat transfer in the first turning and second passage become weak in various extents while the flow rate of outlet 1 increase.

But in the case of the discharge ratio is 40:24:36, the flow rate of outlet 1 is maximum, the heat transfer in the second passage is not the weakest. This means the flow pattern in the turning zone affects the second passage obviously too. The change of discharge ratio may also lead to the change of pressure distribution in the channel, then alter the enter condition of three inlets. The heat transfer in the second turning zone may be changed. In the third passage, whether bleed in the top of passage affects the heat transfer obviously. The effect of bleeding ratio on heat transfer is relatively small. The effect on average heat transfer performance is also small.

## 4. Conclusions

The flow character and heat transfer distribution are investigated in a real shape internal cooling channel. The conclusion can be drawn:

1) The general characteristics of heat transfer distribution in the real channel are similar to those in the ideal unit structure, but the detailed distribution characteristics are obviously different due to the influence of channel shape, inlet state and other factors.

2) The heat transfer characteristics at the junction of the channel and the mid-chord passage are most affected by the flow deflection and outflow. When the ribs in the turning zone are vertical with airflow, the heat transfer distribution in the deflection area is more uniform. The heat transfer enhancement in the second channel is also better. The bleeding at the top of the trailing edge passage is helpful to improve the heat transfer in the passage.

3) The influence of discharge ratio on heat transfer presents a general trend. But it not clear for some discharge ratio because of the complex influence factors.

4) The measurement in actual structure of internal cooling channel by transient liquid crystal testing technique can provide the detailed heat transfer distribution of the complex wall surface, which is helpful to understand the heat transfer characteristics and the heat transfer enhancement mechanism.

## References

[1] J.C.Han,et al. Heat transfer and friction in channels with two opposite rib-roughened walls. Journal of Heat Transfer, 1984, 106(4):774-781.

[2] Casarsa L, et al Arts T (2005) Experimental investigation of the aerothermal performance of a high blockage rib-roughened cooling channel. ASME J Turbomach 127:580–588.

[3] Graham, A.et al, Thole, K. A.. Flowfield measurements in a ribbed channel relevant to internal turbine blade cooling. Proceedings of ASME Turbo Expo 2004, GT2004-53361.

[4] Son, S. Y.et al. PIV flow measurements for heat transfer characterization in two-pass square channels with smooth and 90° ribbed walls. Journal of Heat Transfer, 2002,45(24):4809 -4822.

[5] Matthew A. Smith, et al. Dunn. Heat transfer for high aspect ratio rectangular channels in a stationary serpentine passage with turbulated and smooth surfaces. ASME Journal of Turbomachinery, 2013,136(5): 51002-51012.

[6] Akira Murata, et al. Aiding and opposing contributions of centrifugal buoyancy on turbulent heat transfer in a two-pass transverse- or angled-rib-roughened channel with sharp 180° turns. Journal of Heat Transfer, 2004,47(17-18):3721-3743.

[7] Marco Schüler, et al. The Effect of Side Wall Mass Extraction on Pressure Loss and Heat Transfer of a Ribbed Rectangular Two-Pass Internal Cooling Channel. ASME Journal of Turbomachinery, 2010,133(2):21002-21012.

[8] Schabacker, J.et al. PIV Investigation of the Flow Characteristics in an Internal Coolant Passage with Two Ducts Connected by a Sharp 180° Bend. TheAmerican Society of Mechanical Engineers 1998, GT1998-544.

[9] Akira Murata, et al. Large eddy simulation of turbulent heat transfer in a rotating two-pass smooth square channel with sharp 180° turns. Journal of Heat Transfer,2004,47(4):683-698.

5th International Workshop on Heat/Mass Transfer Advances
for Energy Conservation and Pollution Control
August 13-16, 2019, Novosibirsk, Russia

# Heat transfer at evaporation/boiling in the thin horizontal liquid layer on microstructured surfaces under low pressures

**V. I. Zhukov[1,*], A. N. Pavlenko[2], and V. P. Bessmeltsev[3]**

[1] Novosibirsk State Technical University, K. Marx Prosp., 20, Novosibirsk, Russia, 630073
[2] Kutateladze Institute of Thermophysics, SB RAS, Lavrentiev Prosp., 1, Novosibirsk, Russia, 630090
[3] Institute of Automation and Electrometry, SB RAS, Koptyug Prosp., 1, Novosibirsk, Russia, 630090

E-mail : *vizh@inbox.ru

**Abstract.** The results of the experimental study of heat transfer during boiling of n-dodecane on a capillary-porous surface are presented. The surface is coated using a 3D laser printer. The dependence of the heat transfer coefficient on the heat flux density under different pressures and liquid layer heights is studied. The obtained experimental data were compared with the values obtained during evaporation/boiling on a plain surface under the same conditions.

## 1. Introduction

The capacity of technical equipment depends at a larger extent on the cooling system operation under conditions of significant heat flows removal to keep the temperature below the certain level, thereby preventing damages and maintaining performance. At present, two-phase cooling systems using latent heat of vaporization are recognized to be the most efficient. Preference is given to passive cooling schemes, which apply pool boiling of the working liquid [1, 2]. The use of modified surfaces is a practical way to reduce temperature difference and increase critical heat flux (CHF). The techniques of modified porous surface construction are described in [3] and focused on heat transfer intensification at pool boiling. The experiments [4, 5] were carried out to study heat transfer processes in thin liquid films at low pressure on a capillary-porous surface in operating conditions of flat-plate heat pipe systems. The experiments showed the difference between heat transfer regimes under atmospheric and low pressures. For instance, nucleate boiling did not occur under low pressure. Although the steam-saturation temperature was changing, it tended to grow with the increase of the heat flux [4, 5]. In [6], the evaporation/boiling regime on capillary fed copper particle sintered porous wicks is experimentally investigated at the same reduced pressure to compare the heat transfer performance of wicks with a different particle size, particle type and the wick thickness. The experiments were done at constant pressure of 9.6 KPa; that corresponds to the water-saturation temperature of 45°C. The liquid level may have receded into the wick when the heat flux was high enough to induce boiling heat transfer, therefore, the bubble nucleation cannot be observed experimentally. This is particularly true for the case of a large heating area, such as in the study [6]. The experiments described in the studies [7, 8] demonstrate the absence of bubble nucleation boiling under variation of the heat flux, up to crisis phenomena development on the plain horizontal surface of the thin liquid film. The paper [9] presents the systematic study of evaporation and boiling regimes in a thin horizontal layer of the liquid on the plain surface at large variations of layer heights and reduced pressures. The diagrams of hydrodynamic regimes have been constructed, where the boiling regions or their absence are indicated against each layer height, reduced pressure and the heat flux. This paper presents the results of the heat transfer experimental study of low-pressure evaporation and nucleate boiling in horizontal, different-height layers of n-dodecane on a capillary-porous surface.

## 2. Experiments

The experimental setup consists of a process chamber, a cooling system, a system for measuring pressure and temperature, a system of collecting and processing test data, and a system of heaters power control and supply. The process chamber was designed as a thermosyphon (figure 1). It is a cylindrical vessel of 12Cr18Ni10Ti steel with the internal diameter of 120 mm, the height of 300 mm and wall thickness of 1 mm. A 30 – mm thick high heat – conducting brass plate was enclosed between the

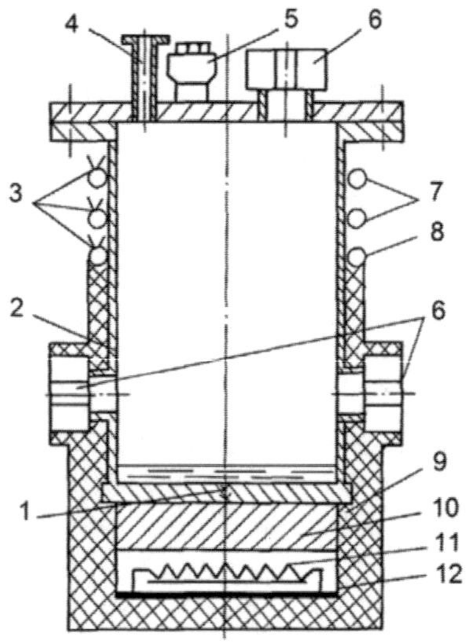


**Figure 1**. The process chamber of experimental setup: 1 – heated surface with coating, 2 – body frame, 3 – thermocouples, 4 – branch pipe for fixing the pumping system and measuring the pressure, 5 – vacuum inlet, 6 – inspection windows, 7 – cooling coil, 8 – heating coil, 9 – thermal insulation layer, 10 – brass plate, 11 – electric heater, 12 – electric heater cover.

bottom and a heater in order to attain uniform of the heat flux on the heated surface. To reduce contact thermal resistance, the gap between the chamber bottom and the plate was filled with a special high heat-conducting paste. There is a cooling coil on the outer surface of the upper part of the chamber. The chamber is cooled by water flowing through the coil. To reduce heat losses due to the leakage along the chamber walls from the bottom to the cooling coil and to provide uniform temperature distribution at the chamber bottom, an optional coil for heating the chamber walls was fixed below the cooling coil. At the top and sides, the process chamber was fitted with the windows for visual observations. To measure the temperature of the heating surface, the copper-constantan thermocouples in stainless capillaries were inserted in five holes of 1.5 mm in diameter into the bottom at different heights. The heat flux was obtained from the temperature gradient measured along the centerline of the upper portion of the bottom using a linear fit from the output of five thermocouples by Fourier equation. The calculated uncertainty was about ±16% at heat flux of $q = 10^3$ W/m$^2$, nearly ±10% at heat flux of $q = 10^4$ W/m$^2$, and ±4% when the heat flux was $q = 10^5$ W/m$^2$. The higher the heat flux, the lower is the uncertainty. The surface temperature was determined by linearly extrapolating the measured temperature profile to the surface. The overall uncertainty in the surface temperature measurement was found to be about ±0.3°C at 200°C. The pressure in the process chamber was measured by a deformation-ionization vacuum gauge and maintained constant. The uncertainty of the measured pressure is ±0.5% of the reading. A capillary-porous coating (figure 2(a)) was applied to the bottom of the process chamber using a 3D laser printing SLS – method (selective laser sintering) technology [10]. The similar technique of surface coatings production is described in [11]. However, opposed to the technique in [11], solid melt of powder particles was not used in the present study. Judging by its properties, the capillary-porous coating is close to the coating made from a sintered metallic powder. The bottom of the process chamber (figure 2(b)) was 12 mm thick and used as a heating surface with the diameter of 120 mm.

(a) (b)

**Figure 2.** Heating surface of the experimental facilities. (a) The photograph of the microstructured capillary-porous coating; (b) The bottom of the process chamber of the experimental setup with the microstructured capillary-porous coating.

The study of heat transfer at evaporation and boiling in thin horizontal liquid layers was conducted with the use of microstructured 2D modulated capillary-porous coatings, which contour could be approximated by the equation:

$$z = \frac{A}{2}\sin\left(\frac{2\pi}{\lambda_m}x\right) + \frac{A}{2} + \delta_0,$$

where $z$ – a vertical coordinate, m; $x$ – a horizontal coordinate, m; $A = \delta - \delta_0 = 0.5 \cdot 10^{-3}$ – the rib height, m; $\delta$ – maximum height of the coating, m; $\delta_0 = 50 \cdot 10^{-6}$ – thickness of the continuous coating without contouring or shaping, m; $\lambda_m = 2l_\sigma = 3.5 \cdot 10^{-3}$ – distance between the ribs, m; $l_\sigma = (\sigma/g(\rho_l - \rho_v))^{1/2}$ – Laplace constant, m; σ – surface tension, N/m; $\rho_l$, $\rho_v$ – liquid and vapour density, respectively, kg/m$^3$; g – acceleration of gravity, m/s$^2$. The coating was made by applying the powder LPW 155 (15-5PH) from stainless steel, with the particles size of 20-40μm, and porosity $\varepsilon = 0.44$.

Prior to the experiments, a certain amount of liquid (n-dodecane), required to form a layer of a desired thickness, was poured onto the bottom. When decreasing pressure in the process chamber, the liquid was deaerated; that was recorded by termination of characteristic bubbling caused by removal of bubbles of dissolved air. For more complete deaeration, the working liquid was boiled under vacuum for several hours.

The experiments were carried out at the pressure of $P$ ($P/P_{cr}$): 0.133 ($7.4\cdot10^{-5}$), 1 ($5.5\cdot10^{-4}$), 10 ($5.5\cdot10^{-3}$), 20 ($1.1\cdot10^{-2}$), where $P$ and $P_{cr}$ are pressure and critical pressure, respectively, KPa; ($P/P_{cr}$) – reduced pressure. Studied layer heights $h$ ($h/l_\sigma$): 1.4 (0.81), 1.7 (0.99), where $h$ – layer height, mm; ($h/l_\sigma$) – dimensionless layer height.

During the experiments, a number of steady heat transfer conditions were achieved. At these stationary regimes, the temperatures over the heated wall thickness, as well as pressure over the liquid layer in the chamber were registered. The data of visual observations were recorded by high-speed video camera. The CHF was determined by the moment of the sharp increase in the heating surface temperature at a constant heat flux.

## 3. Results and discussion

The heat flux density – $q$, W/m$^2$, the temperature of the heating surface – $T_w$, K at constant pressure over the layer were experimentally determined. The saturated-steam temperature in $T_s$, K was calculated from the steam pressure. The heat transfer coefficient was calculated by the formula $\alpha = q/(T_w - T_s)$, W/(m$^2$·K). The results were compared with the experimental data on heat transfer [9, 12] obtained for the given experimental setup during evaporation/boiling of n-dodecane on a plain surface. Figure 3 presents the results obtained for the regime of nucleate boiling. The curve of nucleate boiling with the negative slope was constructed for the 1.4-mm high layer (see figure 3(a)). The similar type of dependence was traced in most studies; in [11] it was observed for the pool boiling of the liquid. This type of the curve behavior is presented and discussed in the review [1]. The nucleate boiling tends to start at lower heat flux on the surfaces with the capillary-porous coating, compared with the coating-free surfaces. The temperature difference is 3-4 time less for the coated surfaces. Figure 3(b) presents data on heat transfer coefficients on the plain surface and on the surface with a capillary-porous coating for two heights of the liquid layer. It's a well-known fact that the heat transfer

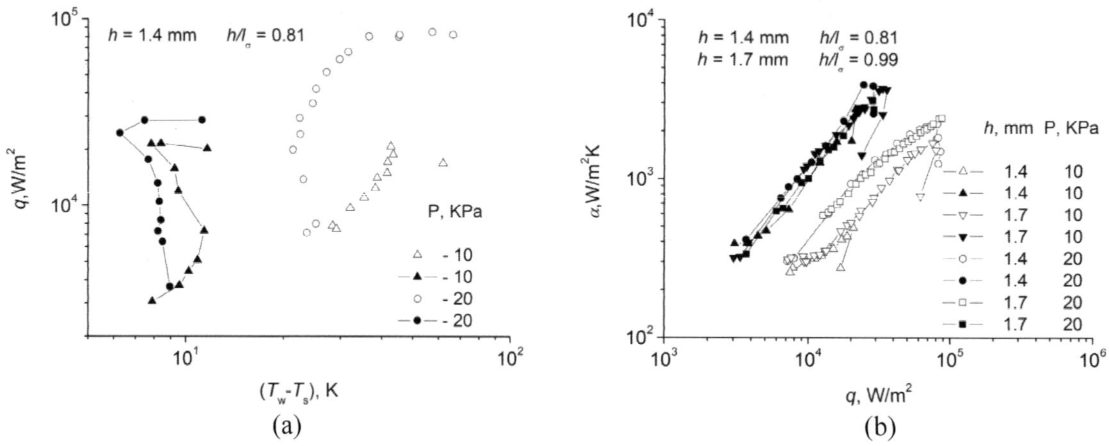


**Figure 3**. The regime of nucleate boiling. (a) The heat flux density versus the temperature difference; (b) The heat transfer coefficient versus the heat flux density. Coloured symbols correspond to the data on the coated surface, open points indicate data on the plain surface.

coefficients tend to increase with a rise of pressure on the plain surface. The pressure is seen to have less effect on the heat transfer coefficients on the capillary-porous coating. The heat transfer coefficients are 3-5 times higher on the capillary-porous coating compared with the uncoated surface. As shown in figures 3(a) and 3(b), heat transfer intensification is higher for the lower pressure.

The nucleate boiling was absent in the range of low pressures – $P$ ($P/P_{cr}$): 0.133 ($7.4 \cdot 10^{-5}$), 1 ($5.5 \cdot 10^{-4}$) KPa both on capillary-porous and plain surfaces [9]. The liquid evaporated from the surface coating at relatively low heat fluxes. The liquid level over the total surface, apart from the area near the chamber walls, tended to decrease with a rise of the heat flux. The ribs were dampened with the liquid and projected above the surface. The liquid was fed along the troughs between the ribs. At some moment in time, pressure pulsations occurred within the working chamber, the surface became partially dry, the liquid level was falling, and a dry spot emerged. While moving round the surface, the dry spot was shrinking or increasing in size. Large-scale low-frequency pulsations of pressure were observed in the total volume depending on the change of the dry spot size at 10-second intervals. The pulsation frequency tended to rise with the increase of the heat flux. The heat transfer coefficient increased sharply with the emergence of the dry spot, and the heat transfer intensified by a factor of 4 - 5 (figure 4). When the pressure reached $P = 0.133$ KPa, the heat transfer coefficient increased at the heat flux density of $q = 9.8 \cdot 10^3$ W/m$^2$. However, at $P = 1$ KPa, the heat transfer coefficient increased at the heat flux density of $q = 2 \cdot 10^4$ W/m$^2$. The moving objects, called "craters" [7-9], form on the plain surface within this range of pressure variations in the thin liquid layer. The "craters" surface is covered with a residual layer of the liquid. The absence of nucleate boiling in the thin liquid layer on the capillary-porous surface at low pressure was observed in the experiments [4–6] with the use of water as a working liquid. As suggested in these studies, the liquid layer decreased or passed into the capillary-porous coating. The movement of the forming objects was not noticed, because much smaller heating surfaces were used. In [4, 5], the heating surface was a square with a side of 11 mm, and in [6] it was a square with a 30 mm side. In the present study, the diameter of the heating surface with the capillary-porous coating was several times larger.

Figure 5 presents a comparison of the experimental data on the CHF values obtained on the capillary-porous surface in the present study with the results obtained in [9] for the plain surface.

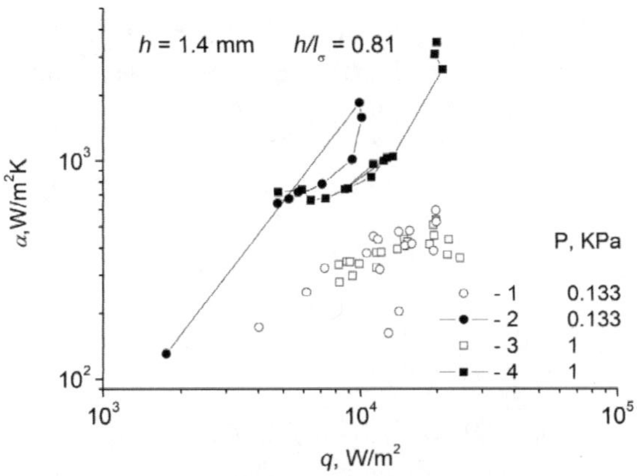


**Figure 4.** The heat transfer coefficient versus the heat flux for the evaporation regime. Coloured symbols indicate data on the coated surface, open symbols indicate data on the plain surface.

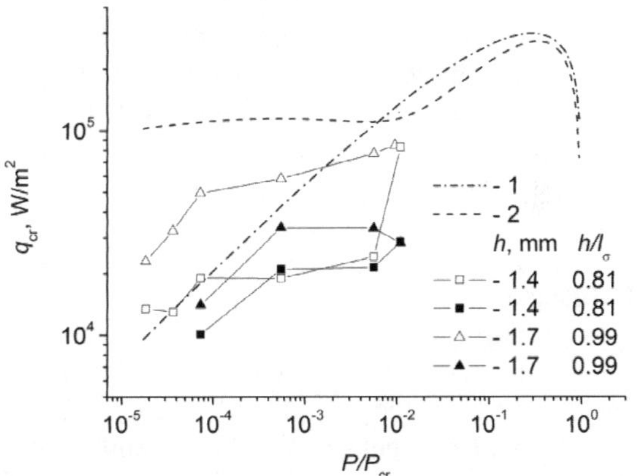


**Figure 5.** The critical density of heat fluxes versus reduced pressure. The calculated dependences: 1-[13], 2-[14]. Coloured symbols indicate data on the coated surface, open symbols indicate data on the plain surface.

The graphs demonstrate that the CHF value obtained on the capillary-porous surface within the studied pressure range significantly depends on the layer height and reduced pressure. The calculated dependences determined by the Kutateladze [13] and Yagov [14] formulae are given on the graph for comparison. Significant deviations from the Kutateladze calculated dependence for the plain surface within the reduced pressure range are explained by the change of the heat transfer mechanism (absence of nucleate boiling), and by the impact of intensive liquid evaporation from the upper free layer boundary. It also confirmed that CHF for evaporation/boiling on the capillary-porous surface don't exceed the values calculated by the Kutateladze formula. The change of the heat transfer mechanism was observed in comparison with conditions of pool boiling. It should be noted that the layers of small heights were studied in this research. Consequently before and during the crisis, the liquid on the most portion of the heat transfer surface was inside the capillary-porous coating, where the free boundary of the liquid film was absent. Evaporation processes took place inside the coating and caused some reduction of CHF compared with the evaporation process in the layers with upper free boundary on the plain heating surface. Similar to the process of boiling/evaporation on the plain heating surface, there exists the pressure region, where CHF considerably depend on the height of the liquid layer and slightly influenced by pressure. This region is biased towards higher pressures for heat transfer on the capillary-porous surface.

## 4. Conclusion

In this study was fabricated 2D modulated capillary-porous surface for evaporation and boiling heat transfer using 3D laser printing SLS – method (selective laser sintering) technology.

The nucleate boiling tends to start at lower heat flux on the surfaces with the capillary-porous coating, compared with the coating-free surfaces. The heat transfer coefficients are 3-5 times higher on the capillary-porous coating compared with the plain surface. Pressure has less effect on the heat transfer coefficients on the capillary-porous coating.

The CHF value obtained on the capillary-porous surface within the studied pressure range significantly depends on the layer height and reduced pressure.

## Acknowledgement

This study was supported in part by the Russian Science Foundation (project no. 19-19-00180 for the block of experimental investigations of heat transfer, critical heat flux, high-speed video monitoring of transition processes and studying structure formation in horizontal liquid layers on the plain surface) and by the Russian Foundation for Basic Research (project no. 18-08-00545-a for the block of experimental investigations of heat transfer in thin horizontal liquid layers on the stainless capillary-porous surface at reduced pressures).

## References

[1] Liang G and Mudawar I 2019 *Int. J. Heat Mass Transfer* **128** 892.
[2] Lin L and Kedzierski M A 2019 *Int. J. Heat Mass Transfer* **131** 1279.
[3] Patil C M and Kandlikar S G 2014 *Heat Transfer Eng.* **35** 887.
[4] Liou J H, Chang C W, Chao C and Wong S C 2010 *Int. J. Heat Mass Transfer* **53** 1498.
[5] Wong S C, Liou J H and Chang C W 2010 *Int. J. Heat Mass Transfer* **53** 3792.
[6] Hong F J, Cheng P, Wu H Y and Sun Z 2013 *Int. J. Heat Mass Transfer* **63** 389.
[7] Gogonin I I, Dorokhov A R and Zhukov V I 1989 *Izv. Sib. Otd. Akad. Nauk SSSR, Ser. Tech. Nauk* **3** 8. [in Russian].
[8] Zhukov V I 1996 *Technical Physics Letters* **22** 879.
[9] Zhukov V I and Pavlenko A N 2018 *Int. J. Heat Mass Transfer* **117** 978.
[10] Bayev S G, Bessmeltsev V P, Goloshevsky N V, Goryaev Ye P, Kasterov V V and Smirnov K K 2017 *Proceedings of the International Scientific Conference "SibOptics - 2017" (Novosibirsk)* vol 1 (Novosibirsk: SSUG&T), p 29.
[11] Zhang C, Zhang L, Xu H, Li P and Qian B 2019 *Int. J. Heat Mass Transfer* **128** 570.
[12] Zhukov V I and Pavlenko A N 2018 *Technical Physics Letters* **44** (6) 508.
[13] Kutateladze S S 1950 *Zhurnal Tekhnicheskoy Fiziki [Zh. Tekh. Fiz.]* **20** 1389 [in Russian].
[14] Yagov V V 2014 *Int. J. Heat Mass Transfer* **73** 265.

5th International Workshop on Heat/Mass Transfer Advances
for Energy Conservation and Pollution Control
August 13-16, 2019, Novosibirsk, Russia

# Research Inside a Gas Heating Device with Different Heat Transfer Medium

**Yun Guo[1,*], Zhongwei Zhou[1], Hao Chen[1], Yeyu Ouyang[1] and Zhiqiang Huang[2]**

[1] School of Mechanical and Automotive Engineering, Shanghai University of Engineering Science, Shanghai 201620, China
[2] University of Shanghai for Science and Technology, Shanghai 200093, China

[*]graceguo1980@126.com

Abstract: Natural gas heating device occupies an important position in natural gas industry, and the most widely used one is the cylindrical heating device with heat transfer medium. The key factors affecting the thermal efficiency of the cylinder gas heating device with heat transfer medium are the type of heat transfer medium and the flow field organization. In order to study the influencing factors of thermal efficiency of the heating device, a visualized practical gas heating device experiment system was established, and a two-dimensional heat transfer model fitting the real object was established, too. The unstructured fitting mesh was used to divide the physical model, and the finite volume method was used for discrete solution, while discrete coordinate (DO) radiation model is used to solve radiation equation. In this paper, water and ethylene glycol are used as heat transfer media to simulate and measure the heat flow field in the cylinder. The results show that ethylene glycol can broaden the application range of gas heating device. Under the same experimental conditions, the heat efficiency of the heating device with ethylene glycol as heat transfer medium is at least 3.8% higher than that of the heating device with water as heat heat transfer medium. The thermal efficiency of gas heating devices is generally less than 87%.

## 1. Introduction

Gas heating device is an indispensable equipment in the application of natural gas. In the process of using natural gas, heating is often needed. For example, in gas-fired generating units, the process has a high requirement on the temperature of fuel gas, so the natural gas used in power plants must undergo heat treatment. LNG also needs to be heated in the gasification process [1]. Therefore, it is of great significance to design and manufacture gas heating devices with high efficiency and energy saving.

## 2. Characteristic and experimental system of gas heating device

The structure of gas heating device is shown in figure 1(a). Fire tube and flue tube bundle are filled with flame and smoke from fuel combustion. The convective tube bundle is filled with natural gas to be heated. The intermediate heat transfer medium fills the inner space of the whole cylinder, and transfers the heat generated in the fire tube and flue tube bundle to the natural gas in the convective tube bundle to complete the heating process of natural gas. In this process, the type of heat transfer medium and its flow field organization are the key factors affecting the thermal efficiency of gas heating device [2]. At present, water and ethylene glycol are the main heat transfer medium used in heaters [3].

The experimental system of gas heating device is composed of test body, gas burner, smoke exhaust system, heated gas system and data acquisition system [4]. Nine-joint armored thermocouple and single-joint thermocouple are installed on A-A section to test the temperature distribution of heat transfer medium in cylinder, as shown in figure 1(b). During experiments, the fuel rates were set at 8 $m^3/h$, 10 $m^3/h$, 12 $m^3/h$, 14 $m^3/h$, 16 $m^3/h$, and the flow rate of the heated gas was set at 1.67 m/s; the pressure and inlet temperature of the heated gas are 5.5 MPa and 280 K, respectively.

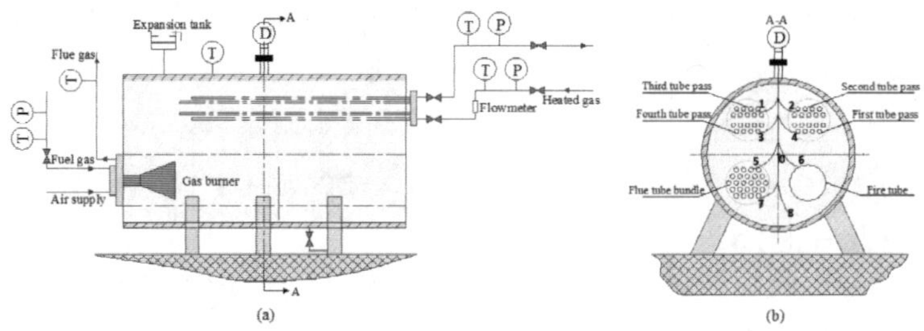


Figure 1. The experimental system of gas heating device.

## 3. Mathematical model

The object of numerical simulation in this paper is the two-dimensional circular section in the middle of the heater cylinder. The structure is shown in figure 2. Among them, D1 is 1300 mm, D2 is 325 mm, b1 is 540 mm and b2 is 510 mm. The diameter of flue tube is 42 mm and the distance between them is 68 mm. The diameter of the convective tube is 38 mm, b3 is 90 mm, b4 is 65 mm, and the angle is 60°.

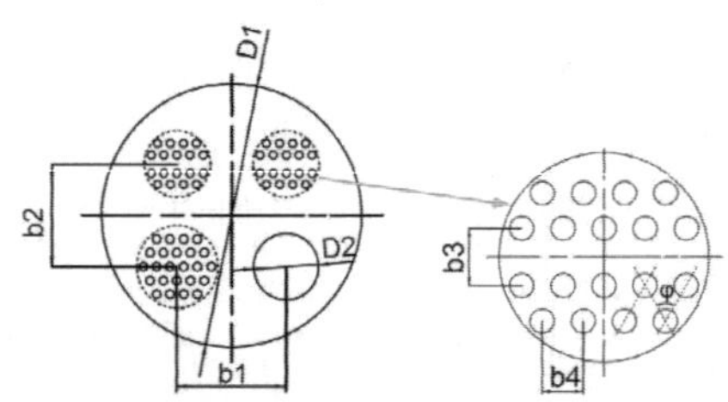


Figure 2. Numerical model simulating a cross-section of the heating device.

Based on the boussinesq assumption[5], the governing equations of two-dimensional flow and heat transfer in combination of natural convection and participating medium radiation are listed below[6].

Continuity equation:

$$\frac{\partial u}{\partial x}+\frac{\partial v}{\partial y}=0 \tag{1}$$

Momentum equations:

$$\rho\left(u\frac{\partial u}{\partial x}+v\frac{\partial u}{\partial y}\right)=-\frac{\partial p}{\partial x}+\mu\left(\frac{\partial^2 u}{\partial x^2}+\frac{\partial^2 u}{\partial y^2}\right) \tag{2}$$

$$\rho\left(u\frac{\partial v}{\partial x}+v\frac{\partial v}{\partial y}\right)=-\frac{\partial p}{\partial y}+\mu\left(\frac{\partial^2 v}{\partial x^2}+\frac{\partial^2 v}{\partial y^2}\right)+\rho g\beta(T-T_\infty) \tag{3}$$

Energy equation:

$$\rho c_p\left(u\frac{\partial T}{\partial x}+v\frac{\partial T}{\partial y}\right)=\alpha\left(\frac{\partial^2 T}{\partial x^2}+\frac{\partial^2 T}{\partial y^2}\right)-Q_s \tag{4}$$

Among them, $u$, $v$ are the velocity components on $x$, $y$ directions, respectively; $p$ is the pressure; $\mu$ is the fluid dynamic viscosity; $\beta$ is the fluid volume expansion coefficient; $\rho$ is the fluid density; $T$ is the temperature; $T_\infty$ is the fluid reference temperature; $\alpha$, thermal diffusivity; and $c_p$ is the specific heat; $Q_s$ is the source term of radiative heat transfer. The radiation transfer equation is solved by discrete coordinate method.

The wall of the cylinder is treated as an adiabatic boundary; the average temperature of the fire tube wall is set as 368K, and the average temperature of the flue tube bundle wall is set as 356K; the boundary condition of the third kind is applied to the boundary of the convective tube bundle, and it is expressed as:

$$-k_e \frac{\partial T_w}{\partial n} = h(T_w - T_g) \tag{5}$$

where $k_e$ is the thermal conductivity of heat-transfer medium; $n$ indicates the wall normal direction; $T_w$ is the wall temperature; and $T_g$ is the heated gas temperature. The average temperature of the heated gas in the first and second tube passes is 290 K, and the average temperature of the heated gas in the third and fourth tube passes is 310 K. $h$ is the heat transfer coefficient calculated by

$$h = \frac{Nu \cdot k_g}{d} \tag{6}$$

in which $k_g$ is the thermal conductivity of gas inside the convective tube bundle; $d$ is the diameter of the convective tubes; and the Nusselt number is calculated by a correlation[7]:

$$Nu = 0.023 Re^{0.8} Pr^{0.4} \tag{7}$$

Where $Re$ is the Reynolds number for the gas flow inside the convective tubes, and $Pr$ is the gas flow Prandtl number.

This paper uses the commercial software Fluent 6.3 to simulate, and uses the finite volume method to discretize. The DO radiation model is adopted because this radiation model can not only solve the S2S problem, but also solve the participating radiation problem of the medium [8]. The physical model uses unstructured Cartesian body-fitted grid [9-10]. The grid is refined near the boundary surface and the grid independence test is carried out [11].

## 4. Results and discussion

The isothermal cloud images and velocity vector distributions of water and ethylene glycol are shown in figure 3 and figure 4, respectively.

The temperature distribution on the cross section of heater is uneven. Generally speaking, the highest temperature is at the top of the fire tube and flue tube bundle, followed by their outer wall. The temperature at the bottom of the cylinder is usually the lowest.

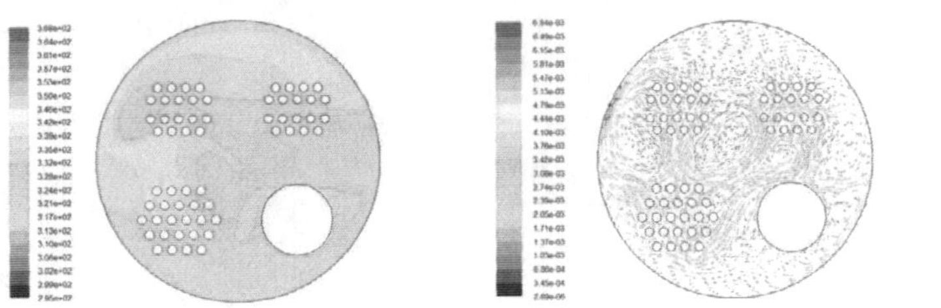

(a) Isothermal contour (K) (b) Flow vectorgram (m/s)

Figure 3. Temperature and flow fields with medium water.

As can be seen from the temperature distribution of water in figure 3(a), the high temperature region at the top of the flue tube bundle presents a wide trunk shape due to the heat transfer behavior of the high-temperature medium water when it flows upward. Due to the strong influence of the low temperature fluid around the convection tube, the high temperature area around the fire tube moves to the left and right. As can be seen from the flow vectorgram in figure 3(b), the water near the fire tube is heated to form upwelling. The water near the first and second convection bundles falls because of lower temperatures. In a traditional heating furnace, heating surface and cooling surface are arranged symmetrically. If there is no proper flow channel between the two surfaces, the relative flow of medium water will be in conflict, as shown in figure 5(a). The chaotic flow field of circular cross section is consistent with the actual working behavior.

Figure 4 (a) shows the isothermal contour of ethylene glycol. As can be seen from the figure, the top of the flue tube bundle appears to be a wide treelike region of high temperature, which is similar to the situation when the medium is water. But at the top of the fire tube, there is a long and narrow high temperature zone, which is shifted to the right and left by the strong influence of the cold fluid around the first pass of the convective tube bundle. According to the velocity vector distribution in figure 4(b) that the ethylene glycol on the top of the fire tube flows upward and to the left and right. At this time, there is still collision between the heating surface of the fire tube and the cooling surface of the convection tube bundle, as shown in figure 5(b), which forms an opposition zone and hinders the development of natural convection flow field, thus greatly reducing the heat transfer efficiency.

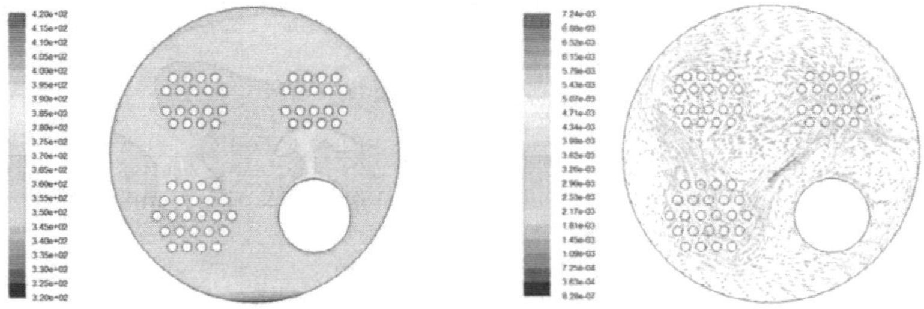

(a) Isothermal contour (K) (b) Flow vectorgram (m/s)

Figure 4. Temperature and flow fields with medium ethylene glycol.

Although the viscosity of ethylene glycol is larger, the velocity of ethylene glycol is a little faster than that of water under the effect of coupled natural convection and participating radiation, and the impact phenomenon of ethylene glycol has been improved slightly than that of water, as shown in figure 5.

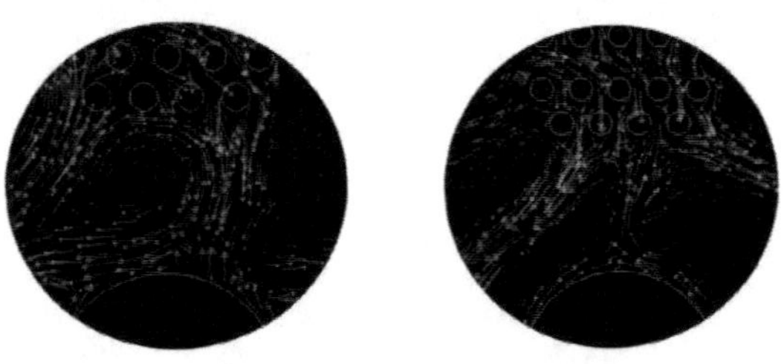

(a) Water (b) Ethylene glycol

Figure 5. Local flow hedging phenomenon.

Figure 6 shows the simulated and experimental temperature distributions of water and ethylene glycol on section A-A under the fuel rate of 10 $m^3/h$. The temperature distribution trend of numerical results and experimental results is consistent. Because the cylinder wall is considered as adiabatic wall, the heat dissipation of the cylinder wall is ignored. Furthermore, the simulation of two-dimensional circular cross section also ignores the axial flow. Therefore, the simulated temperature is usually 1~ 2.5k higher than the corresponding temperature measured in the experiment.

It can be seen from figure 6 that whether water or ethylene glycol is used as the heat transfer medium, the temperature distribution pattern on the circular section is basically consistent.From the temperature values of each measurement point, it can be seen that at different heights of the circular section, the temperature stratification phenomenon is obvious. However, the temperature difference between the upper part and the lower part of the flow field is not large, which indicates that the driving force of natural convection is very small. This defect result in a slow start-up of the heater, and it takes 4 to 6 hours for each test condition to reach a stable state. This leads to a great reduction in the efficiency of gas heating device and high energy consumption.

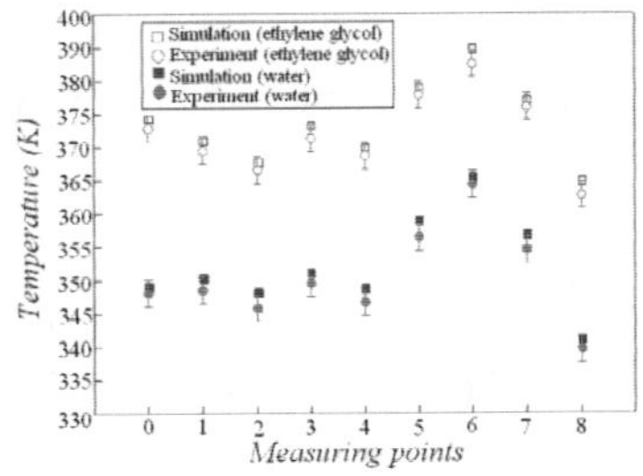


Figure 6. Temperature distributions on section A-A under fuel rate of 10 $m^3/h$.

Figure 6 also shows that, under the same experimental conditions, the temperature at each measurement point of ethylene glycol is 15~ 30k higher than that of water. There are two reasons. First, the heat capacity of ethylene glycol is much smaller than that of water. Second, the high boiling point and low vapor pressure of ethylene glycol make its operating range much larger than that of water.

Figure 7 shows the heat transfer efficiency of heating device with water and ethylene glycol as heat transfer media respectively under different fuel rates. Gas heating device have the highest thermal efficiency with fuel rates of 12 $m^3/h$ of ethylene glycol and 10 $m^3/h$ of water. When the experimental conditions were the same, the heat transfer efficiency of the heating device using ethylene glycol as the heat transfer medium was at least 3.8% higher than that of water. As mentioned above, under all corresponding experimental conditions, the operating temperatures of ethylene glycol are higher than that of water. The highest operating temperature of ethylene glycol can reach 440 K, so the temperature difference will be larger and the thermal efficiency is relatively high. It is one of the reasons to choose ethylene glycol as the intermediate heat carrier medium.

However, due to the structural defects of the traditional heating device itself and the high viscosity of ethylene glycol, it is difficult to form an effective heat flow field inside the cylinder, and the thermal efficiency is usually less than 87%..

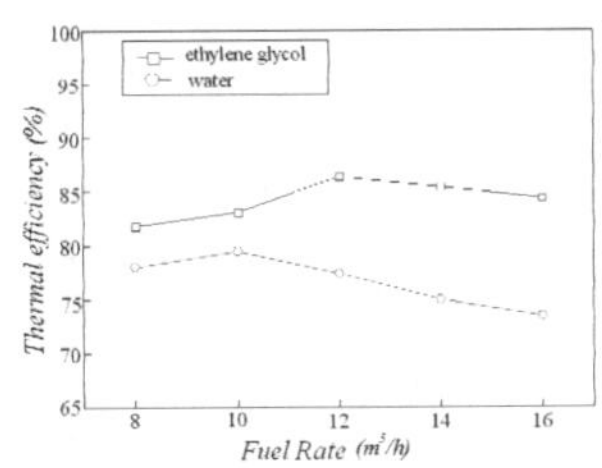


Figure 7. The thermal efficiency under different fuel rates.

## 5. Merits and demerits with ethylene glycol as the heat transfer medium

It is important to choose suitable heat transfer medium for gas heating device. The choice of heat transfer medium mainly depends on the heat transfer performance, pressure and temperature.

When water is used as a heat transfer medium, its operating temperature range is very narrow (323~373 K). Steam or ethylene glycol must be used as heat transfer medium when operating temperature exceeds 373K.When steam is used as a heat carrier, it must be observed frequently, but ethylene glycol is not required. Therefore, ethylene glycol is safe and reliable as a heat transfer medium.

The high boiling point and low vapor pressure of ethylene glycol help its heat absorption and exothermic process at operating temperature without phase transition. In addition, ethylene glycol operates in a wide range of temperatures and pressures as long as the temperature is below 450 K.

## 6. Conclusions

- Under the effect of coupled natural convection and particpating radiation, the heat flow field of ethylene glycol has been improved slightly than that of water, although the visocity of ethylene glycol is larger.
- The operating temperatures of ethylene glycol are higher than that of water under all corresponding operating conditions. Gas heating device have the highest thermal efficiency with fuel rates of 12 $m^3$/h of ethylene glycol and 10 $m^3$/h of water.
- Numerical calculation and experimental results show that the symmetrical arrangement of heating and cooling components in the conventional gas heating device can easily cause the flow field to collide, resulting in the low efficiency of the heating device (less than 87%)..
- When the experimental conditions were the same, the heat transfer efficiency of the heating device using ethylene glycol as the heat transfer medium was at least 3.8% higher than that of water. For gas heating device, the advantages of using ethylene glycol as the heat transfer medium is dependent on its operating temperature and the design of heat transfer components.

**Acknowledgements**
This project is supported by National Natural Science Foundation of China (Grant No. 51606116).

**References**

[1] Li E J, Tang W H and Zheng C 2013 Modular design of process systems of LNG gasification station, Gas and Heat. 33(8): 50-54.
[2] Guo Y, Cao W W and Yan P 2011 Application and research progress of heater in natural gas industry, Heat Transfer Eng. 132: 1003-1008.
[3] Umesh V, Balavignesh S and Raja B 2018 Single-phase convective heat transfer of water and aqua ethylene glycol mixture in a small-diameter tube, J. Eng. Thermophy. 27(1):98-105.
[4] Guo Y, Cao W W, Yan P, Yu C X and Qian S Y 2009 Visualizing experimental investigation on the natural gas heating device, J. Univ. Shanghai Sci. Technol. 31: 47-51.
[5] Cherkasov S G, Anan'ev A V and Moiseeva L A 2018 Limitations of the Boussinesq model on the example of laminary natural convection of gas between vertical isothermal walls, High Temp. 56(6): 878-883.
[6] Hassani M and Karimipour A 2018 Discrete ordinates simulation of radiative participating nanofluid natural convection in an enclosure, J. Therm. Anal. Calorim. 134(3): 2183-2195.
[7] Bejan A 2013 Convection Heat Transfer, 4th ed. (Wiley, New Jersey).
[8] Prieler R, May B and Demuth M 2016 Prediction of the heating characteristic of billets in a walking hearth type reheating furnace using CFD, International J. Heat and Mass Transf. 20: 675-688.
[9] Mathur S R and Murthy J Y 1997 A pressure-based method for unstructured meshes, Num. Heat Transf. B 31: 195-215.
[10] Aghanajafi C and Abjadpour A J 2016 Discrete ordinates method applied to radiative transfer equation in complex geometries meshed by structured and unstructured grids, Braz. Soc. Mech. Sci. Eng. 38(3): 1007-1019.
[11] Guo Y and Guo Z X 2016 Flow and heat transfer inside a new diversion-type gas heating device, Num. Heat Transf. 70: 1-13.

5th International Workshop on Heat/Mass Transfer Advances
for Energy Conservation and Pollution Control
August 13-16, 2019, Novosibirsk, Russia

# Analysis of power and energy management system for new energy dining car

**Yun Guo*, Yeyu Ouyang, Yufei Gao and Bingde Wei**

College of Mechanical and Automotive Engineering, Shanghai University of Engineering Science, Shanghai 201620, China

*graceguo1980@126.com

Abstract: Mobile dining car is very popular in Europe and USA because of its advantages of convenient and fast. However, most of dining cars still use traditional power and energy systems, which is not good for the environment. In recent years, the mobile dining car is developing towards light, convenient, and multi-functional. In large and medium-sized cities where business is concentrated, it is worth studying how to make the dining car cleaner and lower carbon, and make it a kind of green and healthy diet culture. In this paper, the green energy is introduced into the dining car power system and cooperated with the flywheel to realize the effective use of photovoltaic cells, achieve zero emission of the dining car, and put an end to the environmental pollution from the source. The social value and the actual effect will be great. Based on the analysis of the actual operating conditions of the dining car, the key parameters such as electrical equipment and motor load of the dining car are obtained, and then the matching calculation of the power system of the new energy dining car is carried out, including the power generation of photovoltaic cells and the storage capacity of flywheel batteries. Then, according to the unique working mode of the dining car, the energy distribution strategy and control system are analysed. Finally, a mathematical model is established and simulated with MATLAB to verify the effectiveness of the design. The results show that the system can complete the day work of the dining car.

## 1. New energy dining car power system parameter matching

### *1.1. Photovoltaic cell matching calculation*

According to the power consumption of the dining car and the power generated by photovoltaic cells, the number of photovoltaic panels can be calculated.

Set the series-parallel efficiency of the battery as 0.9, and set the number of solar cells required to meet the use of the vehicle at rated power as follows:

$$n = \frac{P_a \cdot 1000}{0.9P_m} \tag{1}$$

Calculate n=31.8, value n=32.

In this chapter, 32 photovoltaic cell modules with a total power of 6.91 kW were selected in the dining car. According to the illumination of Shanghai and other variable factors, the power of solar cells can be calculated [1]. Assume that the dining car on the photovoltaic cell components fully expanded. According to the daily lighting hours, working ten hours per day, the average illumination intensity is one third of the maximum illumination intensity, which can be calculated as, photovoltaic cell modules receive an average of 3 hours of light per day, according to the formula to calculate power $Q = P_s \times t$.

According to the calculation, the daily output of photovoltaic cell modules Q=20.7 kWh, far exceeding the normal energy consumption. Therefore, in order to match the power generation of the photovoltaic cell, a set of flywheel cells should be added.

### *1.2. Flywheel battery matching calculation*

In this paper, the battery is designed according to the power parameters of the dining car. In the case of batteries as the only power supply, at this time, the battery should satisfy the dining car to travel 40km at a speed of 50km/h, and ensure that the highest 7kw electrical appliances work normally for 2 hours.

It is known from the analysis of working conditions that the dining car consumes 15.97 kWh of energy per day. However, considering that the self-discharge of the carbon-cellulose flywheel battery in the latest study is 5% in a day, and the energy conversion efficiency is about 95%, and the specific energy of the existing flywheel battery is not higher than 200 W·h/kg, the flywheel batteries need to store energy more than 200 W·h/kg, so the overall weight of the flywheel battery is at least 88.7 kg.

According to the formula, the mass, radius size and angular velocity of the flywheel battery can be calculated. Because the flywheel battery is a mechanical battery, its energy is calculated according to the mechanical energy. But in actual use, there will be a certain loss. So the actual parameters are shown in the table below:

Table 1 Flywheel battery parameters of this dining car

| Storage capacity | Maximum power | Quality of the battery | Maximum speed | Radius of the rotor | The rotor material | The quality of rotor |
|---|---|---|---|---|---|---|
| 18 kW·h | 150 kW | 90 kg | 450 r/s | 0.3 m | Composite | 60 kg |

### *1.3. Flywheel - photovoltaic system analysis and wiring control*

The flowchart of energy conversion of the flywheel - photovoltaic system composed of photovoltaic wing and flywheel is shown in figure 1:

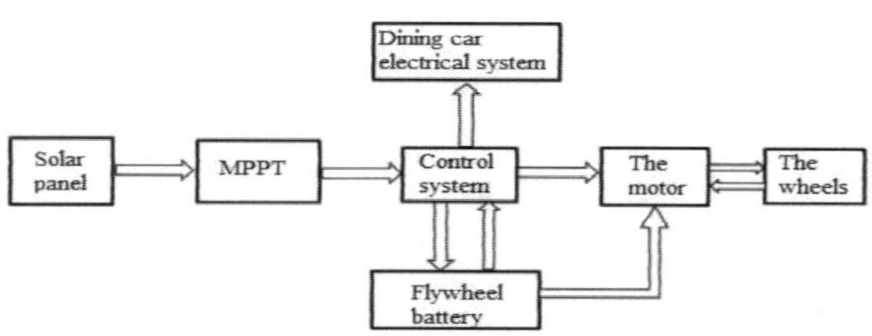


Figure 1. Flow chart of flywheel photovoltaic wing.

In the design of the new energy dining car, the photovoltaic wing is combined by 16 in series and 2 in parallel. The photovoltaic wing is controlled by MPPT controller to track the maximum power point, and its power and output voltage is monitored in real time by the monitoring end of the electronic control system[2]. After the energy distribution device is fed into the flywheel to stabilize the voltage, it is provided to the electric equipment and motor in the car. In this period, inverter is needed for Dc to Ac conversion. Finally, for the sake of safety, Dc circuit breaker and lightning protection switch are installed in the system circuit to protect the safe operation of the system[3] (as shown in figure 2).

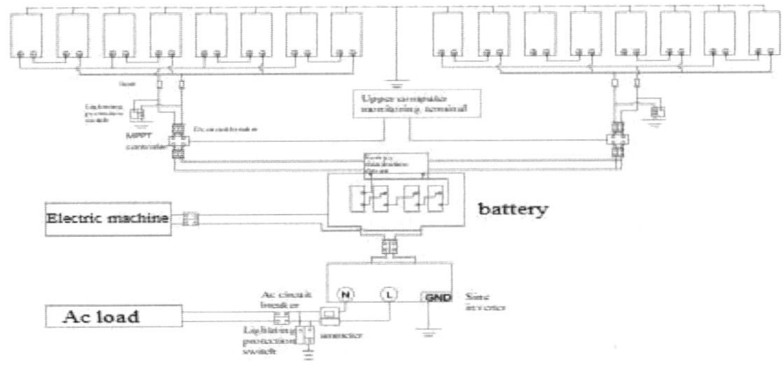


Figure 2. The overall wiring of the vehicle control system.

The energy storage effect of the flywheel is quite strong. After charging in the daytime, there is no problem for the power system of the cooker to work in the evening. Therefore, it will not affect the use of the flywheel because the energy will be transformed into new energy dominated by solar energy[4].

## 2. Work mode and control strategy analysis

### *2.1. Analysis of dining car working mode*

Control strategy of this paper is mainly according to the different working condition of concrete analysis, improve the service life and battery life of the dining car effectively, to put it simply, energy distribution consists of five modes: photovoltaic battery powered separately, flywheel battery powered separately, photovoltaic battery powered for load and charging for flywheel battery ,photovoltaic flywheel battery powered mixed and charging for flywheel battery . Under different conditions, how to use the five kinds of mode, is the focus of energy distribution.

The core of the whole energy distribution system is the flywheel battery, when to charge the flywheel battery, when the flywheel battery discharge, control this one, you can properly deal with the problem of energy distribution. And the diners practical block 32 solar photovoltaic panels, tightening state when driving, as shown in figure 3, the power is small, cannot satisfy the drive motor, electrical appliances, and give the flywheel battery, may also need to the flywheel battery discharge in auxiliary drive, when diners in a fixed mode, all the solar photovoltaic panels as shown in figure 4, the output power depending on light intensity and temperature of the outside world will have different change, the overall power is can satisfy a used electrical appliances and for flywheel battery.

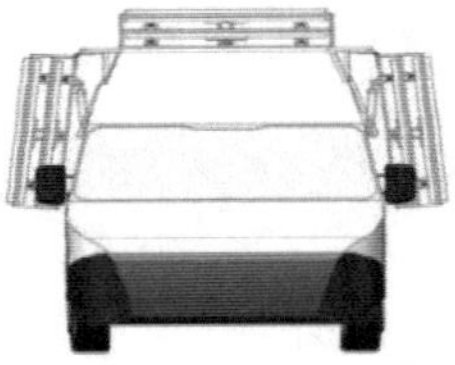

Figure 3. photovoltaic panels contracted state.

Figure 4. photovoltaic panels expanded state.

Now the power distribution of the whole vehicle is listed as an equation:

$$P_{me} + P_{fe} = P_{se} + P_{be} \tag{2}$$

According to the above situation, the following two aspects should be considered when formulating the energy control and distribution strategy of solar energy dining car:

- To consider the SOC value of a flywheel battery, you need to put it between $SOC_{Low}$ and $SOC_{Top}$. That is the best operating range of the flywheel battery, which can maintain its service life and stable output. To do this, reserve a SOC*, which is the highest value that must be charged and the lowest value that can be charged into the flywheel battery.

- The output power of solar photovoltaic panel, because its output power varies greatly with the external environment, so even when the photovoltaic panel alone to power, also need to be in the working state of the battery. According to the different proportions of batteries, the working conditions are divided into the following four types:

When the solar energy is supplied separately without charging the battery:

$$P_{me} + P_{fe} = P_{se} \tag{3}$$

When the solar energy is supplied separately and the battery needs to be charged:

$$P_{me} + P_{fe} = P_{se} - P_{be} \tag{4}$$

When no light, battery independent power supply:

$$P_{me} + P_{fe} = P_{be} \tag{5}$$

When the dining car braking, there is surplus electricity:

$$P_{me} = P_{be} \tag{6}$$

## 3. Simulation

### *3.1. Power source engineering model of new energy dining car*

The above solar cell model is to model the principle of solar power generation under ideal conditions, mainly to analyze its characteristic curve, and the research results can better match the calculation of the battery. According to the solar cells are commonly used several important parameters, such as the short circuit current $I_{sc}$, open circuit voltage $V_{oc}$, when the most high power voltage of maximum power point of current $I_m$, $V_m$, maximum power Pm, these are available according to the battery factory, on this basis, can be simulated under certain precision solar diners a day's work, the first step first according to the principle of photovoltaic cell model established engineering practice[5]. The model is as follows:

$$I = I_{sc}(1 - C_1)\left[exp\,(\frac{V}{C_2V_{oc}}) - 1\right] \tag{7}$$

At $P_m$, $V=V_m$, $I=I_m$, so:

$$I_m = I_{SC}(1 - C_1)\left[exp\left(\frac{V_m}{C_2V_{OC}}\right) - 1\right] \tag{8}$$

Because at normal temperature, $\exp\left(\frac{V_m}{C_2V_{OC}}\right) \gg 1$, So the"-1" term can be ignored, work out $C_1$ :

$$C_1 = (1 - \frac{I_m}{I_{oc}})exp\,(-\frac{V_m}{C_2V_{oc}}) \tag{9}$$

In the open circuit, when I=0, $V=V_{OC}$, because $exp\left(\frac{1}{C_2}\right) \gg 1$, Therefore, the term "-1" can also be ignored, and $C_2$ can be solved as:

$$C_2 = \left(\frac{V_m}{V_{oc}} - 1\right)\left[ln\left(1 - \frac{I_m}{I_{oc}}\right)\right]^{-1} \tag{10}$$

However, this model is still to idealistic. In the practical application of the dining car, the influence of external environmental factors on the model should also be considered. For this reason, more accurate simulation should be carried out according to the changes of the external environment.

### *3.2. The engineering model when the external environmental conditions change*

According to the above research on the *I-V* characteristic curve of solar cells, it shows that solar power and current are related to solar intensity and battery temperature. According to the environment of the dining car, the change range of solar intensity on the ground in Shanghai area is 0 ~1000 W/ $m^2$, and it is the strongest at noon. And the temperature change of the solar cell is bigger, so the change of the design range is 10~70 °C. The characteristic curve of solar photovoltaic panel is simulated [6].When the sunshine intensity and the battery temperature are inconsistent with the standard conditions, the influence of external environmental conditions on the characteristics of solar cells must be considered. Therefore, $G$ is set as any light intensity, and $T$ is set as the solar cell temperature under any ambient temperature $T_{air}$. According to the experimental data and some data searched, the following formula is proved to be of sufficient accuracy in engineering sense [7]:

$$T = T_{air} + K + G \tag{11}$$

In the formula above, K is determined by the slope of the line T (s), According to the common characteristics of the panel as well as access to some information, K = 0.03 °C · $m^2$/W.

According to the derivation of the model and the change of the characteristic curve, two models can be selected as the actual operation of the solar power model of the dining car.

According to I−V characteristic curve of reference sunshine intensity and reference battery temperature, the change curvature of any point ($V$, $I$) can be obtained, and the movement of any point ($V$, $I$) can be used to obtain the new sunshine intensity and any point ($V'$, $I'$) on the i-v characteristic curve under the new battery temperature. The specific algorithm is as follows [8]:

$$I' = I + \Delta I \tag{12}$$

$$V' = V + \Delta V \tag{13}$$

$$\Delta T = T - T_{ref} \tag{14}$$

$$\Delta = -\beta \times \Delta T - R_s \times \Delta I \tag{15}$$

$$\Delta I = \alpha \frac{G}{G_{ref}} \times \Delta T + I_{sc}\left(\frac{G}{G_{ref}} - 1\right) \tag{16}$$

For monocrystalline and polycrystalline silicon solar cells, the test values are:

$$\alpha = 0.0012 I_{sc} \tag{17}$$

$$\beta = 0.005 V_{oc} \tag{18}$$

Thus, the engineering model of solar panels under changing external conditions is as follows:

$$V' = V + \Delta V \tag{19}$$

$$I' = I_{sc}\left(1 - C_1\left[\exp\left(\frac{V - \Delta V'}{C_2 \times V_{oc}}\right) - 1\right] + \Delta I\right) \tag{20}$$

### *3.3. Simulation of photovoltaic cell under actual operation condition of dining car*

Since the key values of this model are all calculated above, and the main influencing factors are the input light intensity and temperature, MATLAB software is used to simulate several key input values, which can be substituted into the model to obtain relatively accurate simulation results.

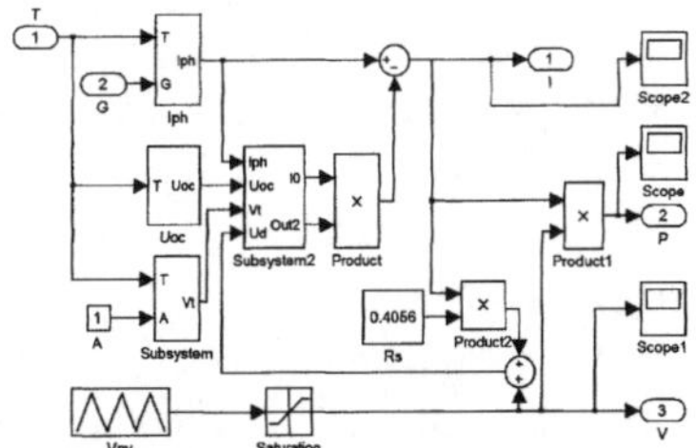

Figure 5. MATLAB simulation model.

All the coefficients calculated above (open circuit voltage $V_{oc}$, peak voltage $V_m$, short circuit current $I_{so}$, peak current $I_m$) and the variables (the change of simulated voltage, light intensity, temperature changes) simulated are substituted into the model, the simulation results of output current and power are shown in figure 7:

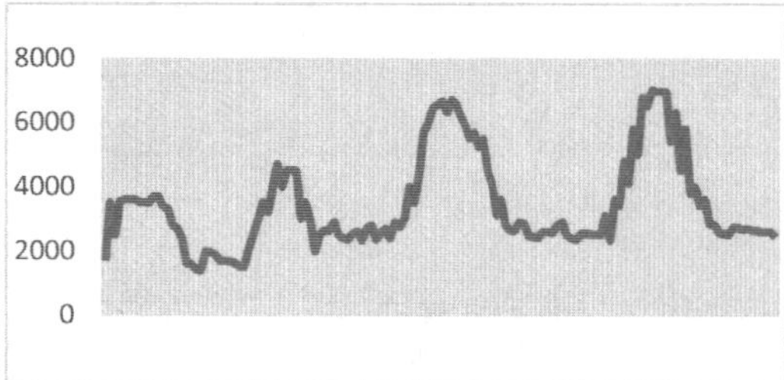


Figure 6. Load power variation curve.

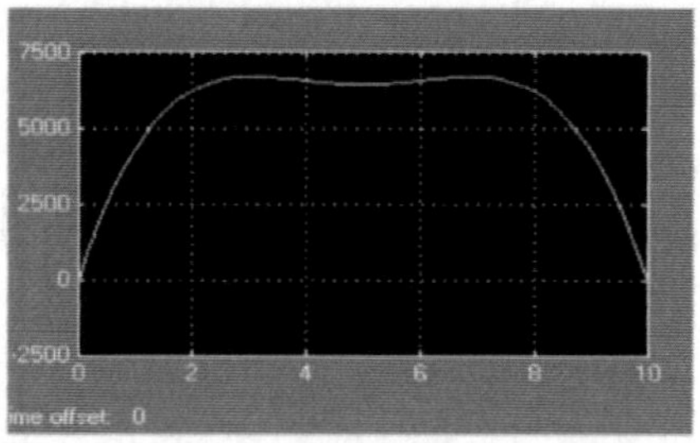


Figure 7. Generation power variation curve.

According to the comparison between the solar power generated by the simulation and the load power obtained from the experiment above, the energy change balance of the new energy dining car studied can continue to cycle. The power of the dining car is all below the power of the power generation, and there are only three wave peaks. The remaining electricity is underestimated. The photovoltaic cell can power the flywheel battery to ensure the power supply when the light is weak. According to the comparison of figure 6 and 7, in this paper, the design of photovoltaic panels slightly larger, because its simulation model is based on light intensity at the time of the theoretical value, therefore should be performed according to the actual climate situation photovoltaic panels and battery matching calculation, different regions, different latitude and longitude of the dining cars, it should be personalized customization, in order to ensure maximum use of energy.

**Acknowledgement**
This paper is supported by National Natural Science Foundation of China (Grant No. 51606116) and Key Scientific and Technological Research Projects of Shanghai Science and Technology Commission (Grant No. 19195810800).

## References

[1] Levasseur A and Cavalett O 2016 *Enhancing life cycle impact assessment from climate science, Review of recent findings and recommendations for application to LCA*, Ecological Indicators.71(12):163-174.
[2] Yan H L and Guo Y. *An intelligent thermoelectric nozzle for a food 3D printer*[P]. ZL201810448387.5
[3] Wang B, *Analysis of PM2.5 sources in Beijing. Motor vehicle exhaust emissions are the main source*.2016-10-31
[4] Chai Y 2014 *Europe is moving towards an era of electric cars*, Seeking Knowledge. 3(05):60-61.
[5] Li G 2011 Research on the key factors affecting the development of China's electric vehicle industry, *Wuhan Univ. Technol. school of management.*
[6] Dai Y D and Zhu Y Z 2013 Bulletin of Chinese Academy of Sciences. *Reshape energy for sustainable development,* 8(02):239-246.
[7] Su L Y, Wang Y and Chen Q 2013 Analysis of energy conservation and emission reduction benefit of pure electric vehicles in China in the future, *Climate Change Research*. 4(11):97-89.
[8] Kang L P and Earley Robert 2014 Do electric cars really save energy and reduce emissions? *New Energy Automobile*. 4(11):87-89.

5th International Workshop on Heat/Mass Transfer Advances
for Energy Conservation and Pollution Control
August 13-16, 2019, Novosibirsk, Russia

# Influence of the thermal parameters on the bubble heat balance at transient boiling of subcooled water

**A A Levin and P V Khan**

Melentiev Energy Systems Institute (ESI) SB RAS, Irkutsk, 130 Lermontova Street, Russian Federation

lirt@mail.ru

**Abstract.** This paper considers the development of the approach presented [1–3] to the determination of heat trasnfer in subcooled boiling flow additively from the components of forced convection and heat fluxes associated with the presence of bubbles on the surface. The case of non-stationary heat release in a vertical channel with an uprising fluid was numerically studied on the basis of the experimental data from [5]. We have studied the effect of the fluid subcooling and the heat flux on the characteristics of transient boiling.

## 1. Introduction

The subcooled liquid is widely used in heat transfer and cooling devices, since it allows removing heat with greater intensity than in case of saturated flow. The formation, growth and collapse of bubbles on the heater surface contribute to the heat transfer enhancement, creating additional heat fluxes. Recently, many papers have been published [3,4] that address calculations of such fluxes based on new experimental data on the bubble growth dynamics and nucleation density. When constructing numerical models of boiling, many authors obtained highly reliable results, but for a limited range of conditions. The reason for this is a large number of closure relations included in the models that describe such key parameters such as the bubbles lifetime, their maximum or lift-off diameter, the size of dry spots and thickness of the superheated liquid layer etc. Given the high variability of the wall boiling conditions (temperature of the flow and heat wall, flow rate, pressure in the system, heat flux, surface wettability etc.), we can hardly expect that the use of semi-empirical models will be successful. Therefore, in recent years, researchers are motivated to focus their studies on the influence of boiling conditions on the simplest individual submodels of heat transfer and/or mechanical equilibrium. In this paper, we consider such components of an individual bubble heat flux as heat flux of the bubble initial evaporation, $q_{ei}$, microlayer evaporation heat flux, $q_{eml}$, heat flux of the heat conduction through the microlayer, $q_{cml}$, heat flux of the heat conduction through the superheated layer, $q_{cs}$, and their sum, $q_b$. We study the influence of such parameters such as subcooling, $\Delta T_{sub}$, heater surface temperature, $T_w$, overall heat flux, $q_{tot}$, and wall heating rate $\partial T/\partial t$ on the calculation of heat fluxes. When calculating the heat fluxes, we used the experimental data presented in Levin and Khan [5].

## 2. Heat transfer model for transient boundary layer

In this paper, we further elaborate on the application of the numerical model presented in [6]. The system of differential equations describing the coupled problem of heat transfer under conditions of forced

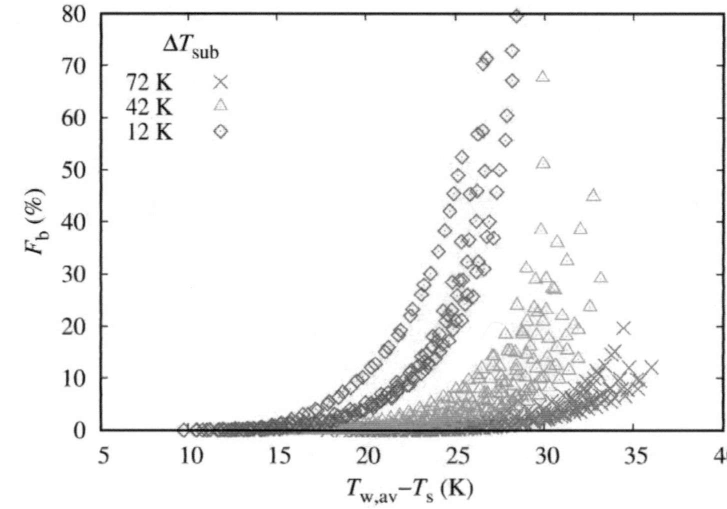


**Figure 1.** Influence of bubble boiling on heat transfer under unsteady heat generation.

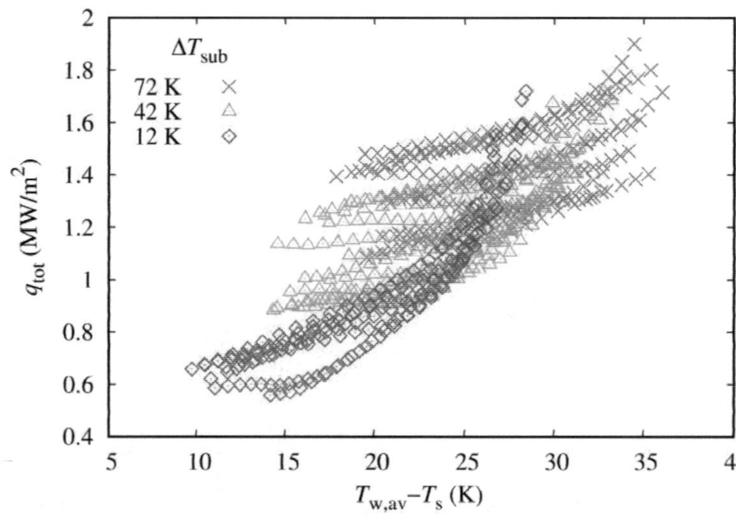


**Figure 2.** Boiling curve for the unsteady heat release.

convection is complemented by closing relations for the heat exchange with bubbles on the surface. The bubble heat flux consists of the thermal fluxes of the initial evaporation of the bubble, $q_{ei}$, the evaporation of the microlayer, $q_{eml}$, the thermal conductivity of the microlayer $q_{cml}$ and the thermal conductivity of the superheated layer, $q_{cs}$:

$$q_b = q_{ei} + q_{eml} + q_{cml} + q_{cs}. \quad (1)$$

Here, the heat balance components are defined as:

$$q_{\mathrm{ei}} = h_{\mathrm{lg}}\rho_{\mathrm{g}}\frac{\pi D_{m}^{3}}{6} f N_{a} \tag{2}$$

$$q_{\mathrm{eml}} = h_{\mathrm{lg}}\rho_{l}\delta_{\mathrm{ml}}\frac{\pi D_{\mathrm{ml}}^{2}}{6} f N_{\mathrm{a}} \tag{3}$$

$$q_{\mathrm{cml}} = \frac{N_{\mathrm{a}}\pi k_{\mathrm{l}}(T_{\mathrm{w}}-T_{\mathrm{s}})D_{ml}^{2}}{4\delta_{\mathrm{ml}}} \tag{4}$$

$$q_{\mathrm{cs}} = N_{\mathrm{a}}\int_{r_{\mathrm{w}}+\delta_{\mathrm{ml}}}^{r_{\mathrm{w}}+\delta_{\mathrm{s}}} k_{\mathrm{l}}(T_{\mathrm{w}} - T_{\mathrm{s}})\,\pi\left(\frac{D_{\mathrm{m}}}{r-r_{\mathrm{w}}} - 2\right) dr, \tag{5}$$

where $D_{\mathrm{ml}}$ is the diameter of the microlayer; $h_{lg}$ – latent heat; $T_w$ – surface temperature; $T_s$ – saturation temperature; $D_m$ – maximum bubble diameter; $k_l$ – thermal conductivity; $N_a$ – nucleation density; f – nucleation frequency; ρ– density; $\delta_s$ – thickness of superheated liquid layer; $\delta_{\mathrm{ml}}$ – thickness of microlayer.

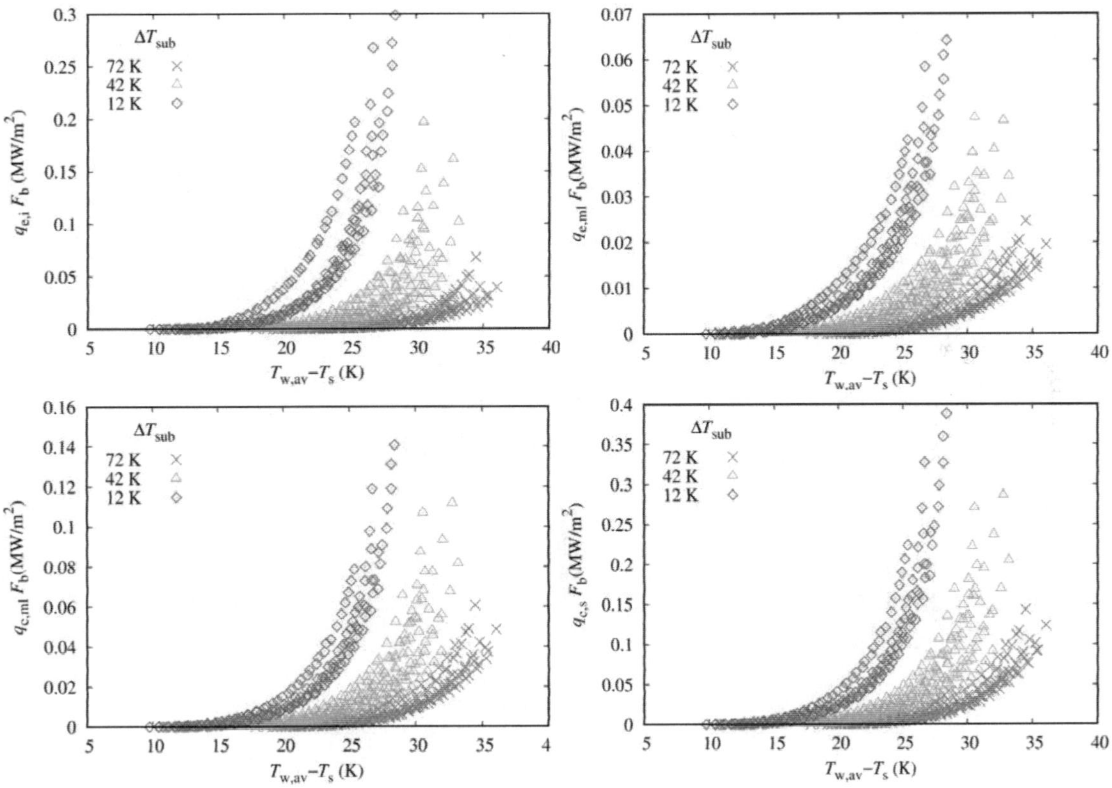


**Figure 3.** Individual components of the heat balance in the wall liquid layer under unsteady heat generation depending on the surface overheating.

Nucleate density for the unsteady case was satisfactory predicted by following correlation [6]:

Journal of Physics: Conference Series **1369** (2019) 012010 doi:10.1088/1742-6596/1369/1/012010

$$N_a = 10^6 \cdot [\exp(0.14(T_{\mathrm{w}} - T_{\mathrm{ONB}}) - 1]. \tag{6}$$

Relative influence area $F_b$, within which equation (1) is responsible for taking into account the heat balance in the wall liquid layer, is defined as a part of the heating surface fully covered by bubbles of diameter $D_{\mathrm{m}}$:

$$F_{\mathrm{b}} = \frac{\pi}{4} N_a D_{\mathrm{m}}^2. \tag{7}$$

Figure 1 presents the influence of the liquid temperature and the heater surface on the surface area with nucleate boiling. The increase of $\Delta T_{\mathrm{sub}}$ from 12 K to 72 K leads to decrease of $F_{\mathrm{b}}$ from 80 to 20%. The exponential nature of the dependence of the area influencing the wall temperature is reflected by formula (6). A significant decrease in the share of the heater surface covered with steam should diminish the role played by the components of the heat balance (2)–(5) when we reduce the initial liquid temperature $T_0$.

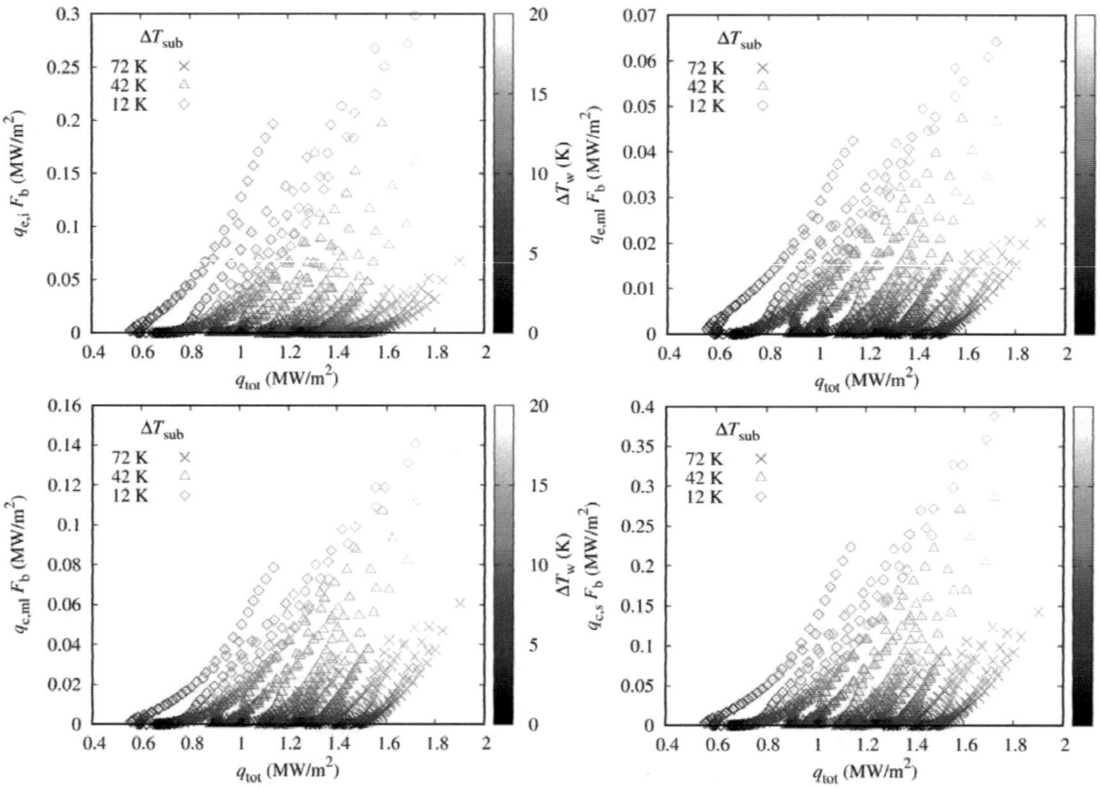


**Figure 4.** Dependence of individual components of the heat balance in the wall liquid layer under unsteady heat release on the heat flux $q_{\mathrm{tot}}$.

The dependence of the calculated values of the heat flux $q_{\mathrm{tot}}$ on the temperature head $\Delta T_w$ demonstrates an increase in the intensification of heat transfer with a decrease in the liquid temperature of the liquid $T_0$ (figure 2). The influence of the liquid subcooling on the heat flux exceeds the results obtained by other authors for stationary boiling. This can be explained by the significantly smaller thickness of the superheated liquid layer in the case of a rapidly increasing surface temperature.

Analyzing the influence produced by the heat releasing surface temperature and the total value of the heat flux into the liquid on the components of the heat balance in the boundary liquid layer (figures 3 and 4), we can observe some common features: the greatest contribution to the overall heat transfer from heat conduction through the overheated layer of fluid $q_{cs}$; the second most important (about 80% of $q_{cs}$) term in the balance is the heat transfer associated with the final condensation of the vapor bubble ($q_{ei}$); the thermal conductivity through the liquid microlayer under the bubble $q_{cml}$ is about 50% of $q_{ei}$; the heat of evaporation of the liquid microlayer $q_{eml}$ is approximately equal to 25% of $q_{ei}$. As expected, an increase in the liquid subcooling leads to a significant decrease in the components $q_{eml}$, $q_{cml}$, $q_{cs}$, $q_{ei}$ under the increase of the general heat flux $q_{tot}$, because the area covered by bubbles $F_b$ decreases.

Using empirical estimates for such nucleate boiling characteristics such as the maximum bubble diameter, nucleation frequency, and nucleation site density, the heat balance components can be written as follows:

$$q_{\mathrm{ei}} = C_{\mathrm{ei}}\left(e^{0.14\Delta T_{\mathrm{w}}} - 1\right)\Delta T_{\mathrm{w}}^{2} \quad (8)$$

$$q_{\mathrm{eml}} = C_{\mathrm{eml}}\left(e^{0.14\Delta T_{\mathrm{w}}} - 1\right)\Delta T_{\mathrm{w}} \quad (9)$$

$$q_{\mathrm{cml}} \approx C_{\mathrm{cml}}\left(e^{0.14\Delta T_{\mathrm{w}}} - 1\right)\Delta T_{\mathrm{w}}^{2} \quad (10)$$

$$q_{\mathrm{cs}} \approx C_{\mathrm{cs}}\left(e^{0.14\Delta T_{\mathrm{w}}} - 1\right)\Delta T_{\mathrm{w}}^{2}\ln \Delta T_{\mathrm{w}} \quad (11)$$

where $\Delta T_{\mathrm{W}} = T_{\mathrm{W}} - T_{\mathrm{ONB}}$, and $C_i$ are the values that do not depend on the wall overheating. It follows from (8)–(11) that the heat balance components can be ordered with the respect to the increase in the heat flux growth when the wall overheating increases: $q_{eml}$ has the smallest size, then $q_{ei}$ and $q_{cml}$ follow, while $q_{cs}$ has the biggest growth rate. Estimating the influence of factors from equations (8)–(11), we can also conclude that the nucleation site density dominates the heat transfer process in the wall liquid layer.

## Conclusions

Under the conditions of unsteady heat release into a subcooled liquid flow, variations of the wall temperature, the initial temperature and the value of the heat flux, the density nucleation centers are influenced more than the size of bubbles. We can assume that this effect will become stronger with the increase in pressure. The increase in the total heat flux from the wall into the liquid under increasing subcooling is caused by a significant decrease in the superheated layer thickness. In this case, the values of heat flux components associated with the presence of the vapor phase are significantly reduced, since a decrease in nucleation density has a greater effect on heat transfer in general than an increase in heat transfer in the neighborhood of a single bubble. It is worth noting that our research does not cover coalescence nucleate boiling, for which a more complex heat transfer model should take into account the mutual influence of the nucleation spots.

## Acknowledgments

The research was performed at Melentiev Energy Systems Institute SB RAS under the support of the Russian Foundation for Basic Research, Grant No. 18-29-10019, using the equipment of the High-Temperature Circuit Multi-Access Research Center.

## References

[1] Rohsenow W M 1953 Heat transfer with evaporation *Heat Transfer—A Symposium held at the University of Michigan During the Summer of 1952* (University of Michigan Press) pp 101–150

[2] Steiner H, Kobor A and Gebhard L 2005 A wall heat transfer model for subcooled boiling flow *International Journal of Heat and Mass Transfer* **48** 4161–73
[3] Gilman L and Baglietto E 2017 A self-consistent, physics-based boiling heat transfer modeling framework for use in computational fluid dynamics *International Journal of Multiphase Flow* **95** 35-53
[4] Mohanty R L and Das M K 2017 A critical review on bubble dynamics parameters influencing boiling heat transfer *Renewable and sustainable energy review* **78** 466–94
[5] Levin A A and Khan P V 2018 Experimental observation of the maximum bubble diameter in non-stationary temperature field of subcooled boiling water flow *International Journal of Heat and Mass Transfer* **124** 876–83
[6] Levin A A and Khan P V 2019 Characteristics of nucleate boiling under conditions of pulsed heat release at the heater surface *Applied Thermal Engineering* **149** 1215-1222

5th International Workshop on Heat/Mass Transfer Advances
for Energy Conservation and Pollution Control
August 13-16, 2019, Novosibirsk, Russia

# Comparative analysis of heat supply options for small and middle-sized settlements of Eastern Siberia by using uncertain and fuzzy information

**V V Khan[1], N P Dekanova[2] and P V Khan[3]**

[1]Irkutsk National Research Technical University, Lermontov str. 83, Irkutsk, Russia
[2]Irkutsk State Transport University, Chernyshevskogo str., bld. 15, Irkutsk, Russia
[3]Melentiev Energy System Institute, Lermontov str. 130, Irkutsk, Russia

*khan@istu.edu*

**Abstract.** The problems of heat supply in small and medium-sized settlements of Eastern Siberia are considered. Variants of using cogeneration heat plant (CHP) or boiler plant on coal or wood waste, the use of liquid petroleum gas (LPG) and liquid natural gas (LNG) as a fuel, the use of electric boiler plant or stand-alone electric boilers, applications of solar collectors and heat pump installations are considered based on a systems approach and fuzzy modelling, by taking into account the uncertainty of some factors.

## 1. Introduction

One of the most important tasks facing the national economy of Russia is to increase the efficiency of heat supply for small and medium-sized settlements in Eastern Siberia. The low efficiency of heat supply systems is due to quite complex problems: considerable distance from fuel sources, centralized electricity and heat supply; long distances between settlements; the duration of the heating period (2/3 of a year); low outdoor temperatures. Traditionally, it is considered that the centralized heating systems are the most efficient [1]. Moreover, East Siberia is characterized by the presence of relatively cheap coal reserves, therefore coal heat sources are most prevalent in this region.

In large cities, due to the combined generation of heat and electricity and the use of highly efficient technologies, the use of coal-fired CHP plants ensures a relatively low cost of thermal energy and an acceptable degree of purification of combustion products. In small and medium-sized settlements, coal-fired boiler houses are mainly used, which are less efficient than CHP plants and, as a rule, are less environmentally friendly. All these factors lead to a high cost of thermal energy and an increase in tariffs for the population and business. Attempts by the tariff service to restrain the rise in prices for thermal energy lead to the fact that the resource supplying organizations lack the motivation to modernize and increase the energy efficiency of equipment [1-3]. The low level of remuneration, the shortage of funds for the necessary materials does not make it possible to ensure quality equipment operation. This, in turn, leads to significant losses due to low equipment efficiency and high wear of heat networks. This vicious circle needs to be broken. At present, it becomes obvious that the existing methods of state regulation in heating systems require serious changes. One of the most important points of the heat market liberalization planned by the Ministry of Energy is the so-called principle of "alternative boiler house"[1]. On the one hand, it is assumed that the development of the market will lead to the introduction of the most efficient technologies. On the other hand, the "alternative boiler house" tariff should set a limit price for heat at the level of the amount that a consumer could spend on heating using his own gas boiler room, taking into account the cost of its purchase, installation and operation. It will be prohibited to sell heat at a price that exceeds this level. Thus, the key criterion for choosing the ways and means of modernization of heat supply systems is the economically reasonable tariff [3, 4].

The modern level of development of technology creates the possibility of applying a wide variety of methods and means to the organization of heat supply to populated areas. In small and medium-sized settlements cogeneration can be effectively used in mini-CHP. In such settlements low population density and low level of operation increase losses in the transport of thermal energy which leads to low efficiency of the centralized heat supply from boiler plants. The analysis shows that in many cases autonomous energy sources using local types of fuel, as well as secondary and renewable energy sources are more profitable.

In determining the paths for the development of heat supply schemes, the final decision is made not only on the basis of the criterion of an economically reasonable tariff. When making a decision, a multi-criteria system analysis is necessary, taking into account the uncertainty of a number of parameters and the lack of clarity of judgment. Environmental constraints are becoming increasingly important. Many areas of Eastern Siberia, in particular, the southern Baikal region, are among the specially protected natural territories, where strict regulations are imposed on the emissions of harmful substances and even moderate risk of man-made disasters are unacceptable [5].

The experience of developing heat supply schemes shows that along with economic, environmental criteria, danger of emergencies, risks of interruptions in fuel supply, a significant role is played by such criteria as the attitude of the government and the population to particular technologies, prospects for the development of the settlement , etc. Such criteria are of a fuzzy nature. Often they are expressed in the following way: "the population will not accept it .."; "..this is not tested ....", "there is not enough experience." Such different nature of factors that must be considered when deciding on the direction of development of the heating system, the technical impossibility of measuring some indicators, and the fundamental unclearness of others led to the choice of fuzzy logic and optimization methods in this study. Inherent in the estimates of the operation of heating systems, fuzziness has led to the widespread use of fuzzy logic and probabilistic approaches in their analysis and the development of automated control systems [6,7].

## 2. Models and research methods

When making decisions on the future development of heat supply systems for small and medium-sized settlements in the southern part of the Baikal region, the determining factors are the factors influencing the formation of economically sound tariffs. When assessing the cost of fuel and materials, it is necessary to take into account systemic factors, including climatic conditions, environmental restrictions, logistics, etc. [2-4]. When choosing the direction of development of the heat supply system of populated areas, the main role is given to such criteria as: the cost of production and transportation of thermal energy, energy efficiency, environmental friendliness, safety and investment.

The algorithm for selecting the option of heating the settlement is divided into three main stages. At the initial stage, all possible sources of energy are analyzed. For further analysis, only those are selected whose use is possible without violating environmental restrictions and is expedient from a logistics point of view. At the second stage, heat supply schemes for the selected sources are formed taking into account the corresponding technologies. At the third stage, the choice of the heat supply scheme is carried out, based on the criterion of the minimum cost.

This paper uses materials obtained in the development of a heat supply scheme for the Baikal urban settlement [9], which was conducted with the participation of the authors of the article. During the development and subsequent updating of the heat supply scheme, special attention was paid to the choice of heat sources and the degree of centralization of the system. Options for centralized and decentralized heat supply and fuels such as coal, LNG, LPG, renewable and secondary energy sources were considered. The following options were investigated as possible heat supply schemes: modernization of the existing CHP plant; construction of a heat source focused on the use of various types of fuel (LNG, LPG, coal, biofuel) using cogeneration (mini CHP); construction of a boiler plant designed to use various types of fuel (LNG, LPG, coal, biofuel); construction of electric boiler; the device of the autonomous heat sources working at electric energy (EE); use of renewable and secondary energy sources.

The first stage of the research is the collection of information on each of the possible variants. The second stage is the selection of options that meet the restrictions on harmful emissions, and are accepted by the population and government officials. The final stage is the selection of options for which the cost of heat produced does not exceed a certain threshold value. The criterion for selecting an acceptable heat supply option at the second stage is as follows:

$$MAX(R_{NOX}, R_{SOX}, R_{ash}, R_{ac}, 0.5\,R_{pop}, R_{gov}) < 0.5,$$

where $0<R_i<1$ is the value of the membership function of the class of solutions that violate the ith constraint. $R_{NOX}, R_{SOX}, R_{ash}$ show the degree of violation of standards for emissions of nitrogen oxides, sulfur oxides and ash content, respectively; $R_{ac}, R_{pop}$, $R_{gov}$ are numerical representations of the linguistic parameters reflecting the danger of an emergency, the attitude to the heat supply variant of the population and the government, respectively. A coefficient of 0.5 reflects the relative importance of the indicator: from the population the absence of pronounced protest is enough while the options that are not accepted by the government cannot be accepted in principle. At the first stage, ten possible heat supply options for the city of Baikalsk are presented for consideration. At the second stage, options based on the use of coal are excluded due to environmental constraints, options based on the use of gas are excluded for high risks of accidents; heat pumps and solar collectors are excluded for negative attitudes of population and government. As a result, biofuel-related options, electric boiler houses and autonomous electric boilers were found to satisfy all the limitations and pass to the third stage of selection.

The main results of the study on heat supply of Baikalsk city are summarized in table 1. In columns 6-11, the indicators of compliance with the restrictions are given in the form of fractions: in the numerator the absolute or linguistic values are given, and in the denominator the degrees of violation of the restrictions are given. In the derivation of the thermal energy cost when using LPG and LNG the fuel price of 20 thousand rubbles per ton is assumed. It should be noted this price might be higher if we take into account the trends of the world market and the positions of leading Russian companies towards the prospects for gas supply in Eastern Siberia.

The high risk of using LPG and LNG is due to the possible consequences of the gas release in case of accidents during transportation or storage. The government' distrust of heat pump technology and solar collectors is due to previous negative experiences. The biofuel-based technologies meet no confidence because of the lack of experience of large-scale use of the technology. The options based on the use of LNG and LPG are excluded due to the lack of confidence in the reliable gas supplies at reasonable prices. Serious concern regarding the construction of the electric boiler plant is the likelihood of a significant increase in electricity tariffs.

The negative attitude of the population towards heat pumps and solar collectors is caused by the need for serious reconstruction of building heating systems and increasing personal responsibility for the operation of the systems. A significant part of the population does not want to take any responsibility for their dwelling heating and express the position of the mere user: "the government must provide ...". However, increasing the degree of a consumer control over the consumption of energy resources increases his/her motivation to save energy. Owners of houses with autonomous heat sources and energy accounting are much more thrifty about their consumption. From this point of view, the use of autonomous heat sources (house electric boilers and/or heat pump installations) should stimulate consumers to save energy.

## 3. Results and discussion

Comparative analysis of the cost of heating options and the required investment is given in figure 1. From the point of view of economic and energy efficiency, the most attractive are options for the use of biofuels, heat pumps and solar collectors, that is, options using secondary and renewable energy resources. The cheapest energy give options with solar collectors. However, considering other criteria, including the amount of necessary capital investments, this option cannot be accepted as a base case. However, the use of solar energy as a supplement to other heat supply options can improve the efficiency of heating systems.

The lowest cost of production of heat energy is provided by CHP due to the joint production of heat and electric energy and due to the low cost of fuel. But due to severe environmental constraints [5], the use of CHP as a base case has also been rejected. As a result, for Baikalsk city, the base option was chosen for heating from boiler houses operating on biofuel such as dry chips and/or pellets [9]. The use of combined heat and power generation using local biofuels provides a lower cost of energy supply than the boiler houses, but requires higher investments. The choice of a cheaper option with regard to investments was influenced by the low assessment of the development prospects of Baikalsk city. At the same time, from the standpoint of energy efficiency, environmental friendliness, as well as in terms of investment volumes, heat sources using electricity are more preferable. In this case, the rejection of options associated with city electric boiler or with autonomous electric boilers is caused by a high estimate of the risk of increase to an unacceptable level of electricity tariffs. When evaluating a variant using heat pump installations, the lack of experience in large-scale use of heat pumps in Russia has played a decisive role. In addition, several years ago in Baikalsk city, a heat pump of 500 kW was installed. The decision to introduce this installation was made in a purely bureaucratic way, without taking into account the fact that the wastewater treatment plant, where the heat pump was installed, was remote from consumers. As a result, the use of the heat pump installation did not justify itself, and, apparently, this formed a long term negative attitude towards innovation on the part of decision makers.

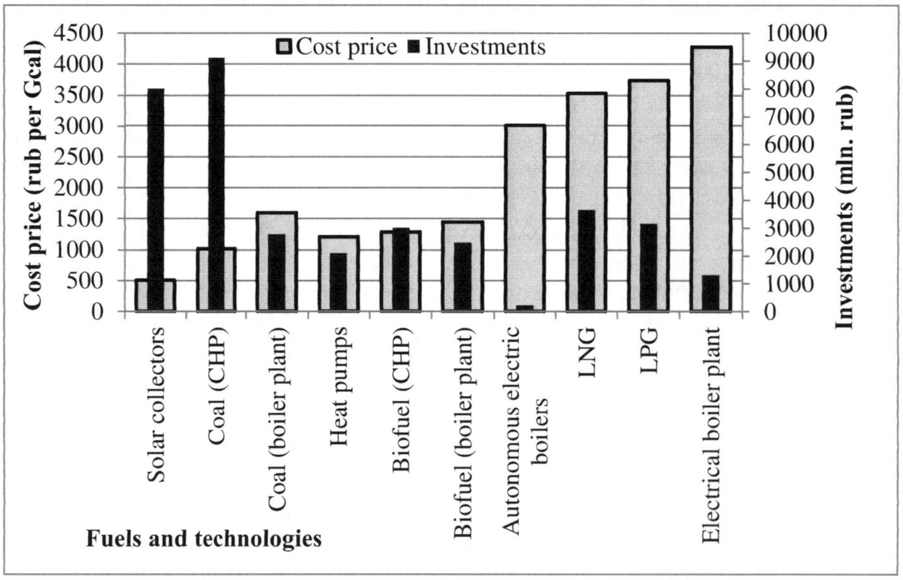


Figure 1. Comparison of options for the cost of production and transport of heat energy and the required investment

Table 1. Characteristics of heat supply development options for Baikalsk city

| Option | Fuel and/or technology | Investments, million rubbles | Cost price, rubbles/ Gcal | Results of restrictions violation accessment | Constraint violation indicators [a] | | | | | |
|---|---|---|---|---|---|---|---|---|---|---|
| | | | | | $NO_x$ emission, $R_{NOX,}$ ton/year | $SO_x$ emission, $R_{SOX,}$ ton/year | Ash emission, $R_{ash}$, ton/year | Accident danger, $R_{ac}$ | Population attitude, $R_{pop}$ | Government asttitude, $R_{gov}$ |
| 1 | 2 | 3 | 4 | 5 | 6 | 7 | 8 | 9 | 10 | 11 |
| 1 | LNG | 3650 | 3527 | declined because of accident danger | 121/0.5 | 0.0 | 0 | high/ 0.9 | neutral/0.2 | neutral/0.2 |
| 2 | LPG | 3150 | 3736 | declined because of accident danger | 121/0.5 | 0.0 | 0 | high / 0.9 | neutral/0.2 | neutral/0.2 |
| 3 | Biofuel (CHP) | 3004 | 1288 | declined because of investments | 6.5/0.03 | 25.0/0.1 | 30.5/0.4 | low/0.1 | neutral/0.2 | neutral/0.2 |
| 4 | Biofuel (boiler plant) | 2474 | 1446 | **accepted** | 6.5/0.03 | 25.0/0.1 | 30.5/0.4 | low/0.1 | neutral/0.2 | neutral/0.2 |
| 5 | Coal (boiler plant) | 2 768 | 1595 | declined because of ecology | 231/1.0 | 251.7/1.0 | 35/0.5 | medium/0. 5 | neutral/0.2 | positive/0.0 |
| 6 | Coal (CHP) | 9 100 | 1020 | declined because of ecology | 320/1.0 | 380/1.0 | 51/1.0 | medium/0. 5 | neutral/0.2 | positive/0.0 |
| 7 | City electrical boiler plant | 1 300 | 4275 | declined because of high risk of EE price growth | 0 | 0.0 | 0 | low/0.1 | positive/0.0 | neutral/0.2 |
| 8 | Autonomous electrical boilers | 209 | 3012 | declined because of high risk of EE price growth | 0 | 0.0 | 0 | low/0.1 | positive/0.0 | neutral/0.2 |
| 9 | Heat pumps | 2100 | 1209 | declined because of population and government attitude | 0 | 0.0 | 0 | low/0.1 | negative/1.0 | negative/1.0 |
| 10 | Solar collectors | 8000 | 505 | declined because of population and government attitude | 0 | 0.0 | 0 | low/0.1 | negative/1.0 | negative/1.0 |

[a] Indicators of violation of restrictions are given in the form of fractions, where the numerator shows the value in its original representation, and the denominator - the value of the corresponding membership function.

Journal of Physics: Conference Series **1369** (2019) 012011 doi:10.1088/1742-6596/1369/1/012011

## 4. Conclusions

Given the full range of system factors, coal-fired boilers in the short term are no longer the cheapest heat sources for many small and medium-sized settlements in the Baikal region. It is advisable for small and medium-sized settlements to individually optimize heat supply schemes. Inflated risk assessments are caused by lack of experience. This, in turn, is due to the lack of pilot projects. To reduce the negative impact of overestimated risk assessments, it is important to intensify pilot projects on the use of heat pumps, solar collectors, biogas, mini CHP.

As for Baikalsk city, finally the biofuel boiler plant was found to be the optimal choice considering all constraints.

## References

[1] Weinsicher B 2014 http://www.forbes.ru/mneniya-column/konkurentsiya/255219-kak-regulirovanie-rynka-tepla-pobezhdaet-zakony-fiziki.

[2] Gasho E G, Guzhov S V and Taktashev R N 2017 Bulletin of the South Ural State University. Ser. Power Engineering 17(4) 14–23

[3] Khan V V, Dekanova N P, Romanova T A and Sharaeva S A 2017 A comprehensive analysis of the effectiveness of energy saving measures for social infrastructure in Eastern Siberia on the basis of system approach *Pr. of Universities. Investments. Construction. Real Estate* 7(1) 101-12

[4] Khan V V, Dekanova N P and Khan P V 2017 Heat supply of villages and towns of Eastern Siberia. Problems and ways of development. *IOP Conf. Ser.: Mater. Sci. Eng.* 262 012081

[5] Order of the Ministry of Natural Resources of Russia dated March 5, 2010 N 63 "On approval of standards for maximum permissible impacts on the unique ecological system of Lake Baikal and the list of harmful substances, including substances belonging to the categories of especially dangerous, highly dangerous, dangerous and moderately dangerous for the unique ecological system of Lake Baikal"

[7] Kobersi I S and Finaev V I 2013 Control of the Heating System with Fuzzy Logic World Applied Sciences Journal **23**(11) 1441-7

[8] Babiarz B 2010 Reliability assessment of heat supply systems in their operational process RT&A **1** 01(16)

[9] 2018 Update of the heat supply scheme completed in Baikalsk, Irkutsk Region: https://irkutskmedia.ru/news/673161/

5th International Workshop on Heat/Mass Transfer Advances
for Energy Conservation and Pollution Control
August 13-16, 2019, Novosibirsk, Russia

# An analytical model of subcritical and critical vapor-liquid flow through a granular bed

**E A Tairov and P V Khan**

Melentiev Energy System Institute, Lermontova 130, Irkutsk, Russia

tairov@isem.irk.ru

**Abstract.** This paper presents a computational and experimental justification of a theoretical model for the two-phase vapor-liquid flow through a fixed bed of solid particles. The theoretical description is based on the equations of gas dynamics of the granular bed and the homogeneous model of a single-component two-phase flow, given the difference between the velocities of liquid and vapor phases. Relationships between the phase slip ratio and polytropic coefficient of isenthalpic expansion of vapor-liquid flow were obtained using the multi-parameter nonlinear regression. We used the experimental data on the vapor-liquid flow with an inlet pressure, $P_1$, of 0.6-1.2 MPa and an inlet flow quality, $x_1$, of 0.016-0.178 through a $H$=50-355 mm bed consisting of steel spheres, 2 and 4 mm in diameter, $d$. An expression for the critical pressure ratio in the granular bed is proposed. The developed model makes it possible to predict the values of mass velocity of the mixture over the entire range of experimental data with an average error of 3%. Based on the model equations, we obtained expressions for dimensionless mass velocity depending on the pressure ratio that can be applied to two-phase flows through both granular beds and other porous media.

## 1. Introduction

The study of a two-phase flow through porous media is of interest in such technical fields as design of chemical catalytic reactors, high-efficient heat exchangers, reactors with spherical coated fuel particles; analysis of nuclear reactor safety; technological advancement of oil and gas production. A major task of fluid mechanics and heat exchange is to establish relation between the flow rate and pressure drop. Most of the theoretical models of the two-phase flow are based on generalization of Darcy's law [1, 2] and the binomial Forchheimer-Ergun equation [3, 4], as well as the equations including interfacial resistance [5, 6]. In all cases the two-phase flow is considered at the mass velocities below their maximum values. The maximum vapor-liquid flow rates, which lead to two-phase flow choking, are calculated, as a rule, as an independent problem. The main results were obtained on the basis of theoretical and experimental studies of critical two-phase flows through long and short pipes, orifices, nozzles, gates and safety valves. The applied theoretical models, which differ in the detail of phenomenon description, are presented in the state-of-the-art papers [7–9]. The slip models for rather long channels are noted to agree well with the experiment in the critical flow rate. The problem is reduced to selection of the method for determination of the specific mixture volume.

The problems of gas-dynamic flow choking in granular media were studied in the monograph by M.A. Gol'dshtik [10], who obtained an analytical expression for the maximum ideal gas flow rate. The flow to vacuum was considered as a condition for the critical flow rate. Generalization of Gol'dshtik 's model to a two-phase vapor-liquid flow is presented in [11]. It successfully described the only available experimental data on the critical flow through fixed beds of spherical particles [12].

This paper generalizes the results presented in [11] to describe subcritical and critical conditions of two-phase vapor-liquid flows through the close-packed beds of solid particles on the basis of the unified theoretical model.

## 2. Mathematical model of vapor-liquid flow

The critical and subcritical vapor-liquid flow through the bed of fixed spherical particles are described using the equations of gas dynamics of the granular bed [10]. The main equations comprise the motion equation

$$\left[1+\frac{4}{3}(1-m)\right]\rho w_m \frac{dw_m}{dy} = -\frac{dP}{dy} - \frac{3}{2}\frac{m(1-m)}{\psi d}\rho w_m^2 , \quad (1)$$

the continuity and polytropic equations

$$\rho w_m = \rho_l w_{1,m} , \quad (2)$$

$$P / P_1 = (\rho / \rho_1)^n . \quad (3)$$

Here $m$ is the average porosity, $w_{\mathrm{m}}$ is the volume-averaged velocity of vapor-liquid flow in the pore space, $d$ is the diameter of a spherical particle, $\psi = 0.508 + 0.56(1-m)$ is the relative minimum flow cross-section.

Integration of these equations by the bed height from 0 to $H$ on the assumption that $d << H$ results in the following expression for the mass velocity:

$$\rho w_m = \left[\frac{2n}{3(n+1)}\frac{d}{H}\frac{\psi}{m(1-m)}P_1\rho_1\left(1-\left(\frac{P_2}{P_1}\right)^{\frac{n+1}{n}}\right)\right]^{0.5} . \quad (4)$$

Here $P_1$, $\rho_1$ are the pressure and density at $y=0$; $P_2$ is the pressure at $y=H$ . The maximum velocity can be obtained by setting $P_2$ to zero.

In view of generalization of the presented model in [11] to a two-phase flow, we assume that the vapor-liquid flow through the bed is isenthalpic, and that the thermodynamic mixture state is subject to the polytropic equation. Phase temperatures are set equal and velocities of vapor and liquid are different. A vapor component moves faster than a liquid one, and the ratio of their velocities, $s = w''_m / w'_m$, called a slip ratio, has a significant effect on the void fraction:

$$\varphi = \left[1 + s\frac{\rho''}{\rho'}\frac{1-x}{x}\right]^{-1} , \quad (5)$$

which in turn determines the mixture density:

$$\rho = \rho'(1-\varphi) + \rho''\varphi . \quad (6)$$

The polytropic coefficient $n$ and the slip ratio $s$ in equations (3)–(5) are determined by generalization of experimental data obtained in [12] in the range of inlet pressures from 0.6 to 1.2 MPa, inlet flow quality $x_1$ from 0.016 to 0.178, and mass velocity from 150 to 1200 kg/(m$^2$s). They are represented by the formulas:

$$n = 0.43 + 0.57(1\text{-}\exp(\text{-}x_1/0.16)), \quad (7)$$

$$s = 1 + a_s(P_1)\exp\left(-\left(\frac{b_s(P_1)}{x_1} + \frac{c_s(P_1)}{1-x_1}\right)\right), \quad (8)$$

where

$$a_s(P_1) = 16.0 + 39.5P_1 - 42.4P_1^2;$$

$$b_s(P_1) = 0.0175 - 0.0123\,P_1 - 0.00249\,P_1^2;$$

$$c_s(P_1) = -2.30 + 11.5\,P_1 - 8.04\,P_1^2 .$$

As follows from equation (7), the polytropic coefficient does not depend on bed parameters and the inlet pressure, and is a function of the inlet flow quality solely.

## 3. Results and their discussion

In this work the model presented by equations (3)–(8) was adapted to description of both the critical and subcritical mass velocity values, which were calculated in a set of experiments with the beds of steel spheres, 2 and 4 mm in diameter and a bed height of 50, 100, 250, 355 mm. The pressure at the bed inlet was 0.6, 0.9 and 1.2 MPa, the inlet flow quality changed from 0.016 to 0.178, the measured mass velocity values were within the range from 150 to 1200 kg/(m$^2$s). Dependence of the mass velocity on the pressure drop in the bed was determined based on 54 processed conditions. The critical mass velocity was obtained by gradual decrease of the outlet pressure to the air pressure at the fixed values of inlet pressure and inlet flow quality. With decrease in the ratio between the outlet and inlet pressures, $\beta = P_2 / P_1$, below the critical pressure (or otherwise, with increase in the pressure drop, $\Delta P = P_1 - P_2$. above the critical pressure) the mass velocity took its maximum value and became insensitive to further decrease of the pressure $P_2$ beyond the bed.

Analysis of the experiments reveals that the critical pressure ratio $\beta_{cr} = P_{2,cr} / P_1$, depending on experiment conditions, can take different values ranging from 0.15 to 0.36. Still, $\beta_{cr}$ does not practically depend on bed parameters, and decreases with increase of $P_1$ and inlet flow quality $x_1$. The least value, $\beta_{cr} = 0.15$, is obtained at $P_1 = 1.2$ MPa and $x_1 = 0.178$. At the same pressure and lower flow qualities that are characterized by the values of $x_1$ = 0.033–0.055, $\beta_{cr} = 0.25$. The decreasing effect of $x_1$ on the critical pressure ratio is also observed in the study of highly wet vapor flow rates through the Laval nozzle [13, 14]. As a result of processing available experimental data on reaching a critical flow through the beds the approximate formula for $\beta_{cr}$ is written as:

$$\beta_{cr} = 0.145 P_1^{-0.822} x_1^{-0.0813}. \qquad (9)$$

Formula (4) for critical conditions using $\beta_{cr}$ takes the form

$$(\rho w_m)_{cr} = \left[ \frac{2n}{3(n+1)} \frac{d}{H} \frac{\psi}{m(1-m)} P_1 \rho_1 \left( 1 - \beta_{cr}^{\frac{n+1}{n}} \right) \right]^{0.5}. \qquad (10)$$

Constitutive relations (7, 8) for the polytropic coefficient and the slip ratio are obtained by generalization of experimental results using formula (10). Note that formulas (7), (8) are related, as a rule, to the known mixture parameters at the bed inlet and applied in the range: $0.016 < x_1 < 0.178$; $0.6 \le P_1 \le 1.2$ MPa.

The developed mathematical model is used to generalize experimental results on the subcritical and critical vapor-liquid flows through different packed beds of spherical particles. A typical behavior of the mass velocity when reaching a critical condition with decrease in the pressure $P_2$ beyond the bed of particles is presented in figure 1. It indicates both experimental values and calculated relationships $\rho w_m (\Delta P)$ determined by formulas: (4) – in the subcritical condition, (10) – in the critical condition. The dashed line in figure 1 shows critical pressure drops – $\Delta P_{cr} = P_1 (1 - \beta_{cr})$ corresponding to the predictions by formula (9). The theoretical model is seen to describe the studied relationships adequately enough.

Agreement between the calculation results and the experimental data at higher pressures are illustrated in figure 2. It is seen that the theoretical model generalizes experimental data obtained at different values of pressure $P_1$, particle diameter $d$ and bed height $H$ in the subcritical and critical conditions with high accuracy. The determination coefficient $R^2$ of the model is equal to 97% for the critical mass velocity, and 98% for the subcritical mass velocity.

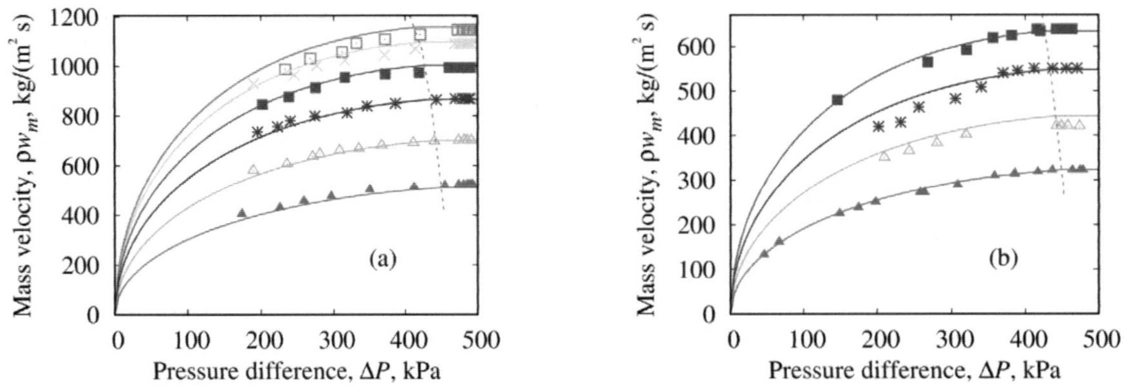


Figure 1. Mass velocity depending on the pressure drop for $P_1$=0.6 MPa (a) $d$ = 2 mm, $H$ = 100 mm; (b) $d$ = 4 mm, $H$ = 250 mm. Dots are experiment with the flow quality $x_1$: □ – 0.016; × – 0.022; ■ – 0.033; * – 0.055; Δ – 0.096; ▲ – 0.178. Solid lines show calculated values; dashed line shows critical condition boundary.

We introduce a relative dimensionless mass velocity via the ratio of expressions (4) and (10). For any fixed solid beds at equal parameters of the two-phase flow at the inlet this indicator depends only on $\beta$, $n$ and $\beta_{cr}$ :

$$\mathrm{PW}_m = \frac{\rho w_m}{\left(\rho w_m\right)_{cr}} = \left( \frac{1-\beta^{\frac{n+1}{n}}}{1-\beta_{cr}^{\frac{n+1}{n}}} \right)^{0.5}. \tag{11}$$

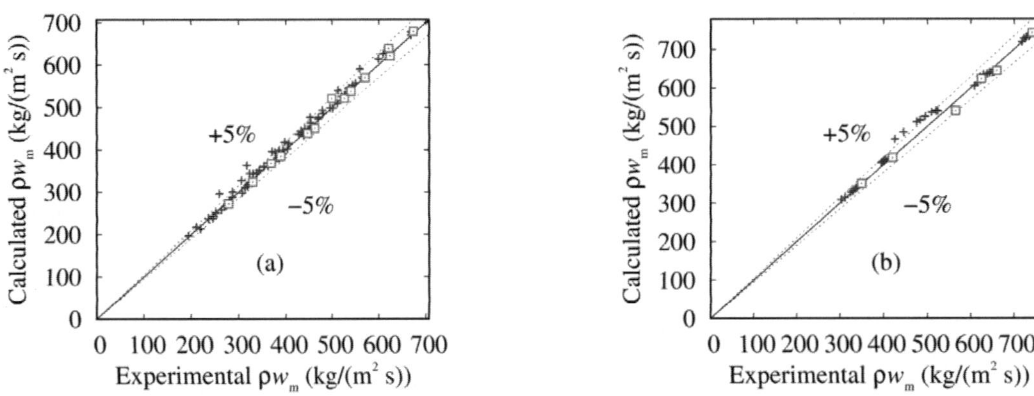


Figure 2. Calculated and experimental values of the mass velocity for $d$ = 2 and 4 mm, $H$ = 250 and 355 mm, $0.016 < x_1 < 0.178$ (a) $P_1$ = 0.9 MPa; (b) $P_1$ = 1.2 MPa: + – subcritical values, □ – critical values.

Besides, $n$ and $\beta_{cr}$ at the fixed pressure $P_1$ at the bed inlet are functions of only the inlet mixture flow quality. In this case, formula (11) may be applied to generalize mass velocity data obtained for different types of beds at equal flow qualities $x_1$. Generalization of experimental results to all beds and some values of the inlet flow quality and the pressure is shown in figure 3. The left margin of the lines is limited by the corresponding value of $\beta_{cr}$.

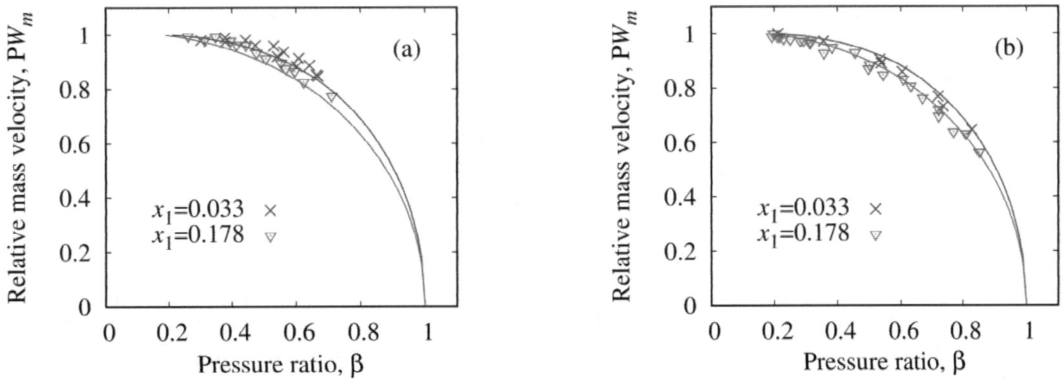


Figure 3. Dimensionless mass velocity (a) $d = 2$ mm, $H = 50$ and 100 mm $P_1 = 0.6$ MPa; (b) $d = 2$ and 4 mm, $H = 250$ and 355 mm $P_1 = 0.9$ MPa: markers show experimental values, lines show calculated values.

In the experiments, the critical pressure ratio is essentially lower than unity and decreases with the increasing inlet pressure $P_1$. The polytropic coefficient is also lower than unity. Both stipulate a very low value: $\beta_{cr}^{(n+1)/n} << 1$. Neglect of its low value in equation (11) yields the formula:

$$\left(\rho w_m\right)_{cr} = \frac{\rho w_m}{\left(1-\beta^{\frac{n+1}{n}}\right)^{0.5}} . \tag{12}$$

The value of $\left(\rho w_m\right)_{cr}$ calculated by the approximate formula (12) is only by 0.3–2.3% higher than by formula (11). Availability of several measurements of $\rho w_m\left(\Delta P\right)$ allows us to eliminate the critical condition parameters using formula (12) and to construct the entire curve for the mass velocity in the subcritical flow.

Formula (12) is suitable for the experimental study of two-phase flows in granular media with a complex internal structure, whose porosity and dominant diameter of grains are difficult to determine.

## 4. Conclusions

We developed a mathematical model of the two-phase vapor-liquid flow through the random bed of solid particles. The model is an extension of the theoretical model, proposed by M.A. Gol'dshtik to describe the ideal gas flow, to a two-phase flow. The average mixture density is calculated subject to the phase slip. The slip parameter and the polytropic coefficient are determined using the experimental data. The developed model allows generalizing the experimental results on the critical and subcritical vapor-liquid flows through different spherical beds with reasonable accuracy. The proposed approach may be applied to generalization of the experimental data on flows through granular beds with a complex internal structure.

**Acknowledgements**

The research was performed at the High Temperature Circuit of Melentiev Energy Systems Institute under the project III.17.1.3 of the Basic Research Program of Siberian Branch of the Russian Academy of Sciences, No. AAAA-A17-117030310443-5 with the partial financial support of the Russian Foundation for Basic Research (Project No. 17-08-00709a).

## References

[1] Stubos A and Buchlin J-M 1993 *Int. J. Heat Mass Transfer* **36**(5) 1391–401
[2] Avdeev A A and Soziev R I 2008 *High Temperature* **46**(2) 251–6
[3] Li L, Zou X, Lou J, Li H and Lei X 2015 *Annals of Nuclear Energy* **85** 290–5
[4] Tung V and Dhir V 1988 *Int. J. Multiphase Flow* **14**(1) 47–65
[5] Li L, Wang H, Zou X and Kong L 2018 *Annals of Nuclear Energy* **112** 62–70
[6] Taherzadeh M and Saidi M S 2015 *Int. J. Multiphase Flow* **69** 115–27
[7] Kim S-M and Mudawar I 2015 *Int. J. of Heat and Mass Trans.* **87** 497–511
[8] Eliasi E and Lellouche G S 1994 *Int. J. Multiphase Flow* **20** 91–168
[9] Fisenko V V 1978 *Kriticheskie dvukhfaznye potoki* [Critical two-phase flows] (Moscow: Atomizdat Publ.) p. 160 (In Russian)
[10] Gol'dshtik M A 2005 *Protsessy perenosa v zernistom sloe* [Transport processes in a granular layer] (Novosibirsk: Kutateladze Institute of Thermophysics at Russian Academy of Sciences) p. 358 (In Russian)
[11] Tairov E and Khan P V 2017 *MATEC Web of Conf.* **115** 05007
[12] Tairov E A, Pokusaev B G and Bykova S M 2016 *High Temperature* **54**(|2) 277–86
[13] Deich M E, Danilin V S, Tsiklauri G V and Shanin V K 1969 *High Temperature* **7**(2) 327–33
[14] Starkman E S, Schrock V E, Neusen K F and Maneely D J 1964 *J. Basic Eng.* **86** 247–54

5th International Workshop on Heat/Mass Transfer Advances
for Energy Conservation and Pollution Control
August 13-16, 2019, Novosibirsk, Russia

# Flow structure and heat transfer in a turbulent vertical bubbly flow downstream of a sudden duct expansion

**A.V. Chinak[1], P.D. Lobanov[1] and M.A. Pakhomov[2]**

[1]Laboratory of Problems of Heat and Mass Transfer, Kutateladze Institute of Thermophysics, Siberian Branch of Russian Academy of Sciences, 630090, Novosibirsk, Russia

[2]Laboratory of Thermal and Gas Dynamics, Kutateladze Institute of Thermophysics, Siberian Branch of Russian Academy of Sciences, 630090, Novosibirsk, Russia

pakhomov@ngs.ru

The flow structure and heat transfer in bubbly flows in a sudden duct expansion were studied. The data was obtained both in a pipe and a flat channel with a sudden expansion. Measurements are performed using shadow photography and PIV/LIF system. For modelling two-phase bubbly flow the set of RANS equations is used. Reynolds stress transport model is used to predict the turbulence of the liquid phase. The influence of sudden channel expansion on the distribution of bubbles and it`s velocity is shown. It is shown, that in the pipe with sudden expansion there are not a strong influence of bubbles on the hydrodynamical structure and heat transfer in the flow recirculation zone. In a contrast, in the vertical channel with back facing step the strong increasement of void fraction were found in similar region. It can result in increasement of heat transfer.

## 1. Introduction

Two-phase bubbly flows are common in different practical applications. Most of studies in this field are dedicated to investigations in vertical pipes. It was found in literature that recirculation flow determines the turbulent flux structure and the intensity of the transfer for momentum, mass, and heat. It can be used in different types of exchangers. But just some works can be found in literature on the topic of two-phase bubbly recirculation flows [1-4]. Experimental investigations of flow structure in ducts with sudden expansion were performed in [1, 2]. The numerical model for predictions of the flow structure in pipe sudden expansion were presented in [3]. In [4] the experimental study of heat transfer in bubbly flow downstream of a pipe sudden expansion and validation of numerical model [3] was performed.

The purpose of this paper is to present a data about flow structure of bubbly flow in a pipe and a flat channel with a sudden expansion. The data about hydrodynamical structure and heat transfer performance behind the pipe`s sudden expansion is presented. The distribution of bubbles over a duct cross section is very important parameter. The distributions of individual velocities of bubbles and void fraction distribution is presented in the last part of the paper. It can be used to obtain another hydrodynamical parameters of two-phase flow for safe operation and for the prediction of various emergency scenarios in heat generators.

## 2. Experimental setup and measurement methods

The experimental setup consists of a closed circuit in a liquid and an open circuit in the gas phase. The test liquid (distilled water) was pumped from the main tank to the test section. The pipe with sudden expansion (pipe diameters before and after the expansion $D_1$=15 mm, $D_2$=42 mm, step height h =13.5 mm) and the flat channel with the height before the expansion $H_1$=8 mm and the step height H=12 mm were used as test sections (Fig. 1a, b). In both cases bubbles were generated by feeding air into the liquid flow through capillaries with inner diameter of 0.7 mm. The capillaries are uniformly distributed across the duct cross-section.

The gas flow rate is determined using a gas flow controller, the liquid flow rate is measured by ultrasonic flow rate meter. The measurement error of the liquid and gas flow rates is ±2% of the measured value.

Direct heating and the thermography using infrared camera were used to measure heat transfer characteristics. In order to perform the measurements of liquid phase velocity distributions a Particle Image Velocimetry (PIV) "Polis-PIV" system was used. In order to obtain velocities and positions of bubbles in a flat channel a high-speed shadow method of imaging was used.

## 3. Numerical model

For the modeling of dispersed phase (gas bubbles) Eulerian two-fluid approach was used. The Eulerian approach is based on kinetic equations for a one-point PDF of bubbles coordinates, velocity, and temperature in the turbulent Gaussian fluid flow fields. The bubbles' behavior in turbulent liquid and their back action on the flow is determined by drag, gravity lift, virtual mass, wall lubrication forces, turbulent transport, and turbulent diffusion [3].

The carrier liquid (water) flow is treated as a steady-state, incompressible and axisymmetrical. Mass conservation continuity, two-momentum and energy equations with taking into account the effect of bubbles presence were used. Axisymmetric approach was applied for the system of RANS equations, the turbulence model and the equations for the dispersed phase (gas bubbles) [3]. The turbulence of the liquid phase is predicted using the model of Reynolds stress transport (second moment closure-SMC) [5]. The SMC model predicts the turbulent Reynolds stresses directly from partial differential equations and allows us to compute the anisotropic flow. The low-Reynolds number elliptic blending second-moment closure of [5] is utilized in the present study. It is modified for the presence of bubbles by the model [6].

In the present study the effect of break-up and coalescence of bubbles is not considered as well as the effects of pressure drop of the liquid and heat up of the gas phase on the bubble size.

## 4. Results and discussion

Radial distributions of the mean axial velocity of the liquid behind the sudden expansion of the pipe in the single-phase (*1*) and two-phase bubbly (*2*) flows are shown in Fig. 2. In the recirculation region negative velocities can be found both for single and two-phase flows ($x/h$ = 4). The maximal difference between the measured and predicted results for both single-phase and carrier fluid in the two-phase bubbly flow is up to 10%.

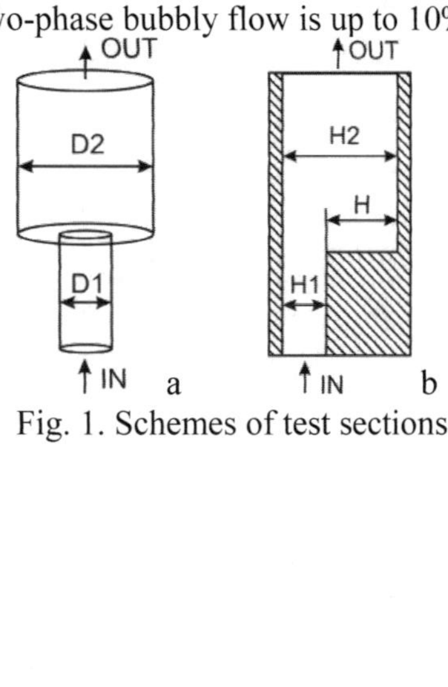


Fig. 1. Schemes of test sections: a - pipe with sudden expansion; b - channel with back facing step.

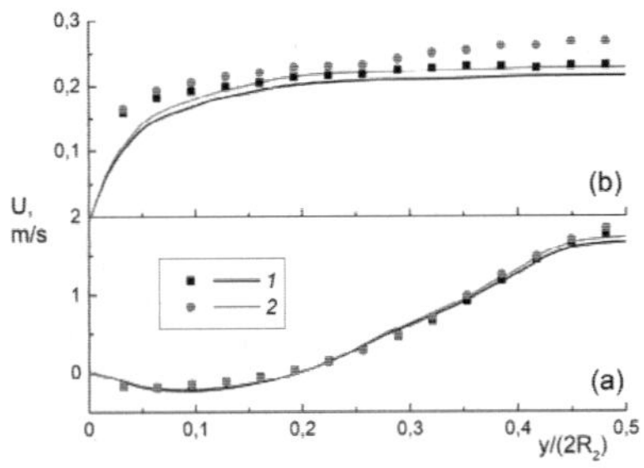


Fig. 2. Profiles of mean axial velocities of liquid phase downstream of the sudden pipe expansion at $x/h = 4$ (a) and 16 (b). Points are measurements, lines are simulations. *1* – single-phase flow at $\beta$ = 0%, *2* – liquid in two-phase flow at $\beta$ = 5%.

Distributions of the parameter for heat transfer enhancement $ER_{max} = Nu_{max}/Nu_{0,max}$ in bubbly flow with various gas volumetric flow rate ratios $\beta$ are shown on the Figure 3. The figure shows the results for two stations downstream of the sudden pipe expansion at $x/h = 20$ and 28. Here $Nu_{max}$ and $Nu_{0,max}$ are the maximal value of the Nusselt number for the bubbly flow and single-phase flow under other identical conditions. Points and lines are the results of measurements and numerical simulations performed by the authors.

A significant increase in the heat transfer rate occurs upon adding air bubbles into the liquid. According to the measurements and simulations, the increase in heat transfer reaches almost three times for the case of $Re_H = 1.02\times10^4$ and $\beta = 10.1\%$. The increased heat transfer coefficient is caused by significant deformation in the velocity profiles of the two-phase flow compared to the single-phase flow and by the increase in the liquid velocity gradient near the pipe wall. The heat transfer enhancement was found mainly in the flow relaxation zone. Actually, these conclusions are in agreement with the measured one. Only a few bubbles penetrate into the flow recirculation region and they are available only in the flow core and shear layer. We obtain a small effect of the gas bubbles on the measured and predicted heat transfer in the flow recirculation region and around reattachment point due to the relatively large bubble size ($d > 1.7$ mm). A recent numerical work [3] showed that small bubbles ($d < 1.5$ mm) caused heat transfer intensification over the entire length of the recirculation zone, while the larger ones caused intensification mostly in the flow relaxation region. The bubbles migrate towards the wall after the reattachment point of the flow and accumulate there due to the action of the lift force; it leads to an increase in fluid phase turbulence and heat transfer in this region.

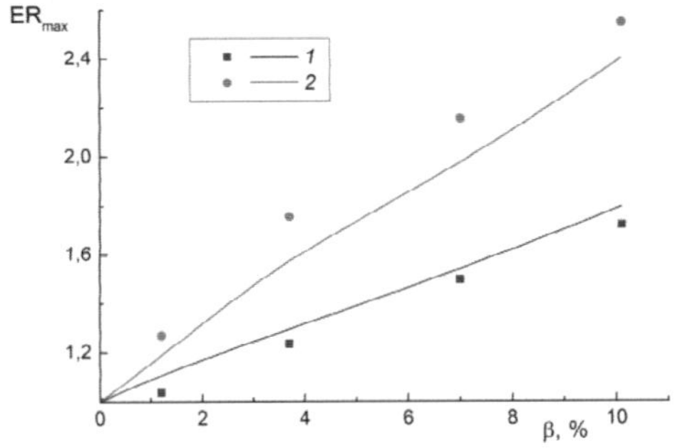


Fig. 3. The distributions of heat transfer enhancement ratios for various gas volumetric flow rate ratios $\beta$ at $Re_H = 1.02\times10^4$. Points and lines are authors' measurements and computations respectively. *1* – $x/h = 20$, *2* – 28.

Figure 4a shows data on the velocity distribution of individual bubbles along the length and width of the sudden expansion of a flat vertical channel. Here L is a line along the channel width (L=0 m is a channel wall position L=0.1 m is the channel centerline). A decrease in the velocity of the bubbles, as well as the carrier liquid was found behind the sudden expansion. It can be pointed that in the recirculation zone of the flow bubbles move along more curved trajectories than before, which is apparently caused by a high degree of flow perturbation.

The local void fraction distribution behind the step is shown in Figure 4b. There is a tendency to increase the local void fraction at some distance from the step. As shown by the visualization of the streamlines in a single-phase flow, the location of areas of highest gas content approximately corresponds to the end of the recirculation area (x/H =7-8).

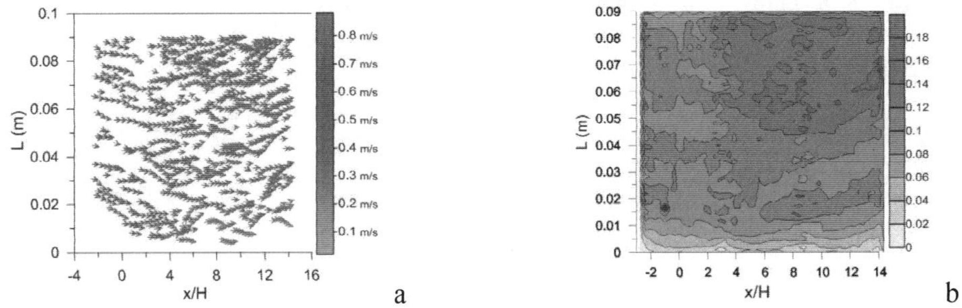


Fig. 4. The distributions of individual bubble velocities (a) and void fraction distribution (b) behind the channel`s sudden expansion. $Re_H = 6\times10^3$.

## Conclusions

The results of experimental and numerical simulations of the flow structure and heat transfer in bubbly flows in a sudden duct expansion are presented. Measurements of bubble sizes, velocities and void fraction distribution are performed using shadow photography, liquid phase velocities are measured by PIV/LIF system. For modelling two-phase bubbly flow the set of RANS equations is used. Reynolds stress transport model is used to predict the turbulence of the liquid phase. The influence of sudden duct expansion on the distribution of bubbles and it`s velocity, liquid phase velocities and heat transfer is shown. It is pointed, that in the pipe with sudden expansion there are not a strong influence of bubbles on the hydrodynamical structure and heat transfer in the flow recirculation zone. In a contrast, in the vertical channel with back facing step the strong increasement of void fraction were found in similar region. It can result in increasement of heat transfer. This topic will be the purpose of the further studies.

## Acknowledgements

The study was performed by partially financial support of the Russian Foundation for Basic Research (Project No. 18-08-00477), heat transfer measurements were performed under the state contract with IT SB RAS (AAAA-A18-118051690120-2).

## References

[1] Bel Fdhila R 1991 Ph.D. thesis, Institut National Polytechnique de Toulouse, France. 1991
[2] Pecherkin NI et. al. Proc. of. 3d Conference on Multiphase Flow, Lion, France, 1998
[3] Pakhomov MA and Terekhov VI 2016 *Int. J. Heat Mass Transfer* **101** 1251
[4] Lobanov PD and Pakhomov MA 2017 *J. Eng. Thermophys.* **26** 377
[5] Fadai-Ghotbi A, Manceau R and Boree J. 2008 *Flow, Turbulence and Combust.* **81** 395
[6] Lopez de Bertodano M, Lee SJ, Lahey RT, Jr. and Drew DA 1990 *ASME J. Fluids Eng.* **112** 107

5th International Workshop on Heat/Mass Transfer Advances
for Energy Conservation and Pollution Control
August 13-16, 2019, Novosibirsk, Russia

# Optimization of phase change material component and its application in buildings

**M Y Wang[1,2], H Z Cui[3], Y Liu[1,2], J Y Bao[1,2] and L Yang[1,2*]**

[1] State Key Laboratory of Green Building in Western China, Xi'an University of Architecture and Technology, 31 Middle Section of Yanta Road, Xi'an, Shannxi 710055, P.R. China

[2] School of Architecture, Xi'an University of Architecture and Technology, 31 Middle Section of Yanta Road, Xi'an, Shannxi 710055, P.R. China

[3] Guangdong Provincial Key Laboratory of Durability for Marine Civil Engineering, College of Civil Engineering, Shenzhen University, Shenzhen, Guangdong 518055, P.R. China

*Corresponding author e-mail address: yangliu@xauat.edu.cn

**Abstract**. Improving the thermal transfer performance of phase change material (PCM) component has attracted more and more attention in building the energy-saving field. This paper presented an optimized PCM component which is encapsulated in a metal container, to promote the thermal performance of the wall. Thermal transfer of PCM component is optimized by adding internal fin through numerical simulation software COMSOL Multiphsics. The simulation model was verified by data from publicly available published literature. The temperature difference of PCM component in front and back heating surface ($\Delta T_s$) was employed to obtain the optimal geometric parameters including spacing ($S_{fin}$) and length ($L_{fin}$) of the fin. Thermal performance of the wall with the optimized and non-optimized PCM component was compared under the summer typical climate. The wall with optimized PCM component has the best capacity for heat storage and release. The interior surface temperature reduction of the wall with optimized PCM component is lower by 2.1 °C than reference wall.

## 1. Introduction

Taking advantage of latent heat into the building envelope, can significantly reduce the building energy consumption and improve the temperature regulation for the indoor thermal environment. Among the different combinations of PCM applied in building, macro-encapsulation is the simplest and most economical method. Package PCM in the container as a PCM component combined with building envelopes is considered a promising option in building energy conservation [1, 2]. The superiority of this technique embodied at various packaging shapes and size, are safe and avoid leakage of material. Sufficient thermal storage capacity and the function of the construction structure can be less affected [3]. However, unreasonable application and poor thermal conductivity would cause the regulating ability untimely and incomplete for the indoor thermal environment [4]. Metal fins are widely used to increase the thermal transfer areas of PCM component. This method could compensate PCM for low thermal conductivity and relatively easy to fabricate [5]. The fin type can be divided into partitioning fin and partial fin according to the objective application [6]. Many types of research mainly focused on optimizing the thermal performance of fin number and geometry in PCM component.

From the above, thermal transfer enhancement of PCM component with fin would promote the charging and discharging rate of PCM. Few pieces of research investigate this technology applied in building envelope and evaluate the application effects under the actual indoor and outdoor thermal environment. In the present work, the optimal geometric parameters of fin were obtained and the thermal performance of the composite wall with optimized PCM component was analyzed. The conclusion can provide references for PCM application in building.

## 2. Numerical simulation method

### 2.1. Physical model

The two-dimensional physical model of the multi-layer wall filled with PCM is shown in Fig. 1. The aerated concrete structure is selected, which is widely used in Southern China. The aerated concrete has good thermal insulation but poor thermal storage capacity. The wall consists of 20 mm thick plaster layer on the exterior and interior surface, and 200 mm aerated concrete layer inside. The dimension of the PCM component are as follows: the net depth, height, and thickness are 200 mm, 200 mm and 10 mm. Aluminum is selected as shell and fin material for good thermal conductivity. The thickness of the shell is set as 1 mm. The partial fin in the PCM component is added to the indoor side (right side), which can adjust the indoor thermal environment directly. The thermo-physical properties of PCM used in the simulation are shown in Table 1. The different spacings ($S_{fin}$) and lengths ($L_{fin}$) of the fin are employed to determine the optimal geometric parameters.

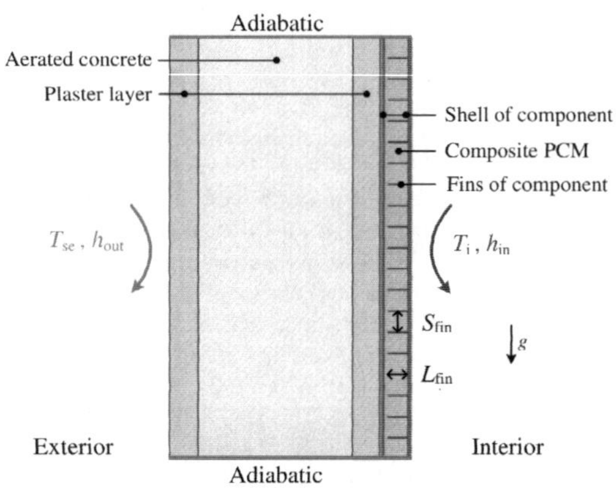


**Figure 1.** Schematic diagram of wall with the optimized PCM component

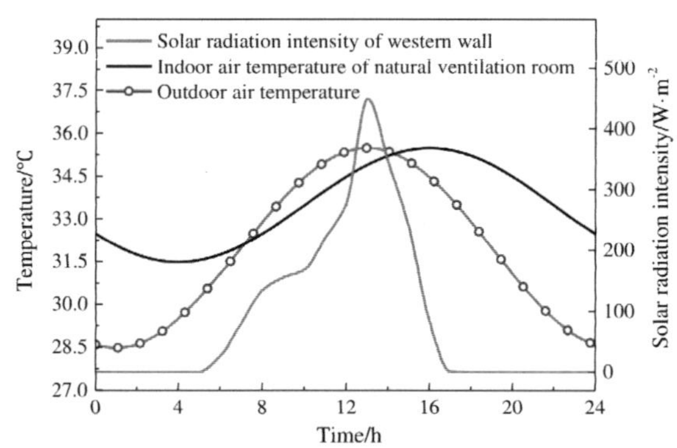


**Figure 2.** Boundary conditions variation over the typical summer day in Shenzhen

**Table 1.** Thermo-physical properties of the materials used in the multi-layered wall

| Martials | $k$ (W·m$^{-1}$·K$^{-1}$) | $\rho$ (kg·m$^{-3}$) | $c$ (kJ·kg$^{-1}$·°C$^{1}$) | $L$ (kJ·kg$^{-1}$) | $T_{pc}$ (°C) |
|---|---|---|---|---|---|
| PCM | 0.23/0.21 | 833.8/786.7 | 1.91/2.03 | 122 | 33.7 |
| Aerated concrete | 0.22 | 700 | 1050 | / | / |
| Plaster layer | 0.93 | 1800 | 1050 | / | / |
| Fin and shell | 211 | 2675 | 9.03 | / | / |

Note: $k$, $\rho$, and $c$ are solid/liquid phase of PCM at each property.

### 2.2. Governing equations

To perform multiple physical fields simulation, the thermal transfer module for solid, liquid and laminar flow physical fields are employed [7]. The assumptions are as following:

(1) One-dimensional thermal transfer is assumed. The boundary conditions of the top and bottom of the PCM component are set as adiabatic and the initial temperature is uniformly distributed.

(2) All the solid materials are homogenous.

(3) The liquid PCM is incompressible Newton fluid, and density change during the phase change process conforms to Boussinesq approximation.

Based on the above assumptions, the effective heat capacity method is used to simulate thermal transfer in the phase change process and the transient energy equations are as follows:

$$\rho C_{\mathrm{p}} \frac{\partial T}{\partial t} + \nabla \cdot (-k \nabla T) + \rho C_{\mathrm{p}} u \nabla T = 0 \tag{1}$$

where $\rho$ is the density of materials; $C_{\mathrm{p}}$ is specific heat of materials; $T$ is transient temperature; $k$ is thermal conductivity of materials; $u$ is a vector of the liquid PCM velocity field. The liquid fraction ($\theta$) of PCM can be defined with $T$:

$$\theta = \begin{cases} 0 & T < T_{\mathrm{m}} - \frac{\Delta T}{2} \\ \frac{T - T_{\mathrm{m}} + \Delta T / 2}{\Delta T} & T_{\mathrm{m}} - \frac{\Delta T}{2} < T < T_{\mathrm{m}} + \frac{\Delta T}{2} \\ 1 & T > T_{\mathrm{m}} + \frac{\Delta T}{2} \end{cases} \tag{2}$$

where $T_{\mathrm{m}}$ is the melting point of PCM; $\Delta T$ is the phase change temperature range; $\theta$ varies linearly from 0 to 1 through the transition zone over $\Delta T$. Based on the definition of $\theta$, density ($\rho_{\mathrm{PCM}}$), thermal conductivity ($k_{\mathrm{PCM}}$) and specific heat ($C_{\mathrm{p,PCM}}$) of PCM can be expressed as follows (Subscripts s and l are solid and liquid properties for PCM):

$$\rho_{\mathrm{PCM}} = \theta \rho_{\mathrm{s}} + (1 - \theta) \rho_{\mathrm{l}} \tag{3}$$

$$k_{\mathrm{PCM}} = \theta k_{\mathrm{s}} + (1 - \theta) k_{\mathrm{l}} \tag{4}$$

$$C_{\mathrm{p,PCM}} = \frac{1}{\rho_{\mathrm{PCM}}} \left( \rho_{\mathrm{s}} C_{\mathrm{p,s}} \theta + \rho_{\mathrm{l}} C_{\mathrm{p,l}} (1 - \theta) \right) + L \frac{\partial \alpha_{\mathrm{m}}}{\partial T} \tag{6}$$

where $L$ is the latent heat of PCM; $\alpha_{\mathrm{m}}$ is volume fraction in the phase change process:

$$\alpha_{\mathrm{m}} = \frac{1}{2} \frac{\rho_{\mathrm{l}}(1\text{-}\theta) - \rho_{\mathrm{s}} \theta}{\rho_{\mathrm{s}} \theta + \rho_{\mathrm{l}}(1\text{-}\theta)} \tag{7}$$

The changes of $\rho$ in phase change drive buoyancy in liquid PCM. Based on the Navier-Stokes equation, the momentum conservation equation is as follows:

$$\rho_{\mathrm{PCM}} (u \cdot \nabla) u = -\nabla p + \nabla \cdot \mu (\nabla u + (\nabla u)^T) + \rho_{\mathrm{PCM}} g \alpha_{\mathrm{m}} (T - T_0) \tag{8}$$

where $p$ is the fluid pressure of PCM; $g$ is gravity; $\mu$ is the dynamic viscosity of PCM.

### *2.3. Validation of the Model*

Before proceeding further, the grid quantity and timesteps are checked. Simple linear free triangular elements are applied to generate a grid. Four grid quantities (39433, 49793, 72623 and 91396) and four timesteps (1, 0.5, 0.25 and 0.1 hours) are studied. Considering the simulation accuracy and computational resources, 72623 grid quantity and timestep of 0.25 h are chosen. The reliability of the model is validated based on the data of Huang et al [5]. The average temperature of the PCM component heating surface ($T_{\mathrm{s}}$) and PCM ($T_{\mathrm{PCM}}$) are selected for validation. The initial temperature and ambient environment temperature are set as 20 C. The front and back surface heat transfer coefficients of PCM are set as 10 $\mathrm{W \cdot m^{-2} \cdot C^{-1}}$ and 5 $\mathrm{W \cdot m^{-2} \cdot C^{-1}}$. The radiating surface is set as a constant value of 1000 $\mathrm{W \cdot m^{-2}}$. The top and bottom boundaries of the system are adiabatic. $T_{\mathrm{s}}$ and $T_{\mathrm{PCM}}$ are compared which are shown in Fig. 3. The maximum relative error of $T_{\mathrm{s}}$ and $T_{\mathrm{PCM}}$ are 4.58 % and 1.67 %, the average relative error of $T_{\mathrm{s}}$ and $T_{\mathrm{PCM}}$ are 0.51 % and 1.62 %. It indicates that the model is credible to investigate the thermal transfer of the PCM component.

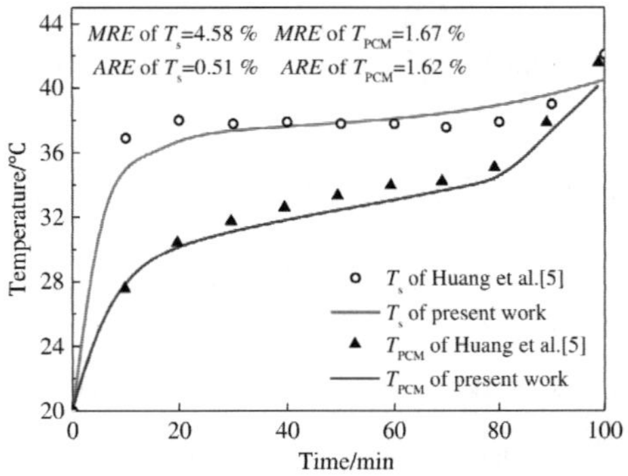


**Figure 3**. Model validation with $T_s$ and $T_{PCM}$

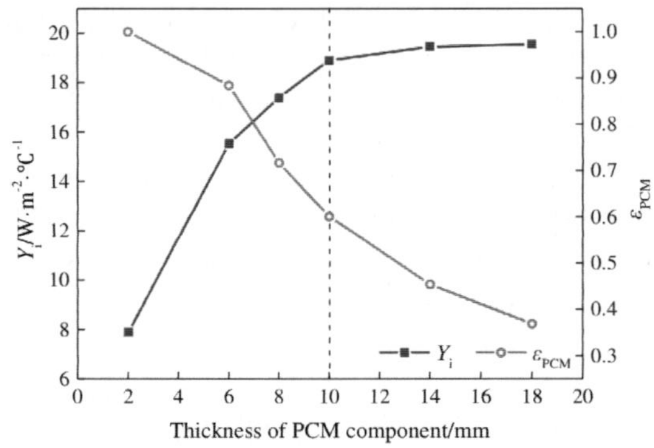


**Figure 4.** Determination of the optimal PCM thickness with $Y_i$ and $\varepsilon_{PCM}$

## *2.4. Evaluation index of the composite wall with PCM*

To avoid the influence of periodical temperature variation on the optimization of the fin, constant boundary conditions of the PCM component are employed to evaluate the PCM thermal behavior with different geometric parameters. The initial temperature is set as 20 °C, the heat transfer coefficient and temperature of the heating surface of the PCM component are set as 8.7 W/m$^2$·K and 40 °C. The temperature difference in front and back heating surface of the PCM component in front and back heating surface ($\Delta T_s$) at heat storage stage is used to assess the efficiency of the fin. The integer multiples of fins' spacing ($S_{fin}$) and length ($L_{fin}$) are used as simulation cases. The values of $S_{fin}$ are 20, 10, 5 and 4 mm, the values of $L_{fin}$ are 8, 4, 6, and 2 mm.

The optimal quantity of PCM can fully achieve the heat storage capacity of the PCM composite wall. Latent heat utilization of PCM ($\varepsilon_{PCM}$) and interior surface coefficient of heat accumulation for the PCM composite wall ($Y_{if}$) are used to determine the optimal PCM thickness:

$$\varepsilon_{PCM} = f_{max} - f_{min} \tag{9}$$

$$Y_{if} = \frac{A_q}{A_0}$$

(10)

In Eq. (9), $f_{max}$ and $f_{min}$ are the maximum and minimum value of $\theta$. In Eq. (10), $A_q$ is the wave amplitude of interior surface heat flux and $A_0$ is the temperature amplitude of the interior surface.

As shown in Fig. 4, $Y_{if}$ and $\varepsilon_{PCM}$ are influenced by PCM thickness. $\varepsilon_{PCM}$ decreases with the increase of PCM thickness. $Y_{if}$ increases with PCM thickness from 2-10 mm, then tend to be stable with 0.6 of $\varepsilon_{PCM}$. The optimal thickness of PCM of the non-optimized PCM component is determined as 10 mm and the thickness of PCM in optimized PCM component is calculated as 10.3 mm.

# 3. Results and discussion

## *3.1. Optimization of fin-based on thermal behavior*

Figure 5 presents the comparison of $\Delta T_s$ with different fins' geometric parameters at heat storage stage. The lower value of $\Delta T_s$ means the better thermal behavior of the PCM component. From Fig. 5 (a) and (b), three stages of phase change process are observed: sensible heat of solid phase, latent heat of phase change process and sensible heat of liquid phase [8]. At the sensible heat of the solid phase, $\Delta T_s$ varies linearly from 0-10 min due to the thermal resistance of the PCM component. By comparing all cases, $L_{fin}$ has a significant impact on the sensible heat of the solid phase. At the latent heat of the phase change process, $\Delta T_s$ tends to be stable and natural convection becomes dominant in thermal transfer. Also, $L_{fin}$ has a significant impact on the latent heat of the phase change process. In addition,

$\Delta T_s$ of the PCM component with no fin is higher than that with a fin. The PCM melting rate decreases due to the PCM is not melted completely. At the sensible heat of the liquid phase, $\Delta T_s$ increases when the PCM melts completely. From the above, the optimized geometric parameters of the fin in the PCM component are determined: 5 mm of $S_{fin}$ and 8 mm of $L_{fin}$.

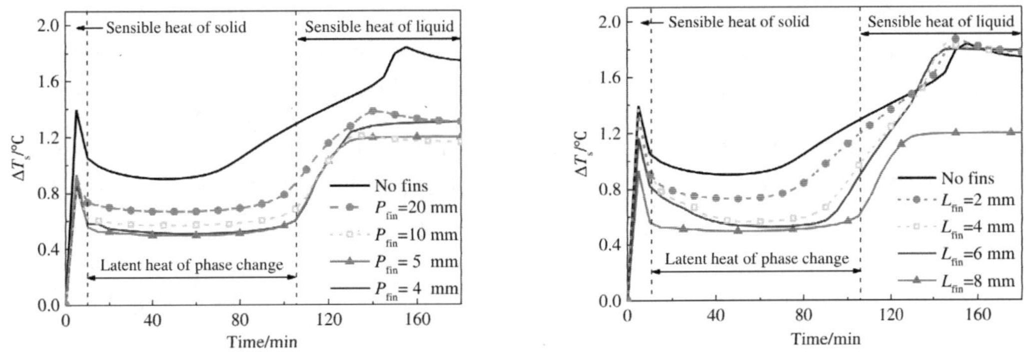


(a) (b)

**Figure 5.** (a) $\Delta T_s$ comparison with different $S_{fin}$, (b) $\Delta T_s$ comparison with different $L_{fin}$

*3.2. Thermal performance of the PCM composite wall*

According to the optimization of the fin, a multi-layered wall without PCM component is defined as "Reference wall". The wall with installed non-optimized PCM component and the wall with installed optimized PCM component is defined as "Wall with non-optimized PCM component" and "Wall with optimized PCM component". To investigate the thermal performance of the PCM wall, interior surface temperature and interior surface heat flux of each wall are compared.

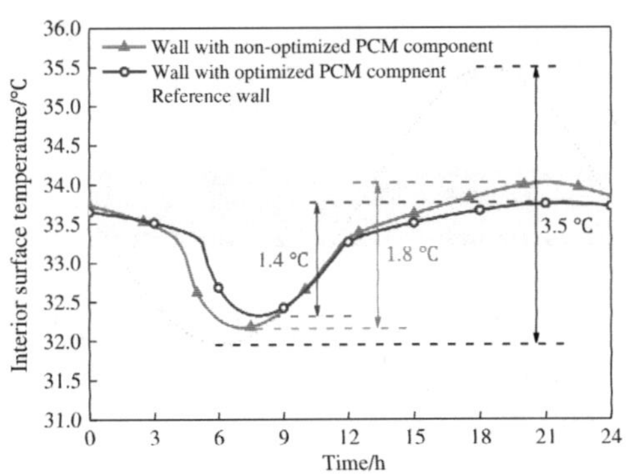


**Figure 6.** Interior surface temperature comparison of different walls

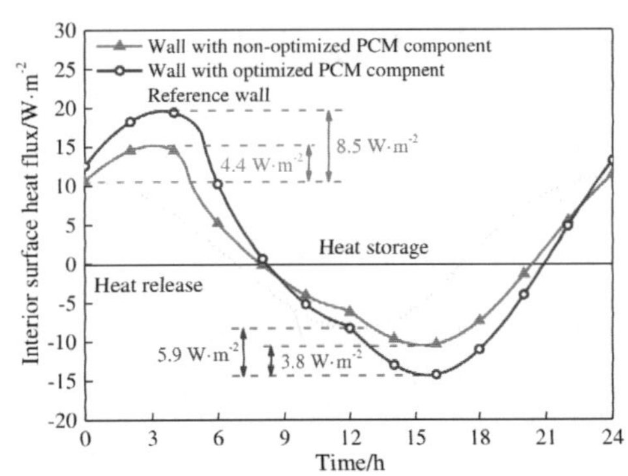


**Figure 7.** Interior surface heat flux comparison of different walls

Interior surface temperature is presented in Fig. 6. The results demonstrate that two types of PCM walls can regulate the peak interior surface temperature. Comparing with reference wall, the average lag time of the peak interior surface temperature is 2.5 h for the wall with non-optimized PCM component and 3 h for the wall with optimized PCM component. The interior surface temperatures reduction of two walls with non-optimized and optimized PCM component are lower by 1.7 °C and 2.1 °C than reference wall, respectively. These results indicate that the wall with the optimized PCM component has the best thermal stability.

Interior surface heat flux is shown in Fig. 7. When heat flux is greater than 0, it means the heat is transferring from interior surface to indoor (heat release at night). The heat is absorbing from indoor to interior surface (heat storage in the daytime) when the heat flux is less than 0. The thermal storage/release capacity of the wall is reflected in two aspects: the duration and intensity. The duration of the reference wall is 10 h (7:00-17:00). The wall with the non-optimized PCM component and optimized PCM component are 12.25 h (8:00-20:15) and 13 h (8:15-21:00). The intensity of heat storage/release can be express by the fluctuation of heat flux. The peak value of the wall with non-optimized and optimized PCM component is 3.8 $W \cdot m^{-2}$ and 5.9 $W \cdot m^{-2}$, which are higher than the reference wall during the daytime. It means that the PCM component promotes the heat storage capacity of the wall and increases the heat storage capacity. Similarly, PCM component increases the heat release capacity at night. The peak heat flux of the wall with the non-optimized and optimized PCM component are higher by 4.4 $W \cdot m^{-2}$ and 8.5 $W \cdot m^{-2}$ than reference wall. According to the above analysis, the wall with optimized PCM component has the best capacity in terms of heat storage and release.

## 4. Conclusions

In present work, optimization of PCM component was conducted based on COMSOL Multiphysics considering the geometric parameters of the fin. The temperature difference in the front and back heating surface of the PCM component heating surface ($\Delta T_s$) was used to assess the fin efficiency. Suitable geometric parameters including spacing ($S_{fin}$) and length ($L_{fin}$) of the fin are a benefit for thermal transfer of PCM component. The optimal $S_{fin}$ and $L_{fin}$ of the fin are determined as 5 mm and 8 mm, respectively. In addition, the optimal thickness of PCM was determined with the actual indoor and outdoor environment. The thermal performance of the wall with the non-optimized and optimized PCM component was compared with the interior surface temperature and heat flux. The wall with optimized PCM component presents the best thermal stability. Results show that the interior surface temperature reduction of the wall with the optimized PCM component is lower by 0.4 °C than the wall with the non-optimized PCM component.

## Acknowledgments

This work was supported by the National Science Foundation of China (No. 51808429), Shaanxi Province Key Research and Development Plan (No. 2017ZDXM-SF-076) and Young Elite Scientists Sponsorship Program by CAST, YESS (No. 2018QNRC001).

## References

[1] Liu Y, Yang L, Zheng WX, Liu T, Zhang XR and Liu JP 2018 *Energy Convers. Manage* **166** 522-33
[2] Yang L, Qiao YH, Liu Y, Zhang XR, Zhang C and Liu JP 2018 *Build Environ* **144** 194-207
[3] Salunkhe PB, Shembekar PS 2012 *Renew. Sust. Energ. Rev* **16** 5603-16
[4] Rajesh B, Balaji C 2013 *Int Commun Heat Mass* **46** 27-30
[5] Huang M.J, Eames P.C, Norton B 2004 *Int J Heat Mass Transf* **47** 2715-33
[6] Kamkari B, Shokouhmand H 2014 *Int J Heat Mass Transf* **78** 839-51
[7] Soares N, Gaspar A.R, Santos P, Costa J.J 2016 *Energ Buildings* **116** 18-34
[8] Biwole PH, Groulx D, Souayfane F, Chiu T 2018 *Int J Therm Sci* **124** 433-46

5th International Workshop on Heat/Mass Transfer Advances
for Energy Conservation and Pollution Control
August 13-16, 2019, Novosibirsk, Russia

# Numerical modelling and optimization of pulverized biomass gasification process

**I G Donskoy**

130 Lermontova st., Irkutsk, Russia, 664033

donskoy.chem@mail.ru

**Abstract.** Biomass, such as wood, forestry and agricultural wastes, etc., is considered to be a fuel with better environmental characteristics than fossils. The main prospects for the increase in the energy use of biomass are associated with the combustion and gasification in small energy systems, including those being a part of hybrid power plants. One of the ways to improve the efficiency of the gasification process is to increase the temperature by the use of oxygen-enriched gasification agent, as well as grinding the fuel to intensify transfer processes and chemical transformations. Thermodynamic estimates show the possibility of achieving the gasification process efficiency at the level of 80-90%, however, the experimental values rarely exceed 70%. In the present work, by means of mathematical modeling, the possibilities of increasing the efficiency of the biomass gasification process are investigated. The range of optimal conditions is evaluated.

## 1. Introduction

The replacement of coal and hydrocarbon fuels with renewable sources allows improving the environmental characteristics of energy systems. Among renewables, plant biomass has a significant advantage. Biomass is widely used for energy production (mainly thermal). The current consumption of plant biomass in the energy sector is about 1.3 billion toes [1], and, according to some estimates, will increase by 2-3 times by 2040. In addition to incineration, gasification, and pyrolysis, biomass can be used [2] in combined production of heat and electricity on the basis of small plants, including gas engine [3]. In areas with high prices for imported fuel, the use of local biofuels may be more profitable than the widely used diesel generators [4, 5]. In addition, biomass gasification can be used for chemical production [6]

The overall efficiency of gas engines operating on low-calorific gas rarely exceeds 30% [7]. Therefore, the technical efficiency of biomass gasification plants is largely determined by the efficiency of thermochemical conversion stage (cold gas efficiency, CGE). Cold gas efficiency is influenced by a large number of parameters: the composition of the fuel (proximate, ultimate, fractional), the composition of the gasification agent, process type, gasification plant capacity, gas cleaning conditions, etc. Thermodynamic analysis of biomass gasification [8] gives an upper estimate for cold gas efficiency at 80-90%. However, as the analysis of the available experimental data [9, 10] shows, these values are difficult to achieve in practice. Relatively low temperatures, which are achieved by air gasification of biomass, cause a high yield of tarry products containing a significant portion of the calorific value of the fuel. Tar conversion processes are reviewed in [11]: they are usually associated with high-temperature post-conversion reactors or with expensive catalytic systems. In the papers [12, 13] the options of the staged gasification process were reviewed. In particular, the authors of [14] reported the achievement of CGE values that were close to thermodynamic limits.

In addition to the fluidized- and fixed-bed gasification processes that are appropriate for particles with a size of the order 0.1-1 cm, there are also entrained-flow processes for the gasification of fuel with a particle size of 0.1-1 mm [15]. An oxygen-enriched gasification agent is commonly used to raise the temperature in the combustion zone. In the works [16, 17, 18, 19, 20, 21, 22, 23] these processes were investigated at laboratory-scale units. Experimental reactor with a capacity of 1 MW is described in [24, 25, 26]. Gasification of biomass in a swirling flow was studied in [27, 28]. However, the achieved CGE remains at the same level (about 70%). In order to understand process behavior, it is necessary to investigate the process efficiency using kinetic models.

The authors of [29] proposed a one-dimensional stationary model of gasification process of pulverized biomass in entrained flow, built as a set of submodels for different zones, which are presented as stirred-tank or plug flow reactors (similar to the reduced-order models for coal gasification processes [30, 31]). The paper [32] examines the behavior of biomass particles in a laboratory reactor taking into account the detailed kinetics of pyrolysis processes and gas-phase oxidation. Models of coal and biomass co-gasification processes are proposed in [33, 34, 35]. The CFD-models for biomass particles gasification in the drop tube reactors are proposed in [36, 37, 38].

In the present work, a simplified model is used to study the efficiency of the biomass gasification process in a wide range of conditions, which allows us to find the optimal gasifier operating modes and ways to improve the process characteristics.

## 2. Input data

Mathematical model described earlier in the previous works of the author [39, 40] was used for calculations. This model is a system of one-dimensional stationary heat and mass transfer equations with a combined kinetic-thermodynamic block describing the processes of pyrolysis and gasification of fuel particles. The reactor is considered to be close to the plug-flow reactor.

The parameters of the reactor are taken from [24]. We consider a cylindrical reactor with fuel consumption of 30-120 kg/h, the working pressure in the reactor is 1-10 bar. Geometrical dimensions of the reactor are as follows: the length of the reaction zone is 2 m; inner diameter is 0.45 m (slightly less than published data; this value was matched for better agreement with residence time). The temperature of the fuel entering the reactor is 25°C; steam temperature is 100°C; the temperature of the gasification agent is 25°C.

The model was verified using [24] data for pressures of 2-7 bar. The results of the comparison between the experimental and calculated data for 28 modes are shown in Fig. 1. Residence time is estimated by the model with good accuracy (using correction for inner diameter). Although the correlation coefficient obtained for all components is 0.91, local errors can reach 50%. Chemical efficiency of gasification is overestimated by an average of 11% (in relative values). The methane content in product gas is underestimated by the model, the content of CO and $H_2$, on the contrary, is higher than the measured ones. Nevertheless, one may conclude that the qualitative correspondence of the modeling results and literature data was observed.

Optimal cold gas efficiency measured in experiments [24] is close to 75%. Gasification agent here is a mixture of $O_2/N_2/H_2O$ with a nitrogen concentration up to 40% vol. Theoretical calculations show the possibility of achieving cold gas efficiency about 87% (for a pressure of 7 bar). It is interesting to find the reasons for this difference.

**Table 1**. Fuel composition and properties.

| **Property** | **Value** |
|---|---|
| $W^r$, % | 6.7 |
| $A^r$, % | 0.34 |

| | |
|---|---|
| $V^{daf}$, % | 82.7 |
| $C^{daf}$, % | 51 |
| $H^{daf}$, % | 6.22 |
| $N^{daf}$, % | 0.1 |
| $S^{daf}$, % | 0.01 |
| $Q^{r}$, MJ/kg | 15.96 |
| Density, kg/m$^3$ | 900 |
| Mean particle diameter, μm | 120-180 |

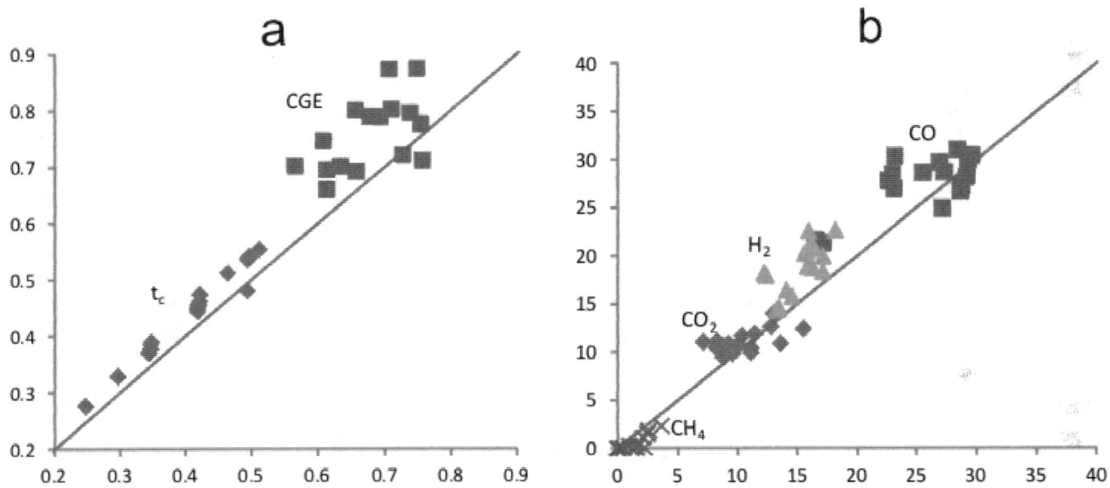


**Fig. 1.** Comparison of measured [24] and calculated values: (a) CGE – cold gas efficiency; $t_c$ – mean residence time, s; (b) dry gas composition, % vol. Solid lines correspond to equation y = x.

## 3. Results of modeling

Further calculations are related to the optimization of the pulverized biomass gasification process in the steam-oxygen entrained flow. The geometrical parameters of the reactor and the fuel reactivity remain the same as for the previous calculations. Fuel consumption is maintained at 50 kg/h. In the calculations, the specific consumption of oxygen and water vapor, as well as the pressure in the reactor (from 1 to 10 bar) varies. Effect of steam consumption within the selected range (0-0.2 mol/mol of carbon) was weak (compared to other parameters). The following results are presented for a fixed steam consumption (0.2 mol/mol of carbon).

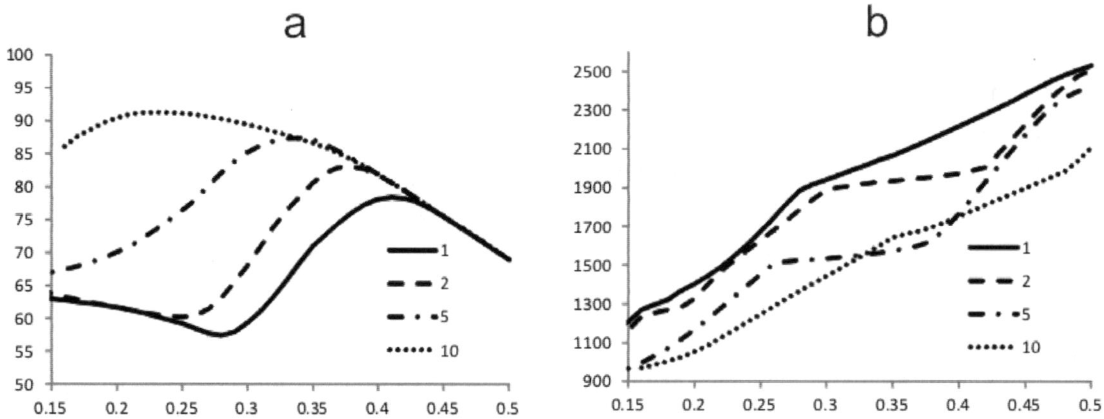


**Fig. 2.** The dependence of the cold gas efficiency, % (a) and combustion zone temperature, K (b) on the stoichiometric ratio (the numbers on the legend – reactor pressure, bar).

Calculations show that the increase of reactor pressure improves the cold gas efficiency, and maximum values of cold gas efficiency are achieved at lower stoichiometric ratios. These effects are associated with heterogeneous kinetics: an increase in pressure leads to an increase of the residence time for fuel particles. As a result, the gas-fuel system is closer to equilibrium state: thermodynamic estimates show that the theoretical value of the stoichiometric ratio for full biomass gasification is within the range of 0.2-0.3. The decrease in the residence time leads to the fact that, with the fixed reactor length, devolatilization is the only completed stage of conversion. In order to gasify the char residue, either high temperature or sufficiently high residence time is required. At low pressures, the first scenario is realized: the maximum cold gas efficiency for 1 bar is achieved at a stoichiometric ratio of 0.41, that corresponds to a combustion zone temperature about of 2250 K (Fig. 2, b). Cold gas efficiency for stoichiometric ratios below 0.3 is obviously determined by the oxidation of pyrogas, which is the main product. With increasing pressure, the gasification reactions complete due to an increase in the particle residence time, as a result, even at low temperatures, the process requires less oxygen.

## 4. Conclusion

Entrained-flow biomass gasification process was investigated using mathematical modeling. The model was tested on published data and showed good agreement with the experiment. Variant calculations were performed to search for optimal conditions for biomass gasification. Calculations show that different modes of fuel conversion are realized at different pressures. At low pressures, the efficiency of the process is determined by the temperature level. With increasing pressure, the optimum value of specific oxygen consumption decreases due to an increase in particles residence time in the reaction zone.

## Acknowledgements

The work was carried out with financial support of the Russian Fund for Basic Research (project No. 18-29-24047) and using equipment of the multi-access scientific center "High-Temperature Circuit".

## References

[1] Bhuiyan A A, Blicblau A S, Sadrul Islam A K M, Naser J 2018 *Journal of the Energy Institute* **91** 1
[2] Bridgwater A B 2003 *Chemical Engineering Journal* **91** 87
[3] Sansaniwal S K, Pal K, Rosen M A, Tyagi S K 2017 *Renewable and Sustainable Energy Reviews* **72** 363.
[4] Kozlov A, Svishchev D, Marchenko O, Solimin S, Shamansky V, Keiko A 2017 *Proc. 25th*

*European Biomass Conf.* (ETA-Florence Renewable Energies) p 776
[5] Bisht A S, Thakur N S 2019 *Renewable Energy Focus* **28** 112
[6] Molino A, Chianese S, Musmarra D 2016 *Journal of Energy Chemistry* **25** 10
[7] Susastriawan A A P, Saptoadi H, Purnomo 2017 *Renewable and Sustainalbe Energy Reviews* **76** 989
[8] Prins M J, Ptasinski K J, Janssen F J J G 2007 *Energy* **32** 1248
[9] Svishchev D A, Kozlov A N, Donskoy I G, Ryzhkov A F 2016 *Fuel* **168** 91
[10] Sansaniwal S K, Rosen M A, Tyagi S K 2017 *Renewable and Sustainable Energy Reviews* **80** 23
[11] Abdoulmoumine N, Adhikari S, Kulkarni A, Chattanathan S 2015 *Applied Energy* **155** 294
[12] Heidenreich S, Foscolo P U 2015 *Progress in Energy and Combustion Science* **46** 72
[13] Mednikov A S 2018 *Thermal Engineering* **65** 531
[14] Gadsboll R O, Clausen L R, Thomsen T P, Ahrenfeldt J, Henriksen U B 2019 *Energy* **166** 939
[15] Tremel A, Becherer D, Fendt S, Gaderer M, Spliethoff H 2013 *Energy Conversion and Management* **69** 95
[16] Kobayashi N, Tanaka M, Piao G, Wu X J, Kobayashi J, Hatano S, Itaya Y, Mori S 2007 *Journal of the Japan Institute of Energy* **86** 904
[17] Kobayashi N, Tanaka M, Piao G, Kobayashi J, Hatano S, Itaya Y, Mori S 2009 *Waste Management* **29** 245
[18] Hernandez J J, Aranda-Almansa G, Bula A 2010 *Fuel Processing Technology* **91** 681
[19] Zhao Y, Sun S, Zhou H, Sun R, Tian H, Luan J, Quan J 2010 *Fuel Processing Technology* **91** 910
[20] Qin K, Lin W, Jensen P A, Jensen A D 2012 *Fuel* **93** 589
[21] Goktepe B, Umeki K, Hazim A, Lundstrom T S, Gebart R 2017 *Fuel* **201** 111
[22] Schneider J, Grube C, Herrmann A, Ronsch S 2016 *Fuel Processing Technology* **152** 72
[23] Sripada P P, Xu T, Kibria M A, Bhattacharya S 2017 *Fuel* **203** 942
[24] Weiland F, Wiinikka H, Hedman H, Wennebro J, Pettersson E, Gebart R 2015 *Fuel* **153** 510
[25] Ogren Y, Toth P, Garami A, Sepman A, Wiinikka H 2018 *Applied Energy* **226** 450
[26] Wiinikka H, Toth P, Jansson K, Molinder R, Brostrom M, Sandstrom L, Lighty J S, Weiland F 2018 *Combustion and Flame* **189** 240
[27] Risberg M, Ohrman O G W, Gebart B R, Nilsson P T, Gundmunsson A, Sanati M 2014 *Fuel* **116** 751
[28] Kostyunin V V, Potapov V N 2014 *Modern Science: Researches, Ideas, Results, Technologies* **14** 82
[29] Andersson J, Umeki K, Furusjo E, Kirtania K, Weiland F 2017 *Energy Technology* **5** 1
[30] Monaghan R F D, Ghoniem A F 2012 *Fuel* **91** 61
[31] Sahraei M H, Duchesne M A, Hughes R W, Ricardez-Sandoval L A 2017 *Fuel* **196** 520
[32] Billaud J, Valin S, Peyrot M, Salvador S 2016 *Fuel* **166** 166
[33] Valero A, Uson S 2006 *Energy* **31** 1643
[34] Jeong H J, Hwang I S, Park S S, Hwang J 2017 *Fuel* **196** 371
[35] Donskoi I G 2019 *Soild Fuel Chemistry* **53**
[36] Chen W H, Chen C J, Hung C I, Shen C H, Hsu H W 2013 *Applied Energy* **112** 421
[37] Ku X, Li T, Lovas T 2014 *Energy & Fuels* **28** 5184
[38] Adeyemi I, Janajreh I, Arink T, Ghenai C 2017 *Applied Energy* **185** 1007
[39] Donskoi I G 2016 *Solid Fuel Chemistry* **50** 191
[40] Donskoy I G, Shamansky V A, Kozlov A N, Svishchev D A 2017 *Combustion Theory and Modelling* **21** 529

Journal of Physics: Conference Series **1369** (2019) 012016 doi:10.1088/1742-6596/1369/1/012016

5th International Workshop on Heat/Mass Transfer Advances
for Energy Conservation and Pollution Control
August 13-16, 2019, Novosibirsk, Russia

# Investigation of the features of polyolefins and wood biomass thermochemical conversion for their energy utilization

**I G Donskoy, V A Shamanskiy, A N Kozlov, M V Penzik, M A Kozlova**

130 Lermontova st., Irkutsk, Russia, 664033

donskoy.chem@mail.ru

**Abstract**. Combustion of solid combustible wastes is the most common way of their disposal. In developed European countries, up to 50% of all municipal waste is burned; in Russia, it is estimated to be less than 2.5%. The efficiency of modern thermal plants operating on MSW and biomass, as well as the capacity of individual combustion units are continuously increasing, approaching the average values that are close to fossil fuels plants. Replacing coal with combustible waste will make it possible to reduce the cost of energy production and reduce the load on waste landfills. However, to this end, it is necessary to solve the problems of organizing efficient combustion as well as cleaning of combustion products. Of particular interest is the co-processing of municipal waste along with lignocellulosic waste from agriculture and the forest industry. In the present work, the thermal behavior of woody biomass and polyethylene mixtures (in different mass ratios) when heated in an oxidizing environment is investigated using methods of thermal analysis. The distribution of the products of pyrolysis and oxidation is investigated by means of mass spectrometry.

## 1. Introduction

The problems of municipal waste disposal are urgent. A significant proportion of them are polymeric materials. Production of plastics in the world reaches 300 million tons per year, of which about 40% are packaging materials, which for the most part form waste. Almost 30% of the total production of industrial polymers is polyethylene (PE). In developed countries, in recent years, the specific production of plastic decreases, and the capacity of waste-processing units increases [1, 2]. This is achieved through appropriate environmental legislation and social policies [3]. There are countries importing waste, the international market for waste recycling is continuously developed [1].

Plastic waste can be used by sorting and recycling [4], using thermal decomposition to obtain valuable raw materials (monomers, combustible gases, fuels) [5-7], or by complete combustion producing heat [2]. The choice of technology depends on the content of polymeric materials in the waste and their component composition. Most of the recyclable municipal waste is currently utilized by combustion.

Thermochemical conversion of polymers faces a number of problems, primarily, the cleaning of thermal decomposition gases to prevent the emission of halogens and polyaromatic compounds. The addition of chlorine-containing polymers affects the evaporation of mineral components [8]. The formation of a significant amount of liquid products can be both a positive and a negative circumstance, depending on the target products of processing.

The development of new efficient methods for processing plastics requires the study of the physicochemical features of their thermochemical transformations [9, 10]. The polymers have a high calorific value, but their complex thermal behavior (melting, endothermic destruction, swelling [11]) leads to difficulties in combustion associated with the mechanical stability of materials and the unpredictable composition and complex chemical behavior of thermal decomposition products [12, 13]. Therefore, the most common reactors for thermochemical conversion of polymers are rotary kilns and augers, and the processes are carried out in allothermic and staged modes [14-17], including post-conversion of liquid tarry products [18, 19]. Promising ways to recycle waste are plasma and catalytic processing [20, 21].

Gasification and incineration of polymeric waste often require the addition of higher-grade fuel. For example, biomass can be used, which is usually also a waste. It has better mechanical and reactive properties, but less calorific value. Their co-conversion in some cases is more effective than separate processing [21].

Gasification and combustion can be carried out in fluidized bed reactors [22-24] (including those with two-stage reactors [25-28]) and in the fixed-bed reactors [29-31]. In the works on thermochemical conversion of mixtures of biomass and polymers, sintering and clinkering problems are often noted, which are associated both with sticking of particles due to the adhesive properties of pyroliquid [32], and because of a temperature increase in the combustion core leading to melting of ash [33]. In this regard, the proportion of plastic in the fuel mixture is often limited [13, 34].

Thermal oxidation of polyolefins proceeds through the radical reactions concerned with peroxide compounds [35], so plastic decomposition in an oxidizing environment begins at lower temperatures compared to pyrolysis [36]. Measurements (for example, carried out in [37]) show that the set of oxidation products is very diverse. These are mainly hydrocarbons, alcohols, aldehydes and ketones, the number of carbon atoms in molecules of products can reach 20 or more units. Aromatics may form at high temperatures [38]. Wood contains a significant amount of oxygen, so its thermal destruction begins at lower temperatures compared to plastics. Composition and particle size of polyolefin and wood could significantly affect the kinetics mixture decomposition due to averaging of thermophysical properties, reaction chains interaction and interfacial phenomena [39-41]

## 2. Experimental study

In this work, thermal decomposition of wood biomass and low pressure PE was studied at different mass ratios when heated at a rate of 30°C/min in an oxidizing environment using the complex thermal analysis unit STA449F1 combined with the QMS403C quadrupole mass spectrometer. Calibrated pine sawdust (lignin content 29.6% wt., tar content 4.4% wt.) with a size of 0.6-1.0 mm, and particles of PE with a size ~1 mm were used as raw materials. Mixtures with PE content ranging from 0 to 100% wt. were investigated with increments of 20%. The environment gas mixture was formed from 10 ml/min of argon and 70 ml/min of air and was fed into the analyzer's heat chamber. The obtained thermogravimetric and DSC curves are shown at Fig.1. First curve at Fig. 1a corresponding to a sample consisting of pure wood (0% PE) has a characteristic section, where it is possible to distinguish intervals with sequential decomposition of cellulose and lignin components of wood. The final section of the curve corresponds to the oxidation of char, which is formed during oxidative pyrolysis. For a pure PE sample, the rate of mass loss monotonously increases, and thermogravimetric curve has no characteristic sections. That indicates that in the process of oxidative pyrolysis of pure PE char is not formed.

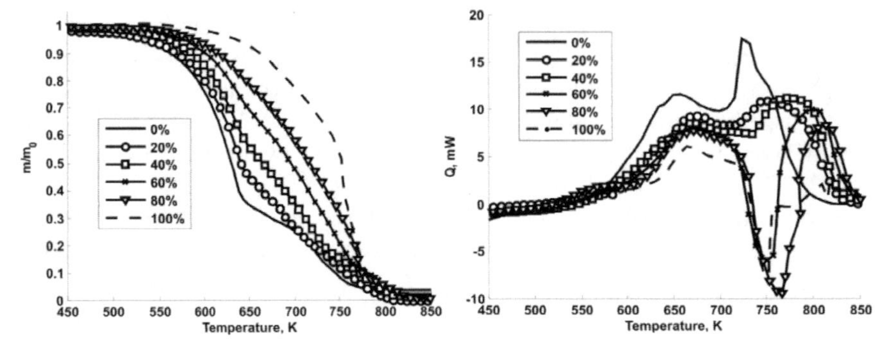


**Fig. 1.** TG-curves (a) and DSC-signal (b) measured during oxidation of wood and PE (legend shows PE content).

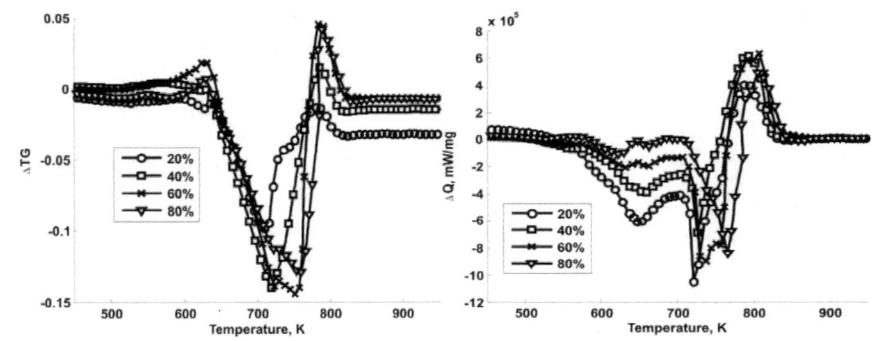


**Fig. 2.** Differences between actual and additive thermal behaviour of samples (legend shows PE content).

The main effect that is expected from the thermal co-conversion of various organic polymers is the mutual influence on the rate and products of decomposition. Fig. 2b shows deviation curves obtained as the difference between actual thermogravimetric curves and additive curves calculated for present mixture compositions. One can see two weak positive peaks in the vicinity of temperatures of 630 and 820 K, which are the higher the greater the amount of PE in the mixture, and a significant negative peak in the region of 720-750 K.

The first two peaks occur in areas of the most rapid burning of wood, so it can be concluded that the presence of PE to a small extent contributes to the oxidation of both the carbohydrate components of the wood and the char conversion. In this case, the pure PE in the region of ~620 K is slightly oxidized, which is also confirmed by the relatively small value of the DSC signal at this temperature for pure PE (Fig. 1b). At the same time, in the 820 K zone, the deviation from the additivity of the DSC signal is considerable and positive, and when the content of PE in the mixture exceeds 20%, the difference does not grow significantly (Fig. 2b). This indicates that the char conversion rate and the PE molecules destruction rate both increase, and decomposition products also could react in the area above the sample.

Figure 1b also shows that in this temperature range, for pure PE, noticeable endothermic processes take place, which, apparently, is connected with the thermal destruction of the hydrocarbon polymer. Thus, the char particles, whose surface is covered with molten products of incomplete decomposition of PE, during combustion could contribute both to the acceleration of the endothermic destruction of hydrocarbon chains and their partial oxidation.

The region of negative deviations of the measured mass of mixtures from additivity (Fig. 2a) indicates a significant decrease in the overall burning rate as a result of the mutual influence of the components. For samples with a PE content of more than 60%, the total thermal effect of oxidative thermal conversion at temperatures of 850–950 K drops to the region of the endothermic process, which is certainly due to the predominance of PE in the mixture and incomplete oxidation of thermolysis products. Figure 3 shows the mass spectra of the products of thermal-oxidative decomposition of the studied mixtures.

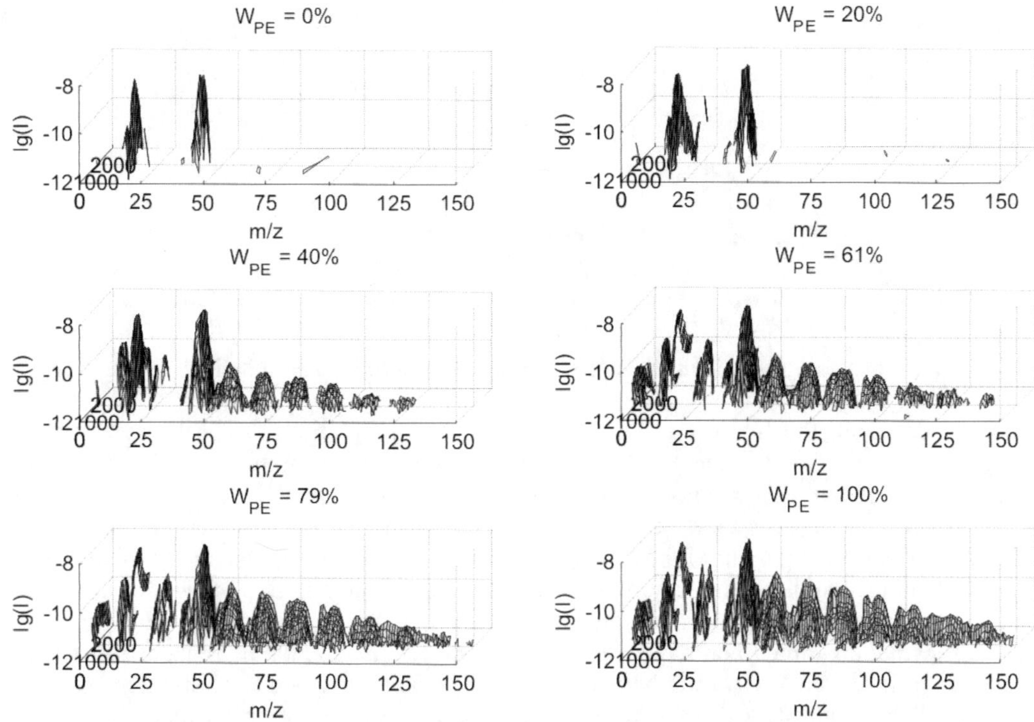


**Fig. 3.** Time-resolved mass spectra of volatile oxidation products.

As expected, the main products of oxidation are carbon dioxide and water. Moreover, they are practically the only ones for mixtures up to 20% PE content. When the content of PE is 40% and higher, an increase in the amount of incomplete oxidation products appears. It is extremely difficult to analyze the composition of these products. For the case of pure PE, starting with a mass number of 50, mass spectra signals appear on almost every mass number up to 150. As the amount of wood increases, "gaps" appear between the mass numbers. According to the experimental conditions, the actual amount of oxygen in the thermoanalyzer chamber is much higher than stoichiometric; therefore, high content of oxidized products leaving the reaction space (in the case of pure PE) indicates first of all the existence of kinetic restrictions obstructing complete oxidation. Since the development of the combustion process proceeds by a radical mechanism, it is reasonable to assume that in the case of thermal decomposition of PE, the resulting hydrocarbon radicals terminate the development of the oxidation process, in which the OH, H and $HO_2$ radicals play a role in the formation of alcohols, aldehydes and carboxylic acids (similar to inhibiting pyrolysis process in [42]). Since the mass numbers corresponding to the hydroxyl (OH), methoxyl ($CH_3O$) and carboxyl (COOH) groups overlap with more intense peaks of water, oxygen and carbon dioxide, respectively, this fact can be observed most clearly at ion currents corresponding to a m/z = 29 (aldehyde group, CHO), shown at Fig. 4.

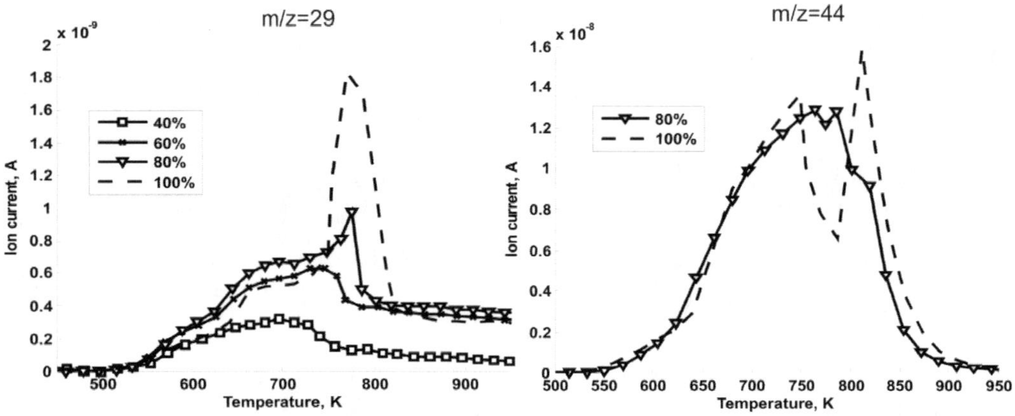


**Fig. 4.** Ion currents obtained for m/z = 29 and 44 of pure PE and wood-PE mixtures.

It can be seen from the figure that with an increase in PE fraction, the peak values of the ion currents grow (as well as their areas). In the case of PE content in a mixture is 20% and below, the aldehyde peaks in the mass spectra are not observed (Fig. 3). Comprehensive analysis of the oxidation kinetics is beyond the scope of this article, but it should be noted that the increase in the yield of aliphatic aldehydes decreases within the temperature range, where an increase in the mixture burnout rate (Fig. 1a) and the exothermic deviation of the DSC signal from additivity (Fig. 2b) are observed. This indicates that in this temperature region the presence of wood prevents the formation of oxygenates polymeric hydrocarbon molecules. Figure 4 shows also the ion currents of carbon dioxide mass spectra, from which it is clear that even a small amount of wood in the system reduces the formation of aliphatic aldehydes in the combustion products towards complete oxidation of organic carbon. Perhaps this is explained by the decomposition and oxidation of lignin components taking place in the same interval in the case of pure wood (Fig. 1). Heat release during their oxidation contributes to the development of deeper oxidation processes of hydrocarbon fragments produced by the degradation of PE. As for the peak on the aldehyde mass spectra corresponding to pure PE and the decrease in its area during carbon dioxide formation, both of these circumstances indicate an increase in the decay rate of carbon chains, and a slight decrease in the oxidation rate of the terminal carbon atoms in polymer molecules, which contributes to the formation of carbon dioxide.

## Conclusions

The processes of combined oxidative thermopyrolysis of wood and PE have a significant non-additive effects, both due to a change in the chemical mechanism of the processes taking place and due to the interaction of dissimilar thermal and kinetic effects. In our opinion, processes of initiation and inhibition of radical chemical reactions in the gas phase and at the surface of particles covered by the molten plastic play important role.

When the content of PE is 40% and higher, products of incomplete oxidation are intensively emitted, and in order to eliminate them, it is necessary to organize an additional oxidation zone. The optimal consumption of wood, PE and gas, as well as the geometry of the secondary reaction space needs to be studied using detailed kinetic and hydrodynamic models.

One of the problems of plastic-rich waste incineration is swelling and agglomeration of particles, which are associated with the formation of high-adhesive liquid substances [51]. In this regard, it is interesting to search for conditions in which the yield of such products is minimal. The results of the study show that, with a plastic fraction of 20% and below, no essential formation of long hydrocarbons and oxygenates are observed. However, the local fraction of plastic may be high enough due to incomplete mixing of raw materials and due to the heterogeneity of such systems: the interaction between biomass and plastic particles occurs at the interface, where the proportions of the components are comparable even with a low gross fraction of plastic. Therefore, in addition to the physicochemical characteristics of the conversion of mixtures, it is necessary to investigate the behavior of particles under fixed bed conditions.

**Acknowledgements**
This work was supported by the Russian Fund for Basic Research (project number 19-08-00744) and was carried out using equipment of the multi-access scientific center "High Temperature Circuit".

**References**
[1] Malinauskaite J, Jouhara H, Czajczynska D, et al. 2017 *Energy* **141** 2013
[2] Castaldi M, van Deventer J, Lavoie JM, et al. 2017 *Waste Biomass Valor.* **8** 1875
[3] Mwanza BG and Mbohwa C 2017 *Procedia Manufacturing* **8** 649
[4] Singh N, Hui D, Singh R, et al. 2017 *Composites Part B: Eng.* **115** 409
[5] Arena U 2012 *Waste Manag.* **32** 625
[6] Wong SL, Ngadi N, Abdullah TAT and Inuwa IM 2015 *Renew. Sust. Energy Rev.* **50** 1167
[7] Ardolino F, Lodato C, Astrup TF and Arena U 2018 *Energy* **165B** 299
[8] Zhang Z, Liu J, Shen F, Zhang Z and Dong Y 2018 *Fuel* **221** 249
[9] Horton SR, Zhang Y, Mohr R, Petrocelli F and Klein MT 2016 *Energy Fuels* **30** 7904
[10] Ranzi E, Faravelli T and Manenti F 2016 *Adv. Chem. Eng.* **49** 1
[11] Zhou C, Zhang Q, Arnold L, Yang W and Blasiak W 2013 *App. Energy* **107** 173
[12] Salganskaya MV, Glazov SV, Salganskii EA and Zholudev AF 2010 *Russ. J. Phys. Chem. B.* **4** 928
[13] Donskoi IG 2018 *Solid Fuel Chem.* **52** 121
[14] Malkow T 2004 *Waste Manag.* **24** 53
[15] Alvarez J, Kumagai S, Wu C, et al. 2014 *Int. J. Hydrogen Energy* **39** 10883
[16] Chen D, Yin L, Wang H and He P *Waste Manag.* 2014 **34** 2466
[17] Sipra AT, Gao N and Sarwar H 2018 *Fuel Proc. Tech.* **175** 131
[18] Ohmukai Y, Hasegawa I and Mae K 2008 *Fuel* **87** 3105
[19] Jin Q, Wang X, Li S, et al. 2019 *J. Energy Inst.* **92** 108
[20] Al-Salem SM, Antelava A, Constantinou A, et al. 2017 *J. Env. Manag.* **197** 177
[21] Lopez G, Artetxe M, Amutio M, et al. 2018 *Renew. Sust. Energy Rev.* **82-1** 576
[22] Ephraim A, Minh DP, Lebonnois D, et al. A 2018 *Fuel* **231** 110
[23] Van Kasteren JMN 2006 *J. Mater. Cycles Waste Manag.* **8** 95
[24] Mastelone ML, Zaccariello L, Santoro D and Arena U 2012 *Waste Manag.* **32** 733
[25] Narobe M, Golob J, Klinar D, et al. 2014 *Biores. Tech.* **162** 21
[26] Cho M-H, Mun T-Y, Choi Y-K and Kim J-S 2014 *Energy* **70** 128
[27] Robinson T, Bronson B, Gogolek P and Mehrani P 2016 *Fuel* **178** 263
[28] Lin C-L and Weng W-C 2017 *Renew. Energy* **109** 135
[29] Nunes SM, Paterson N, Herod AA, et al. 2008 *Energy Fuels* **22** 1955
[30] Garcia-Bacaioca P, Mastral JF, Ceamanos J, et al. 2008 *Biores. Tech.* **99** 5485
[31] Vonk G, Piriou B, Dos Santos PF, et al. 2019 *Waste Manag.* **85** 106
[32] Ouadi M, Brammer JG, Kay M and Hornung A 2013 *App. Energy* **103** 692
[33] Madadian E, Crowe C and Lefsrud M 2017 *J. Cleaner Prod.* **164** 137
[34] Couto ND, Silva VB and Rouboa A 2016 *J. Cleaner Prod.* **139** 622
[35] Iring M and Tudos F 1990 *Prog. Polymer Sci.* **15** 217
[36] Peterson JD, Vyazovkin S and Wight CA 2001 *Macromolecular Chem. Phys.* **202** 775
[37] Bravo A and Hotchkiss JH 1993 *J. App. Polymer Sci.* **47** 1741
[38] Goncalves CK, Tenorio JAS, Levendis YA and Carlson JB 2008 *Energy Fuels* **22** 372
[39] Oyedun AO, Gebreegziabher T and Hui CW 2013 *Chem. Eng. Trans.* **35** 883
[40] Sophonrat N and Yang W 2017 *Energy Procedia* **142** 315
[41] Xiong S, Zhuo J, Zhou H, et al. 2015 *J. Analyt. App. Pyrolysis* **112** 66
[42] Kumagai S, Fujita K, Kameda T and Yoshioka T 2016 *J. Analyt. App. Pyrolysis* **122** 531

Journal of Physics: Conference Series **1369** (2019) 012017 doi:10.1088/1742-6596/1369/1/012017

5th International Workshop on Heat/Mass Transfer Advances
for Energy Conservation and Pollution Control
August 13-16, 2019, Novosibirsk, Russia

# Numerical study on the influence of slit shapes and sizes between the lower wall and the plate on turbulent flow and heat transfer characteristics

**D H Kong, V N Afanasiev, P A Semenev**

Bauman Moscow State Technical University, 2 Baumanskaya ul., 5, Moscow 105005, Russia

kongdehai2013@gmail.com

Abstract. The results of numerical study of flow structure and heat transfer in turbulent flow of air over a flat plate at $q_w$ = const with a rectangular rib with slit channel of different shapes: confusor, diffusor and const cross-section are presented. The slit channel is located between the plate and the lower wall of the rib. Numerical simulation is performed for an incompressible fluid in the framework of three-dimensional Reynolds-averaged Navier-Stokes (RANS) equations with a Kato-Launder $k$-$\varepsilon$ turbulence model for turbulence closure. The parameter effects, including the slit channel sizes and shapes (confusor, diffusor and plane-parallel), on turbulent flow structure and normalized local heat transfer coefficients on the heat transfer surface are examined. It is established that the rib with a slit, especially with a confusor, substantially reduces the size of the recirculation region behind the rib due to the effect of the generated near-wall jet on main flow. In addition, in the presence of a slit between the wall and the rib, in transverse direction, a pair of symmetrical extended spots with high values of heat transfer coefficients relative to the middle of the plate is observed. Rib with a consfor slit having a longer confusor part (Case B2) provides the best overall heat transfer performance.

## 1. Introduction

It is an extremely urgent problem to design and create a highly efficient and compact cooling system, which is often and successfully solved with the use of the methods of heat transfer enhancement. Many studies have focused on methods of enhancing heat transfer by affecting the flow and primarily the boundary layer by means of various intensifiers – ribs of different geometry [1]. As every additional turbulization of the flow is bound to associate with additional energy loss, the selection of location and the method of the additional turbulization of the flow are crucial in the development of effective methods for heat transfer intensification. The most effective method is the intensification of flow, which may affect the near-wall layers of the fluid at a distance, which is of the order of y+ < 30...100 from the wall, without affecting the flow core [2-4]. It can be assumed that such a method of heat transfer intensification can significantly increase the heat transfer coefficient together with a moderate increase in the friction coefficient, i.e. to achieve an increase in heat transfer faster than the growth of friction resistance. The interest in ribs, which are immersed in a boundary layer and somewhat larger than the height of the laminar sublayer has increased.

As is known, when flowing around a solid rib, in front of it and behind it, low-speed separated zones are formed, leading to a significant deterioration of heat transfer in its vicinity. As shown in [5], reduction or elimination of these zones can be achieved through perforation of the rib with plane-parallel slit channels. Comparison of the thermal-hydraulic efficiency of the slit rib (horizontal and vertical slit) and the perforated rib in a turbulent channel flow was studied in [6]. It is shown that such a slit rib is able to eliminate separation zones due to the throttling effect [7]. The studies of [8-10] under the guidance of Panigrahi P. K. are devoted to experimental studies of the processes of flow around permeable ribs with different slit geometries. In addition, in many experimental and numerical studies [11-14] much attention is paid to the study of the flow features of the detached ribs located at a small distance from the heating surface. The study of [11] presents experimental results of the distribution of local heat transfer coefficients in a rectangular tube with detached ribs. The influence of the distance between the pipe wall and the rib, as well as the angle of its installation on the separated flow structure and heat transfer are numerically studied in [12, 13]. In [14] it is shown that the organization of near-wall jets formed by the flow of air from the external flow to the bottom region of the rib can be used to control hydrodynamic and thermal processes in flow around the rib.

Recently, an experimental study of turbulent flow around transverse slit ribs located on a flat plate was carried out in [4]. The confusor, diffuser and constant cross-section slits of finite length between the plate and the lower wall of the rib are considered. It is noted that the structure of the vortex zones before and after the slit rib significantly depends on the shape of the slit. Based on this, in this paper, a detailed numerical study is conducted to investigate the influence of three shapes of a slit (confusor, diffuser and constant cross-section slit) and different slit sizes on the turbulent flow and heat transfer.

## 2. Computational investigation

### *2.1. Formulation of the problem*

Spatial low-velocity turbulent air flow and heat transfer in the vicinity of a single transverse rectangular rib mounted on the plate in the presence and in the absence of a profiled slit are computed. The computational domain size corresponds to experimental section sizes in [4]. Figure 1 depicts the computational domain. The length of the computational domain in longitudinal, transverse and vertical directions is 195, 300 and 80 mm respectively. A square rib with the height of 3.2 mm is located at a distance of 35 mm from the inlet boundary. The ratio of the boundary layer thickness δ in the inlet section to the height of the rib is 4.2.

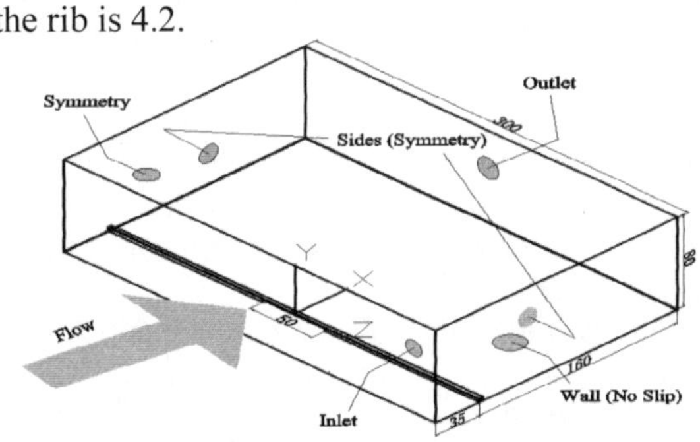


**Figure 1.** Scheme of the computational domain. All dimensions are in millimeters.

The present work considers ribs with a three-type slit (eight cases). The detailed configurations of different slit channels between the wall and the lower wall of the rib with the indication of the geometric dimensions are shown in Figure 2 and in table 1.

Case A1: a single transverse square rib of the height h and the length B = 300 mm is mounted on the heated plate.

In Cases B1-B3, unlike a solid rib (Case A1), a confusor slit is made between the plate and the lower wall of the rib. The slit length in the central part of the rib is 50 mm. In Case B1, the slit consists of confusor and plane-parallel parts. In Case B2, the slit looks like a complete confusor. In Case B3, the inclination angle of the confusor slit is smaller. The vertical size at the slit outlet is somewhat larger than in Case B2 (Figure 2 and Table 1).

Cases C1-C2 are similar to Case A1 but have a diffuser slit between the plate and the lower wall of the rib. In Case C1, the slit consists of plane-parallel and diffuser parts. In Case C2, the slit looks like a complete diffuser.

Cases D1-D2 are similar to Case A1 but have a constant cross-section slit between the plate and the lower wall of the rib. In Cases D1 and D2, the gap between the plate and the rib wall is different; in Case D2, the gap is smaller than that in Case D1.

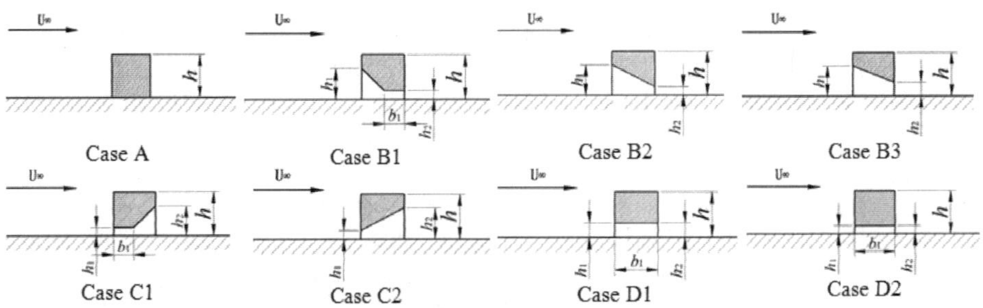


**Figure 2.** Location of square ribs with different-shape and different-size slits on the plate for all cases

**Table 1.** Size of studied ribs

| Case | Type of slit | Rib height, $h$ (mm) | Slit size $h_1$ (mm) | $h_2$ (mm) | $b_1$ (mm) |
|---|---|---|---|---|---|
| A | Absent | 3.2 | 0 | 0 | 0 |
| B1 | Confusor | 3.2 | 2.2 | 0.6 | 1.5 |
| B2 | Confusor | 3.2 | 2.2 | 0.6 | 0 |
| B3 | Confusor | 3.2 | 2.2 | 1.0 | 0 |
| C1 | Diffuser | 3.2 | 0.6 | 2.2 | 1.5 |
| C2 | Diffuser | 3.2 | 0.6 | 2.2 | 0 |
| D1 | Constant section | 3.2 | 1.0 | 1.0 | 3.2 |
| D2 | Constant section | 3.2 | 0.6 | 0.6 | 3.2 |

*2.2. Numerical method and Boundary conditions*

The Reynolds number calculated by $U_\infty$ and $h$ is taken to be 3000, which corresponds to the experimental conditions of [4]. The computational software package ANSYS Fluent 17.2 (license number: 339001) coupled with Kato-Launder $k$-$\varepsilon$ model with enhanced wall function [15] was applied in the analysis of the three-dimensional turbulent flow.

In the present study, the working medium (air with the temperature $T_\infty$ = 297.15 K) and the flow thermophysical properties were assumed constant. Experimental obtained velocity and temperature profiles in [4] were assigned at the computational domain inlet. Turbulence kinetic energy and dissipation rate profiles were determined from velocity and temperature profiles with the use of the Plandtl mixing length hypothesis and the local equilibrium conditions [16].

Outlet boundary conditions are assigned at the computational domain outlet. At the lower wall and rib surface, the no-slip condition is satisfied and the constant heat flux of $q_w$=350 W/m$^2$ is assigned. The conditions of symmetry of governing parameters are satisfied at the upper and side boundaries of the computational domain.

## 3. Results and discussions

Figure 3 shows the flow structure in the vicinity of the longitudinal-vertical ($x$-$y$) plane along the middle of the plate as well as the predicted dimensionless field of the longitudinal velocity component and the streamlines. In flow around the solid rib (Case A) behind it there develop an extended recirculation zone ($0 \leqslant x/h \leqslant 8$) and a secondary vortex characteristic of it. The attached flow providing the high heat transfer intensity is observed in the sections (from $x/h$ = 8 to $x/h$ = 10).

The flow structure undergoes change in the presence of a profiled gap between the plate and the lower wall of the rib. The presence of a near-wall jet significantly changes the structure of the recirculation zone. Figure 3 illustrate that the stagnation and recirculation zones in front of and behind the rib with a slit are suppressed in comparison to the solid rib. In the case of the rib with a confusor slit (Cases B1 and B2), a very thin separation zone is seen ($4 \leqslant x/h \leqslant 6$). When the gap between the plate and the back wall of the rib (Case B3) is increased, the separation zone disappears due to a larger flow rate of the accelerated near-wall flow issuing from the slit. Also, in Cases B1, B2 and B3, behind the rib ($h - h_2$) the backflow zone develops. It represents a pair of vortices rotating towards each other and their scale decreases a little at $h_2 = 1$ mm (Case B3).

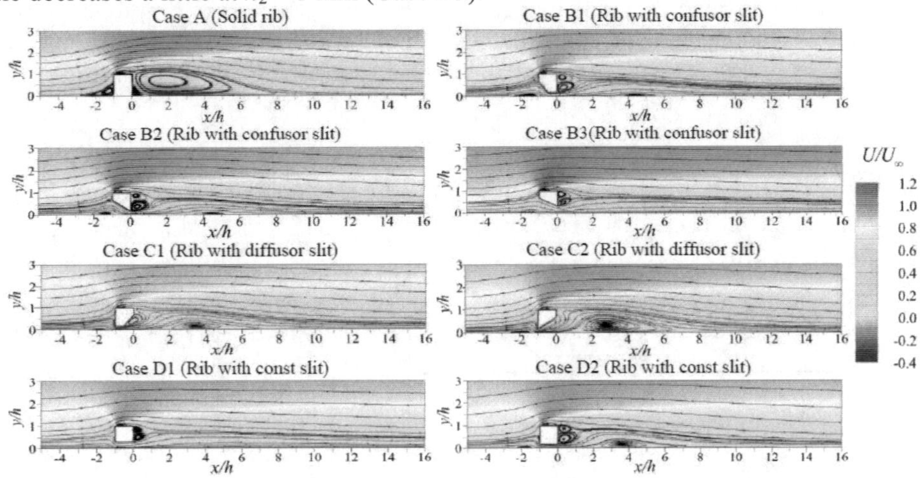


**Figure 3.** Dimensionless longitudinal velocity component field and streamlines around the rib in the longitudinal-vertical ($x$-$y$) plane along the middle of the plate for different cases

In the case of the rib with a diffuser slit (Cases C1 and C2), some part of the near-wall flow rushes into the slit where it slows down in the expanding part of the slit. A small recirculation zone is seen within the range ($1 < x/h < 7$). Its size appears to be significantly smaller in comparison to the similar zone for the solid rib (Case A). When the size of the expanding part of the slit between the wall and the rib is increased, the separated bubble grows in size. It should be noted that in this case, only one vortex is observed directly near the back wall of the rib ($h_1 - h_2$). Behind the rib with a constant cross-section slit (Cases D1 and D2), flow is a lot like flow around the rib with a confusor slit and a pair of vortices is also observed behind the rib ($h - h_2$). However, at $h_2 = 0.6$ mm (Case D2) the separated bubble forming within the range ($2 < x/h < 7$) significantly increases in size and at $h_2 = 1$ mm (Case D1) the separation zone behind the rib disappears as in Case B3. From the aforesaid it follows that in the case of the rib with a confusor slit the accelerated near-wall jet flow entrains the flow separated from the upper surface of the rib and more effectively suppresses the recirculation zone in comparison to ribs with a diffuser slit and a constant cross-section slit.

Figure 4 shows the distribution of the relative heat transfer coefficients ($St/St_0$) in front and behind each type of the rib on the lower ribbed wall. Here the parameter $St_0$ describes heat transfer in the developed turbulent boundary layer in flow around a flat plate at $q_w = \text{const}$ [17].

The classical pattern of heat transfer is observed in flow around a solid rib (Case A) - the region with an extremely low heat transfer coefficient appears immediately behind the rib ($0 \leqslant x/h \leqslant 2$). Then, in the process of the flow movement, there is an area ($2 \leqslant x/h \leqslant 12$) with a moderate increase in the heat transfer level ($St/St_0 \approx 1.1$) as a result of recirculation and reattachment of the flow, respectively. Further, the heat transfer gradually decreases to the level of a flat wall.

In the case of ribs with a confusor slit (Cases B1 and B2), a high-speed near-wall jet is formed in the gap between the plate and the rib, leading to a significant heat transfer intensification on the heat

Journal of Physics: Conference Series **1369** (2019) 012017 doi:10.1088/1742-6596/1369/1/012017

transfer surface immediately in front ($x/h \approx -1$) and behind ($0 \leqslant x/h \leqslant 2$) the rib. Further downstream, for Cases B1 and B2 along the longitudinal direction ($2 \leqslant x/h \leqslant 6$), the heat transfer deteriorates sharply in the zone of interaction of the near-wall jet with the recirculation region. For Case B2, due to the flow reattachment on the heat transfer surface, an extensive region ($6 \leqslant x/h \leqslant 20$) of slight heat transfer intensification is observed ($St/St_0 \approx 1.05$). Compared with Case B2, for Case B1, the value of the heat transfer coefficient in this zone is slightly lower. When the size of the near-wall jet (Case B3) increases, as in the Cases B1 and B2, a near-wall jet is formed, but with a greater intensity of injection. In so doing, the region ($0 \leqslant x/h < 4$) of heat transfer intensification observed immediately behind the rib is bigger. Further downstream from the rib, the heat transfer rate decreases slightly and then slowly increases downstream.

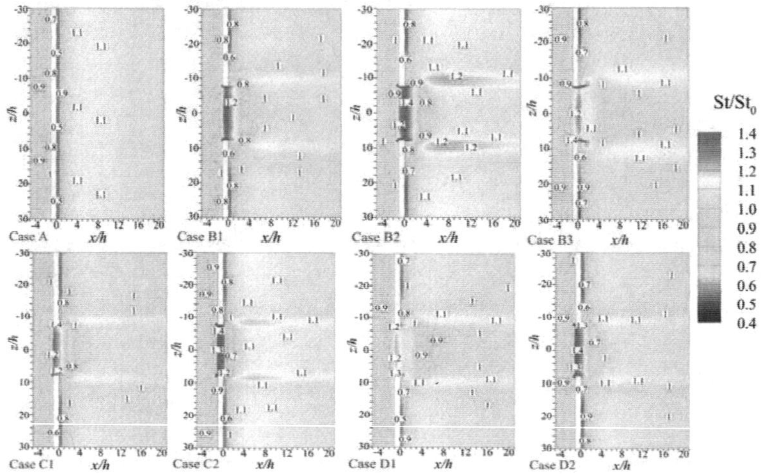


**Figure 4.** Local relative Stanton number distribution on the heat transfer surface for different cases

In the case of ribs with a diffuser slit (Cases C1 and C2), a low-speed near-wall jet is formed. In doing so, immediately behind the rib, the area ($0 \leqslant x/h < 1$) of heat transfer intensification is significantly reduced. However, further downstream, large differences are observed between Cases C1 and C2, and the heat transfer efficiency for Case C2 is significantly higher than that for Case C1. In the case of rib with a constant cross-section slit with a smaller gap size (Case D2), the heat transfer pattern on the heat transfer surface is largely similar to the pattern observed in the case of the rib with a diffuser slit (Case C1). With further increase of the gap size (Case D1), the heat transfer pattern in this case is largely similar to the case of the rib with a confusor slit having the same size at the outlet of the slit (Case B3). But behind the rib, the area of heat transfer intensification is slightly reduced.

It should be noted that in the presence of a gap between the wall and the rib, in the transverse direction, a pair of symmetrical extended spots with high values of heat transfer coefficients is observed, especially in the case of ribs with a confusor and diffuser slit (Cases B2 and C2), relative to the middle of the plate.

Figure 5 shows the averaged Nusselt number in spanwise direction (Z-axis direction) for each rib. Comparing with solid rib, ribs with a slit considerably improve the heat transfer upstream and immediately downstream from the ribs.

Figure 6 illustrates the calculation results of the integral relative Stanton number $St/St_0$ for different geometry ribs. It is seen that the ribs with confusor and diffuser slits (Cases B2 and C2) provide a higher mean value of the Stanton number than the solid rib. In Cases B1, B3, C1, D1 and D2, the heat transfer intensity appears to be somewhat less than that for the solid rib. A smaller value of the Stanton number is seen in flow around the rib with a diffuser slit (Case C1). Upstream of the rib in the selected region, the mean value of the Stanton noticeably increases for the rib with a slit, especially with a confusor slit (Cases B2 and B3) in comparison to the solid rib.

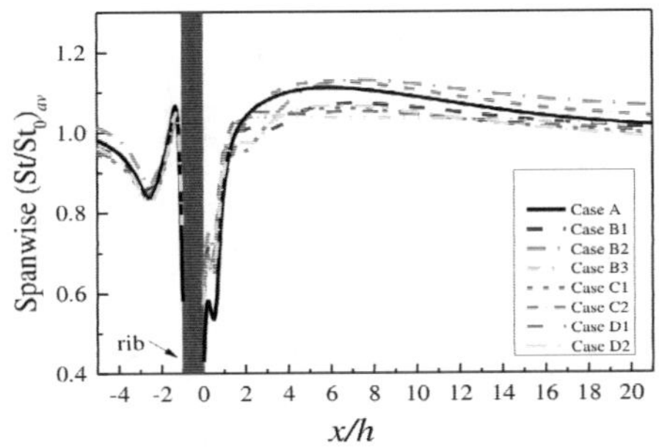


**Figure 5.** Spanwise average relative Stanton number

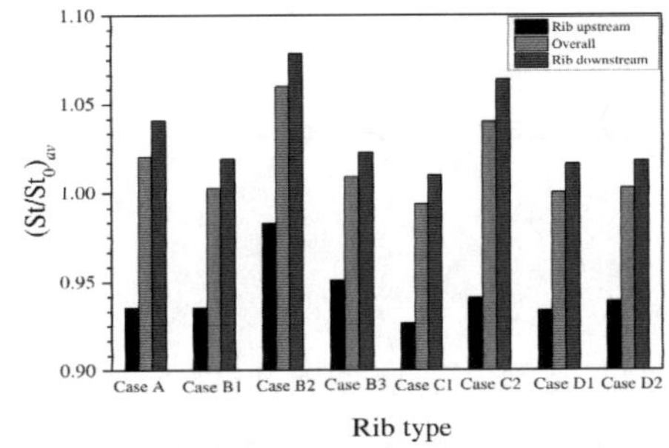


**Figure 6.** Average relative Stanton number on the heat transfer surface (upstream of the rib, downstream the rib and overall value)

When considering the overall mean values of the heat transfer coefficient (Figure 6) a maximum value is seen in the case of the rib with a confusor slit (Case B2). So the overall relative value of $St/St_0$ is by 4% higher than in the case of the solid rib.

## 4. Conclusions

Hydrodynamics and heat transfer processes are numerically investigated in the case of turbulent air flow around the plate surface heated at $q_w$ = const when rectangular ribs ($\delta/h \approx 4$) with confusor, diffuser and constant cross-section slits are mounted across the flow. It is shown that the near-wall jet forming in the slit substantially affects the separated flow structure behind the rib. Behind the rib with a slit, especially with a confusor slit, the separation zone is substantially smaller than that behind the solid rib. When intense jets are formed (Cases B3 and D1) separation zones do not develop on the plate. It is found that the rib with a confusor slit substantially enhances heat transfer directly behind it in comparison to the solid rib. A maximum heat transfer intensity is achieved in the case of the rib with a confusor slit having a longer confusor part (Case B2).

## References

[1] Dreitser G A 2006 *Therm. Eng.* **53** 279–287
[2] Kalinin E K, Dreitser G A and Yarkho S A 1990 *Enhancement of the Heat Transfer in Channels* (Moscow: Mashinostroenie) pp. 18–29 (In Russ.)
[3] Afanasiev V N and Kong D H 2017 *J. Phys. Conf. Ser.* **891** 012140
[4] Afanasiev V N and Kong D H 2018 *Proc. 16th Int. Heat Transfer Conf. (Beijing)* (Connecticut: Begell House) pp 5057–64
[5] Huang J J and Liou T M 1997 *J. Turbomach.* **119** 617–623
[6] Isaev S A, Vatin N I, Guvernyuk S V, Gagarin V G, Basok B I and Zhukova J V 2015 *High Temp.* **53** 873–876
[7] Liou T M, Chen S H and Shih K C 2002 *Int. J. Heat Mass Transf.* **45** 4493–4505
[8] Tariq A, Panigrahi P K and Muralidhar K 2004 *Exp. Fluids* **37** 701–719
[9] Panigrahi P K and Tariq A 2003 *J. Visual-Japan* **6** 407–416
[10] Panigrahi P K, Schröder A and Kompenhans J 2006 *Exp. Fluids* **40** 277–300
[11] Liou T M and Wang W B 1995 *Int. J. Heat Mass Transf.* **38** 91–100
[12] Bogatko T and Terekhov V 2017 *MATEC Web Conf.* **92** 01063
[13] Yongsiri K, Eiamsa-ard P, Wongcharee K and Eiamsa-ard S 2014 *Case Stud. Therm. Eng.* **3**
[14] Molochnikov V M, Mikheev N I, Davletshin I A and Paereliy A A 2008 *Proceedings of the Russian Academy of Sciences. Power Engineering.* **1** 137–144 (In Russ.)
[15] Valencia A 2000 *Numer. Heat Tr. A-Appl.* **37** 289–306
[16] Benodekar R W, Goddard A J H, Gosman and A D and Issa R I 1985 *AIAA J.* **23** 359–366
[17] Kutateladze S S and Leontiev A I 1990 *Heat transfer, mass transfer, and friction in turbulent boundary layers* (New York: Taylor and Francis)

5th International Workshop on Heat/Mass Transfer Advances
for Energy Conservation and Pollution Control
August 13-16, 2019, Novosibirsk, Russia

# Development and Prospect of the Pumped Hydro Energy Stations in China

**B S Zhu*, Z Ma**

Department of Energy and Power Engineering, Tsinghua University, China

bszhu@mail.tsinghua.edu.cn

Abstract. Pumped hydro energy storage (PHES) has been recognized as the only widely adopted utility-scale electricity storage technology in the world. It is able to play an important role in load regulation, frequency and phase modulation and black starts in power systems. Due to its outstanding functions, this technology has been widely used worldwide.

This paper introduces PHES including its functions and the technologies briefly to supply a comprehensive cognition of this technology. The key technologies of PHES, such as design and manufacture of reversible pump-turbine runners, design of heavy-duty thrust bearings and shaft sealings, startup of unit etc., are important for the PHES. In addition, more and more new technologies, such as operation with seawater, underground lower reservoirs, adjustable speed units etc., were put forward and used in PHES to make it perform better.

This paper focuses on the development of PHES in China. Thanks to a rapid development of China's economy in the recent years, PHES has gained a fast development. At present, China has owned the biggest installed capacity (29.9 GW by the end of 2018) of PHES the world. PHES has been an indispensable part of the power grid to increase the stability of the grid and improve the penetration of sustainable energy such as wind power, solar energy etc. With the employment of new technologies. China's new PHES stations, such as Guangzhou, Tianhuangping, Xilongchi PEHS etc., have become the representatives of the world's most advanced level. Based on the requirement of the current power grid for PHES, some new PHES have been under construction and more PHES are in the planning and design stage.

Keywords: Pumped hydro energy station (PHES), function, China, development

## 1. Introduction

Pumped hydro energy stations (PHES) is the only proven large-scale (>100 MW) energy storage technology [1]. Apart from the energy storage, it can also help in the load regulation, promoting grid-connected clean energy and maintaining the security and stability of the electric power system. The benefits of PHES on electricity system operations are well documented in textbook and journals [2].

PHES has been widely used worldwide, especially in the developed countries and regions [3]. Until 2018, pumped storage hydropower has remained the dominant source of large-scale energy storage, accounting for an estimated 96% of global energy storage capacity [4]. Global pumped storage capacity rose by more than 7 GW in 2018, for an estimated year-end total of 160 GW [5].

As the world's second economy, China has a huge generation of electricity. However, the coal-based power grid is difficult in peak shaving and village filling, fast startup and accommodating the renewable energy. To guarantee the stable operation and to improve the renewable energy, an effective power storage technology is urgently needed in China. Due to its satisfactory performances, PHES is expected to be China's primary peaking power source in the future [6].

In the present paper, a brief introduction of PHES are presented firstly. Some of the key technologies of PHES are classified and discussed. The development of Chinese PHES are introduced with showing the distribution of the PHES and the updating of the technology in detail. Meanwhile, the prospect of PHES in China is also depicted based on the policy.

## 2. Description of PHES

The main motivation of PHES is to compensate the random nature of consumption of thermal power. It is implemented by storing energy in excess by pumping water to a higher reservoir or delivering energy in consumption peak by releasing water to push the hydraulic turbine. In Figure 1, a sketch of a PHES are depicted showing the flow direction of the water in different modes.

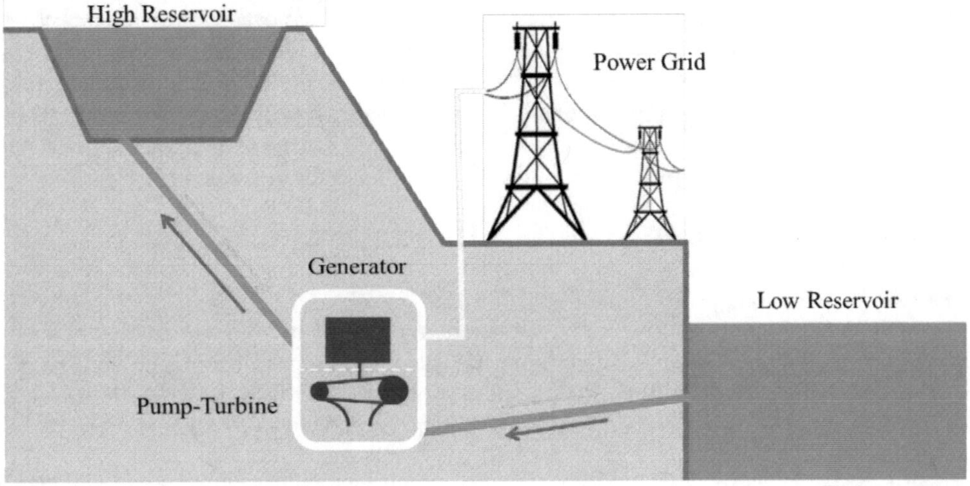


Figure 1. Sketch of the pumped hydro energy station as well as the pump mode and turbine mode.

Koutnik et al. [7] claimed that PHES play an important role in improving the development of the renewable energy, such as wind power, solar power etc. PHES can help to reduce the impact of renewable energy on the power system by smoothing their generation and suppressing their variability [7]. Although, large scale energy storage systems such as underground compressed air energy, battery storage and large flywheels can also store the renewable energy, the energy storage in elevated water remains the most economical method of load leveling for the power stations [8]. PHES also have some other advantages including flexible start/stop and fast response; tracking load changes and drastic load changes adaption; frequency modulation as well as maintenance of the voltage stability [2].

## 3. Technology of PHES

### *3.1. Key technology of PHES*

The unit in PHES plays the most important role as energy storage/generation. The whole unit can be divided into four parts including flow passage, bidirectional shaft, generator, and the relevant systems. A summary of the key technologies is displayed as the following shows:

a. The technology of pump-turbine runners:

1) Hydraulic design and optimization of pump-turbine runners; 2) Manufacture of the runners.

b. The technology of bidirectional thrust bearing and shaft:

1) Design calculation and experiment technology of large scale bidirectional and heavy-duty thrust bearing for the generator; 2) Analysis technology of shaft stability, stiffness and dynamic characteristics for large generators; 3) Technology of shaft seal.

c. The technology of startup.

d. The technology of the relevant system:

### *3.2. Potential technologies of PHES*

Some potential technologies are summarized according to the new trend.

1) Seawater PHES: PHES can operate with seawater. Seawater PHES is a solution of coastal energy storage, especially in an insular system. The quantity of water required for the operation of a PHES is often not available in insular systems due to the low annual rainfalls [9]. Therefore, in such a lacking fresh water region, seawater PHES shows its advantage.

2) Adjustable speed unit: The inherent limitations which may reduce PHES ability to contribute to the balancing of the rapidly varying output of new renewable energy and ancillary services provision [10]. Compared with the fixed-speed unit, adjustable speed unit offers several advantages for both pump and generation modes such as (1) the possibility of an almost instantaneous active power control, (2) reactive power control and (3) higher efficiency and a wider range of operation [11].

3) Underground reservoirs: Some sites are not suitable to build PHES only due to the flat topography. However, underground PHES could be one alternative [12]. Underground PHES also has two reservoirs, but the lower one is underground while the upper one is located at the surface or at shallower depth [13]. That makes these PHES possibly be built in the flat region [14].

## 4. Development of PHES in China

### *4.1. Power system structure*

Coal-based stations developed quickly and have become the main way to generate electricity in China. However, this electricity grid lacks an economic means of providing peaking power. To solve this problem, pumped hydro energy stations were paid more attention. From the late 1960s, many PHES were built in China and China has owned the greatest installed capacity of PHES at present.

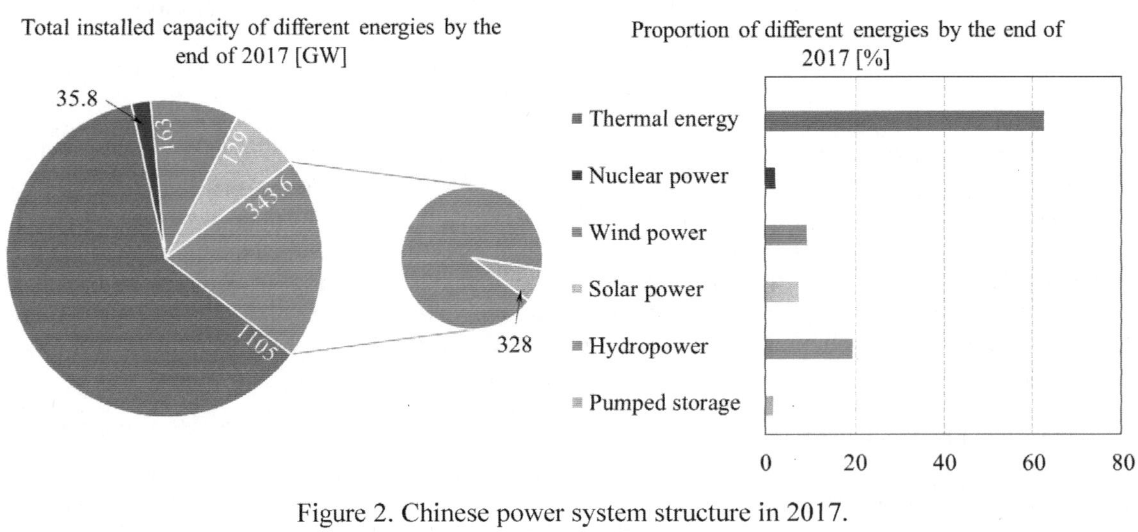


Figure 2. Chinese power system structure in 2017.

In China's power system, thermal energy is no doubt the dominant occupying about 62% of capacity at end of 2017. Hydropower including pumped storage, nuclear energy, wind energy and solar energy take the second to fifth place as Figure 2 shows [15]. With the rapid increase of power system, the stability and environment friendliness were paid more attention. Reduction of the CO2 emission, peak shaving and valley filling have been paid more and more attention.

At present, China is zealously popularizing renewable energies. Development of renewable power can provide obvious environmental benefits, but renewable power may reduce the power system's security and stability. Therefore, an effective and economic energy storage method is needed in China.

*4.2. Significance of PHES in China*

The functions of PHES are summarized as:

1) PHES's peak shaving and valley filling help the coal-based station to save fuel, to avoid restarting, to smooth the output and to improve the load efficiency [16].
2) PHES can adapt quickly load changes and modulate the frequency as well as maintain the voltage [17]. PHES can just be used as an emergency backup to prevent system collapses.
3) PHES is a complement with balancing the disequilibrium of renewable power generation and regulating the frequency of grid.

*4.3. Development of PHES in China*

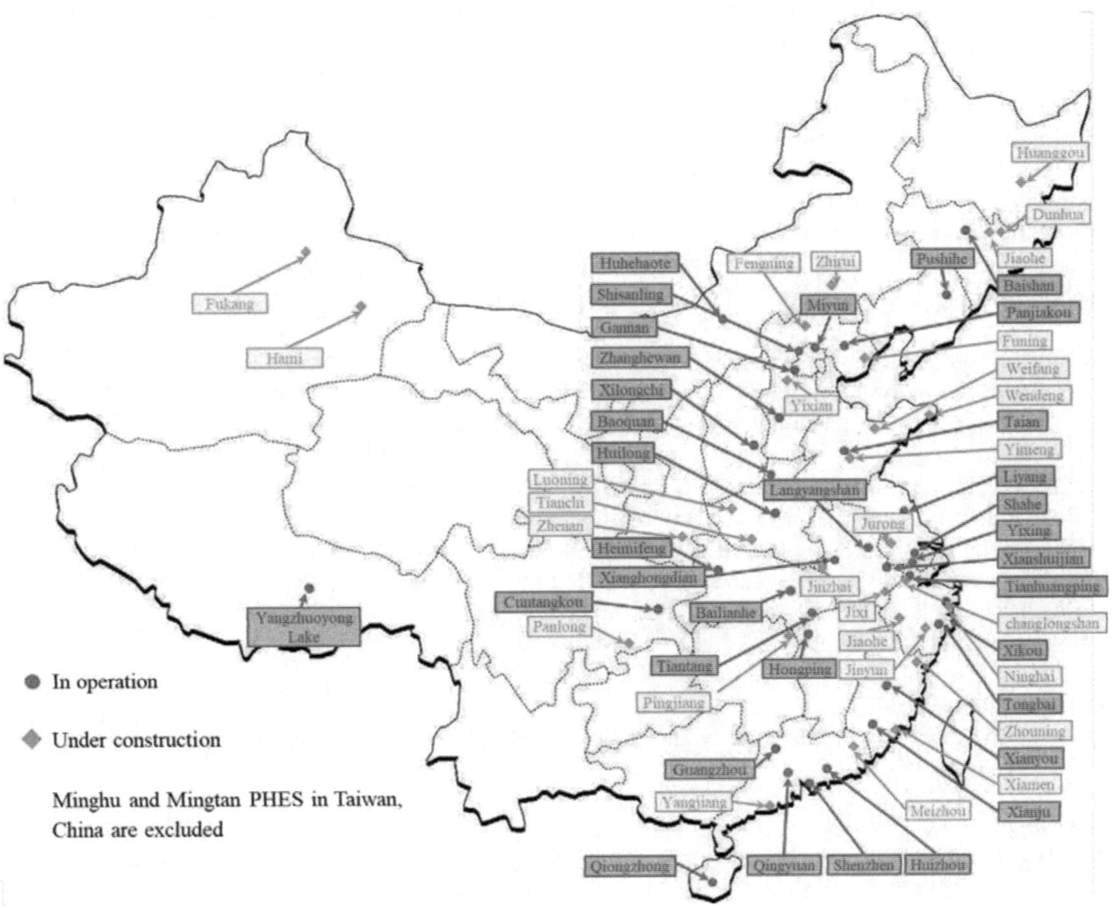


Figure 3. Locations of PHES in operation and under-construction by the end of 2018.

Gangnan station is the beginning of the development of PHES in China in 1968 with 11 MW capacity. The second PHES is Miyun station with installed capacity of 22 MW. After these two PHES, the development remained dormant until 1990s. The end ten years of the last century is a period of rapid progression for PHES station. Coming to the 21st century, the second climax of PHES construction appeared. Several large-scale PHES stations were built, such as Taishan, Langyashan, Yixing, Zhanghewan, Xilongchi, Baoquan, Bailianhe, Heimifeng, Pushihe, Xiangshuijian. Figure 3 show the operating and under construction PHES at the end of 2018 [18, 19].

The development of PHES equipment in China can be divided into three stages. At the beginning, only small and middle size units can be manufactured in China and the stations were designed by foreign companies; Secondly, new technologies were absorbed from the foreign companies; Finally, the capability of design, manufacture and site installation for large scale PHES units were grasped.

China's stations built in recent years have already employed the forefront technologies. Guangzhou and Huizhou stations were the second PHES in terms of installed capacity only less than the bath country pumped storage station in USA (maximum capacity is 3030 MW) [20]. Fengning PHES in Hebei province will rank the first place with a total installed capacity of 3600 MW [21] and it will be the first adjustable speed unit in China [22]. The maximum head of the single stage reversible pump-turbine in Changlongshan PHES is 710 m [23], which ranks only second to Kazunogawa PHES in Japan (714 m) [24]. As to civil engineering, the Shisanling PHES station utilized the reinforced concrete in the whole reservoir to reduce the permeability, which is also the leading level in the world. The technology of asphalt concrete face slab used in whole reservoir bottom against penetration for the Tianhuangpin, Zhanghewan and Xilongchi PHES stations is on the world advanced level.

### *4.4. Prospect of PHES in China*

China has planned to build more PHES in the future. Total installed capacity of pumped storage is set to reach at least 40 GW by 2020 from its current 28.49 GW, and with 60 GW either under construction or in the planning stages according to Chinese "13th Five Year Plan" [25]. Some new technologies will be employed. A nationwide general investigation of seawater PHES has been finished which shows an installed capacity of 42083 MW. China also has begun to employ adjustable speed units in the world's largest PHES in Hebei province [22].

## 5. Conclusions

PHES plays an important role in the power grid. PHES is made of many complex subsystems. Each of them consists of some key technologies and some potential technologies.

The development and the prospect of China's PHES are introduced in the present paper. Some conclusions can be drawn as follow: 1) PHES is important in China's power grid in improving the stability and promoting the development of the renewable energies such as wind power and solar energy; 2) The installed capacity and generation of PHES increased rapidly, in the same time, China's PHES had been in the world advanced level and the relevant design and manufacture technologies are improved rapidly; 3) PHES in China has a promising future with building more stations and employing advanced technologies.

## Acknowledgement

This work was partially supported by the national natural science foundation of China (Grant NO. 51679122, 51736008).

## References

[1] Wang Z. J., Zhu B. S., Wang X. H., et al., Pressure Fluctuations in the S-Shaped Region of a Reversible Pump-Turbine. Energies, 2017, 10(1):96-.

[2] Hino T., Lejeune A., Pumped storage hydropower developments. Compr Renew Energy 2012, 6:405-434.

[3] Fujihara T., Iman H., Oshima K., Development of pump turbine for seawater pumped storage power plant. Hitachi Rev, 1998, 47(5):199-202.

[4] Zuo Z, Liu S, Sun Y, et al., Pressure fluctuations in the vaneless space of High-head pump-turbines—A review. Renewable & Sustainable Energy Reviews, 2015, 41:965-974.

[5] IHA. Hydropower status report 2019. International Hydropower Association, http://www.hydropower.org/.
[6] Renewables 2018 Global Status Report, http://www.ren21.net/gsr-2018/.
[7] Koutnik, J., Foust, J., Nicolet, C., Saiju, R. and Kawkabani, B., 2010, Pump-Storage Integration with Renewables – Meeting the Needs Using Various Concepts, Proceedings of Hydro Vision International, Charlotte, NC, USA.
[8] Hasmatuchi, V., Roth, S., Botero, F., Farhat, M., and Avellan, F., , Hydrodynamics of a Pump-Turbine at Off-Design Operating Conditions: Numerical Simulation, Proceedings of the 7th International Symposium on Pumping Machinery, 2011, Hamamatsu, Japan
[9] Katsaprakakis D. A., Christakis D. G., Stefanakis I., et al., Technical details regarding the design, the construction and the operation of seawater pumped storage systems. Energy, 2013, 55(1):619-630.
[10] Moreira C., Fulgêncio, N., Silva B., et al., Identification of Dynamic Simulation Models for Variable Speed Pumped Storage Power Plants[J]. Journal of Physics: Conference Series, 2017, 813(1):012006.
[11] Martínez-Lucas, G., Pérez-Díaz, J. I., Sarasúa, J. I., et al., Simulation model of a variable-speed pumped-storage power plant in unstable operating conditions in pumping mode. J Phy: Conf Series, 2017, 813(1):012028.
[12] Elena P., Schüttrumpf Holger., Reflection Phenomena in Underground Pumped Storage Reservoirs. Water, 2018, 10(4):504-.
[13] Barnes, Frank S., and Jonah G. Levine, Large energy storage systems handbook, 2011.
[14] Pujades E., Orban P., Jurado Elices A., et al., Water chemical evolution in Underground Pumped Storage Hydropower plants and induced consequences, Egu General Assembly Conference. EGU General Assembly Conference Abstracts, 2017.
[15] Pujades E., Orban P., Bodeux S., et al., Underground pumped storage hydropower plants using open pit mines: How do groundwater exchanges influence the efficiency. Applied Energy, 2017, 190:135-146.
[16] Chen H. S., Cong T. N., Yang W., et al., Progress in electrical energy storage system: a critical review. Review Article Progress in Natural Science, 3(19),291-312.
[17] Wang F., Jiang J. G., A novel static frequency converter based on multilevel cascaded H-bridge used for the startup of synchronous motor in pumped-storage power station. Energy Convers Manage, 52(5), 2085-2091.
[18] China pumped storage plants networks. Statistical tables of pumped storage power stations have been built in China (by the end of December 2018). http://www.psp.org.cn/upload/news/n2019050913041273.pdf.
[19] China pumped storage plants networks. Statistical tables of pumped storage power stations are under construction in China (by the end of December 2018). http://www.psp.org.cn/upload/news/n2019050913071363.pdf.
[20] The inside story of the world's biggest 'battery' and the future of renewable energy. https://thinkprogress.org/the-inside-story-of-the-worlds-biggest-battery-and-the-future-of-renewable-energy-8984e81283c/.
[21] Fengning pumped storage power station. https://en.wikipedia.org/wiki/Fengning_Pumped_ Storage_Power _Station.
[22] GE's variable speed technology to power the world's largest pumped-storage hydro plant. https://www.gepowerconversion.com/press-releases/ge%E2%80%99s-variable-speed-technology-power-world%E2%80%99s-largest-pumped-storage-hydro-plant.
[23] http://www.chinanews.com/gn/2015/11-27/7645555.shtml.
[24] Kazunogawa pumped storage power station.https://en.wikipedia.org/wiki/Kazunogawa _Pumped_ Storage _Power_Station.
[25] National Energy Administration, Development of hydropower in the "13th Five Year Plan". http://www.nea.gov.cn/2016-11/29/c_135867663.htm.

5th International Workshop on Heat/Mass Transfer Advances
for Energy Conservation and Pollution Control
August 13-16, 2019, Novosibirsk, Russia

# Flow and heat transfer of fluid in a pulsating heat pipe

**Megha Sharma and Basant Singh Sikarwar**

Department of Mechanical Engineering, Amity University Uttar Pradesh, Sector 125, Noida 201313, India

msharma@amity.edu

**Abstract:** Cooling of electronic components are preferred heat pipes imbedded in heat sink because of high heat transfer through them. The oscillating motion of the working fluid due to continual phase change facilitates both sensible and latent heat transfer in pulsating heat pipes. This enables transfer of high heat flux efficiently over long distances. In this work, a mathematical model is developed to optimize the parameters which involves in heat pipe design. The parameters are filling ratio, size of tubes and degree of temperature. In House Finite Volume based CFD solver is used for simulation. The obtained simulation results are validated against experimental results present in literature. These results are useful for understanding thermos-hydrodynamics of pulsating heat pipe for achieving maximum thermal performance

## 1. Introduction

With the advent of technology and use of portable systems, microelectronics has become a significant part of our day-to-day lives. Advanced electronic components are subjected to high heat loads due to increased electrical demand and smaller sizes. A major concern is the inefficient heat transfer in these devices which causes over-heating of the components. As a result, there are various cooling systems used in electronic components such as air/liquid cooling, synthetic jet cooling, spray cooling, solid cooling, passive cooling using PCM, cold plates, heat pipes etc. Among these cooling techniques, heat pipes have been widely employed in various industries, since they can carry high heat loads[1]. Out of the various heat pipes used for heat transfer, Oscillating heat pipes (OHPs) are very popular due to their simple design and low fabrication cost as compared to conventional designs. Moreover, they do not require an additional wick structure to transport condensed liquid back to the evaporator.

An Oscillating Heat Pipe (OHP) utilizes phase change for effective heat transfer over long distances. It was invented by Akachi in 1996 [2] and since then it has found numerous applications such as solar energy collection, thermal energy storage, cryogenics, drying, electronics cooling, etc. A typical OHP consists of a working fluid inside a meandering channel, which passes through a heat reception and a heat rejection area[3]. The high heat input from the reception area (evaporator section) triggers nucleate boiling of the working fluid and forms vapor bubbles inside the heat pipe. The energy from the evaporator provides the driving force for fluid motion, which causes the vapor bubbles to move towards the heat rejection area (condenser section). The vapor condenses as it encounters the cold surface of the condenser and the temperature is reduced below saturation temperature. The continuous evaporation and condensation of the working fluid creates a circulatory and oscillatory motion of vapor and liquid plugs which facilitates both sensible as well as latent heat transfer through the OHP[4]. Hence, OHPs can transport a high amount of heat flux over long distances.

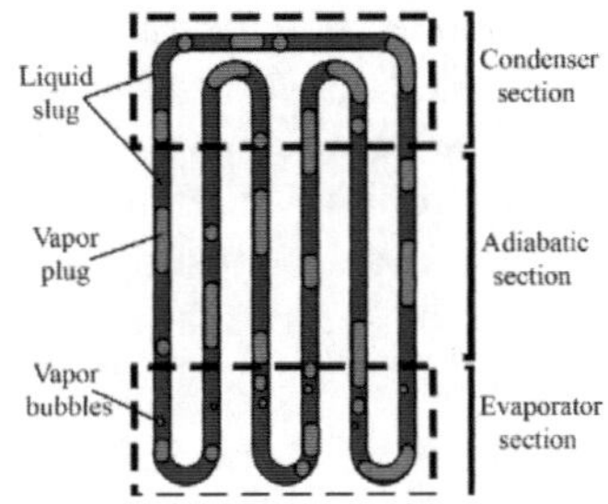


**Fig. 1** Illustration of liquid-vapour slug flow in an oscillating heat pipe

The hydrodynamics and thermal performance of heat pipes have been studied extensively over the years. Tong et al. [5] conducted an experimental study on a vertically oriented Pyrex glass OHP with Methanol as the working fluid. They observed nucleation boiling at the evaporator, coalescence of bubbles and movement of liquid-vapour in a "slug-train" arrangement inside the heat pipe creating an oscillating flow inside the heat pipe. Some other researchers such as Pouryoussefi[6] have developed a numerical model to simulate chaotic flow in a closed loop pulsating heat pipe under several operating conditions. Lin et al.[7] compared Volume of Fraction (VOF) and Mixture models to study the heat transfer mechanism of miniature oscillating heat pipes via simulations. They reported that the mixture model was more suitable for predicting the thermal performance. However, the VOF model was more suitable for flow visualization as it does not allow interpenetration of phases and liquid-vapor slug flow can be clearly seen. Although several analytical and experimental studies have been carried out to analyse the performance of an OHP[8–10], computational analysis demands more attention. In this background, numerical simulations are performed in this paper using Volume of Fluid (VOF) multiphase model to analyse the effect of heating wall temperature (393-413K) for water as working fluid at 50% filling ratio inside the oscillating heat pipe.

## 2. METHODOLOGY

**2.1 Computational model:** The transport phenomenon inside an OHP has been simulated to analyse the effect input heat flux and filling ratio of water on the flow behaviour and thermal performance of the OHP. The 2D computational model of a three-turn closed loop OHP is created. The OHP has an evaporator and a condenser each of length 20mm separated by a 110mm adiabatic section with an inner diameter of 1.3 mm as shown in Fig 2. It is partially filled with the working fluid. The computational domain is discretised using a structured grid with approximately 30,000 elements as shown in Figure 2.

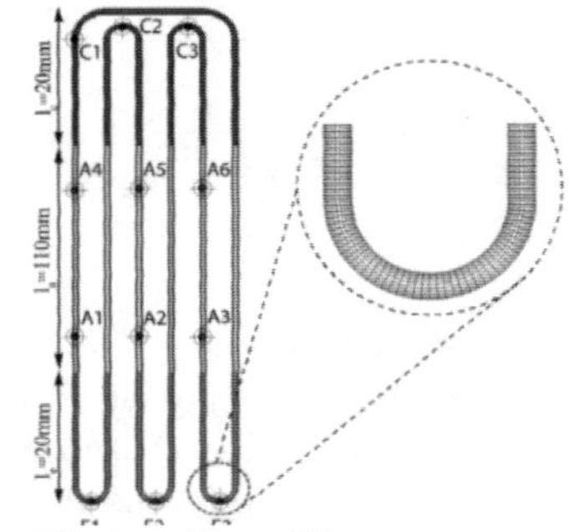


**Fig. 2** Structured grid on a 2D Oscillating Heat Pipe geometry with 30,000 elements. Location of temperature monitors is shown in evaporator, condenser and adiabatic sections.

A In House Finite Volume based CFD solver was used for modelling transport phenomenon inside the Closed Loop Oscillating Heat Pipe. The details and basic layout of CFD solver was reported

elsewhere[11]. The phase change of water into its vapor and vice-versa was modelled using the in-built multiphase Volume of Fluid (VOF) model, which uses the Euler-Euler approach to calculate phasic volume fraction and track the interface between two phases. An additional parameter 'α' whose value lies between 0 and 1, determines the volume fraction of one or more phases in each cell. The value of α in each cell for a phase, say liquid, corresponds to,
α = 0, absence of liquid phase in the cell, 0 < α < 1, interface between liquid and the other phase and α =1, cell is entirely filled with liquid

**2.2 Governing equations:** The solution of the continuity equation for the secondary phase is used to obtain the value of α which takes the following form,

$$\frac{\partial \alpha_l}{\partial t} + \nabla \cdot (\alpha_l \vec{u}) = \frac{S_m}{\rho_l} \quad (1)$$

For the two-phase vapor-liquid oscillating heat pipe, the primary phase was set as water vapor and secondary phase as water. The density and viscosity of the mixture phase is calculated as using volume fraction and properties of individual phases as:

$$\rho = \alpha_l \rho_l + (1 - \alpha_l)\rho_g \quad (2)$$

$$\mu = \alpha_l \mu_l + (1 - \alpha_l)\mu_g \quad (3)$$

A single momentum equation is solved for the entire control volume and the resultant velocity is shared among the phases.

$$\frac{\partial}{\partial t}(\rho\vec{u}) + \nabla \cdot (\rho\vec{u}\vec{u}) = -\nabla p + \nabla \cdot \left[\mu(\nabla\vec{u} + \nabla\vec{u}^T) - \frac{2}{3}\mu\nabla \cdot uI\right] + \rho\vec{g} + \vec{F}_{CSF} \quad (4)$$

Where, FCSF in the momentum equation is the Continuum Surface Force term, which is applied in conjunction with Wall Adhesion to model surface tension.

$$F_{CSF} = \sigma_{l-g}\left(\frac{\rho\kappa_l \nabla\alpha_l}{\rho_l + \rho_g}\right) \quad (5)$$

Similarly, the energy equation is also shared between the two phases and is given as:

$$\frac{\partial}{\partial t}(\rho e) + \nabla \cdot (\rho e\vec{u}) = \nabla \cdot (k \cdot \nabla T) + \nabla \cdot (p\vec{u}) + S_E \quad (6)$$

Where, k is the thermal conductivity of the mixture phase which is determined using the thermal conductivities and volume fraction of the phases in each cell.

$$k = \alpha_l k_l + (1 - \alpha_l)k_g \quad (7)$$

The energy term 'e' and temperature T in equation (6) are treated as mass averaged variables by the VOF model as per equations (8-10).

$$e = \frac{\alpha_l \rho_l e_l + \alpha_g \rho_g e_g}{\alpha_l \rho_l + \alpha_g \rho_g} \quad (8)$$

$$e_l = c_{p,l}(T - T_{sat}) \quad (9)$$

$$e_g = c_{p,g}(T - T_{sat}) \quad (10)$$

**2.3 Solution Methodology:** To model the boiling and condensation phenomena inside the heat pipe, a user-defined function reported in the literature[12] was used to incorporate energy and mass source terms in equations (1) and (6). Constant temperature boundary conditions were applied to the evaporator and condenser walls along with zero heat flux condition on the adiabatic walls. All the walls were treated as fully wetting with a contact angle of 50. The problem was initialized with a temperature of 368 K to speed up the boiling process. The evaporator was subjected to a range of temperatures (393-413K) keeping condenser temperature fixed in all cases at 298K. The various conditions can be altered as necessary to suit the requirements of the model [13].

A second order polynomial was used to model the surface tension [20] as per equation (11):

$$\sigma_{l,g} = 0.09805856 - 1.845 \times 10^{-5}T - 2.3 \times 10^{-7}T^2 \quad (11)$$

The density of liquid water was also modeled as polynomial [20] as per equation (12):

$$\rho_l = 859.0083 + 1.252209T - 0.0026429T^2 \quad (12)$$

The pressure-velocity coupling was achieved using the SIMPLE scheme and the discretization schemes used for pressure, momentum, energy and volume fraction were Body Force Weighted, Second Order Upwind, First Order Upwind and Compressive respectively. Under-relaxation factors of 0.2, 0.5 and 1 for pressure, momentum and energy respectively were used. Convergence criterion was set to 1e-06 for energy and 1e-03 for momentum and continuity. This model was validated with experimental results shown in the literature [14] for a thermosyphon and was extended to a multi-turn oscillating heat pipe upon agreement. Post validation, a parametric study was performed to study the effect of input heat flux on the hydrodynamics and thermal behavior of the heat pipe. Volume fraction of vapor and temperature distribution inside the OHP was studied and analyzed for evaporator wall temperatures (393-413K).

## 3. RESULTS AND DISCUSSION

Oscillating flow inside a closed loop three-turn OHP was simulated using volume of fluid multiphase model to study the behavior of OHPs at various operating conditions. The model was successfully able to imitate the oscillating motion of liquid and vapor plugs inside the OHP as shown in Figure 3.

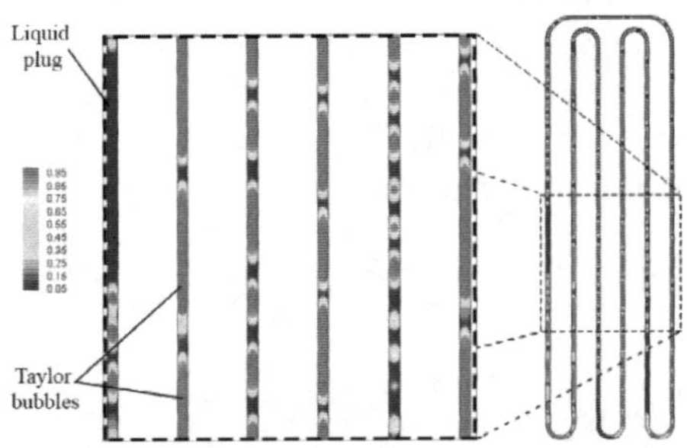


**Fig. 3** Small water vapor bubbles, Taylor bubbles and liquid plugs formed in the OHP at $T_{e,W}$=393 K.

The evolution of volume fraction of vapor during start-up of a multi-turn oscillating heat pipe is illustrated in figure 4. Figure 4 (a) shows the initial state of the OHP filled with 50% water at temperature 368K. Formation of very small vapor bubbles can be seen in figure 4 (b). After 0.3 seconds, the ascent of vapor slugs towards condenser and transition to a vapor-liquid plug flow is observed.

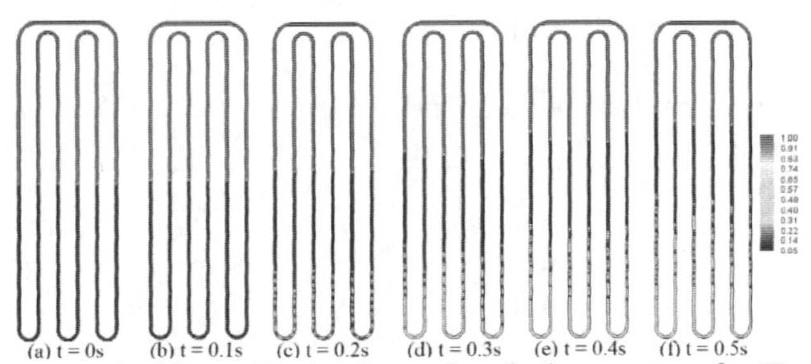


**Fig. 4** Vapor fraction contours inside oscillating heat pipe during start up for $T_{e,W}$ = 393K and $T_{c,W}$ = 298K. Formation of distinct vapor bubbles and transition into plug flow can be seen.

The flow is observed to follow a liquid-vapor slug regime. A circulatory and oscillatory motion of working fluid is observed as shown in Figure 5. Permanent evaporator dry-out is not observed for the range of evaporator wall temperatures tested (398-413K) in the present study. However, signs of temporary evaporator dry-out can be seen in Fig 5(e) as it is subjected to a high evaporator wall temperature of 413K.

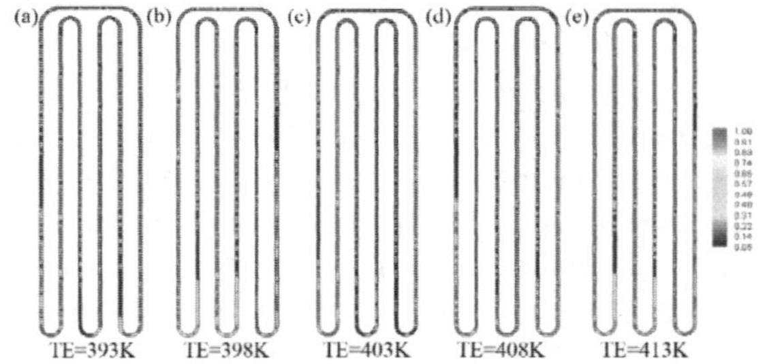


**Fig. 5** Volume fraction contours of water vapor at 15 seconds for various evaporator wall temperatures.

Temperature at various locations (as per fig. 2) was measured and area-weighted average values were obtained at evaporator, condenser and adiabatic regions for a time of 15 seconds. These values of temperature exhibit an oscillating trend as shown in Fig 6. These temperature fluctuations are a characteristic of chaotic flow inside an oscillating heat pipe. Although not much variation is seen in the temperature of fluid in the evaporator section, the adiabatic zone shows an interesting trend of temperature fluctuations. Fig 6(a) and 6(b) show an almost periodic trend after 10 seconds in the adiabatic region. A periodic behavior is also observed in the condenser section for these two cases. Fig 6(a) also shows a similar trend. However, there are no periodic oscillations in the adiabatic section temperature. Due to a low temperature of 393K at the evaporator wall, there could be a delay in the oscillating trend of temperature in the adiabatic region.

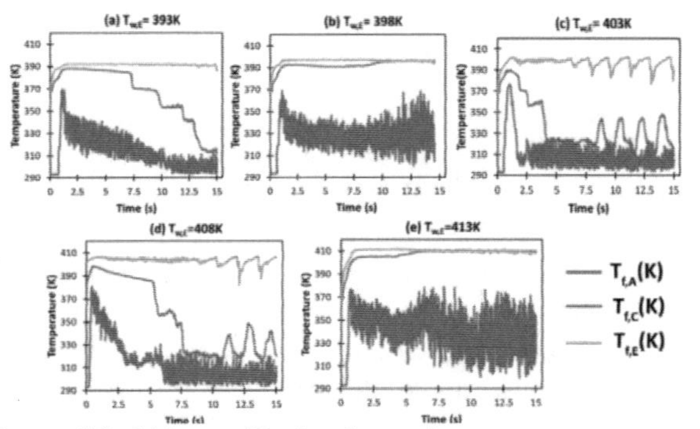


**Fig. 6** Temperature variations of fluid in oscillating heat pipe at evaporator, condenser and adiabatic sections for various evaporator wall temperatures.

**Table 1** Temperature drop across evaporator and condenser sections and average temperatures in evaporator, condenser and adiabatic sections of OHP for various wall heating temperatures.

| Evaporator wall temperature (K) | $T_{f, E}$ (K) | $T_{f, A}$ (K) | $T_{f, C}$ (K) | $\Delta T = T_{f, E} - T_{f, C}$ (K) |
|---|---|---|---|---|
| 393 | 391.37 | 366.09 | 314.45 | 76.92 |
| 398 | 396.06 | 392.31 | 327.97 | 68.09 |
| 403 | 396.98 | 337.72 | 309.16 | 87.81 |
| 408 | 403.50 | 353.30 | 314.53 | 88.97 |
| 413 | 409.63 | 407.29 | 339.61 | 70.01 |

The average values of the temperatures of working fluid and temperature drop across in the evaporator, condenser and adiabatic sections are tabulated in Table 1. It is observed that for evaporator wall temperatures 403K and 408K where periodic fluctuations in temperature are seen in Figure 6(c) and 6(d), the temperature drop across the evaporator and condenser is more than 80K. However, for other cases, a lower value of temperature drop is seen from Table 1. These results can be used to determine the optimum heat load range and can be useful for understanding thermo-hydrodynamics of pulsating heat pipe for achieving maximum thermal performance. Furthermore, a correlation can be obtained for predicting flow behavior in an oscillating heat pipe to prevent evaporator dry-out and maximize its performance.

## 4. CONCLUSION

Numerical simulations have been performed on a three-turn oscillating heat pipe to visualize the oscillating flow of water and its vapor. The heat pipe was initially filled with 50% of water. Constant temperature boundary conditions were applied to the evaporator and condenser section along with a zero-heat flux boundary condition at the walls of the adiabatic section. The temperature of the evaporator wall was varied (393-413K) while keeping the condenser wall temperature constant at 298K. Simulations were run until quasi-steady state was observed and the solution reached convergence. The temperature of the fluid inside the oscillating heat pipe at various locations was monitored. Volume fraction contours of vapor phase show formation of vapor plugs, and no abnormality was observed. Even at high evaporator wall temperatures, pulsatile flow regime was observed with no signs of a permanent dry-out. The graphs obtained for the temperature of the fluid inside the heat pipe revealed a chaotic behavior inside the heat pipe. These results are useful in determining the range of heat load for the proper functioning of the heat pipe. However, the effect of other parameters such as filling ratio, working fluids, and geometry of the heat pipe must be incorporated for a comprehensive optimization study which forms the basis for further research in this domain.

## References

[1] Zuo J, North MT, Wert KL. High Heat Flux Heat Pipe Mechanism. Inter Soc Conf Therm Phenom. 2000.

[2] Akachi H, Polasek F, Stulc P. Pulsating Heat Pipe. In: Proceedings of the 5th International Heat Pipe Symposium,. Melbourne, Australia; 1996. p. 208–17.

[3] Ma H. Oscillating heat pipes. Oscillating Heat Pipes. 2015. 1-427 p.

[4] Khandekar S, Manfred Groll. On the definition of pulsating heat pipes: An overview. In: Proc 5th Minsk International Seminar (Heat Pipes, Heat Pumps and Refrigerators). 2003.

[5] Tong BY, Wong TN, Ooi KT. Closed-loop pulsating heat pipe. 2001;21:1845–62.

[6] Pouryoussefi SM, Zhang Y. Analysis of chaotic flow in a 2D multi-turn closed-loop pulsating heat pipe. Appl Therm Eng. 2017;1–8.

[7] Lin Z, Wang S, Shirakashi R, Zhang LW. Simulation of a miniature oscillating heat pipe in bottom heating mode using CFD with unsteady modeling. Int J Heat Mass Transf. 2013;57:642–56.

[8] Sun Q, Qu J, Li X, Yuan J. Experimental investigation of thermo-hydrodynamic behavior in a closed loop oscillating heat pipe. Exp Therm Fluid Sci [Internet]. 2016.

[9] Zhang XM, J. L. Xu, Z. Q. Zhou. Experimental study of a pulsating heat pipe using FC-72, ethanol, and water as working fluids. Exp Heat Transf A J Therm Energy Gener , Transp , Storage , Convers. 2004;17(December 2012):47–67.

[10] Rao Z, Wang Q, Zhao J, Huang C. Experimental investigation on the thermal performance of a closed oscillating heat pipe in thermal management. Heat Mass Transf. 2017;

[11] Sikarwar BS, Khandekar S, Muralidhar K. Simulation of flow and heat transfer in a liquid drop sliding underneath a hydrophobic surface. Int J Heat Mass Transf. 2013;57(2):786–811.

[12] W.H. Lee. A pressure iteration scheme for two-phase flow modeling. Vol. 1, Multiphase Transport Fundamentals, Reactor Safety, Applications. 1980.

[13] Lenhard R, Jandačka J. Two-phase modeling of interphase heat transport from the condensation to evaporation part of the heat-pipe. AIP Conf Proc. 2013;1558:2138–41.

[14] Fadhl B, Wrobel LC, Jouhara H. Numerical Modelling of the Temperature Distribution in a Two- Phase Closed Thermosyphon. :1–27.

5th International Workshop on Heat/Mass Transfer Advances
for Energy Conservation and Pollution Control
August 13-16, 2019, Novosibirsk, Russia

# Particle Size Measurement of Reaction Product Aerosol of Sodium-Oxygen

**Y. Deguchi and T. Kamimoto**

Graduate School of Technology, Industrial and Social Sciences, Tokushima University, Tokushima 770-8501, Japan

**S. Kikuchi, A. Kurihara, T. Takata and H. Ohshima**

Advanced Fast Reactor Cycle System Research and Development Center, Japan Atomic Energy Agency, Ibaraki, 311-1393, Japan

corresponding author : ydeguchi@tokushima-u.ac.jp

A multi-level scenario simulation system as a safety infrastructure technology is required for design optimization, safety margin adjustment, and innovative technology developments of sodium-cooled fast reactor (SFR). In order to properly implement the Verification and Validation (V & V) of the simulation system, it is indispensable to ensure experimental database of sodium chemistry as specific SFR safety issue. In this study, measurement results of aerosols generated by sodium-oxygen reaction for sodium fire event were reported with the aim of clarifying the radiation heat transport phenomena in the reaction field. The sodium-oxygen counter-flow diffusion flame was formed one-dimensionally above the sodium pool by the reaction between sodium vapour and oxygen. Argon (Ar) including 2% oxygen were introduced to a liquid sodium pool (temperature: 820K) under the reduced pressure condition (0.05MPa). Ar guard flows were employed to stabilize the reaction. The reaction continued more than 600 seconds without any changes in terms of flame shape and position. Aerosol size was measured as a function of Z (the distance from the sodium pool surface) and r (the distance from the center of the sodium pool). Laser Induced Incandescence (LII) and the Mie scattering method using the different wavelength laser beams (405nm, 450nm, 520nm, 532nm, 638nm and 650nm) were employed to ensure the measurement accuracy. Aerosol sizes from several hundred nm to 1 μm were measured in this reaction field and the aerosol size increased toward the sodium pool. This stemmed largely from aerosol growth and polymerization because the flow rate decreased near the sodium pool. It was also confirmed that the size of aerosol measured by LII was in good agreement with the measurement using the Mie scattering method under the same conditions. The refractive index of the aerosol was also evaluated to be 1.42-0.5i.

## 1. Introduction

Liquid sodium is used as coolant in a sodium-cooled fast reactor (SFR) to transfer heat from its reactor core to a steam generator (SG) because of its excellent heat transport capability. On the other hand, sodium has strong chemical reactivity with oxygen and water vapor. The sodium-water and sodium-oxygen reactions become one of design basis accidents [1]. A multi-level scenario simulation system as a safety infrastructure technology is required for design optimization, safety margin adjustment, and innovative technology developments of SFR. In order to properly implement the

Verification and Validation (V & V) of the simulation system, it is indispensable to ensure experimental database of sodium chemistry as specific SFR safety issue [2]. One of the technical difficulties in developing the evaluation technique of this process has been to clarify the phenomena on sodium-water and sodium-oxygen chemical reactions. In particular, it is important to consider the surface and gas-phase reactions because water vapor and oxygen react with gas and liquid sodium at the boundary surface of reaction fields. The surface reaction mechanism has been investigated theoretically and experimentally by Kikuchi et. al. to reveal the sodium-sodium hydroxide reaction as secondary surface reaction of the sodium-water reaction phenomena [3, 4]. On the other hand, the gas-phase and its relevant reactions were studied mainly as a single-step reaction [5-12]. Deguchi et. al. have developed an elementary reaction set of sodium-water and sodium-oxygen gas-phase reactions which consist of 17 $H_2$-$O_2$ and 8 Na-$H_2O$ reactions including 14 species [13]. The theoretical results using it were compared to the measurement by laser diagnostics [14, 15] using the sodium-water counter-flow reaction vessel [13, 14, 15]. In these studies, it was revealed that the major reaction of sodium-water reaction in gas phase was $Na+H_2O \rightarrow NaOH+H$ and OH was produced by $H_2O+H \rightarrow H_2+OH$. The quantitative analysis, however, was not enough to evaluate the products such as aerosols produced by sodium-oxygen reactions in gas phase.

The purpose of this study aims to clarify the sodium-oxygen gas phase reaction path and reaction products. Measurement of aerosols generated by sodium-oxygen reaction for sodium fire event are reported with the aim of clarifying the radiation heat transport phenomena in the reaction field. Aerosol size was measured using sodium-oxygen counter-flow reactions. Laser Induced Incandescence (LII) [16] and the Mie scattering method using the different wavelength laser beams (405nm, 450nm, 520nm, 532nm, 638nm and 650nm) were employed to ensure the measurement accuracy.

## 2. Experimental Setup

Fig. 1(a) - (b) show an overview of the experimental apparatus. The reaction chamber had four windows (diameter: 50mm) for the measurement of aerosols generated by sodium-oxygen reaction using laser diagnostics [16]. Sodium-oxygen counter-flow reactions were applied in this study. 2 and 4 % oxygen concentrations in Ar were introduced from the pipe (diameter:15mm) to a liquid sodium pool (diameter:30mm, temperature: 820 K) under the reduced pressure condition (0.05 MPa). Ar and Ar guard flows (diameters of 65mm and 130mm, respectively) were employed to stabilize the sodium-oxygen reactions. Experimental conditions are summarized in Table 1. The sodium-oxygen counter-flow diffusion flame was formed one-dimensionally above the sodium pool by the reaction between evaporated sodium and oxygen. The reaction continued more than 600 seconds without any changes in terms of flame shape and position. Aerosols generated by sodium-oxygen reaction were measured as a function of Z (the vertical distance from the sodium pool surface) and r (the horizontal distance from the center of the sodium pool).

Fig. 2 shows the experimental apparatus of aerosol measurements. The sodium-oxygen counter-flow diffusion flame was visualized by a CCD camera using the Mie scattering signals from a 532 nm diode laser. Aerosol characteristics were measured by using LII and the Mie scattering method. LII is a method detecting aerosol number density and size using emission from the heated aerosol by the laser radiation. The experimental apparatus comprises an Nd:YAG laser (Spectra physics, Model PS100, laser power : 100mJ/p, pulse width : 5ns), a spectrometer (JASCO, CT-10S) and an ICCD camera (Andor, iStar DH334T-18U-03). Laser wavelengths of 532nm (second harmonic) was used to generate LII signals. 415 nm and 653 nm LII signals were detected two-dimensionally to evaluate the aerosol size using the ICCD camera. Aerosol size and its refractive index were also evaluated using the Mie scattering signals from diode lasers with the wavelengths of 405nm, 450nm, 520nm, 532nm, 638nm, 650nm as listed in Table 2. Each laser beam was combined by beam combiners to have the same laser path as shown in Fig. 3 and was adjusted to be the same laser power at 100 mW. The Mie scattering signals were also detected by the ICCD camera to evaluate the aerosol size and its refractive index. The experimental Mie scattering signal intensity was corrected by the efficiency of experimental apparatus such as the ICCD camera. The experimental accuracy was confirmed by measuring the Mie scattering signals from the latex particles (Magsphere, Inc, particle diameter : 1μm, refractive index : 1.59) in water and compared to the Mie scattering theory.

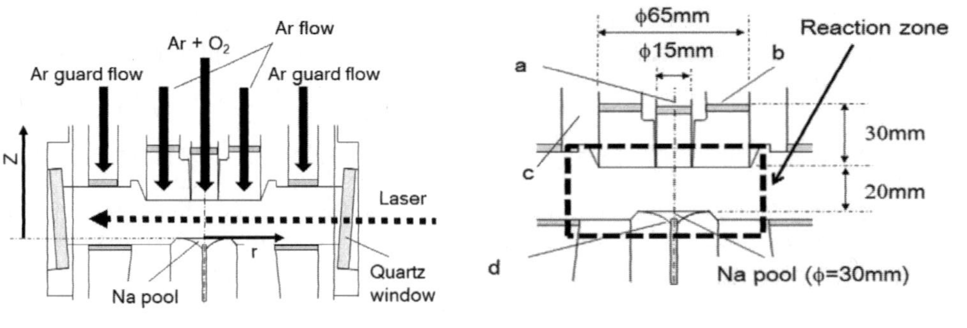


(a) Schematic of reaction chamber (b) Detailed diagram of reaction zone

Fig. 1 Experimental apparatus of sodium-oxygen counter-flow diffusion flame

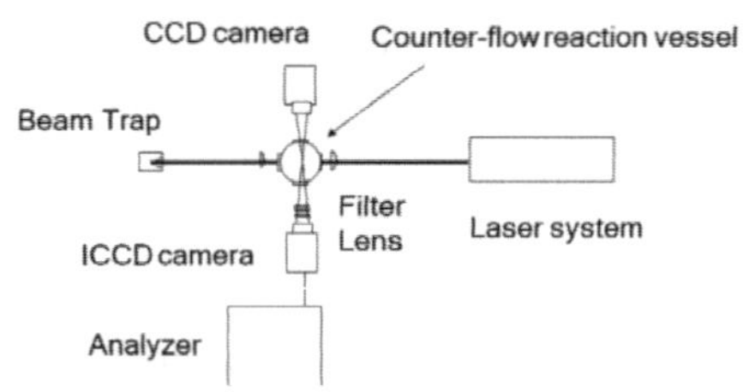


Fig. 2 Experimental apparatus of aerosol measurements

Table 1 Experimental conditions

| Item | Temperature | Position (Fig.1(b)) | Flow rate |
|---|---|---|---|
| Ar + $O_2$ [1)] | 873 K | a | 4.33 x $10^{-6}$ $m^3/s$ |
| Ar flow | 873 K | b | 2.16 x $10^{-5}$ $m^3/s$ |
| Ar guard flow | 340 K | c | 1.33 x $10^{-4}$ $m^3/s$ |
| Na pool | 820 K | d | - |

[1)] Concentration of $O_2$ : 2 and 4 %

Table 2 List of diode lasers

| NO. | Wavelength | Maximum laser power |
|---|---|---|
| 1 | 405nm | 250mW |
| 2 | 450nm | 500mW |
| 3 | 520nm | 700mW |
| 4 | 532nm | 150mW |
| 5 | 638nm | 250mW |
| 6 | 650nm | 250mW |

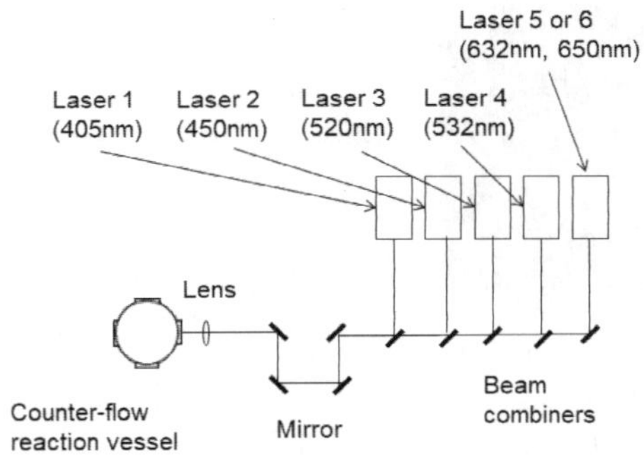


(a) Laser configuration

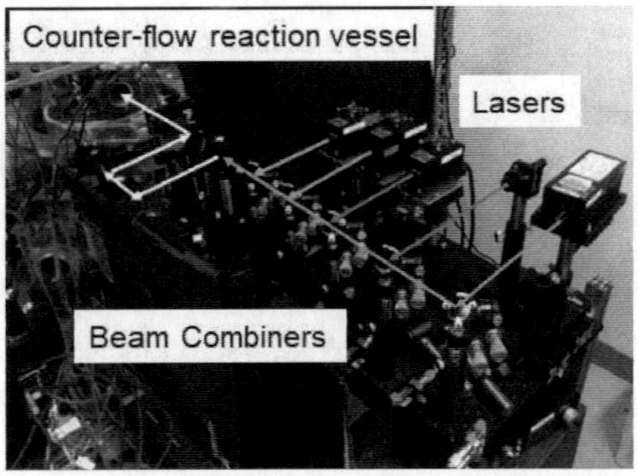


(b) Picture of laser configuration

Fig. 3 Optical setup of Mie scattering experiment

## 3. Results and Discussion

The visualization results of sodium-oxygen reaction fields are shown in Fig. 4. In the case of the reaction between sodium and oxygen, there is no luminescence from the reaction [15] and the Mie scattering signals from the reaction products were observed to identify the reaction zone. In case of 2% oxygen concentration, the strong Mie scattering signals were observed at the reaction front between sodium and oxygen at Z=7-8mm.

Fig. 5 shows the time dependent intensity of LII signals. The LII signal intensity has a relation with the cube of aerosol diameter and it decreases with time because of the decrease of the aerosol temperature [15]. The decrease rate of the LII signal depends on the aerosol size can be measured by measuring the decay of the LII signal intensity. 415 nm and 653 nm LII signals were detected two-dimensionally to evaluate the aerosol size with the correction of aerosol temperature effects. The decrease rates of LII signal intensity at different r positions were almost the same. Fig. 6 shows the particle size distributions measured by LII. Aerosols from several hundred nm to 1 μm were measured in this sodium-oxygen reaction field and the aerosol size increased toward the sodium pool. This stemmed largely from aerosol growth and polymerization. These phenomena were eminent toward the sodium pool because the flow rate decreased near the sodium pool [15]. The reaction front was moved toward the sodium pool with the increase of the oxygen concentration and the aerosol growth rate became bigger with the increase of oxygen concentration as shown in Fig. 6(a) and (b). The number density of aerosol increased with the increase of oxygen concentration and it accelerated the aerosol growth and polymerization. The dependence of aerosol size on r was not eminent and it was consistent with Fig. 5.

Fig. 7 shows the comparison of measured and calculated Mie scattering intensity at different wavelengths and aerosol size. The theoretical Mie scattering intensity between 400-800nm was calculated by the Mie scattering theory with the parameters of the aerosol size and its refractive index. The experimental Mie scattering signal intensity was compared to the theoretical Mie scattering intensity using the least-square technique to determine the aerosol size and refractive index. The aerosol size and refractive index (Z=6mm, r=10mm) generated by sodium-oxygen reaction were determined to be 1.02-1.04 μm and 1.42-0.5i, respectively. The measurement results showed good agreement with the measurement using LII under the same conditions.

The aerosol characteristics such as size and refractive index play an important role in the radiation heat transport phenomena in the reaction fields and the measured results were employed for the evaluation of the SFR simulation system.

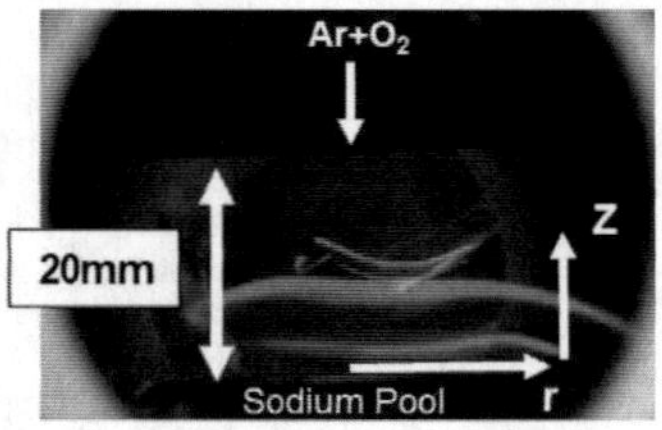


Fig. 4 Visualization of sodium-oxygen reaction ($O_2$ : 2%)

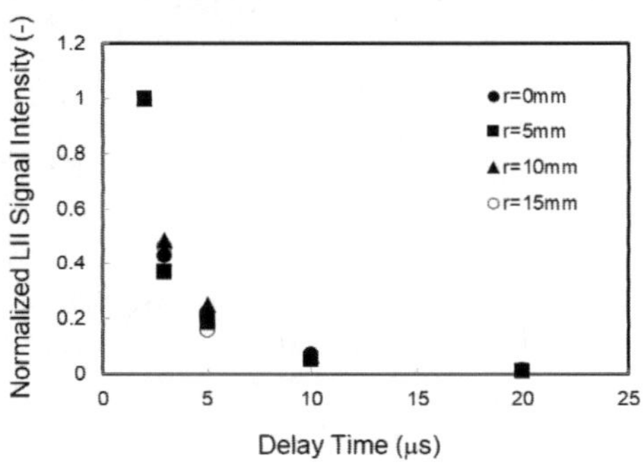


Fig. 5 Time dependence of LII signal intensities ($O_2$ : 2%, Z=7mm, 405nm)

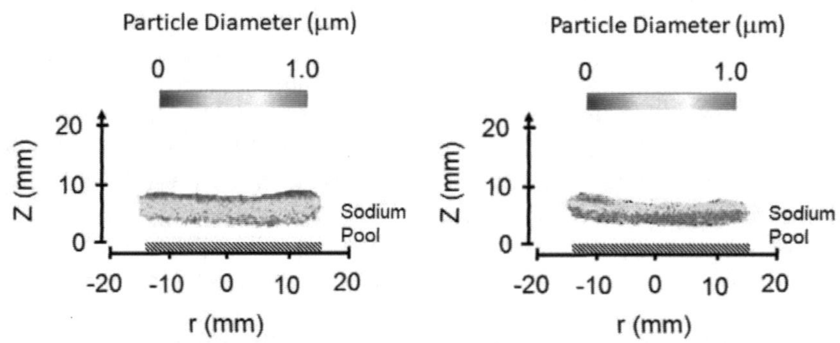


(a) $O_2$ : 2% (b) $O_2$ : 4%

Fig. 6 Particle size distributions Measured by LII.

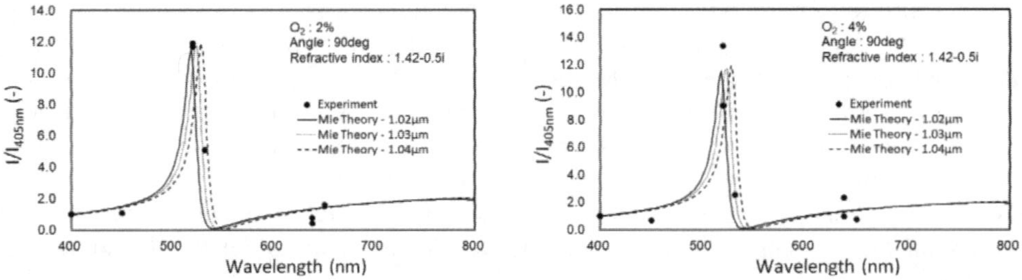


(a) $O_2$ : 2% (b) $O_2$ : 4%

Fig. 7 Comparison of measured and calculated Mie scattering intensity at different wavelengths and size (Z=6mm, r=10mm).

## 4. Conclusions

LII and Mie scattering methods measurement techniques were applied to the sodium-oxygen reaction fields to clarify the aerosol characteristics generated by sodium-oxygen reaction. The aerosol characteristics play an important role in the radiation heat transport phenomena and the following results were obtained.

1) LII was successfully applied to measure the aerosol size. Aerosols from several hundred nm to 1 μm were measured in the sodium-oxygen reaction field. The aerosol diameter increased toward the sodium pool because of the aerosol growth and polymerization.

2) The experimental Mie scattering signal intensity was compared to the theoretical Mie scattering intensity to determine the aerosol size and refractive index. The aerosol size and refractive index (Z=6mm, r=10mm) generated by sodium-oxygen reaction were determined to be 1.02-1.04 μm and 1.42-0.5i, respectively. The measurement results showed good agreement with the measurement using LII under the same conditions.

**Acknowledgement**
This work was supported by a study of "Research and Development of Multi-Level and Multi-Scenario Plant Simulation Systems for Innovative Sodium-cooled Fast Reactors" entrusted to "Japan Atomic Energy Agency (JAEA)" by the Ministry of Education, Culture, Sports, Science and Technology of Japan (MEXT). This study was performed under the contract between JAEA and Tokushima University.

**References**

[1] Ohno S, Kawata K and Miyahara S 1994 PNC Technical Review No.92, PNC TN1340 94-004
[2] Takata T and Yamaguchi A 2003 *Journal of Nuclear Science and Technology* **40**(10) 708.
[3] Kikuchi S, Kurihara A, Ohshima H and Hashimoto K 2012 *Journal of Power and Energy Systems* **6**(2) 76.
[4] Kikuchi S, Seino H, Kurihara A and Ohshima H 2013 *Journal of Power and Energy Systems* **7**(2) 79.
[5] Plane J M C and Husain D 1986 J. Chem. Soc., Faraday Trans. **82**(11) 2047.
[6] Mayer S W and Schieler L 1966 *J. Chem. Phys.* **45**(1) 385.
[7] Vinckier C, Dumoulin A and Jaegere S De 1991 *J. Chem. Soc. Faraday Trans.* **87**(8) 1075,.
[8] Jensen D E and Jones G A 1982 *J. Chem. Soc. Faraday Trans.* **78**(9) 2843.
[9] Patrick R and Golden D M 1984 *Int. J. Chem. Kinet.* **16**(12) 1567.
[10] Cox R M and Plane J M C 1999, *Phys. Chem.* **1**(20) 4713.
[11] Mayer S W , Schieler L and Johnston H S 1967 *Symp. Int. Combust. Proc.* **11** 837.
[12] DeMore W B, Sander S P, Golden D M, Hampson R F, Kurylo M J, Howard C J, Ravishankara A R, Kolb C E and Molina M J 1997 *JPL Publication* **97-4**(12) 1.
[13] Deguchi Y, Takata T, Yamaguchi A, Kikuchi S and Ohshima H 2015 *Mechanical Engineering Journal* **2** (1) 14.
[14] Deguchi Y, Tamura K, Muranaka R, Kusano K, Kikuchi S and Kurihara A 2013 *The Review of Laser Engineering* **41**(11) 927.
[15] Deguchi Y, Muranaka R, Kamimoto T, Takagi T, Kikuchi S and Kurihara 2017 *Applied Thermal Engineering* **114** (5) 1702.
[16] Deguchi Y 2011 *Industrial applications of Laser Diagnostics*, CRS Press.

5th International Workshop on Heat/Mass Transfer Advances
for Energy Conservation and Pollution Control
August 13-16, 2019, Novosibirsk, Russia

# Moist Air Condensation on Inclined Hydrophobic Metallic Surfaces: Simulation & Experiments

**Vishakha Baghel[1] and Basant Singh Sikarwar[1*]**

[1]Department of Mechanical Engineering, Amity University Uttar Pradesh, NOIDA - 201313, India

Corresponding author's Email ID: bssikarwar@amity.edu

**Abstract.** Atmospheric Water Generators (AWGs) are popularly used for harvesting portable water from atmospheric air of hot, humid and arid regions of the world. Hydrophobic metallic surfaces are preferred in AWGs systems because condensation of moist air on/underneath these surfaces have high efficacy. There are several issues to fabricate hydrophobicity on metallic surfaces. Although with the advent of Nanotechnology and thin film coating technologies, the fabrication of hydrophobicity on metallic surfaces has become easier and realizable in the recent era. In this manuscript, a comprehensive mathematical model is developed for simulating the moist air condensation in the form of droplets on various substrates and in different environmental conditions. The experiment is carried out for validation of the present model. Post validation, the effect of surface hydrophobicity, relative humidity and degree of sub-cooling on the condensation rate are addressed. The simulation results show that vertically orientated metallic surfaces having a high contact angle along with low contact angle hysteresis are efficient for condensing unit. Larger condensation rates are observed at higher relative humidity and a high degree of sub-cooling. This research is helpful for designing efficient and effective AWGs for the hot and humid region.

## 1. Introduction

Earth's atmospheric air contains more than 12,800 trillion liters water in the form of moisture (1). This amount of water is enough to fulfill the requirements of entire flora and fauna living on the Earth if it is extracted effectively and efficiently. It is a green technology, feasible, portable, compatible with existing systems and it does not produce any concentrated hypersaline wastewater discharge. However, high energy requirements (2260 Kilojoules/Kg of water) and low transport coefficients of the condensation process hinders its deployment in large scale applications. With the advent of advanced instrumentation and fabrication processes, extraction of water via condensation from moist air is a possible and reliable technology (2). With the acquaintance of the fact that the transport coefficients of dropwise condensation are higher as compared to filmwise condensation (3). Emphases are been laid on the fabrication of hydrophobicity on metallic surfaces because they promote condensation in drop mode and have high thermal conductivity (4). Several studies have been carried out to demonstrate the dropwise condensation of pure vapor however, the literature is limited on moist air condensation in the form of drop for procuring fresh water from the atmosphere.

In contrast to the pure vapor, condensation of moist air is a complex process because of NCGs. The vapor-liquid interface is impermeable to the NCGs and these gases get accumulate during condensation near the interfaces forming NCG diffusion layer (5). Hence, NCGs plays a crucial role in moist air condensation for water harvesting applications. Several researcher and scientists (6, 7) have experimentally and theoretically demonstrated condensation of steam. They reported that the heat transfer coefficient of dropwise condensation process reduces due to the presence of NCGs in the air-steam mixture. Recently, some researchers (8-10) have reported the effect of NCGs in moist air at controlled environmental conditions. Baghel et al. (8) experimentally investigated the effect of static contact angle on the moist air condensation rate. Fouda et al. (9) experimentally and theoretically investigated the condensation of water vapor present in moist air around the horizontal pipe. Their study shows that condensation rate increases with increase in moist air mass flow rate, relative humidity and degree of sub-cooling. Zheng et al. (10) developed a mathematical model for dropwise condensation of moist air for hemispherical droplet using the kinetic theory of gases and law of continuum fluid dynamics formulated by the Knudsen layer and continuum region.

To the best of author's knowledge, a holistic mathematical model for moist air condensation in drop mode is not reported in the literature. In this context, a comprehensive mathematical model is developed to simulate the moist air condensation in the form of droplets on the vertical cold surface under atmospheric conditions. The skeleton of the mathematical model is similar to the one reported by other elsewhere (11) for pure vapor condensation in drop mode. Current moist air condensation model is an extension of the author's previous pure vapor condensation model. Here, the growth rate at individual nucleation sites, the critical size of the drop and terminal velocity are altered based on the kinetic theory of gases and law of continuum fluid dynamics. The experiment is performed to validate the simulation results of moist air condensation. Post validation, the effect of surface hydrophobicity, relative humidity and degree of sub-cooling are addressed. The results of the present model are useful in enhancing the efficiency of AWGs.

## 2. Mathematical Modelling and Experiment of Moist Air Condensation

The moist atmospheric air is a binary mixture of water vapor and the number of NCGs. Therefore, the condensation process in the presence of moist air is distinct from that of pure vapor. This work is an extension of the author's previous work (12) in which the presence of NCG is also taken into consideration in the simulation of dropwise condensation. An additional resistance acts due to the formation of the Knudsen layer near the three-phase contact line due to the accumulation of NCGs at the interface. Author's previous drop growth model is modified as given in Equation 1, considering the various assumptions as reported previously (12).

$$\frac{dr_d}{dt}=\left(\frac{4(T_{sat}-T_w)}{\rho_d(\mathrm{H}_{lv})}\right)\times\left[\frac{\left(1-\frac{r_{min}}{r_d}\right)}{\left(\frac{r_d\beta\lambda_m\left(1-\cos\theta_{avg}\right)}{k_m\left(r_d+\beta\lambda_m\right)}+\frac{2}{h_{int}}+\frac{r_d\theta_{avg}(1-\cos\theta_{avg})}{k_d\sin\theta_{avg}}+\frac{4\delta_{coat}}{k_{coat}(1+\cos\theta_{avg})}\right)}\right]\times\left[\frac{\left(1-\cos\theta_{avg}\right)}{\left(2-3\cos\theta_{avg}+\cos^3\theta_{avg}\right)}\right]$$

(1)

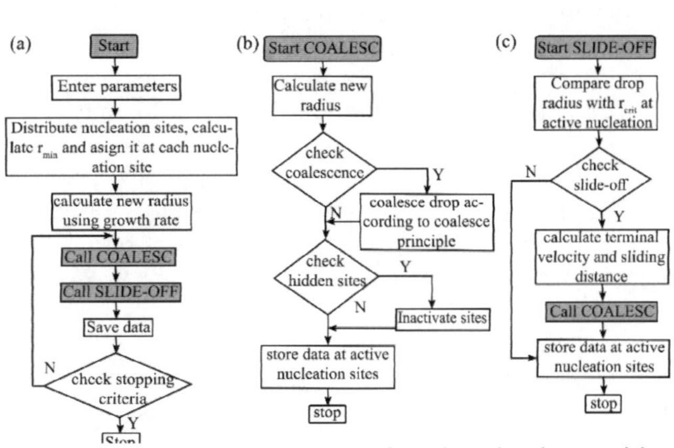


**Figure 1.** Flow diagram of dropwise condensation of moist air along with drop COALESC and SLIDE-OFF subroutines.

The water vapor present in moist air is sensibly cooled till the dewpoint before it starts to condense in unsaturated moist air. The value of the $H_{lv}$ in equation 1 for unsaturated air is as follow;

$$H_{lv} = h_{lv} + C_p (T - T_{sat}) \tag{2}$$

The interfacial heat transfer coefficient $h_{int}$ in equation 1 is,

$$h_{int} = \left[\frac{2\vartheta}{2-\vartheta}\right]\left(\frac{h_{lv}^2(\rho_d - \rho_m)}{T_{sat}}\right)\left(\frac{M_m}{2\pi \overline{R} T_{sat}}\right)^{1/2}\left[1 - \frac{P_v}{2h_{lv}(\rho_d - \rho_m)}\right] \tag{3}$$

where $\vartheta$ is the accommodation coefficient, in this study $\vartheta$ = 0.001 is considered for moist air condensation. $T_{sat}$ is the dew point temperature and $T$ is the temperature of moist air. $M_m$, $K_m$ and $\lambda_m$ are molecular weight, thermal conductivity and mean free path of moist air respectively. The properties of moist air are calculated as reported in the literature (13). $\rho_d$, $K_d$ and $r_d$ are density, thermal conductivity, and radius of the droplet.

The presence of NCGs in moist air also alters the droplet size ($r_{crit}$) and terminal velocity ($U$) during drop sliding on a vertically hydrophobic surface. The modified equations are as:

$$r_{crit} = \left[\left(\frac{k \sin\theta_{avg}}{2 - 3\cos\theta_{avg} + \cos^3\theta_{avg}}\right)(\cos\theta_r - \cos\theta_a)\left(\frac{\sigma}{g(\rho_d - \rho_m)}\right)\right]^{1/2}$$

(4)

$$U = \left[\frac{2(F_{g\parallel} - F_r)}{C_f \rho_d \pi r_d^2 (1 - \cos^2\theta_{avg})}\right]^{1/2}$$

(5)

Where k is a coefficient whose value is considered as 0.5 for moist air condensation, $\theta_{avg}$, $\theta_r$ and $\theta_a$ are static, receding and advancing contact angles for droplet respectively. The $C_f$ is evaluated by the law of continuum fluid dynamics where the droplet is considered stationary with respect to moist air surrounding the droplet. The moist air flow at a velocity equal to the terminal velocity of the droplet.

In the present model, primarily various variables are initialized. These variables are thermos-physical properties of moist air, physio-chemical properties of condensing surfaces, nucleation site density, relative humidity, time step and a total time of the simulation. In this study nucleation sites of $10^6$ $cm^{-2}$ are randomly distributed on the surface by a random number generator. All the nucleation sites are instantaneously occupied by the droplet of minimum radius '$r_{min}$' on the virgin substrate. Subsequently, droplet whose radius is more than or equal to $r_{min}$ further grow by direct condensation as per equation 1 (solved by $4^{th}$ order Runga-Kutta method) and coalesce with the neighboring droplets. However, droplets with a radius less than minimum radius evaporate and vanish. Due to direct condensation and coalescence phenomena the size of droplet increases to a limit that the gravity forces eventually surpass the limiting surface tension force at the triple phase contact line of the condensed droplet. This creates an imbalance in forces and as a result, droplet deforms to achieve necessary static balance. In the course of time, the droplet reaches a critical radius and incipient sliding from the vertical hydrophobic condensing surface. It then re-activates the exposed sites created due to drop coalescence and slide-off and provide a minimum radius drop on newly exposed sites. The aforementioned procedure repeats until a dynamic steady-state is reached. The flow diagram of the numerical algorithm is shown in Figure 1. The C++ program is written to carry out simulation as per the proposed algorithm.

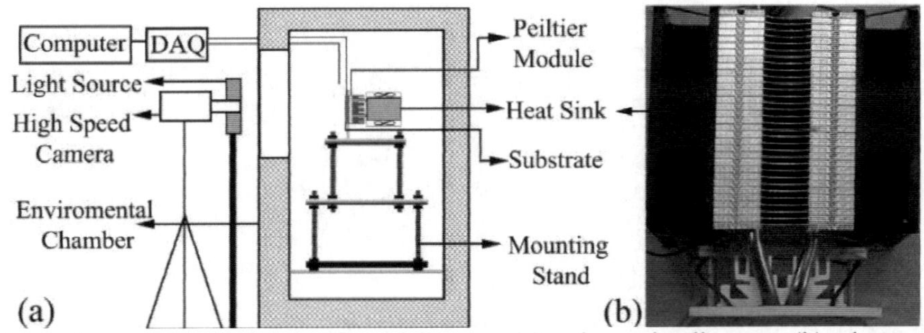


**Figure 2.** Moist air condensation experimental setup (a) schematic diagram (b) photograph of the condensing unit.

The moist air condensation experiments are carried out on chemically etched irradiated metallic surfaces for generating data for code validation. Figure 2 shows the schematic diagram and photograph of the moist air condensation experimental setup. The surface fabrication and experimental procedures are as reported by the author elsewhere (8).

## 3. Results and discussion

The moist air condensation experiments are carried out on chemically treated irradiated surfaces for validation of the code. Figure 3 and 4 shows the qualitative and quantitative comparison of experiment and simulation results respectively. Both the studies were performed at atmospheric pressure and degree of sub-cooling of 10°C. The controlled temperature and humidity of 30°C and 95% are maintained throughout the studies. In this work, a vertically oriented, 40×40 mm surface is considered for moist air condensation. The static contact angle and contact angle hysteresis on this surface are measured as 154° and 7.5° respectively. Validation results show that the simulation results are in good agreement with the experimental data. Hence the model is appropriate and credible. Figure 3 shows the spatiotemporal droplet distribution on the superhydrophobic condensing surface. Droplet imaging at t = 10 min shows droplet growth by direct condensation and coalescence. However, re-nucleation after droplet slide-off is visualized at time t = 60 min. Figure 4 shows the comparison of experimental and simulation condensation rates. The value of simulation condensation rate is higher than the experimental values due to various assumptions made in the simulation study.

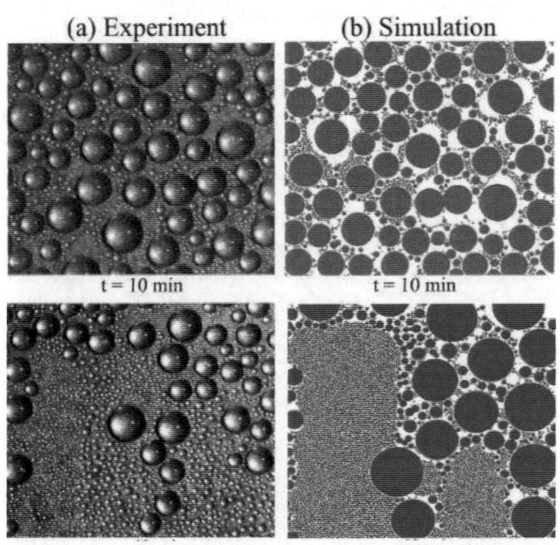


**Figure 3.** Spatio-temporal drop distribution of moist air condensation experiment and corresponding simulation results.

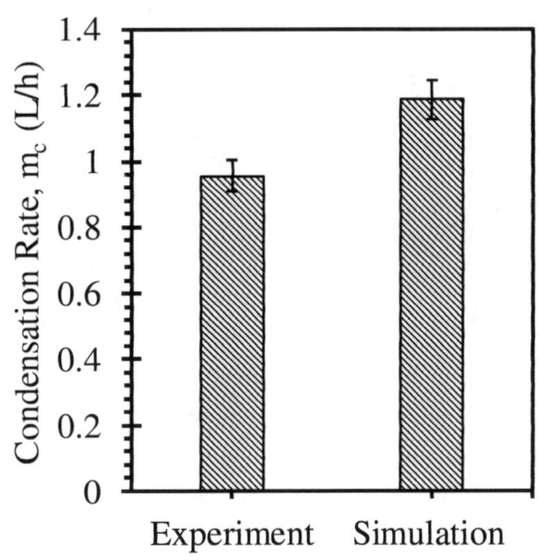


**Figure 4.** Quantitative comparison of experimental and simulation results.

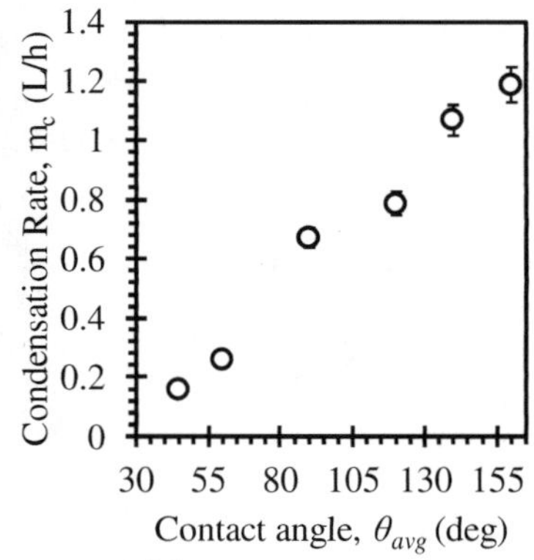


**Figure 5.** Variation of condensation rate with respect to static contact angle of the condensing surface. Simulation carried out considering $T_{sat}$ = 30°C, $\Delta T$ = 10°C and RH = 95%.

Post validation, simulations are carried out for a range of parameters which are not covered in the experimental study. Here the effect of static contact angle, Relative humidity and degree of sub-cooling on condensation rate are investigated. The effect of static contact angle on the condensation rate is shown in Figure 5. The simulation is performed for the controlled operating condition, various values of contact angle and corresponding contact angle hysteresis. The results show that as the static contact angle of surface increases, contact angle hysteresis decreases, drop pinning on surfaces reduces and results into enhanced droplet shedding. Therefore, the condensation rate increases as the static contact angle of the condensing surfaces increases and becomes highest for superhydrophobic surfaces as compared to hydrophilic and hydrophobic surfaces. Figure 6 and 7 shows the effect of relative humidity and degree of sub-cooling on condensation rate respectively. Results show that the condensation rate becomes twice as the relative humidity of moist air increase from 40% to 95%. It is because as the amount of water vapor present in moist air increases, the condensation rate augments with increasing relative humidity. To investigate the effect of degree of sub-cooling the condensation rate is simulated by varying the condensing surface temperature keeping moist air temperature and humidity constant. Results show that the condensation rate linearly increases with increasing degree of sub-cooling.

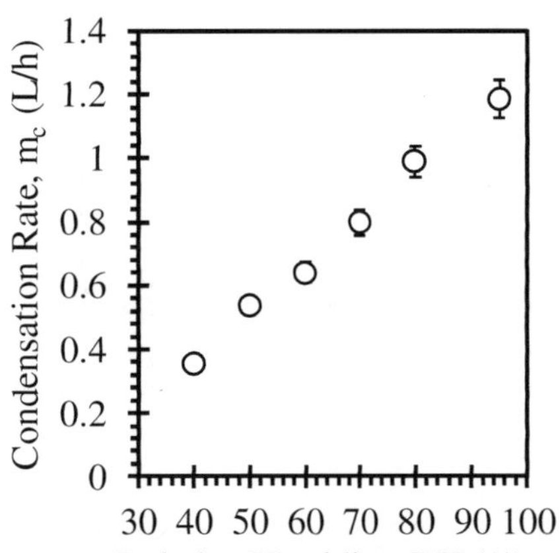


**Figure 6.** Variation of condensation rate with respect to relative humidity of moist air. Simulation performed considering $T_{sat}$ = 30°C, $\Delta T$ = 10°C, $\theta$ = 154°and CAH = 7.5°.

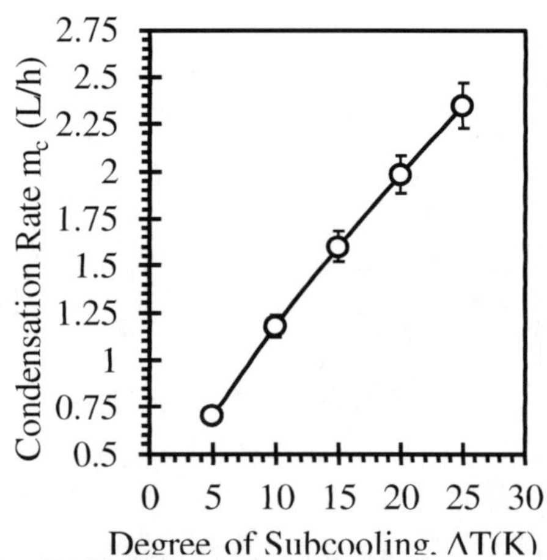


**Figure 7.** Variation of condensation rate with respect to degree of subcooling. Simulation performed considering $T_{sat}$ = 30°C, RH = 95%, $\theta$ = 154°and CAH = 7.5°.

## 4. Conclusion

The moist air condensation is a process for extracting the fresh water from the atmosphere of hot and humid regions. The dropwise condensation of water vapor present in moist air is distinct from the condensation of pure water vapor due to the presence of NGCs in moist air. In this work, a mathematical model is developed to simulate the condensation of water vapor present in moist air in the form of droplets. The simulation results were validated against the data obtained from experimental study and good agreement was found between them. Post-validation, parametric studies were carried out to know the effect of static contact angle, relative humidity and degree of sub-cooling on condensation rate. The study concludes that the surfaces of higher contact angle and low contact angle hysteresis are more efficient for the condensation process. Superhydrophobic surfaces with contact angle hysteresis less than $10^{o}$ promote efficient dropwise condensation and exhibit enhanced droplet shedding. Large condensate rate is estimated at higher relative humidity and a higher degree of sub-cooling.

**Acknowledgment**

The authors acknowledge the financial support from the Science and Engineering Research Board (SERB), Department of Science and Technology (DST), Govt. of India (Project No. ECR/2016/000020).

**References**

[1] Bagheri F. Performance investigation of atmospheric water harvesting systems. Water Resources and Industry. 2018;20:23-8.
[2] Clarke NP, C. Camarillo, Atmospheric moisture collection device1992.
[3] Schmidt VE, W. Schurig, and W. Sellschopps. Versuche uber die Kondensation von Wasserdampf in Film- und Tropfenform. Forsch Ingenieurwes. 1930;1:53-63.
[4] Lee A, Moon MW, Lim H, Kim WD, Kim HY. Water harvest via dewing. Langmuir: the ACS journal of surfaces and colloids. 2012;28(27):10183-91. Epub 2012/06/27.
[5] Luo X, Fan Y, Qin F, Gui H, Liu J. A new model for the processes of droplet condensation and evaporation on solid surface. International Journal of Heat and Mass Transfer. 2016;100:208-14.
[6] Ma X-H, Zhou X-D, Lan Z, Li Y-M, Zhang Y. Condensation heat transfer enhancement in the presence of non-condensable gas using the interfacial effect of dropwise condensation. International Journal of Heat and Mass Transfer. 2008;51(7-8):1728-37.
[7] Lan Z, R. Wen, A. Wang, and X. Ma. A droplet model in steam condensation with noncondensable gas. International Journal of Thermal Sciences. 2013;68:1-7.
[8] Baghel V, Sharma DK, Sikarwar BS, Kumar R, Avasthi DK. Tailoring the hydrophobicity of copper surface with ion beam irradiation. Radiation Effects and Defects in Solids. 2019:1-13.
[9] Fouda A, M. G. Wasel, A. M. Hamed, E.-S. B. Zeidan, and H. F. Elattar. Investigation of the condensation process of moist air around horizontal pipe. International Journal of Thermal Sciences. 2015;90:38-52.
[10] Zheng S, Eimann F, Philipp C, Fieback T, Gross U. Modeling of heat and mass transfer for dropwise condensation of moist air and the experimental validation. International Journal of Heat and Mass Transfer. 2018;120:879-94.
[11] Sikarwar BS, Battoo NK, Khandekar S, Muralidhar K. Dropwise Condensation Underneath Chemically Textured Surfaces: Simulation and Experiments. Journal of Heat Transfer. 2011;133(2):021501.
[12] Sikarwar BS, Khandekar,S., and Muralidhar,K. Mathematical modelling of dropwise condensation on textured surfaces. Sadhana. 2013;38:1135-71.
[13] Tsilingiris PT. Thermophysical and transport properties of humid air at temperature range between 0 and 100°C. Energy Conversion and Management. 2008;49(5):1098-110.

5th International Workshop on Heat/Mass Transfer Advances
for Energy Conservation and Pollution Control
August 13-16, 2019, Novosibirsk, Russia

# Tailoring the Thermal Conductivity of Paraffin and Low-cost Device for Measuring thermal conductivity of Phase Change Material

**Mohit Bhandwal, Amritanshu Verma, Basant Singh Sikarwar***

Department of Mechanical Engineering, Amity University, Uttar Pradesh, Noida-201313

*bssikarwar@amity.edu

Abstract. In this work, experiment is carried out for melting process of paraffin wax dispersed with high conductive nano-fillers such as Carbon Nanotube (CNT) and ZnO. In addition, a simple, economically viable, easy operating experimental setup is proposed to measure the thermal conductivity of altered PCMs. In this setup, the sample is placed in an insulated transparent cavity and its one end maintained at constant temperature. However, other ends are adiabatic. Temperature at four specific location of melting paraffin in the cavity are measured by T-type thermocouples. The digital camera is used to capture the instantaneous images of the moving thermal front of melting at various time steps. ImageJ® is used to know velocity of the melting front of PCMs from captured images. Based on energy balance, a mathematical expression for estimating the conductivity of the altered PCMs is derived. The measurement of proposed device is compared with commercial thermal conductivity measuring device and data available in the literature. The obtained results have 95% confidence level against the commercial device results and literature results. Post-calibration, experimental study of pure paraffin and altered paraffin wax are carried to know insights of Spatio-temporal energy storing in form of latent heat. This research is useful for design the heat sink for thermal energy management of pulsating heating generating device.

Key words: Melting, Solidification, Latent heat Energy, altered Paraffin wax, Carbon nanotubes (CNT).

## 1. Introduction

Energy storage is essential to utilize renewable energy sources effectively and efficiently [1]. Solar-thermal energy is stored in the form of latent, thermo-chemical and sensible heat. Among these methods, Latent heat storage is more effective for its high energy per volume capacity at isothermal nature and this method uses Phase Change Materials (PCMs) as storage medium which exchange the energy in form of latent heat by melting and solidification [2-4]. Any storage medium which have suitable thermo-physical properties reduce the cost of the system and make it more efficient [5]. In this context, several organic, inorganic and eutectic PCMs and their thermo-physical properties have been critically reviewed in literature for optimizing thermophysical properties of PCMs [6]. Among several PCMs, Paraffin wax has been widely employed, as it is chemically stable, readily available, safe and non-reactive, non-corrosive and economical. It is also available in a wide range of melting temperatures, which makes it a desirable thermal storage medium. These properties make paraffin wax

suitable PCM for domestic hot water systems, building cooling, heat exchangers etc. However, low thermal conductivity of it has caused an incessant challenge in its large-scale implementation. To enhance the thermal conductivity of paraffin wax, scientists and researchers [[7-14]] have reported various methods such as addition of nano-fillers, metal-powder, metal matrix, carbon-fibers/graphite. On the other hand, addition of conducting fins and altering surface morphology methods have also been explored for enhancing the performance of latent heat storage.
Several experimental studies [15-18] have been performed on tailoring the thermal conductivity of paraffin wax using various nano-additives. CNT and ZnO are the nano-fillers high intrinsic thermal conductivity for enhancing thermal conductivity of paraffin wax. However, the effect of CNT and ZnO dispersion in paraffin remains unexplored in the literature. In this work, CNT and ZnO are used as nanofiller to enhance thermal conductivity of paraffin wax. In addition, an experimental setup is fabricated to measure thermal conductivity of altered paraffin wax. The thermal conductivity measurement of proposed device is compared with commercial thermal conductivity measuring device and data available in the literature. The prosed device results have 95% confidence level against the commercial device results and literature results. Post-calibration, insights of Spatio-temporal energy storing in form of latent heat of altered paraffin were explored. It is found that convection is main mode of heat transfer in melting zone of altered paraffin wax because large buoyancy force. This research is useful for design the thermal energy management for uneven availability of energy [19].

## 2. Materials and Experiment

Pure paraffin wax which has melting temperature of 62±1 °C was bought from *Chandri Wax Specialities Private Limited India*. Various weight percentage of CNT and Zno are used to alter thermal conductivity of paraffin wax. CNT is multi-walled carbon nanotubes grown by chemical vapor deposition (CVD) method using camphor as the carbon source. It is clear hollow tubular structures of outer diameter 5–20 nm, maximum abundance of diameter ~10 nm, length ~ 1–10 μm and specific surface area ~ 100 $m^2$/g. However, powder form Zno are bought from local vendor (Shakti chemical Delhi, India). These nano-filler is added into Paraffin wax when its temperature of wax is10 °C more than the its melting temperature. It is stirred for 45 minutes at 1500 rpm after pouring. Finally, the liquid form altered wax is ultra-sonicated for 60 minutes. It is noted that the temperature from stirring to sonicating process remains constant (=70 °C). Eventually, the altered paraffin is placed at room temperature for cooling. After cooling, it is placed in the refrigerator at 20 °C for two days.

Figure 1 shows an experimental setup for measuring the thermal conductivity of altered wax. Figure 1 (a) shows the schematic diagram and the photograph of experimental setup is given in Figure 1(b). The setup has rectangular 10 mm thick acrylic material cavity (Dimension: 3 cm × 3 cm × 6 cm) in which paraffin wax is filled up. The four faces of the cavity are insulated Nitrile PVC tape. However, right side face of cavity has aluminum block (3cm ×6 cm) to maintain constant wall temperature for charging and discharging the energy to wax. The left face of the cavity maintains room temperature using heat sink with fins and fan. Constant temperature bath is used to flow water in the block, as shown in Figure 1. Front side of cavity has glass window for visualizing the melting and solidification process wax. Double glass window is used to minimize the loss through the visualization window. At top face of cavity, a port is provided to fill up and drain out the paraffin wax from cavity. T-type thermocouples with a bead size of 0.1 mm and measurement accuracy of ±0.05 C are located at different positions in the chamber. The location of Thermocouples is given in table 1.

Table 1: Position of thermocouple in rectangular cavity

| Thermocouple | *X (cm)* | *Y (cm)* | *Z (cm)* |
|---|---|---|---|
| $T_1$ | 0.00 | 1.5 | 4 |
| $T_2$ | 1.00 | 1.5 | 4 |
| $T_3$ | 2.00 | 1.5 | 4 |
| $T_4$ | 3.00 | 1.5 | 4 |

Journal of Physics: Conference Series **1369** (2019) 012022 doi:10.1088/1742-6596/1369/1/012022

To minimize the convection effect of atmospheric air, the complete setup is placed in another acrylic chamber. Before carrying out experiment, probes and devices are calibrated, and water is heated in constant temperature bath. To ensure that there no temperature gradient inside the Al block, water flow rate is sat up at 40 L/min. The Phantom camera and Nikon (D-3300) camera with macro-lenses are used to capture the melting and solidification process of altered paraffin wax. NI DAQ (9174) with a frequency of five second is used to collect temperature data of various location automatically in a computer. In this way, the temperature variation with respect to time at four locations and spatio-temporal images of melting paraffin wax are recorded.

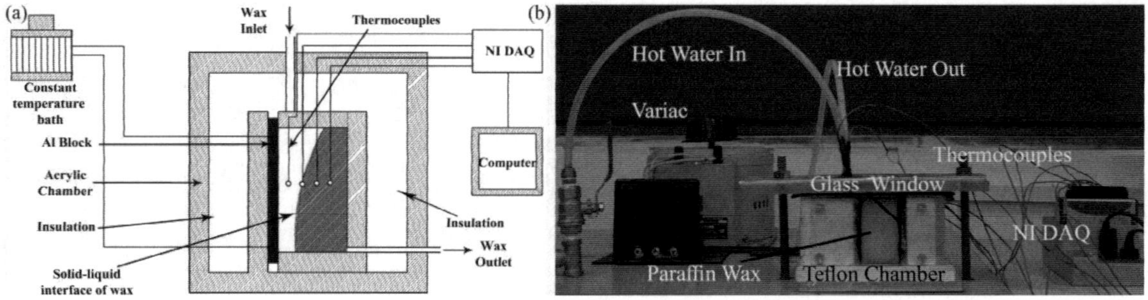


Fig. 1: Experimental setup (a) Schematic diagram and (b) photograph

*Procedure for measuring the thermal conductivity:* The thermal conductivity of wax is estimated by knowing spatio-temporal position melting front and temperature distribution in wax. Figure 2(a) shows the pictorial view melting at various time step. After image processing, a clear image of solid-liquid interface is extracted, Figure 2(b). The mathematical expression is derived by applying energy balance at the solid-liquid interface. Its expression is as:

$$k = \frac{\sigma \rho H_{sl} \delta L_2 L_1}{\Delta t \left[ L_2 \left( T_1 - T_2 \right) - L_1 \left( T_3 - T_4 \right) \right]} \quad (1)$$

Here, $L_1$ and $L_2$ are distances between the thermocouples $T_1$, $T_2$ and $T_3$, $T_4$ respectively. The temperatures were measured at two different points of liquid region ($T_1$ and $T_2$) and two different point of solid region ($T_3$ and $T_4$). The experimental results show that there is negligible convection after three by fourth depth from the top of the surface. For minimizing the convection effect, thermocouples were imbedded at ¾ depth from the top surface. The $\delta = (\delta_1 - \delta_2)$ is estimated by superimposed images captured at $\Delta t = (t_1 - t_2)$time step, as shown in Figure 2. By knowing the measured values from experimental studies, the thermal conductivity was measured using Equation 1

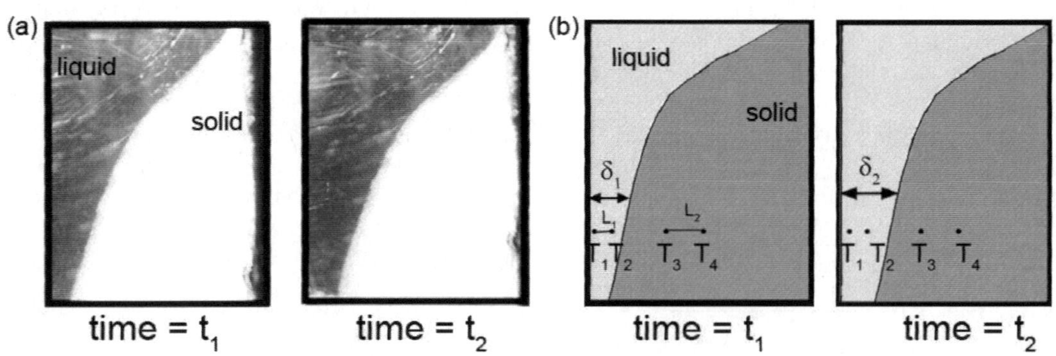


Fig. 2: (a) Photograph of paraffin wax at two-time step (b) Photographs after images process. $\delta_1$ an $\delta_2$ are position of solid-liquid interface at time $t_1$ and $t_2$. However, $L_1$ and $L_2$ are distance between thermocouples in liquid phase and sold phase of melted wax.

## 3. Results and Discussion

The device is calibrated before measuring the thermal conductivity of altered paraffin wax. Figure 3 shows the comparison of present device measurement with value of *k* given by manufacturer and value of *k* measured by commercial device (KD2-PRO). The device measurement is very close to measurement by commercial device and value reported by the literature. Post-calibration, the thermal conductivity of altered paraffin is measurement. Table 2 shows the thermal conductivity of altered paraffin wax.

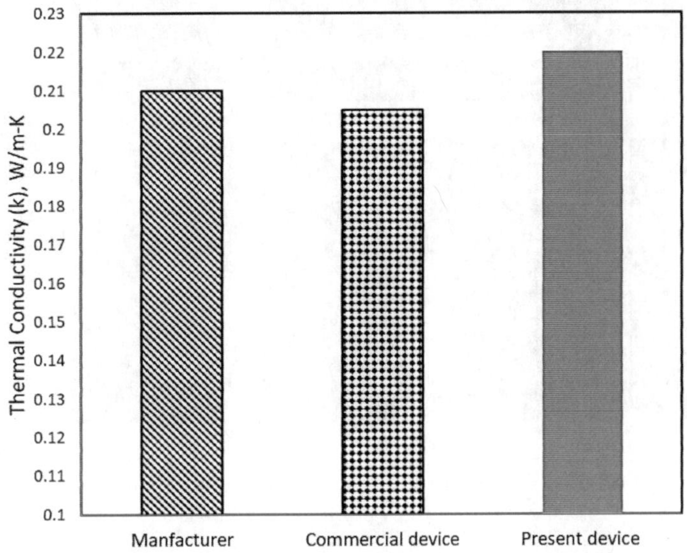


Fig. 3: Solid-liquid interface motion captured during (a) experimental study and (b) simulation study

Table 2: Thermal conductivity (W/m-K) of altered PCM

| PCM+Nano-fillers | 0.15% | 0.5% | 0.75% | 1% |
|---|---|---|---|---|
| Paraffin + ZnO | 0.3659 | 0.4971 | 0.5685 | 0.9362 |
| Paraffin +CNT | 0. 38 | 0.50 | 0.51 | 0.52 |

Figure 5 shows that solid-liquid interface displacement during melting process is faster than the other. At same percentage weight mixing of ZnO has high thermal conductivity as compared to CNT based altered paraffin wax.

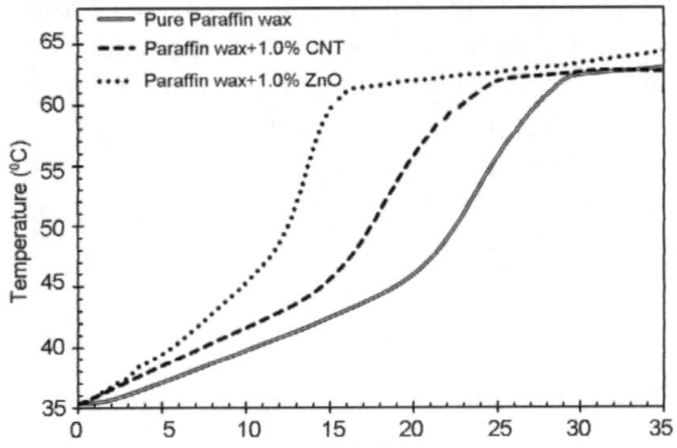


Fig. 4: Temperature variation at location $T_2$ with respect to time of melting process of wax.

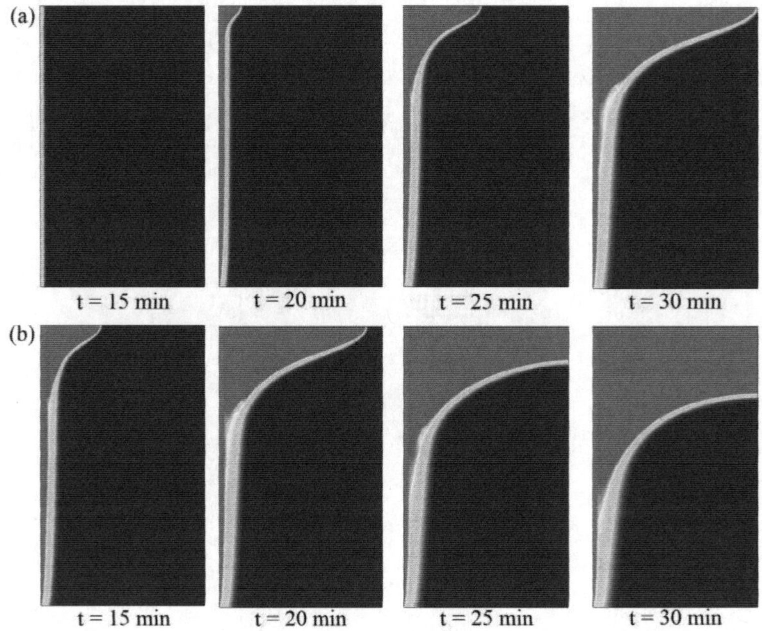


Fig. 5: Spato-temporal location of Solid-liquid interface for altered paraffin wax (a) 1% CNT and (b) 1% ZnO.

Figure 5 shows displacement of liquid-solid interface captured during melting experiment. Initially, the dominant mode of heat transfer in solid paraffin is conduction but as the wax starts melting, a rapid increment in temperature of wax is observed at top because of convection. Experimental results show that the transient response of altered paraffin wax is faster as compared to pure wax as the thermal conductivity of altered paraffin wax increases with increasing the percentage of CNT and ZnO. 1% CNT in paraffin wax enhance 12% speed to heat diffusion in tailored paraffin wax. However, 1% ZnO in wax enhances 10% speed of heat diffusion. Hence, the ZNO is more effective than CNT mixed paraffin wax.

## 4. Summary and Conclusion

An experimental study was conducted to analyze the thermal performance of paraffin wax when it is altered by adding CNT and Zno fillers. The transient response of thermal properties of wax was recorded during the melting process. The comparison of results from both proposed device and commercial device exhibited good agreement. In order to make an effective heat storage device using paraffin wax, it is essential to tailor the thermal properties of material so that the potential of energy storage device is fully utilized. This study provides a framework for designing the effective heat exchanger storing the solar energy in form of latent heat of wax. This research also gives a method to calculate the thermal conductivity of PCM by simple and low-cost device. the ZNO is more effective nanofiller as compared to the CNT.

## 5. Acknowledgement

The Department of Science and Technology (DST), Govt. of India for project No. ECR/2016/000020 is acknowledged for providing fund for buying camera and DAQ.

## References

[1] Miguel Miguel, Teresa Nogueira, Florinda Martins, 2017. "Energy storage for renewable energy integration: the case of Madeira Island, Portugal". Energy Procedia; 136, pp.251-257.

[2] Seddegh, S., Wang, X. and Henderson, A.D., 2016. "A comparative study of thermal behaviour of a horizontal and vertical shell-and-tube energy storage using phase change materials". Applied Thermal Engineering, 93, pp.348-358.

[3] Papadimitratos, A., Sobhansarbandi, S., Pozdin, V., Zakhidov, A. and Hassanipour, F., 2016. "Evacuated tube solar collectors integrated with phase change materials". Solar Energy, 129, pp.10-19.

[4] Kabeel, A.E. and Abdelgaied, M., 2016. "Improving the performance of solar still by using PCM as a thermal storage medium under Egyptian conditions". Desalination, 383, pp.22-28.

[5] Y. B. Tao, V.P. Carey, 2016. "Effects of PCM thermophysical properties on thermal storage performance of a shell-and-tube latent heat storage unit". Applied Energy, 179, pp.203-210.

[6] Jiang, G., Huang, J., Fu, Y., Cao, M. and Liu, M., 2016. "Thermal optimization of composite phase change material/expanded graphite for Li-ion battery thermal management". Applied Thermal Engineering, 108, pp.1119-1125.

[7] Karaipekli, A., Biçer, A., Sarı, A. and Tyagi, V.V., 2017. "Thermal characteristics of expanded perlite/paraffin composite phase change material with enhanced thermal conductivity using carbon nanotubes". Energy conversion and management, 134, pp.373-381.

[8] Wang, C., Lin, T., Li, N. and Zheng, H., 2016. "Heat transfer enhancement of phase change composite material: Copper foam/paraffin". Renewable Energy, 96, pp.960-965.

[9] Liu, L., Su, D., Tang, Y. and Fang, G., 2016. "Thermal conductivity enhancement of phase change materials for thermal energy storage: A review". Renewable and Sustainable Energy Reviews, 62, pp.305-317.

[10] Jiang, X., Luo, R., Peng, F., Fang, Y., Akiyama, T. and Wang, S., 2015. "Synthesis, characterization and thermal properties of paraffin microcapsules modified with nano-Al2O3". Applied energy, 137, pp.731-737.

[11] Li, Y., Li, J., Deng, Y., Guan, W., Wang, X. and Qian, T., 2016. "Preparation of paraffin/porous TiO2 foams with enhanced thermal conductivity as PCM, by covering the TiO2 surface with a carbon layer". Applied energy, 171, pp.37-45.

[12] Guan, W.M., Li, J.H., Qian, T.T., Wang, X. and Deng, Y., 2015. "Preparation of paraffin/expanded vermiculite with enhanced thermal conductivity by implanting network carbon in vermiculite layers". Chemical Engineering Journal, 277, pp.56-63.

[13] Sharma, R.K., Ganesan, P., Tyagi, V.V., Metselaar, H.S.C. and Sandaran, S.C., 2016. "Thermal properties and heat storage analysis of palmitic acid-TiO2 composite as nano-enhanced organic phase change material (NEOPCM)". Applied Thermal Engineering, 99, pp.1254-1262.

[14] Şahan, N., Fois, M. and Paksoy, H., 2015. "Improving thermal conductivity phase change materials—A study of paraffin nanomagnetite composites". Solar Energy Materials and Solar Cells, 137, pp.61-67.

[15] Sikarwar, B.S., Chopra, A., Bhandwal, M., Kumar, M. and Avasthi, D.K., 2017. "Tailoring the thermal conductivity of paraffin wax by nano-fillers for thermal storage applications". In ISHMT Digital Library. Begel House Inc..

[16] Tang, Y., Su, D., Huang, X., Alva, G., Liu, L. and Fang, G., 2016. "Synthesis and thermal properties of the MA/HDPE composites with nano-additives as form-stable PCM with improved thermal conductivity". Applied energy, 180, pp.116-129.

[17] Nourani, M., Hamdami, N., Keramat, J., Moheb, A. and Shahedi, M., 2016. "Thermal behavior of paraffin-nano-Al2O3 stabilized by sodium stearoyl lactylate as a stable phase change material with high thermal conductivity". Renewable energy, 88, pp.474-482.

[18] Khan, Z. and Khan, Z.A., 2018. "Experimental and numerical investigations of nano-additives enhanced paraffin in a shell-and-tube heat exchanger: A comparative study". Applied Thermal Engineering, 143, pp.777-790.

[19] Mekhilef S, Saidur R, Safari A., 2011. "A review on solar energy use in industries". RenewSustain Energy Rev; 15: 1777–1790.

5th International Workshop on Heat/Mass Transfer Advances
for Energy Conservation and Pollution Control
August 13-16, 2019, Novosibirsk, Russia

# Numerical modeling of interface diffusion during heating of spherical metal heterostructure

**E I Golovneva[1], I F Golovnev[1] and A M Igoshkin[2]**

[1] Khristianovich Institute of Theoretical and Applied Mechanics SB RAS, Novosibirsk 630090, Russia

[2] Nikolaev Institute of Inorganic Chemistry SB RAS, 3, Acad. Lavrentiev Ave., Novosibirsk 630090, Russia

elena@itam.nsc.ru

This work is a pilot one and addresses to the development of a physical and mathematical model of alloy formation; this model permits conducting a micro-level detailed study of the influence imposed by the ratio of the thermophysical parameters of the matrix and nanoclusters materials, as well as external controlled parameters on the physical properties of the alloys. The article presents the developed methodological approaches for the numerical simulation of alloys formation with the help of molecular dynamics method (MDM).

## 1. Introduction

Necessity to have nano-modified alloys with prescribed properties causes the need in a developed physical and mathematical model of the atomic-level formation of such alloys and in the analysis of the influence of combination of matrix nanocluster materials, as well as the effect of external conditions on the alloy properties.

In recent decades, interest to the molecular-dynamic modeling (MDM) of alloy formation on the atomic level has begun to grow. There are quite a lot of works where the ready alloy is used as the initial data [1]. The cycle of papers [2 – 5] approaches better to the authors' paper by the nature of investigations. The major difference is in the following: the liquid film formed initially as a stoichiometric mixture, and was then melted through with the aid of the modeling of the impulse atom distribution in accordance with the Maxwell distribution. The system temperature is increased abruptly in 100 K. This approach does not permit modeling the process of heat transfer during the heating and cooling which, as we know from experiment, dictates the final structure of the alloy. In [6 – 7] there is the MD modeling of the structural and dynamic properties of the alloy Cu-Ni on the liquid and solid interface.

The main difference of this paper from the above mentioned ones is that we use the free boundary conditions which permits modeling the effect of the nanoparticle size, taking into account the influence of the nanoparticle surface. Moreover, the purpose of the work is to model the formation of the alloys, not only analysis of the thermophysical properties of the ready (formed) alloys, which is the fundamental difference from the works of other scientific teams.

## 2. Physical system and calculation method

The studied system presents a sphere of one material covered with a spherical layer of another material; such a system can be referred to as a "spherical heterostructure".

Copper and silver were used at the stage of methodology refinement. To model the interaction between atoms in this system, we used the Voter potential [8] obtained by the embedded atom method (EAM), which has successfully used in many tasks. To prepare the system and bring in the stationary equilibrium state in the potential energy minimum, the artificial viscosity method was used [9]. Then, the found values of coordinates and momenta were used as the initial data for the system heating up to the needed temperature. To do this, we used the stochastic momenta method [10]. The following heating parameters were applied for every system atom: the random momentum amplitude_ $\Delta p = 4$ (in program units), and the interval between absorption of momenta by atoms is $\Delta N_\tau = 20$ numerical time pitches. In this calculation, in every 25 K of the system temperature growth, the coordinates and momenta arrays were output.

The velocity Verlet algorithm [11] is used in every calculation of this paper, the chosen time pitch is $\tau = 10^{-16}$ s.

Below there are the results for the case when the inner copper sphere radius is 20 Å, whereas the external silver sphere is 40 Å (i.e. the silver layer thickness is 20 Å). This system consists of 2899 atoms of copper and 12784 atoms of silver. Totally there are 15683 atoms in the system. All of them are located in the nods of a perfect FCC lattice, the constants of the copper ($a_{Cu}$=3.6149 Å) and silver lattice ($a_{Ag}$=4.09 Å) are used. The origin of the laboratory system of coordinates lies in the sphere center.

## 3. Numerical cooling of the system

The cooling procedure results in the fact that the system evolves into the state with the temperature close to the absolute zero, the potential energy of the total system and each Cu and Ag subsystem decreases and takes the constant equilibrium values, the atoms in the Cu-Ag interface shift toward the more energetically advisable position. A layer of 5 Å was cut in the equatorial plane for the better visualization of the processes in the inner volume of the heterostructure.

The system temperature is calculated through the kinetic energy of the chaotic motion of the atoms, the kinetic energy of the mass center is ignored (figure 1).

Figure 1 – (c) presents the change of the potential energy of the copper subsystem $\Delta U_{Cu}$ (the calculation is such as if there is no silver subsystem), change of the potential energy of the silver subsystem $\Delta U_{Ag}$ (the calculation is such as if there is no copper subsystem). Opposite to the full potential energy $U_t$, which decreased during the cooling (figure 1 – b), the potential energies of the subsystems rise. The bound energy between two subsystems is the key characteristic of the heterostructure; it is found as follows:

$$U_b = U_t - \left(U_{Cu} + U_{Ag}\right).$$

Reduction of the full potential energy at the simultaneous increase of $\Delta U_{Cu}$ and $\Delta U_{Ag}$ is possible owning to the reduction of the bound energy (figure 1 – d). Coordinates and pulses of the system resulting from the cooling process, were then used for the system heating.

## 4. Modeling of diffusion during the heating of the spherical heterostructure

### *4.1. Heating of the spherical heterostructure*

After the heating which technique is briefly described in chapter 2., every characteristic of interest was averaged by thermal fluctuations within 1,000 calculation stepes (or 0.1 ps). After the heating, all necessary characteristics were analyzed during the relaxation of the closed and isolated system. It turned out that the heterogeneous system needs much more time to get into the thermodynamically

Journal of Physics: Conference Series **1369** (2019) 012023 doi:10.1088/1742-6596/1369/1/012023

equilibrium state. The system containing only copper atoms gets into the equilibrium state after the heating within approximately 500 calculation pitches (0.05 ps). Whereas the spherical heterostructure Cu-Ag needs about 500,000 calculation pitches to equalize the temperatures of the silver and copper subsystems (see figure 2).

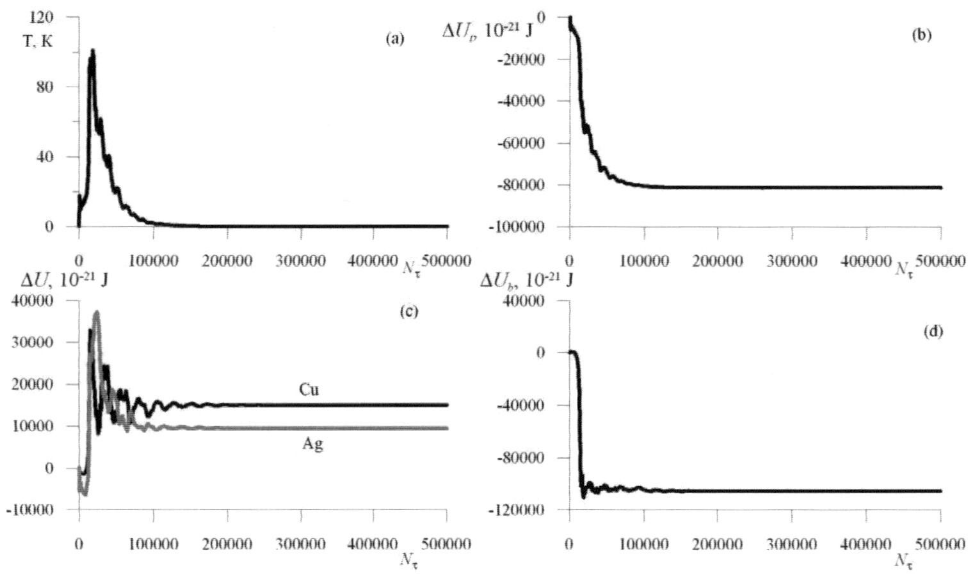


**Figure 1.** System characteristics versus time during the cooling. (a) temperature; (b) variation of the full potential energy of the system; (c) variation of the potential energy of copper (black line) and silver (grey line) subsystems; (d) variation of the bound energy between the copper and silver subsystems.

*4.2. Detection of the diffusion on the interface*

Initially, the diffusion on the interface between Cu and Ag is found visually. For better evidence, the atomic layer of 5 Å thick is shown for different temperatures; it is cut out from the spherical heterostructure in the equatorial plane. Figure 3 presents as an example the physical system at T = 1700 K. This visualization clearly shows both the beginning and the developed diffusion process.

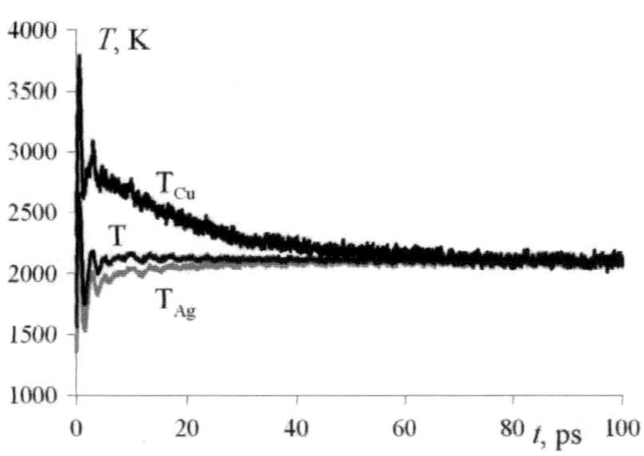


**Figure 2.** Change of the temperature of the heterostructure system component versus relaxation time. T is the average temperature of the total system, $T_{Cu}$ is the copper subsystem temperature, $T_{Ag}$ is the silver subsystem temperature.

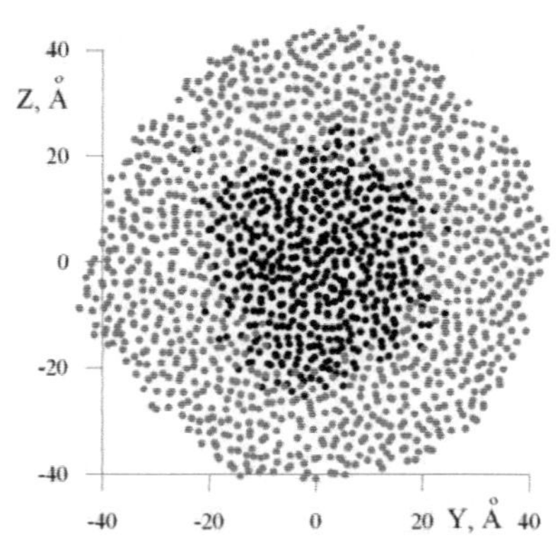


**Figure 3.** Position of atoms in the YZ plane during the heating. *T*=1700 K. Layer thickness is 5 Å.

### *4.3. Meso-analysis*

It is necessary to have the distribution of a number of characteristics inside the spherical heterostructure. To do this, the technique of meso-analysis has been developed. The logics of space division to subsystems depends on both the geometry of the studied object and the processes occurring in it. In the case of spherical heterostructure heating, the optimal division of the space is to split it to spherical layers between two radiuses $r$ and $(r+\Delta r)$, where $\Delta r$ – the layer thickness. The center of these inserted spheres coincides with the studied object center.

The distance between the layers is $\Delta r$ = 5 Å, the radius of the most inner meso-cell is also 5 Å. For the chosen size of the studied system $R$=40 Å, the lattice of spherical inserted meso-cells up to 80 Å is assigned. Such a stock is used to take into account the metal expansion during the heating, and even possible evaporation of the atoms from the surface.

The files with data arrays about the coordinates and momenta of each atom of the system obtained with the aid of the program on the spherical heterostructure heating are used as the initial data in the calculation program of the meso-analysis.

The temperature distribution over the meso-cells has been obtained (figure 4). Certain temperature decrease from the center toward boundary meso-cells is observed, which vindicates the need to increase the time of relaxation of the heterogeneous system during the heating.

The distribution has been calculated about the relative amount of the atoms of each kind (copper and silver in this task) for the meso-cells at different temperatures of the system heating. It is found as the relation of the substance (copper or silver) atom amount in the certain meso-cell at the chosen temperature to the total amount of the atoms (both copper and silver) in the same meso-cell at the same temperature. Figure 5 shows the results for $T$ = 1100 K.

Every graph contains for better visualization the initial (in the cooled system) boundary of the inner copper sphere and external silver sphere $R_{Cu}$, and the external system boundary $R_{Sp}$ (actually, the radius of the external shell from silver).

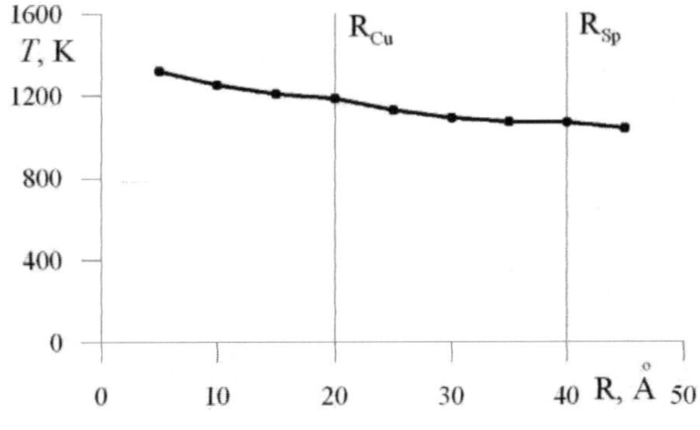


**Figure 4.** Temperature distribution over meso-cells. The heated system temperature T=1100 K.

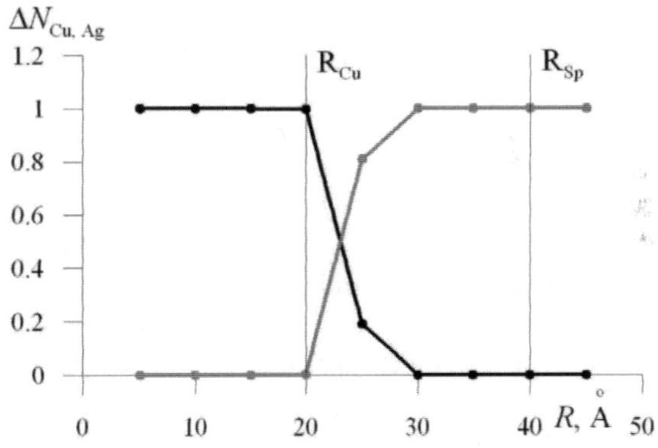


**Figure 5.** Distribution of the relative amount of copper atoms (black line) and silver atoms (grey line) over meso-cells at the heating temperature T=1100 K.

### *4.4. Calculation of the parameters characterizing the diffusion on the interface*

The algorithm of parameters calculation has been developed; these parameters characterize the diffusion through the medium interface and through the external interface (which in fact corresponds to the thermal expansion of the substance). The essence is that in certain moments during the heating we determine the position of copper and silver atoms in respect to the initial position of the interface between the inner copper sphere and external silver layer. In the same way we detect, if copper (and silver) atoms have crossed the external boundary of the system, and what is was before the heating start. The amount of atoms of each substance is summed up and their content in percent from the total amount of the atoms of the same substance is determined. Thus the qualitative development of the diffusion during the heating becomes clear (figure 6). At the temperature of 1100 K, atoms diffusion is still weak, which is proven by the results below.

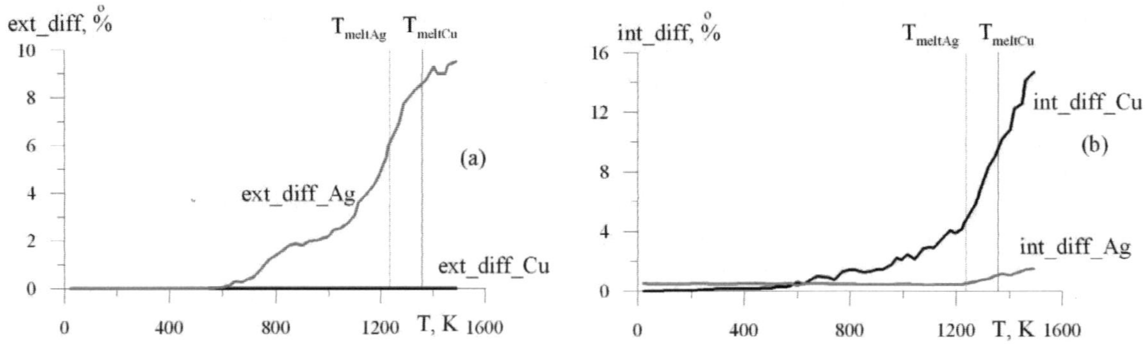


**Figure 6.** Diffusion of copper (black line) and silver (grey line) atoms through the external system boundary (the interface with the external medium) (a) and interface between the substances (b).

It is evident that the diffusion of copper atoms through the inner interface begins from about 600 K. Note that the diffusion of silver atoms through the initial external interface of the system begins also from 600 K. I.e., the thermal expansion becomes noticeable at the temperature about 600 K. Insignificant amount of silver atoms from the external shell get into the inner sphere of Cu (figure 6 – b, grey line), but even at the copper melting temperature which is higher than the silver melting point the diffusion of Ag atoms from the external shell through the inner interface toward the center remains negligibly low (about 1% of all Ag atoms), whereas the thermal expansion of the whole system makes about 10% of the total amount of atoms at the copper melting temperature (figure 6 – a grey line, figure 6 – b black line). I.e., the diffusion of both copper and silver from the spherical heterostructure center dominates over the diffusion toward the system center.

## 5. Conclusions

- The technique of the numerical simulation of the heating of the heterogeneous system has been developed on the example of the spherical heterostructure: the inner sphere from Cu, the external shell from Ag up to the temperatures above the melting point of the heterostructure components.
- The approach has been developed to perform the meso-analysis of the spherical structure for the better visualization of the physical parameters distribution in the system.
- On the base of the meso-analysis, the temperature profile distribution has been found in the spherical heterostructure, as well as the distribution of the relative amount of the atoms of each substance for the meso-cells. The resulting graphs demonstrate how the compound profile changes as the temperature grows.
- The algorithm has been developed for the qualitative analysis of the diffusion in the system. It is shown that the dominating diffusion is that which coincides in direction with the atoms motion at the thermal expansion of a substance.

## 6. Acknowledgments

The work has been supported by the RFBR, Grant number 18-01-00287-a “Investigation of the effect of nanostructuring and the conditions for the formation of nanomodified alloys by the molecular dynamics method”.

Journal of Physics: Conference Series **1369** (2019) 012023 doi:10.1088/1742-6596/1369/1/012023

## 7. References

[1] Yue Q, Tahir Ç, Yoshitaka K and Goddard W A III 1999 *Phys. Rev. B* **59** 3527–33
[2] Levchenko E, Evteev A, Belova I, Murch G 2010 *Comp. Mat. Sc.* **50** 331–7
[3] Levchenko E, Evteev A, Riley D, Belova I, Murch G 2010 *Comp. Mat. Sc.* **47** 712–20
[4] Evteev A, Levchenko E, Hagel Fl, Belova I, Murch G 2011 *Intermetallics* **19** 934–41
[5] Levchenko E, Evteev A, Löwisch G, Belova I, Murch G 2012 *Intermetallics* **22** 193–202
[6] Qi C, Li J, et al 2016 *Comp. Mat. Sc.* **125** 72–81
[7] Lu P, Zhou K, et al 2017 *Comp. Mat. Sc.* **135** 22–8
[8] Voter A 1993 Embedded Atom Method Potentials for Seven FCC Metals: Ni, Pd, Pt, Cu, Ag, Au, and Al *Los Alamos Unclassified Technical Report* # LA-UR 93-3901
[9] Golovnev I, Golovneva E, Fomin V 2003 *Phys. Mesomech.* **6** 41–45
[10] Golovnev I, Golovneva E, Utkin A 2018 *Phys. Mesomech.* **21** 523–8
[11] Allen M, Tildesley D 1987 *Computer Simulation of Liquids* (New York: Clarendon Press) p 385

Journal of Physics: Conference Series **1369** (2019) 012024 doi:10.1088/1742-6596/1369/1/012024

5th International Workshop on Heat/Mass Transfer Advances
for Energy Conservation and Pollution Control
August 13-16, 2019, Novosibirsk, Russia

# Formation of thermocapillary structures in heated liquid film

**E A Chinnov**

Kutateladze Institute of Thermophysics, Novosibirsk, 630090, Russia

E-mail: chinnov@itp.nsc.ru

**Abstract.** Data on the formation of thermocapillary structures and breakdown of a heated liquid film flowing down a vertical surface with Reynolds number variation from 0.1 to 500 have been analyzed and generalized. Small-scale thermocapillary structures are distinguished. It is shown that the distances between temperatures heterogeneities (rivulets) of thermocapillary structures do not depend on the Reynolds number. It is determined that the interaction of waves with thermocapillary structures leads to an increase in the critical heat flux corresponding to the liquid film breakdown as compared with the data of publications.

## 1. Introduction

It is known that two-dimensional waves in isothermal liquid films are unstable to three-dimensional perturbations. When two-dimensional waves decay into three-dimensional ones, the synchronous waves without phase displacement of waves in the transverse direction and subharmonic waves with phase displacement are distinguished in [1, 2]. It is shown that the length of a wave unstable to transverse three-dimensional perturbations decreases with increasing Reynolds number. Propagation of two-dimensional and three-dimensional waves along a vertically flowing isothermal water film at Re = 10-100 and the effect of artificial perturbations are studied in [3]. It was experimentally shown that for Re > 40, two-dimensional waves are unstable to transverse perturbations with a wavelength of approximately 20 mm. They break up into three-dimensional waves that interact strongly and combine with each other. At Re < 20, the length of a wave unstable to transverse three-dimensional perturbations increases to 30 mm and more. It is determined that the transition from regular two-dimensional structures to a three-dimensional flow is accompanied by a significant redistribution of liquid in the longitudinal direction [4, 5]. The characteristic shapes of three-dimensional structures developing in the process of transition are described. It is found that the predominant structures on the film surface at Re > 50 are the short-lived rivulets, which are the chains of not less than 5-8 waves with close values of the transverse coordinate [6].

To date, various mechanisms of rivulet formation in the heated liquid films have been discovered, and two thermocapillary regimes have been identified: A and B [7]. The structures of two types A and B differ in the level of heat flux density, required for their formation, and size and nature of the dependence of interrivulet distance on the heat flux density and Reynolds number. Formation of regular structures in thermocapillary regime A on the surface of a smooth liquid film flowing over a vertical plane with small heaters of 6.5 × 13 mm at low Reynolds numbers was discovered and investigated in [8, 9]. Under the influence of thermocapillary forces directed against the flow, film thickening occurred. When the threshold heat flux density was achieved, the flow divided into vertical rivulets at a certain distance $\Lambda$ with a thin film between them. Data on the formation of structures of this type on the water film surface on extended heaters with the length of 60 mm are presented in [10]. This phenomenon at $0.1 \le Re \le 2$ was studied theoretically and analyzed in detail in [11]. It was determined that the distance between the rivulets is almost independent of Re, and for a fixed Marangoni number, it can even decrease with increasing Reynolds number.

In regime B, the rivulet flow is formed gradually with increasing heat flux density and distance from the upper edge of the heater. It is shown in [7, 10, 12] that in the region of sufficiently high heat fluxes with Re> 15, the distance between the rivulets depends weakly on the Reynolds number and film path length. Temperature fluctuations and wave characteristics of the water film flow over a vertical plate with a heater are studied in [13]. It is shown that two thermocapillary effects appear in-between the rivulets with increasing heat flux density. The transverse temperature gradients cause a significant decrease in the liquid film thickness, and the longitudinal gradients increase the relative wave amplitude as compared with the isothermal conditions. Formation of thermocapillary structures at small Reynolds numbers of up to Re = 0.1 is studied experimentally in [14]. It is shown that the transverse size of structures is determined by the capillary constant and does not depend on the viscosity of the liquid.

Breakdown dynamics is studied in [15] using a fiber-optic sensor. The influence of substrate wettability on thermocapillary breakdown of a liquid film is considered in [16]. In [17], the thermocapillary breakdown of a film is studied with a change in a wide range of Reynolds numbers and heater sizes.

The existence of thermocapillary structures in regime A at Re = 150 was detected in [18]. Formation of the smaller structures in the form of temperature heterogeneities across the flow with a distance between maxima of $2 \pm 0.4$ mm was registered there.

This work is aimed at analysis and generalization of data on the formation of various types of thermocapillary structures and breakdown of the heated liquid film with a variation of Reynolds number from 0.1 to 500.

## 2. Results and discussion

Based on the analysis of experimental data on the hydrodynamics of the flow of a heated liquid film, it has been established that the structures of type A are observed in the range of Reynolds numbers from 0.1 to 150. They appear on the liquid film surface when the threshold of heat flux density is reached. When structures of type A appear, high-temperature gradients of up to 10-15 K/mm are registered in the upper part of the heater. The boundary condition close to T = const took place on the heater surface. Under the conditions of a developed wave flow, thermocapillary structures of type A were formed in the residual layer of the liquid film after propagation of the three-dimensional wavefront. The structures existed for a limited period of time and interacted with the front of the next wave transforming into rivulets that moved along the heater, changing their direction, Fig. 1a.

In regime B, a rivulet flow was formed gradually with increasing heat flux and distance from the upper edge of the heater, Fig. 1a. Structures of type B were registered in the range of Reynolds numbers from 0.1 to 500. Boundary condition q = const was implemented on the heater surface, and temperature gradients on the film surface did not exceed 1 K/mm.

Formation of the smaller structures in the form of temperature heterogeneities across the flow has been found. These structures, designated as type C, were observed, as a rule, in the residual layer of liquid after propagation of a large wave before the formation of structures of type A. Temperature gradients on the liquid film surface at different points of time are shown in Fig. 1b. At zero time, we can see only gradients from the fronts of passing waves. With a decrease in the film thickness behind the wavefront and its heating, temperature heterogeneities appear on the surface with an average distance between them of 1.5 mm (t = 20 ms). Then, they are enlarged and transformed into thermocapillary structures of type A. At lower Reynolds numbers, the stationary structures of type C are registered, Fig. 2. Three maxima appeared on the left side of the heater. The maximal temperature of liquid on the film surface increased to 68°C. The difference between the maximal and initial temperature was 44°C. The distance between small-scale heterogeneities was 2.5–3 mm.

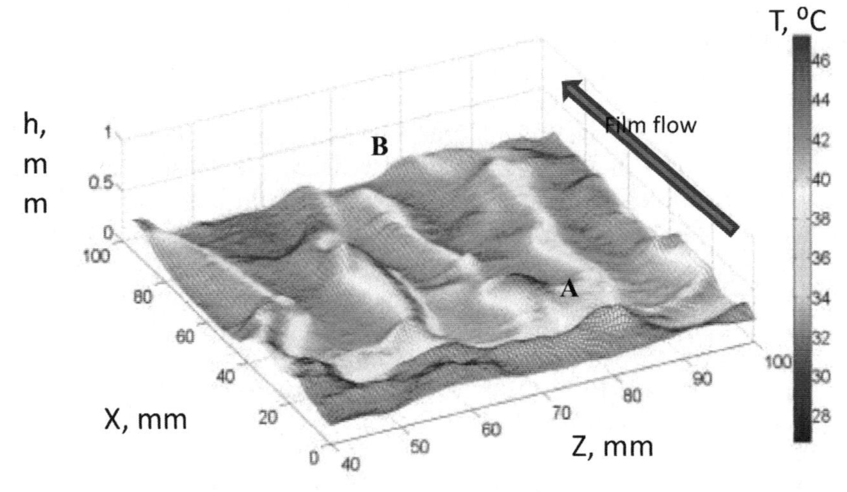


a

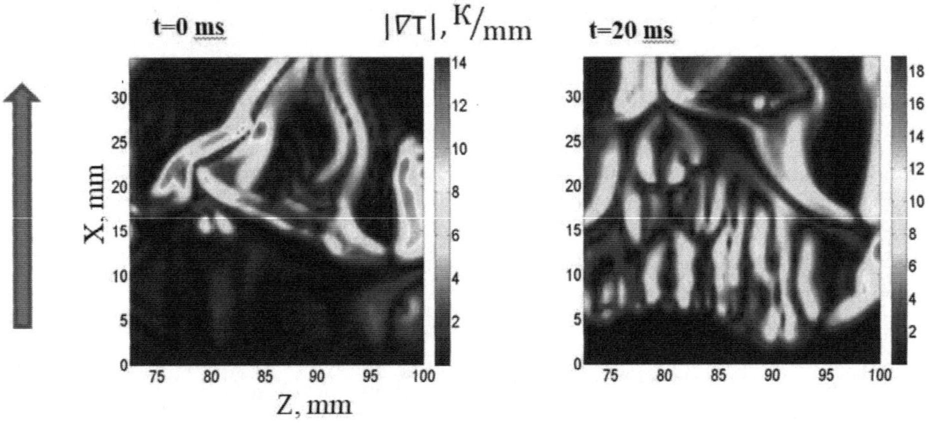


b

Fig. 1. Three-dimensional distribution of thicknesses, temperatures, and gradients over the water film surface, Re=50, q= 6 W/cm$^2$, the heater of 150x100 mm. a - Three-dimensional distribution of thickness and temperature over the liquid film surface. b – Distribution of temperature gradients over the liquid film surface.

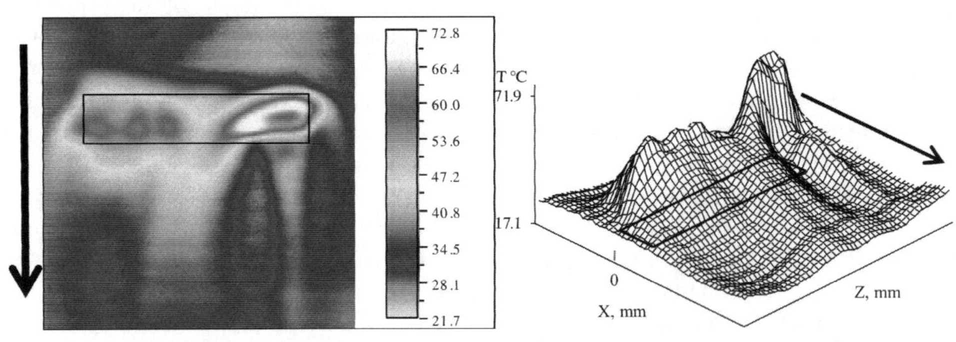


Fig. 2. Thermogram and temperature distribution over the water film surface, Re=11.8, q=9.9 W/cm$^2$, the heater of 6.55 x 22 mm.

Generalization of available experimental and theoretical data on the dimensionless distance between the rivulets of the observed structures is presented in Fig. 3. The characteristic regions of thermocapillary structures A, B, and C are identified. It can be seen that the dimensionless distances between the rivulets in all three cases are self-similar relative to the Reynolds number. Dependence of the distance between the crests of three-dimensional hydrodynamic waves (3D) on Reynolds number, derived from the data of [3], is given. At low Reynolds numbers, the distance between the crests of three-dimensional waves decreases sharply with increasing Re. For Re > 30, it does not depend on Reynolds number and is close to the distance between the rivulets in regime B.

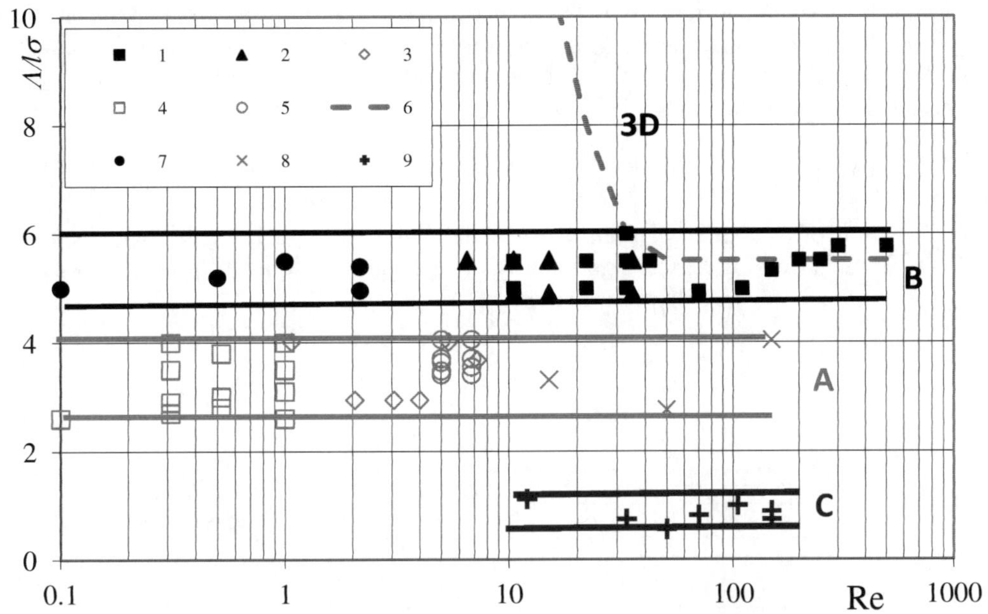


Fig. 3. Generalization of experimental data on dimensionless distance between the rivulets of thermocapillary and wave structures. 1 - structures B, water [14]; 2- structures B, FC-72 [7]; 3- structures A, water, [10]; 4- structures A, data of [11]; 5 - structures A, FC-72, [7]; 6 – dependence of wavelength of unstable formation of three-dimensional waves on Reynolds number according to [3]; 7- structures B, water solution of glycerin [14]; 8- structures A, water; 9 - structures C, water. $l_\sigma$ - capillary constant.

Under the action of thermocapillary forces directed from the hotter to the colder areas, an increase in deformation of the film surface occurred. As a result, when the heat flux density became higher, the liquid film breakdown took place. When forming the structures of type A, the liquid film breakdown occurred in the upper part of the heater, where maximum deformations of the film surface were registered, and during the formation of structures of type B, it occurred in the lower part of the heater. Generalization of experimental data on the heat flux, corresponding to the formation of structures of type A is fulfilled in the form of dependence between criterion $Km_A^*$ and Re, Fig. 4. Criterion $Km_A^*$ characterizes specific thermal power $W_A/B$, released on the heater for heating the film

$$Km_A^* = -q_A(\sigma_T/(c\rho^2 g^{2/3}\nu^{5/3}))L/l\nu = -W_A\sigma_T/(Bc\rho^2 g\nu\, l\nu^2), \qquad (1)$$

where B is heater width, c is heat capacity of liquid, g is acceleration of gravity, L is heater length, lv is scale of viscous-gravitational interaction equal to $(\nu^2/g)^{1/3}$, $q_A$ is heat flux density, corresponding to

formation of structures in regime A, $\nu$ is kinematic viscosity of liquid, $\sigma_T$ is temperature derivative of surface tension coefficient equal to $\partial\sigma/\partial T$, $\rho$ is liquid density.
The condition for structure formation in regime A on the liquid film surface is described by the following relation

$$Km_A^* = 113Re^{1.08} \quad . \qquad (2)$$

Data on the liquid film breakdown during the formation of A and B structures are presented in Fig. 4. If only structures B were formed on the liquid film surface, breakdown always occurred in the lower part of the heater between the formed rivulets. With an increase in the heat flux density in the lower part of the heater, there was a sharp increase in liquid film deformation, which led to its breakdown. A summary of data for different types of liquid is presented in Fig. 2b. Data for the breakdown in regime B are summarized by dependence from [19]

$$Km_{crB}^* = 165Re \quad . \qquad (3)$$

During the formation of structures in regime A on the liquid film surface, the heat flux densities corresponding to breakdown increased significantly. The nature of liquid film breakdown changed, which is explained by a change in film deformation

$$Km_{crA}^* = 290Re \quad . \qquad (4)$$

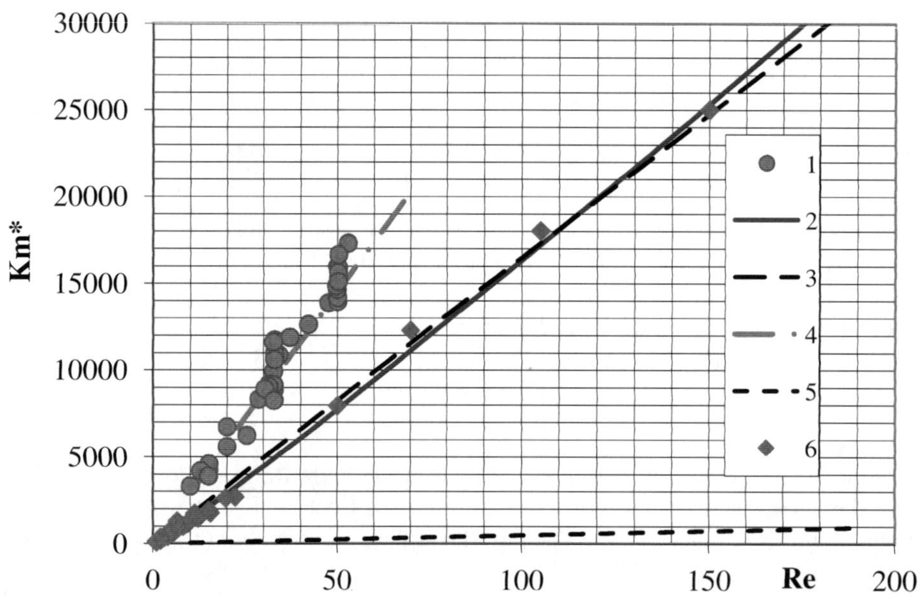


Fig. 4. Generalization of experimental data on the formation of thermocapillary structures and liquid film breakdown. 1 – experimental data on water film breakdown after formation of structures A, 2 – dependence (2), 3 – dependence (3), 4 – dependence (4), 5 – line, above which structures B become perceptible, 6 – experimental data on formation of structures A on the water film surface.

Thus, the data on the formation of thermocapillary structures and breakdown of a heated liquid film flowing down a vertical surface with Reynolds number variation from 0.1 to 500 have been analyzed and summarized. Thermocapillary structures of a small scale were distinguished. It is shown that the distances between the rivulets of thermocapillary structures for these regimes do not depend on the Reynolds number. It has been established that the formation of metastable thermocapillary structures of type A in the upper part of the heater affects the value of the critical heat flux corresponding to liquid film breakdown. It is shown that the critical heat flux increases with increasing film Reynolds number and at high Re numbers, it exceeds significantly the data on a breakdown in the absence of thermocapillary structures of type A.

## Acknowledgments

The study of rivulet formation, liquid films rupture, and data generalization were supported by the Russian Science Foundation (grant No. 18-19-00407). The study of small-scale thermocapillary structures of type C was performed in the framework of the state task to IT SB RAS (AAAA- A17-117022850022-0).

## References

[1] Joo S W, Davis S H 1992 *J. Fluid Mech.* **242** 529.
[2] Liu J, Schneider J B., Golub J P 1995 *Phys. Fluids.* **7** 55.
[3] Park C.D, Nosoko T 2003 *AIChE Journal* **49** 2715.
[4] Alekseenko S.V., Guzanov V.V., Markovich D.M., et al 2012 *Technical Physics Letters* **38** 739.
[5] Alekseenko S.V., Bobylev A.V., Guzanov V.V., et al 2014 *Technical Physics Letters* **40** 1034.
[6] Bobylev A.V., Guzanov V.V., Kharlamov S.M., et al 2017*Technical Physics Letters* **43** 694.
[7] Chinnov E.A., Kabov O.A 2003 *Mechanics and Technical Physics* **44** 708.
[8] Kabov O.A., Chinnov E.A 1997 *Russian J. Engineering Thermophysics* **7** 1.
[9] Kabov O. A 1998 *Thermophysics and Aeromechanics* **5** 547.
[10] Chinnov E.A 2009 *Thermophysics and Aeromechanics* **16** 69.
[11] Frank A.M., Kabov O.A 2006 *Phys. Fluids.* **18** 032107-1.
[12] Chinnov E.A. et al 2007 *Thermophysics and Aeromechanics* **14** 249.
[13] Kabov O.A., Marchuk I.V. et al 2008 *Fluid Dynamics* **4** 108.
[14] Shatskii E.N., Chinnov E.A. et al 2017 *Technical Physics Letters* **43** 1080.
[15] Zaitsev D.V., Rodionov D.A., Kabov O.A 2007 *Micrograv. sci. technol.***19** 100.
[16] Zaitsev D.V., Kirichenko D. P., Kabov O.A 2015 *Technical Physics Letters* **41** 551.
[17] Zaitsev D.V., Semenov A.A., Kabov O.A 2016 *Thermophysics and Aeromechanics* **23** 625.
[18] Chinnov E.A 2012 *Technical Physics Letters* **38** 711.
[19] Chinnov E.A., Sharina I.A 2008 *Thermophysics and Aeromechanics* **15**113.

5th International Workshop on Heat/Mass Transfer Advances
for Energy Conservation and Pollution Control
August 13-16, 2019, Novosibirsk, Russia

# Stefan Problem with Internal Heat Generation: Comparison of Numerical Modeling and Analytical Solution

**J C Crepeau[1*], A Yu Sakhnov[2] and V S Naumkin[2]**

[1]University of Idaho, Department of Mechanical Engineering, 875 Perimeter Drive MS 0902, Moscow, Idaho, USA 83844

[2]Kutateladze Institute of Thermophysics SB RAS, Lavrentyev Ave., 1, Novosibirsk, Russia 630090

*Corresponding author's e-mail: crepeau@uidaho.edu

The Stefan problem with internal heat generation can be found in various industrial applications. In particular, this problem describes melting of nuclear fuel rods during a reactor emergency. The present study reports on numerical modeling of the heat conduction equation and the Stefan equation with an application of the method of catching of the front into a space grid node. To solve numerically the heat conduction equation we realized an implicit four-node scheme using the Thomas method and iteration procedure.

## 1. Introduction

The first published solution to the solid-liquid phase change problem is attributed to Lamé and Clapeyron [1]. Independently, Stefan, for whom the Stefan problem is named, published an analysis of solid-liquid phase change based on data collected from ice formation on ships as they sailed the polar seas. His analysis compared favorably with the experimental data. The solid-liquid phase change is well studied [3,4] and continues to be a fruitful area of research today, especially in light of research involving phase change materials used in cooling systems.

We propose solutions to a non-homogeneous form of the Stefan problem, namely phase change under the influence of internal heat generation. Applications to this problem include geophysical systems, materials processing and cryogenic treatment of biological systems. The primary application of this work is nuclear safety and the analysis of melting of nuclear fuel rods. This paper compares the analytical solutions in a cylindrical geometry with a numerical result using the catching of the front method.

## 2. State of Problem and Solving Methods

### 2.1. Governing equation and conditions

We consider melting of a cylinder with internal heat generation $\dot{q}$ in both phases (Fig. 1). The radius of the cylinder equals $r_0$ and the distance from the cylinder center to the melting front equals $s$. The outer surface of the cylinder has a temperature $T_{out}$ and the temperature at the interphase border equals melting temperature $T_m$. Top and bottom surfaces of the cylinder are assumed to be adiabatic, i.e. there is no heat flux along the cylinder. Heat conductivity $k$ and heat capacity $c_p$ are considered to be the same in both phases. For such conditions the melting process can be described by the equation of heat conductivity and the Stefan condition for interphase border respectively:

$$\frac{1}{\eta}\frac{\partial}{\partial \eta}\left(\eta \frac{\partial \Theta(\eta,\tau)}{\partial \eta}\right) + Q = \frac{\partial \Theta(\eta,\tau)}{\partial \tau}, \tag{1}$$

$$\left.\frac{\partial \Theta_{liq}(\eta,\tau)}{\partial \eta}\right|_{\eta=\zeta(\tau)} + \frac{1}{St}\frac{d\zeta(\tau)}{d\tau} = \left.\frac{\partial \Theta_{sol}(\eta,\tau)}{\partial \eta}\right|_{\eta=\zeta(\tau)}, \tag{2}$$

where $\eta = r/r_0$ – nondimensional radius of the cylinder, $\Theta(\eta,\tau) = (T(r,t)-T_0)/(T_m - T_0)$ – nondimensional temperature, $\tau = \alpha t / r_0^2$ $Q = \dot{q} r_0^2 / k(T_m - T_0)$ – nondimensional internal heat generation, Stefan number – $St = c_p(T_m - T_{out})/\Delta h_f$, $\zeta(\tau) = s(t)/r_0$ – nondimensional coordinate of the melting front.

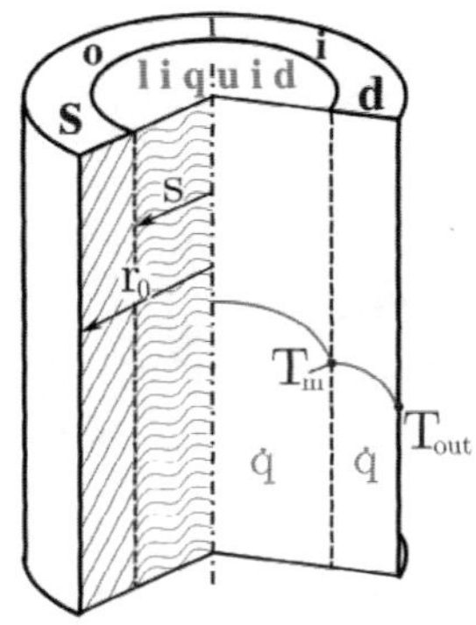


Fig. 1. Schematics of the melting problem with internal heat generation.

In non-dimensional variable the boundary conditions can be written as follows:

$$\frac{\partial \Theta(0,\tau)}{\partial \eta} = 0, \quad \Theta(\zeta(\tau),\tau) = 1, \quad \Theta(1,\tau) = 0. \tag{3}$$

For the initial temperature distribution, we set a constant zero value $\Theta(\eta,0) = \Theta_0 = 0$ excluding Fig.5, where the effect of the initial temperature is presented.

*2.2. The method of catching of the front into a space grid node.*

To solve numerically the heat conduction equation (1) we realized an implicit four-node scheme using the Thomas method [5] and iteration procedure. The Stefan equation (2) was modeled with an application of the method of catching of the front into a space grid node [6]. The sense of the method consists in that the distance $\Delta\zeta \approx d\zeta$ in equation (2), which melting front has gone for time $\Delta\tau$, is set strongly equal to the step of space grid $\Delta\eta$. Then one can express the time step $\Delta\tau \approx d\tau$ from equation (2) as follows:

$$\Delta\tau = \frac{\Delta\eta}{St\left(\left.\frac{\partial \Theta_{sol}(\eta,\tau)}{\partial \eta}\right|_{\eta=\zeta(\tau)} - \left.\frac{\partial \Theta_{liq}(\eta,\tau)}{\partial \eta}\right|_{\eta=\zeta(\tau)}\right)}. \tag{4}$$

This method does not allow direct setting of time step. Nevertheless, the space step can control the time step. It is worth to note that the time step decreases with the increase of Stefan number that means more intensive melting process.

Convergence tests were done for space grid. At 200 nodes, the melting front distribution did not depend on the number of nodes in space grid. However, we used the space grid with 13000 nodes in order to reach the position of the interphase border in the steady-state regime.

## 3. Results

Fig. 2 shows profiles of non-dimensional temperature on the cylinder radius for $St = 1$ and $Q = 5$. At the non-dimensional time $\tau = 0.05$ the temperature profile belongs to the stage of initial heating. At $\tau = 0.25$ temperature in the center of the cylinder is rather close to one. This value of non-dimensional temperature corresponds to the melting temperature. Further, the melting process starts.

Temperature profile at $\tau = 0.6$ shows a characteristic temperature distribution for the active melting process of the cylinder. At $\eta = 0.22$ we can see the kink in the temperature profile that corresponds to the interphase transition.

At time $\tau = 4.16$ melting of the cylinder is finished in fact and temperature distribution corresponds to the steady-state regime. At that, the kink in the temperature profile disappears. It has two reasons which are finish of melting process that gives a zero derivative on the front position in equation (2) and adopted assumption about properties' equality of liquid and solid phases.

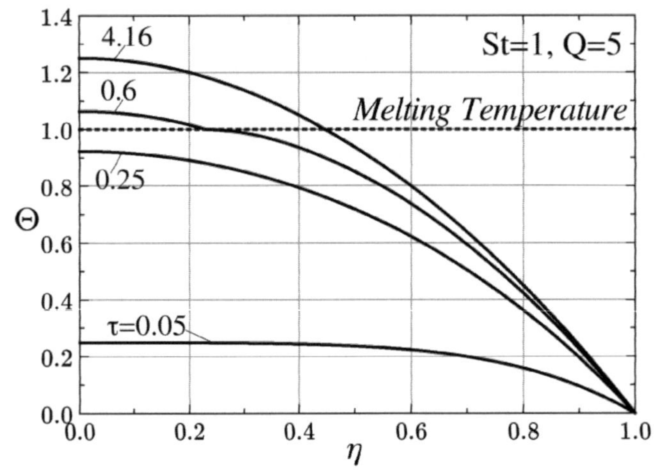


Fig. 2. Temperature profiles at melting process with internal heat generation.

Fig. 3a demonstrates an effect of Stefan number on movement of melting front and comparison of numerical modeling results and analytical solution of the Stefan problem in cylindrical coordinates [7]:

$$
\begin{aligned}
&-\sum_{n=1}^{\infty}\lambda_n A_n \exp\left(-\lambda_n^2\tau\right) J_1\left(\lambda_n\zeta(\tau)\right)+\frac{1}{St}\frac{d\zeta(\tau)}{d\tau}= \\
&=-\sum_{n=1}^{\infty}\tilde{\lambda}_n B_n \exp\left(-\tilde{\lambda}_n^2\tau\right)\left[Y_1\left(\tilde{\lambda}_n\eta\right)-\frac{Y_0\left(\tilde{\lambda}_n\right)}{J_0\left(\tilde{\lambda}_n\right)}J_1\left(\tilde{\lambda}_n\eta\right)\right]+\left(\frac{Q}{4}\left(\zeta^2(\tau)-1\right)+1\right)\frac{1}{\zeta(\tau)\ln\zeta(\tau)}.
\end{aligned}
\tag{5}
$$

As it is seen, the increase of Stefan number accelerates the melting process. One can allocate three-time parts which are initial heating, active melting, and steady-state regime. Analytical and numerical dependencies coincide well with each other at Stefan numbers 0.1 and 0.01. There is some discrepancy between numerical modeling and analytical solution at Stefan numbers 1.0 and 10. We explain this difference by machine error at the calculation of Bessel functions that formula (5) contains. For high Stefan numbers the second term of (5) has lower values than for small ones, consequently, the contribution of the first and third terms of (5) increases.

Fig. 3b shows dependencies of melting front position on the complex $\tau St$. Time parts of initial heating were deleted from the data. Applying this complex allows to compress the time scale and to represent all data within one order. Moreover, dependencies for $St = 0.01 - 1.0$ come down to the unity line practically. On the one hand, it proves that within the pointed range Stefan number and as a consequence material properties have the main influence on the melting process. On the other hand, we can conclude that at $St = 0.01-1.0$ steady-state regime starts from $\tau St \approx 2$. In this case, at known Stefan number one can calculate easy the total time of the active melting.

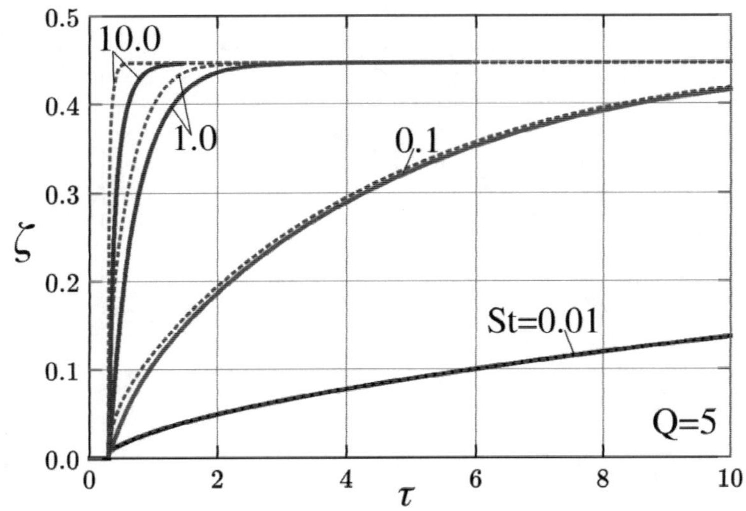


a)

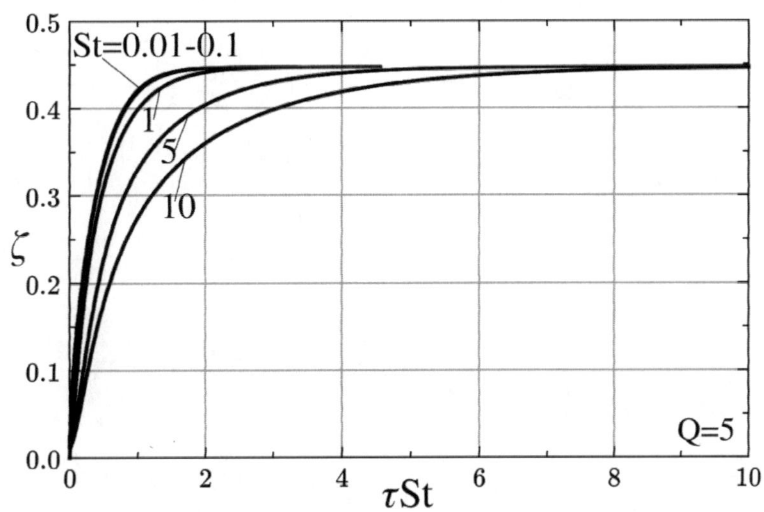


b)

Fig. 3. Effect of Stefan number on melting front movement. a) Comparison results of numerical simulation and analytical solution [3] b) Time scaling

Fig. 4 presents the effect of internal heat generation on the interphase border position in the steady-state regime. There is no melting process at $Q < 4$. In addition, it follows from the analytical solution [7]:

$$\zeta_{\max} = \sqrt{1 - \frac{4}{Q}}. \tag{6}$$

The absence of melting indicates that the total heat generated within the cylinder is less than heat flux through the outer surface of the cylinder.

The other limit of the presented dependence is the outer surface of the cylinder, i.e. that is the case of full melting of the cylinder. As it is seen, it can be realized at rather high values of the internal heat generation.

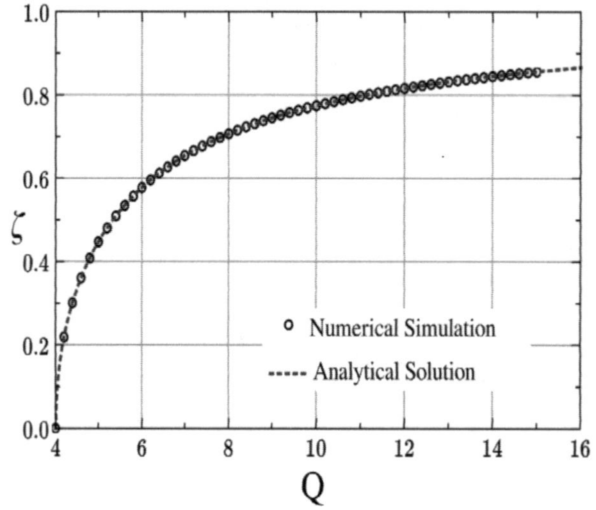


Fig. 4. Effect of internal heat generation on the interphase border at the steady-state regime.

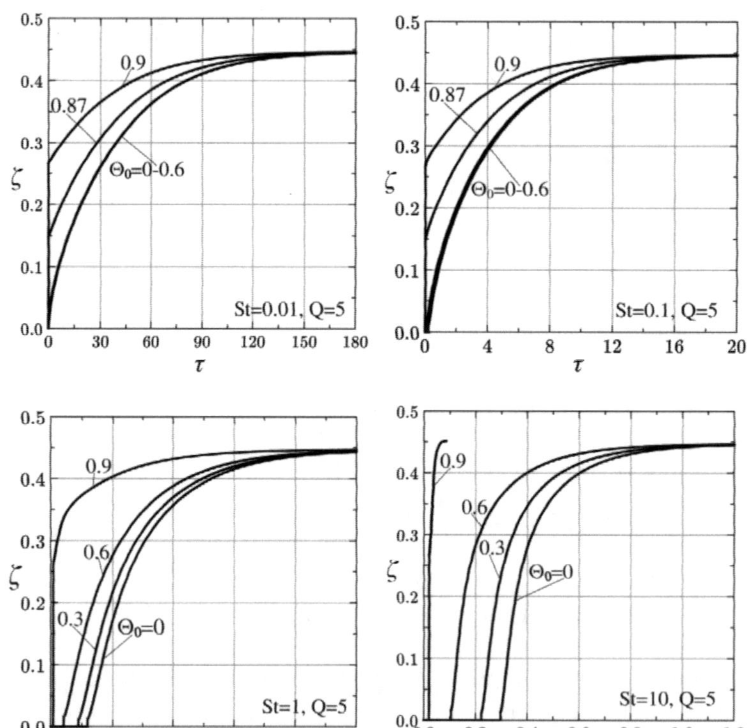


Fig. 5. Influence of initial temperature on the melting front movement at different Stefan numbers.

Fig. 5 shows dependencies of melting front position on non-dimensional time for various Stefan numbers and at values of initial temperature $\Theta_0 = 0 - 0.9$. For low Stefan numbers 0.01 and 0.1 the initial temperature has a weak influence on the melting process. Only beginning from $\Theta_0 = 0.87$ it is changed. At that, on the initial time, the area with non-dimensional radius $\eta = 0.15$ melts instantly. This area of instant melting increases with the increase of the initial temperature.

At Stefan numbers 1 and 10 initial temperature of the cylinder has a strong effect on the melting process. The effect is realized through the decreasing of the time of initial heating. There is an area of instant melting at high initial temperatures as well. At that, its radius does not depend on Stefan number.

## 4. Conclusions

The paper presents results on the numerical modeling of the melting process of a cylinder with internal heat generation. We have demonstrated the following results:

- The method of catching of the front into a space grid node [6] allows a successful numerical model of the rod (cylinder) melting with internal heat generation. Comparison with the results of the analytical solution of Stefan Problem [7] proves it.
- Application of $\tau St$ leads to time scaling of melting front movement at different Stefan numbers. At that, for St=0.01 – 1 steady-state regime starts from $\tau St = 2$.
- Melting process begins at Q > 4, when the total internal heat generation exceeds heat flux through the outer surface.
- The initial temperature of the cylinder has a stronger influence on the melting process at higher Stefan numbers.

## References

[1] Lamé, G., Clapeyron, B.P., Memoire sur la solidification par refroidissement d'un globe liquide, Annales Chimie Physique, 47 (1831), pp. 250-256.

[2] Stefan, J., Uber die Theorie der Eisbildung, insbesondere uber die Eisbildung im Polarmeere, Sitzungsberichte der k.k. Akademie der Wissenschaften in Wien, Mathematische-Naturwissenschaften, Abteilung II (1889), pp. 965-983.

[3] Rubenstein, L.I., The Stefan Problem, AMS Pubs., Providence 1971

[4] Alexiades, V., Solomon, A.D., Mathematical Modeling of Melting and Freezing Processes, Hemisphere 1993

[5] Poulikakos D, *Conduction Heat Transfer*, Prentice-Hall, Englewood Cliffs 1994

[6] Samarskii A A, Vabishchevich P N, *Computational Heat Transfer: Volume 1 Mathematical Modelling*. John Wiley & Sons Ltd. 1995.

[7] L Barannyk, J Crepeau, P Paulus, A Siahpush, Fourier-Bessel series model for the Stefan problem with internal heat generation in cylindrical coordinates. *Proceedings of the 26th International Conference on Nuclear Engineering ICONE26, London, England*, 2018.

# Numerical study of combustion of the diesel fuel sprayed with a steam jet in a prospective burner device

**D V Krasinsky[1], A I Tsepenok[2]**

[1] Kutateladze Institute of Thermophysics SB RAS, Novosibirsk, 630090, Russia
[2] OOO "ZiO-COTES", Novosibirsk, 630049, Russia

E-mail: dkr@itp.nsc.ru

**Abstract.** A prospective laboratory-scale burner device that employs a concept of liquid fuel spraying in the high-speed superheated steam jet is considered. The presented work is devoted to numerical study of aerothermochemical processes occurring in this novel-design burner during the combustion of diesel fuel sprayed with axial jet of water steam. The mathematical model is based on Euler-Lagrange approach for two-phase turbulent reacting flow description applied in 2D axisymmetric steady-state formulation. The vaporized diesel fuel combustion is described via EDC model with chemical kinetics mechanism for *n*-heptane (30 species, 66 reactions). The two-step soot model is also used, as well as NOx prediction at post-processing stage. The fields of velocity, temperature, heat fluxes, gas species and disperse phase concentrations, including soot and NOx, have been obtained from numerical simulation for a typical operational mode with diesel supply rate of 1.2 kg/hour and water steam flowrate of 0.8 kg/hour. Analysis of these flowfield data has allowed to reveal the two-layer coaxial structure of reacting flow (consisting of relatively cold near-axis jet layer and the coaxial layer formed by the hot recirculation zone) inside the burner.

## 1. Introduction

The development of small-scale burners utilizing the low-grade liquid hydrocarbon fuels stipulates the need for fundamental research aimed at investigation of aerothermochemical processes leading to drastic intensification of combustion while matching the environmental safety standards. It is recognized that ad hoc supply of water steam into reaction zone can improve the performance and ecological aspects of heavy-weight hydrocarbon fuel combustion. Such principle of the fuel combustion enhancement by superheated water steam jet has been implemented in a pilot ~10 kW burner developed earlier at the Institute of Thermophysics SB RAS [1]. The steam nozzle is mounted at the burner axis, the vertical cylindrical chamber (main channel of the burner) is located coaxially. There is a separate supply of fuel, air and superheated steam in the burner, with atmospheric air being sucked into the burner through a row of openings made around perimeter of the steel body. The burner reaches the autonomous self-sustained operational mode when the steam pressure inside its boiler part grows up to ~3 bar or higher. In this mode, owing to the supply of steam jet, the formation of bright and short flame is observed with drastic intensification and improved performance of combustion process in the burner (compared to an absence of water steam) characterized by significantly smaller soot and NOx emissions [2]. The processes of the vaporized hydrocarbon fuel combustion in this pilot burner have been recently studied numerically in [3] with mathematical description based on the conserved scalar approach with constrained-equilibrium chemistry model.

Journal of Physics: Conference Series **1369** (2019) 012026 doi:10.1088/1742-6596/1369/1/012026

With further elaboration of this burner's design, carried out recently at the Institute of Thermophysics SB RAS, a novel technique of the liquid fuel spraying by the use of high-speed steam jet has been suggested and studied in experiments [4–6]. In this technique the liquid fuel flows down from the fuel-supply pipe inclined at 30° to horizon, ending with a needle located in the vicinity over the steam nozzle orifice, thus appears a hanging fuel drop releasing the liquid that is sprayed by the steam jet stream. It has been noted in [4] that due to this way of liquid fuel feeding into the burner, as there's no direct contact between the fuel and the nozzle body, the surface of steam nozzle is not subject to carbonization or coke deposition. As a result, the gas-droplet reacting turbulent flow is formed inside the chamber of this modified burner with liquid fuel spraying technique. The body of this novel-design laboratory-scale burner is comprised of the vertical cylindrical chamber of 97 mm height and internal diameter ∅51 mm. A sketch of the burner geometry and its operational principle is given in [5, 6]. The manifold of heat-mass transfer processes takes place there, including the following: the high-temperature vaporization of diesel fuel droplets, the multistage non-equilibrium chemistry reactions in gas-phase multicomponent mixture including soot, the convective and radiative heat transfer. The presented work is devoted to numerical modelling of two-phase aerodynamics, heat transfer and combustion, with account for the aforementioned complex of processes, in the considered novel-design burner and its outer flame.

## 2. Mathematical model

The 2D axisymmetric steady-state formulation in cylindrical coordinates is employed for mathematical model that is based on Euler-Lagrange approach for two-phase flow description and on RANS approach with the use of "realizable $k$-ε" turbulence model [7]. Radiative heat transfer is modelled according to *P-1* approximation with WSGGM [8] and soot absorption taken into account. The two-step soot model of Moss-Brookes [9] has been used. The Eddy Dissipation Concept of Magnussen [10] (EDC-model) has been applied as turbulent combustion model. To represent a real diesel fuel, the *n*-heptane (chemical formula $nC_7H_{16}$) has been chosen as a commonly used model (or surrogate) fuel. The thermophysical properties of *n*-heptane as liquid fuel, including temperature-dependent expressions for the latent heat of evaporation and for the saturated vapour pressure, needed to model vaporization rate of a fuel droplet, are given in [11]. The size distribution of droplet particles at their initial injection points (located near the steam nozzle exit) was taken from measurements [4]. The chemical kinetics mechanism adopted in the presented work and describing the non-equilibrium reactive processes during combustion of *n*-heptane, is based on the one developed in [12] and contains 30 species (including inert nitrogen $N_2$) and 66 reactions. The numerical solution procedure is generally similar to the one described in [3], computations have been performed using CFD package ANSYS Fluent.

## 3. Numerical results

The burner operational mode with diesel fuel feeding rate of 1.2 kg/hour and superheated water steam flowrate of 0.8 kg/hour, that was used as a typical regime in experimental works [5, 6], has been considered in the numerical study. The performed numerical modelling of the whole range of aerothermochemical processes in the studied burner has allowed to obtain a detailed flowfield information including the fields of velocity, temperature, radiated heat fluxes, species and disperse phase concentrations, including soot and NOx, in the burner and its outer flame (torch). These results are shown in figures 1–6 (it should be noted that the field values are represented there with mirror reflection at burner symmetry axis for clearer view) and are also analysed as profile distributions along the burner axis in figures 7–9. The field of velocity magnitude (see figure 1) shows the burner aerodynamic structure formed by the high-speed jet of water steam released from the axially-mounted steam nozzle (orifice ∅0.6 mm). Due to the conditions (temperature $T_t$=250°C and pressure $P_t$=7.8·10$^5$ Pa) of superheated steam supplied at the input tube (internal diameter ∅3 mm) that feeds the nozzle, the steam jet discharged from the nozzle orifice becomes supersonic with Mach number $M_n$ ~1.2 (at nozzle exit plane). Then velocity of the supersonic expanding jet stream grows further, so within the jet initial region (up to the distance of about 12 nozzle diameters, i.e. ~ 7.2 mm from the steam nozzle

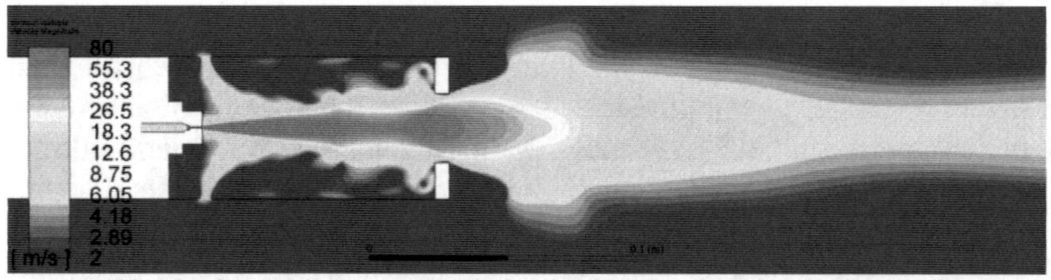


**Figure 1.** Field of velocity magnitude in the burner and torch (logarithmic scale), m/s.

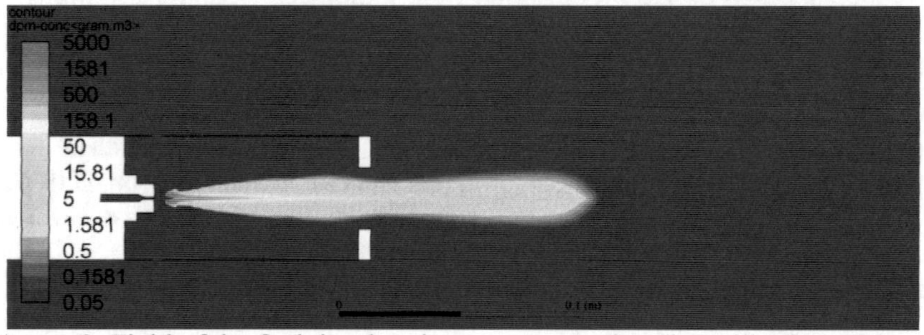


**Figure 2.** Field of the fuel droplet phase concentration (logarithmic scale), g/m$^3$.

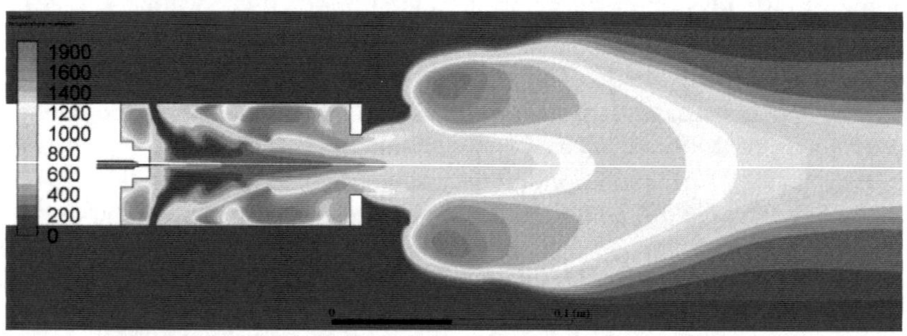


**Figure 3.** Field of temperature in the burner and torch, °C.

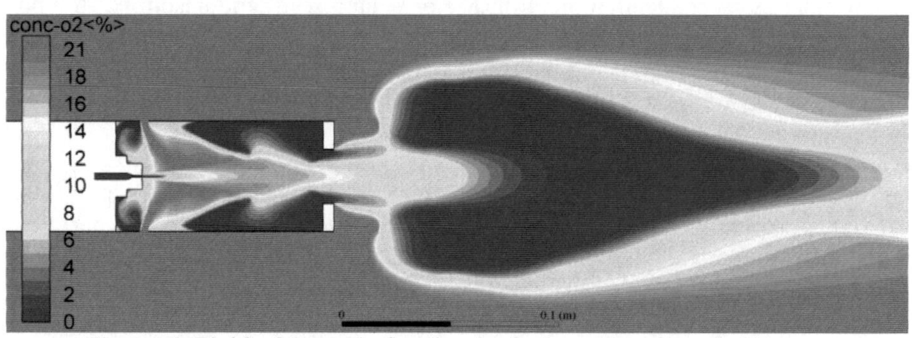


**Figure 4.** Field of $O_2$ mole fraction in the burner and torch, %.

exit) M≈1.47 and the corresponding axial velocity is $U_x$≈721 m/s, as found from the obtained velocity and Mach number profiles along the axis. Further downstreams the jet velocity decays (with transformation to subsonic flow regime) and the jet spreading inside the burner chamber causes the entrainment of external air (see figure 1) that is sucked into the chamber through the air supply openings located along the perimeter of chamber wall. The presence of upper lid (with internal orifice ∅25 mm and width of 5 mm – seen as white rectangle in figure 1), covering the chamber top, results in partial inclination and attachment of the flow towards the lid wall and the chamber wall, thus leading to recirculation of the flow formed between a shear layer of the central jet and the chamber and lid walls. This recirculation zone is also "bounded" by air entrainment from the lower part of the chamber. Therefore the mixing process takes place between the fuel and entrained oxidizer within this recirculation zone.

The distribution of fuel droplets concentration is shown in figure 2 where the locations of initial droplets "seeding" (applied in numerical simulation as 5 injection rakes in a framework of Lagrangian tracking for disperse phase) can be noticed as a region of maximum concentration of disperse phase that "hangs" over the steam nozzle exit. The initial size distribution for the injected fuel droplet particles was specified in the range of 12–35 μm, following the data from experimental study [4]. It is seen from figure 2 that the largest concentration of droplets is reached at the axis line – a more detailed analysis indicates that air entrainment effect can be the cause for such behaviour. Because of evaporation process, the region where the droplets are seemingly present is bounded (note a logarithmic scale of legend in figure 2). Part of the model diesel fuel (*n*-heptane) evaporated inside the burner chamber enters the recirculation zone (mentioned above) as gaseous species $nC_7H_{16}$ and undergoes the reaction processes there. Another part of *n*-heptane in gaseous or disperse phase is moved out of the burner along an axial jet, thus fuelling the torch (i.e. outer flame region).

The temperature field shown in figure 3 demonstrates a complicated thermal structure inside the burner and in the outer torch region. High temperature level is seen inside the chamber recirculation zone where several spots of reaction heat release can be noticed. However the stream of entrained air separates the central jet from hot flame fronts (occurring in the recirculation zone), therefore the near-axis steam jet remains relatively cold, with temperature less than 485°C along axis inside the chamber. As observed from figure 3, the flow exiting the burner chamber through the lid orifice (∅25 mm) towards the torch is comprised of two layers: the near-axis "cold" jet layer, and coaxial layer of hotter stream leaving from recirculation zone. These two layers provide different composition of reactive mixture that fuels the outer torch: the near-axis "cold" jet is carrying mostly $nC_7H_{16}$, $H_2O$ and $O_2$ (see curves *1, 2, 7* in figure 7), while the coaxial layer is carrying the species of gasification process $H_2$ and CO (as seen from figures 5–6) mixed together with complete combustion products $H_2O$ and $CO_2$. The two-layer structure of the flow and chemical composition exiting the burner forms a peculiar feature of outer torch: from temperature field it can be seen that the flame front in the torch is horseshoe-shaped.

The aforementioned aerothermochemical structure in the burner and its torch can also be observed from figures 4–6 representing the fields of $O_2$, $H_2$ and CO mole fractions. It is seen that $O_2$ is consumed inside the chamber recirculation zone and within the horseshoe-shaped bounds of the outer torch. In these low-oxygen zones the fuel gasification prevails, therefore the concentrations of $H_2$ and CO reach ~10% (by volume) there. Profiles of the species mole fractions along the burner axis plotted in figure 7 provide quantitative details for analysis of reactive mixture composition in the torch. It should be noted that in figures 7–9 the value $x=0$ of axial coordinate denotes the burner exit plane (i.e. $x<0$ relates to the region inside the burner chamber and lid, while $x>0$ relates to the outer region). It can be seen from figure 7 that mass fraction of gaseous fuel *n*-heptane (curve *2*) is gradually decreasing along axis and diminishes to almost zero at $x\approx98$ mm. As opposite, the concentrations of intermediate reactants – $CH_4$, $H_2$ and CO (curves *3, 4, 5*) – are increasing and reaching their maxima at similar location $x\approx100$–118 mm. Further downstreams these species concentrations start decreasing, while $CO_2$ mole fraction (curve *6*) grows till $x\approx201$ mm; after that the torch flame zone cools down because of entrainment of outer air – see the rise of $O_2$ mole fraction (curve *1*) at $x>200$ mm. It is worth to note that $H_2O$ mole fraction (curve *7*) remains significant (up to 18%) for the whole range of combustion zone plotted in figure 7, thus enhancing conditions for gasification process.

Profiles of the soot and NO emissions (in mole fraction ppm) along the burner axis are depicted in figure 8. A nonmonotonic behaviour seen in the soot profile (curve *1*) can be attributed to the nonlinear processes of formation (primarily due to the fuel decomposition), growth, coagulation and oxidation of soot. The nitrogen oxides (NOx) in the burner and its torch are mainly formed via the extended "thermal" $N_2$ oxidation mechanism of Ya.B.Zeldovich, as it has been estimated in computations performed at a post-processing stage of main numerical solution. The computed integral value of NO emission in exhaust gases from the burner equals to 44.7 ppm – this value corresponds very well (with 2% accuracy) to NO emission measured in [6] for the studied operational mode of the burner.

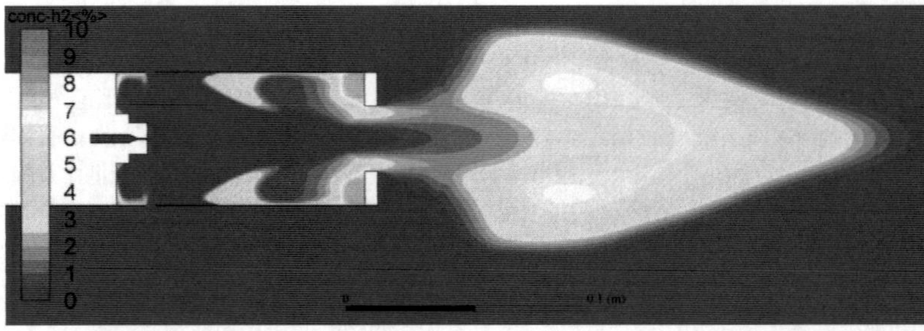


**Figure 5.** Field of $H_2$ mole fraction in the burner and torch, %.

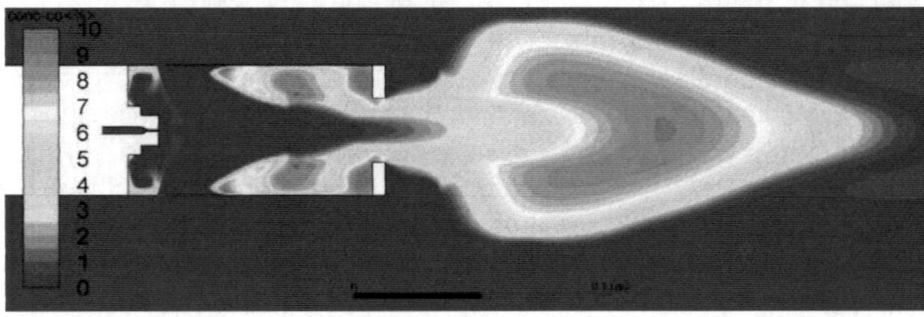


**Figure 6.** Field of CO mole fraction in the burner and torch, %.

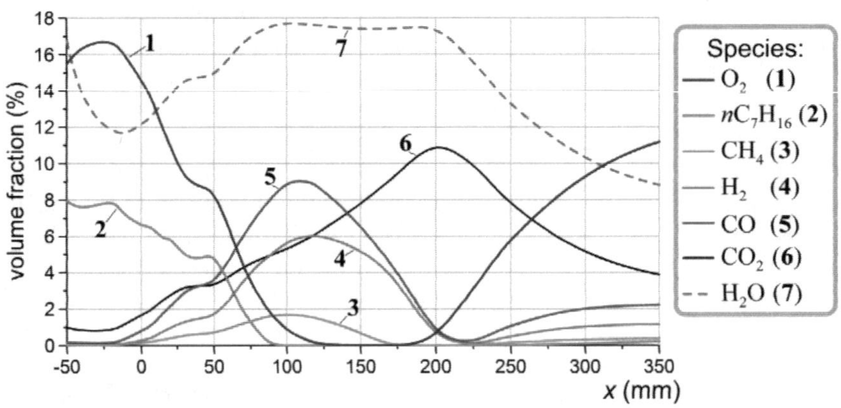


**Figure 7.** Profiles of the species fractions along the burner axis, %
(curve *2* represents the mass fraction of *n*-heptane, other curves represent the mole fractions).

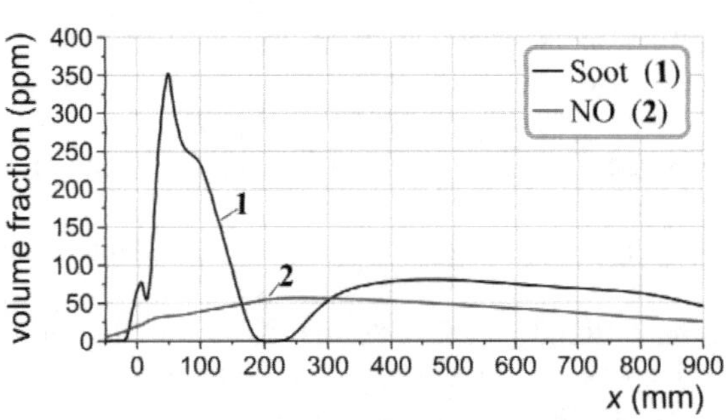


**Figure 8.** Profiles of the soot and NO mole fractions along the burner axis, ppm.

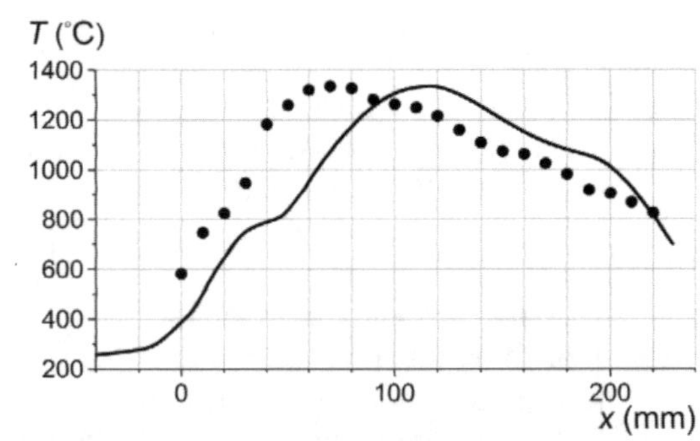


**Figure 9.** Profile of temperature along the burner axis, °C. Solid line: numerical modelling, dots: thermocouple measurements [6].

In addition to analysis of the temperature field structure (figure 3), the temperature profile plotted along the burner axis is compared in figure 9 versus the in-flame thermocouple measurements data reported in [6]. One can see that, notwithstanding some shift in the positions of temperature maxima between two profiles, the values of these maxima are almost equal (~1334°C), and the shapes of both experimental and numerical profiles are similar. Therefore the performed comparison on the level of NO emission and on the temperature profile data indicates to adequacy of the used modelling description and the obtained numerical prediction of thermochemical conditions inside the burner torch.

## 4. Conclusions

The results of numerical investigation of the whole complex of aerothermochemical processes taking place in the novel-design laboratory-scale burner during the combustion of diesel fuel sprayed with a high-speed jet of water steam have been presented. The mathematical model, applied in 2D axisymmetric steady-state formulation, is based on Euler-Lagrange approach for two-phase flow description and allows to take into account the turbulent aerodynamics, the convective and radiative heat transfer processes, the vaporization of liquid fuel droplets, and the combustion of multicomponent gas mixture (with EDC turbulent combustion model using the chemical kinetics mechanism for 30 species and 66 reactions, and also the two-step soot model).

The numerical simulation, performed for the burner operational mode with diesel supply rate of 1.2 kg/hour and water steam flowrate of 0.8 kg/hour, has allowed to obtain and analyse the detailed flowfield information in the burner and its torch, including the fields of velocity, temperature, radiated heat fluxes, species and disperse phase concentrations, including soot and NOx. It has been concluded that the flow inside the burner chamber (and further exiting the chamber towards the torch) is comprised of two layers: the near-axis relatively cold jet layer, and the coaxial layer of hotter stream leaving from the chamber recirculation zone. These two layers also differ in their mixture composition, with near-axis jet layer formed mainly by three components (the injected water steam, the sprayed *n*-heptane (as model diesel fuel) and the entrained air), while the coaxial layer is carrying the syngas species $H_2$ and CO and the combustion products $H_2O$ and $CO_2$. The comparison of the obtained numerical results versus the measurements data [6], performed on NO emission (~45 ppm) and on the temperature profile data, has shown an agreement between experimental and numerical results, thus indicating to adequacy of the used modelling description.

The presented numerical investigation forms the scientific basis for further development of the studied burner device employing a concept of liquid fuel spraying in the high-speed water steam jet.

## Acknowledgements

The studies were carried out with the financial support of the RF Ministry of Education and Science, RFMEFI60417X0185.

## References

[1] Alekseenko S V, Pashchenko S E, Salomatov V V 2010 *J. Eng. Phys. and Thermophys.* **83**(4) 729-741
[2] Alekseenko S V, Anufriev I S, Vigriyanov M S, Kopyev E P and Sharypov O V 2016 *Combust. Expl.& Shock Waves* **52**(3) 286-293
[3] Krasinsky D V 2018 *J. Phys.: Conf. Ser.* **1105**(1) 012035
[4] Anufriev I, Kopyev E and Shadrin E 2017 *MATEC Web of Conf.* **115** 05010
[5] Anufriev I S et al 2018 *J. Phys.: Conf. Ser.* **1105** 012036
[6] Anufriev I S, Kopyev E P 2019 *Fuel Processing Technol.* **192** 154-169
[7] Shih T-H, Liou W W, Shabbir A, Yang Z, Zhu J 1995 *Computers&Fluids* **24**(3) 227-238
[8] Smith T F, Shen Z F, Friedman J N 1981 *Proc. XX-th Nat. ASME-AIChE Heat Transfer Conf.*
[9] Brookes S J and Moss J B 1999 *Combustion and Flame* **116** 486-503
[10] Gran I R and Magnussen B F 1996 *Combust. Science and Technol.* **119** 191-217
[11] Sazhin S S 2006 *Progr. Energy and Combust. Science* **32**(2) 162-214
[12] Patel A, King S-C and Reitz R D 2004 *SAE Tech. Paper Ser.* 2004-01-0558

# Numerical analysis of the water steam jet gasdynamic parameters influence on the hydrocarbon fuel combustion in a pilot burner

**D V Krasinsky[1], O V Sharypov[1]**

[1] Kutateladze Institute of Thermophysics SB RAS, Novosibirsk, 630090, Russia

E-mail: dkr@itp.nsc.ru

**Abstract.** On the basis of the conserved scalar approach with constrained-equilibrium chemistry model, variant numerical simulation of aerodynamics, heat transfer and the vaporized diesel fuel combustion processes, including soot formation and its effect on radiative heat transfer, has been performed in the pilot burner designed according to a concept of a hydrocarbon fuel combustion enhancement by axial injection of superheated water steam high-speed jet. The aerothermochemical structure of multicomponent reacting flow inside the burner has been analysed for the three cases of numerical study, including two regimes of the water steam axial injection with different jet velocity, and one regime of air injection through the same axial nozzle. The dependencies of the burner flame characteristics and its ecological performance on the given gasdynamic parameters of the discharged steam or air jet have been obtained. Also the numerical study has confirmed the low-NOx features of the burner operational regimes with water steam injection.

## 1. Introduction

It is known that ad hoc injection of water steam towards a reaction zone can improve the performance and ecological aspects of combustion of heavy-weight hydrocarbon fuels. This principle of the fuel combustion enhancement by superheated water steam jet has been implemented in a laboratory-scale 10 kW burner device developed at the Institute of Thermophysics SB RAS [1, 2]. In this novel-design device the steam nozzle is mounted at the burner axis, and the vertical cylindrical chamber (main channel of the burner) is located coaxially. The experimental research reported in [1–4] have demonstrated the appearance of intensified low-soot bright flame discharged from the studied burner. This burner device, fed with diesel fuel and, separately, with water steam generated in a boiler unit, reaches autonomous steady-state regime when the steam pressure inside the boiler grows up to ~3 bar or higher. Then, owing to axial injection of high-speed steam jet, the performance of combustion process in the burner changes distinctly (compared to an absence of water steam), with substantially smaller soot and NOx emissions in the burner torch [1, 2]. The detailed description of the burner design and its operational principle has been given in [3, 4].

With aim to investigate the conditions and mechanisms leading to improved performance of this novel burner, the processes of the vaporized hydrocarbon fuel combustion in this burner have been recently studied numerically in [5] with mathematical description based on the conserved scalar approach with constrained-equilibrium chemistry model. The presented paper is devoted to further development of numerical study [5] with an emphasis on comparative analysis of the influence of the water steam high-speed jet gasdynamic parameters on the diesel fuel combustion in the novel burner.

## 2. Mathematical model and inlet conditions

The reacting flow of multicomponent gas-phase mixture is assumed to be steady-state. Within RANS framework, the Favre-averaged governing equations are closed with "realizable" *k*-ε turbulence model [6]. To describe the mixture thermochemical state with account for many (>10) species, the conserved scalar approach (Shvab-Zeldovich approximation) has been used in the work. In this approach the thermochemical solution vector of dependent quantities at each point of the flow is determined from values of independent variables: mixture fraction $\xi_1$ and its variance for which the corresponding transport equations are solved. Enthalpy $H$ equation is also solved to account for heat loss. The chemical constrained-equilibrium model [7] has been used. To add the water steam jet as a separate stream, the secondary conserved scalar (denoted as $\xi_2$ herein) submodel [8] has been applied, therefore the space of independent variables has been specified as $\left(\overline{\overline{\xi_1}}, \overline{\overline{\xi_2}}, H\right)$ in computations. The lookup table of dependent quantities constructed for 32 species has dimensions 40×25×128 in this space [5].

The radiation energy transport equation is based on the known *P-1* approximation of the spherical harmonics method, with this the no-soot gas mixture absorption coefficient $\alpha_g$ is calculated according to WSGGM model [9]. The effect of soot on radiative heat transfer is very important because of soot particles' influence on the total absorption coefficient defined as $\alpha_\Sigma = \alpha_g + \alpha_s$ where for soot absorption $\alpha_s$ the approximation [10] has been used. The two-step non-equilibrium model [11] has been applied for the soot processes modelling.

Due to axial symmetry of the burner, the problem is formulated in 2D cylindrical coordinates $(x, r)$. The burner computational domain is schematically shown in figure 1, where numbers *1–3* denote the inlet boundaries of the studied burner configuration. At the external inlet boundary *4* (surrounding the burner body at $r > 46$ mm), the air coflow velocity of 0.8 m/s has been specified to stabilize computations. More detailed description of the employed mathematical model and numerical algorithm is given in [5]. Computations have been performed with the use of CFD package FLUENT.

The variant numerical simulation, carried out in the work, includes two regimes of the water steam injection through the axial nozzle, denoted herein as cases 'S-jet-1' and 'S-jet-2', that differ by the steam jet velocity provided the same flowrate of steam. The boundary conditions at inlet *1* for these cases are given in table 1. Another regime, when air has been supplied through the same nozzle instead of steam, is denoted as 'A-jet' case, for which the air jet velocity and the flowrate are the same as in case 'S-jet-1', but the inlet air temperature was set to $T_{in}$=640 K to compensate the molar weight inequality of air and water.

The diesel fuel at inlet boundary *2* has been represented by a mixture of the hydrocarbon species $\{CH_4, C_2H_2, C_2H_4, C_6H_5, C_6H_6, C_8H_2\}$ as a model fuel with molar weight of 58.6 kg/kmol, assumed to be vaporized (at inlet temperature $T_{in}$=600 K), with stoichiometric mixture fraction value for this model fuel equal to $\xi_{1,st} \approx 0.092$. Other inlet boundary conditions are also specified in table 1.

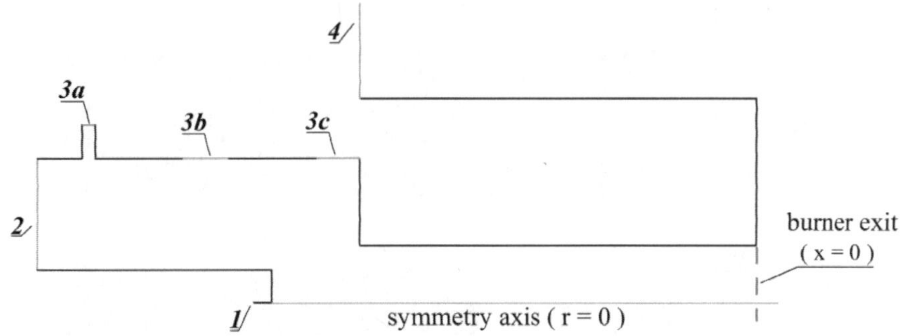


**Figure 1.** Sketch of computational domain (zoom-in view of the burner body).

The flowrate of the diesel fuel feeding into burner, specified in the numerical study, was 0.81 kg/hour, and the superheated water steam flowrate has been set to 0.25 kg/hour. The mass-based ratio $\gamma = G_3 / G_2$ of inlet air and fuel flowrates inside the burner volume, specified in computations, is $\gamma \approx 7.93$, with this the corresponding equivalence ratio $\varphi \approx 1.24$, thus inside the burner channels the fuel-rich conditions are formed and the combustion process completes only in the outer flame (torch) owing to further mixing with external (atmospheric) air.

**Table 1.** Inlet conditions for the steam jet burner regimes.

| B.C. nr. $i$ | Inlet type | $\xi_{1,in}$ | $\xi_{2,in}$ | Mass flowrate $G_i$ (g/s) | Case of steam jet regime | Inlet velocity $U_{in}$ (m/s) | Inlet temperature $T_{in}$ (K) |
|---|---|---|---|---|---|---|---|
| *1* | *$H_2O$-steam* | 0 | 1 | 0.0688 | **S-jet-1** | 637.9 | 400 |
| | | | | | **S-jet-2** | 1188.2 | 745 |
| *2* | *Fuel* | 1 | 0 | 0.2245 | | 0.06 | 600 |
| *3a* | | | | 0.353 | | 0.4 | |
| *3b* | *Air-to-burner* | 0 | 0 | 0.290 | – | 0.12 | 300 |
| *3c* | | | | 1.138 | | 0.48 | |
| *4* | *Air-external* | | | 730.1 | | 0.8 | |

## 3. Numerical results

The results of variant numerical modelling (including the cases 'S-jet-1', 'S-jet-2' and 'A-jet') of the flow and the vaporized diesel fuel combustion processes in the studied burner are demonstrated in figures 2–9. The field of mixture fraction $\xi_1$ distribution inside the burner channels is shown in figure 2 (for the case 'S-jet-2'). The $\xi_1$ scalar essentially determines the thermochemical state of the mixture (excluding soot). The field of secondary conserved scalar $\xi_2$ denoting the evolution of the water steam jet is shown in figure 3 (in logarithmic scale, case 'S-jet-2'). It can be seen from figure 3 that the steam axial jet, due to its momentum effect, is streaming through the main channel, thus forming aerodynamic structure of reacting flow inside the burner. As it has been concluded in [5], a reacting flow inside the main channel of this burner is characterized by a three-layer coaxial structure, with its near-axis layer featuring the low-oxygen zone with conditions facilitating the process of the fuel gasification there, and its middle layer representing the conical flame front. Because the near-axis layer is directly formed by the high-speed axial jet of scalar $\xi_2$ (water steam), it is evident that the given gasdynamic parameters of the steam jet can exert a substantial influence on an overall aerothermochemical structure in the burner. Quantitative estimation of this influence is analysed in

**Figure 2.** Field of mixture fraction $\xi_1$ in the burner (case 'S-jet-2').

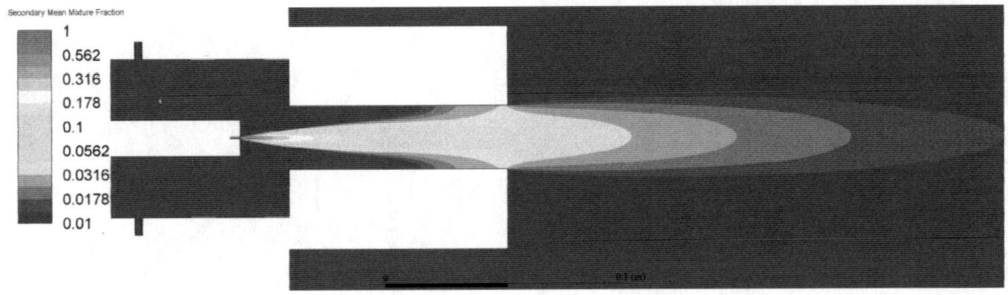


**Figure 3.** Field of secondary mixture fraction $\xi_2$ (logarithmic scale) in the burner, case 'S-jet-2'.

figures 4–9 representing the profiles of variables along the burner axis for the three compared cases. Here it has to be noted that abscissa value $X = 0$ denotes the burner exit plane in these figures, and $X<0$ relates to inner part of the burner, while $X>0$ relates to external region of burner torch (outer flame).

As indicated in table 1, the jet inlet velocity (prescribed at the steam nozzle orifice) is ~1.86 times larger in case 'S-jet-2' than in two other cases (provided the same flowrate) – this difference is clearly seen from the profiles of axial velocity component in figure 4. Then in case 'S-jet-2' the near-axis values of corresponding mixture fraction $\xi_1$ (see figure 5) and soot concentration $C_{soot}$ (see figure 6) appear to be lower inside the burner main channel because of more pronounced air entrainment (due to higher jet velocity) from the bottom chamber into the main channel. Decreasing the values of mixture fraction $\xi_1$ there effectively implies approaching to the stoichiometric value $\xi_{1,st}$, i.e. to the reaction front – this leads to increasing the flow temperature inside the near-axis layer of main channel in case 'S-jet-2' (compared to case 'S-jet-1'), as seen from figure 8. Another factor of importance is the soot influence on radiative heat transfer and, consequently, on the temperature field, because the presence of soot provides the largest contribution to the total absorption coefficient $\alpha_\Sigma$. One can see from figure 7 that the profiles of $\alpha_\Sigma$ are similar to the soot concentration profiles plotted in figure 6. Therefore, when soot concentration is lower (in cases 'S-jet-2' and 'A-jet'), decreasing the values of $\alpha_\Sigma$ results in temperature rising due to weakening the heat removal (by radiative heat transfer) from the reaction front. It should be noted that the temperature maximum in case 'S-jet-2' takes place inside the burner main channel (at $X<0$, see figure 8) – this is a qualitative distinction to the case 'S-jet-1' with lower inlet velocity of steam jet.

In the case 'A-jet', when air is supplied through the axially-mounted nozzle instead of steam, the temperature level along the burner axis becomes notably higher (see figure 8), because in presence of $O_2$ from air jet the vaporized fuel is directly oxidized there (instead of the fuel gasification process at low-$O_2$ conditions in case of steam jet), and also the effect of weakening the radiative heat transfer (see figure 7) plays its role in increasing the temperature in case 'A-jet'. Therefore, as the nitrogen oxide $NO$ formation in the burner and its outer flame torch primarily follows the "thermal" mechanism of Ya.B.Zeldovich, the level of $NO$ emission in case 'A-jet' appears to be an order of magnitude higher (see figure 9), than in the steam jet cases, because of the effects of increasing the temperature and presence of $O_2$ mentioned above. This rise of $NO$ emission in the burner is a principal drawback of the regime 'A-jet' with axial jet of air.

In the performed variant numerical simulations the following ecological characteristics of the burner exhaust have been obtained: the integral value of NOx concentration is equal to 12 mg/m$^3$ in case 'S-jet-1', 20 mg/m$^3$ in case 'S-jet-2', and 150 mg/m$^3$ in case 'A-jet'. Thus, the comparison of the burner ecological performance for two regimes ('S-jet-1' vs 'S-jet-2') with axial injection of superheated water steam allows to conclude that the case 'S-jet-1' with smaller inlet velocity of axial jet appears to be more advantageous. However, compared to the regime with air supply through the axial nozzle, both regimes with water steam axial injection demonstrate a drastic reduction in NOx emission by 7 – 12 times in the studied burner.

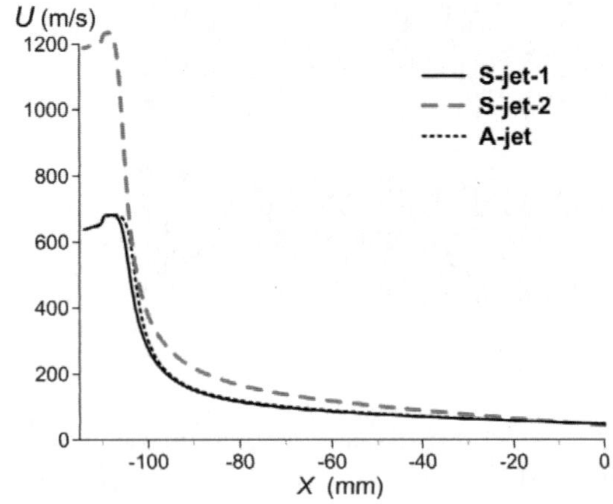


**Figure 4.** Profiles of axial velocity component $U$ along the burner axis, m/s.

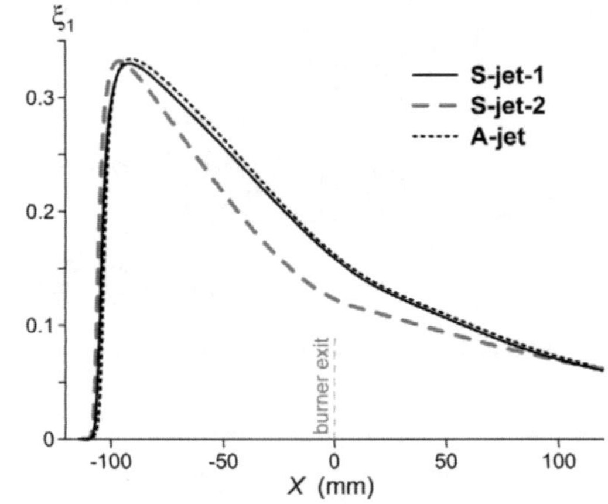


**Figure 5.** Profiles of mixture fraction $\xi_1$ along the burner axis.

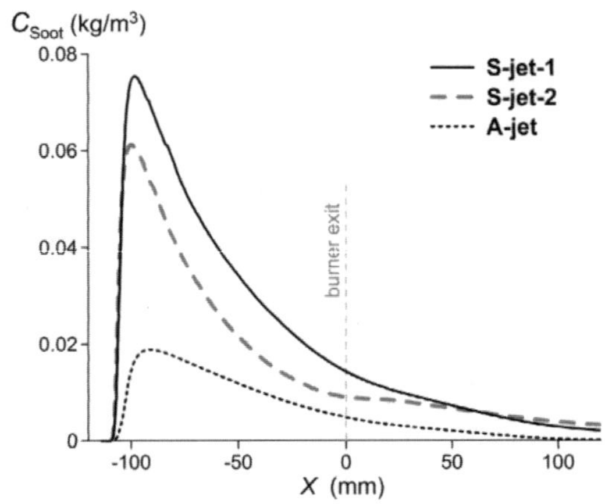


**Figure 6.** Profiles of soot concentration $C_{soot}$ along the burner axis, kg/m$^3$.

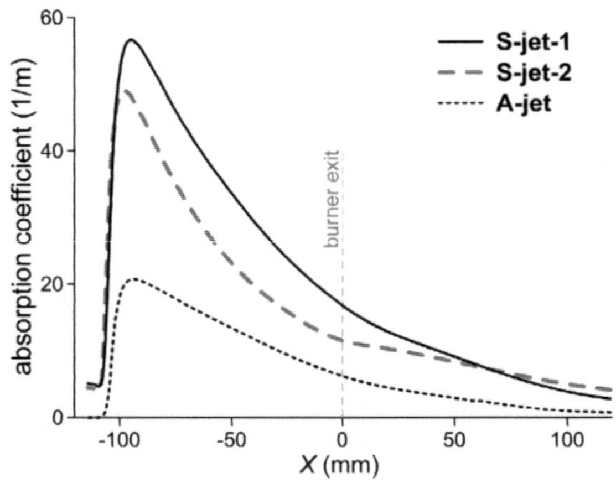


**Figure 7.** Profiles of absorption coefficient $\alpha_\Sigma$ of gas-soot mixture (along the burner axis), 1/m.

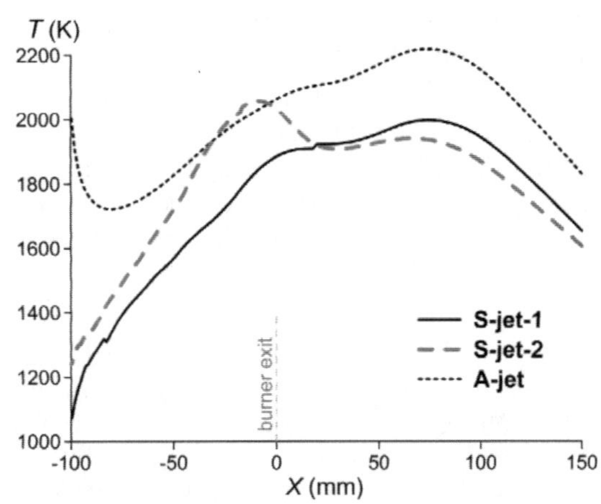


**Figure 8.** Profiles of temperature $T$ along the burner axis, K.

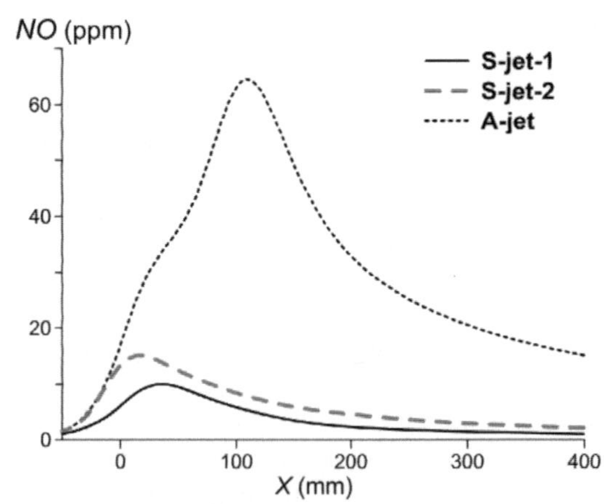


**Figure 9.** Profiles of NO mole fraction along the burner axis, ppm.

## 4. Conclusions

A pilot laboratory-scale burner of prospective design with axial injection of superheated steam jet, utilizing the concept of steam-enhanced combustion of hydrocarbon fuels, has been considered. As a continuation of previous numerical study [5] using mathematical description based on the conserved scalar approach with constrained-equilibrium chemistry model, a variant numerical modelling of aerodynamics, heat transfer and the vaporized diesel fuel combustion processes (including soot formation and its effect on radiative heat transfer) in the considered burner has been performed in 2D axisymmetric formulation. Detailed information on aerothermochemical structure of multicomponent reacting flow inside the burner channels and in its outer flame has been obtained for the three variants of numerical simulation, including two regimes of the water steam high-speed axial injection with different jet velocity, and one regime of air supply through the same axial nozzle. On the basis of comparative analysis of these numerical modelling results it has been shown that the high-speed axial jet is forming a certain aerodynamic and thermochemical structure of reacting flow inside the burner. The dependencies of the burner flame characteristics and its ecological performance on the given gasdynamic parameters of the supplied steam or air jet have been analysed. It has been concluded that, compared to the regime with air supply, both regimes with water steam injection demonstrate a drastic reduction in NOx emission – thus confirming the low-NOx feature of the studied burner.

**Acknowledgements**
This work was carried out under state contract with IT SB RAS (AAAA-A17-117022850029-9).

## 5. References

[1] Alekseenko S V, Pashchenko S E and Salomatov V V 2010 *J. of Engineering Physics and Thermophysics* **83**(4) 729-41
[2] Patent RU2219435 (F23C11/00, F23L7/00) *Method for sootless combustion of fuel* 2003 Vigriyanov M S, Salomatov V V, Alekseenko S V Institute of Thermophysics SB RAS
[3] Alekseenko S V, Anufriev I S, Vigriyanov M S, Dulin V M, Kopyev E P and Sharypov O V 2014 *Thermophys.&Aeromech.* **21**(3) 393-6
[4] Anufriev I S, Arsentyev S S, Agafontsev M V, Kopyev E P, Loboda E L, Shadrin E Yu and Sharypov O V 2017 *J. Phys.: Conf. Ser.* **925** 012014
[5] Krasinsky D V 2018 *J. Phys.: Conf. Ser.* **1105**(1) 012035
[6] Shih T-H, Liou W W, Shabbir A, Yang Z, Zhu J 1995 *Computers&Fluids* **24**(3) 227-38
[7] Bilger R W and Starner S H 1983 *Combustion and Flame* **51**, 155-176
[8] ANSYS Fluent User's Guide. 2018 *ANSYS Inc.*
[9] Smith T F, Shen Z F, Friedman J N 1981 *Proc. XX-th Nat. ASME-AIChE Heat Transfer Conf.*
[10] Sazhin S S 1994 An approximation for the absorption coefficient of soot in a radiating gas *Manuscript* (Fluent Europe, Ltd.)
[11] Brookes S J and Moss J B 1999 *Combustion and Flame* **116**, 486-503

5th International Workshop on Heat/Mass Transfer Advances
for Energy Conservation and Pollution Control
August 13-16, 2019, Novosibirsk, Russia

# Experimental study of local void fraction distribution in bubble flow in rod assembly

**M A Vorobyev, O N Kashinsky**

Kutateladze Institute of Thermophysics SB RAS, 630090, Novosibirsk, Russia

E-mail: vorobyev@itp.nsc.ru

In this work an experimental study of local void fraction distribution in square arrangement rod bundle channel 3 * 3 under bubble flow regime was carried out. Measurements were performed with conductivity probe help, which was introduced into the flow through an angular or side rod at an angle of 15° to the flow. This method made it possible to determine the local void fraction close to the walls of the rods and in the narrow gap between them. Measurements have shown that a dome-shaped profile of local void fraction is realized in the gap between the side rode and the central one, while a saddle-shaped distribution of the void fraction is observed on the line connecting the central and angular rods. Also the effect of the spacer grid on the distribution of the gas phase in the channel was shown.

## 1. Introduction

A number of studies of domestic and foreign authors are devoted to experimental investigations of the dispersed gas-liquid flow structure [1–5], but these experiments were carried out in channels of simple geometry. The development of theoretical and numerical models of gas-liquid flows is hampered by the lack of sufficient experimental information obtained for various flow conditions. In the last 5-10 years, there has been a noticeable tendency of transition of leading groups, which for many years conducted researches on two-phase flows in pipes, to studies in channels used in modern technologies. So, for example, annular channels and assemblies of vertical rods — simulators of fuel elements are used to study flows in relation to the modeling of fluid flow in fuel assemblies of modern reactor installations. For such channels, a detailed experimental study of hydrodynamics and heat transfer has been carried out for the flow of single-phase fluid in the assembly of vertical rods [6–9]. Criterial relationships have been obtained for calculating the hydraulic resistance coefficient, heat transfer coefficient, and critical heat flux. At the same time, the two-phase flow in such channels remains insufficiently explored.

Numerical codes designed to predict emergencies in nuclear reactors are usually based on two-fluid models. Reliable experimental data of gas phase distribution in the fuel assembly channel are necessary both for the compilation of the closure relations and for the verification of the models. There is a number of works devoted to this topic. So in works [10, 11] the void fraction averaged over the volume in the rod assembly was measured using pressure transducers. The paper [12] presents the distribution of gas content in the assembly averaged over the area, obtained using the X-ray scanner system. In [13], the volume-averaged gas content was measured by an impedance sensor. In addition to the works devoted to the measurement of averaged gas content, there are also studies which were devoted to measurements of local void fraction. Such experiments are carried out using optical or conductivity probe. In [14], measurements of the local gas content by an electric resistance probe were

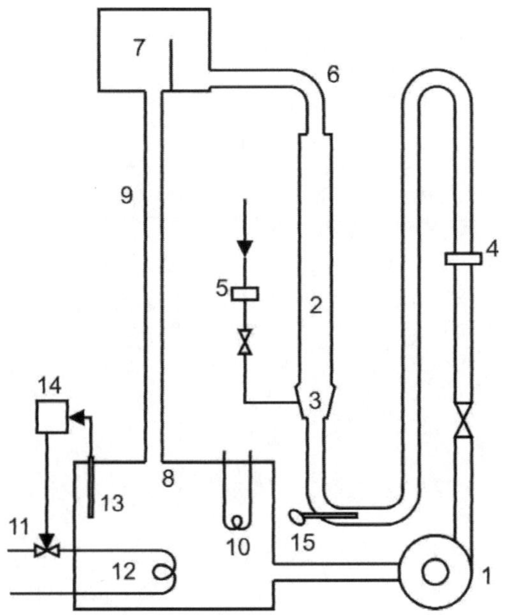


Figure.1. Scheme of setup

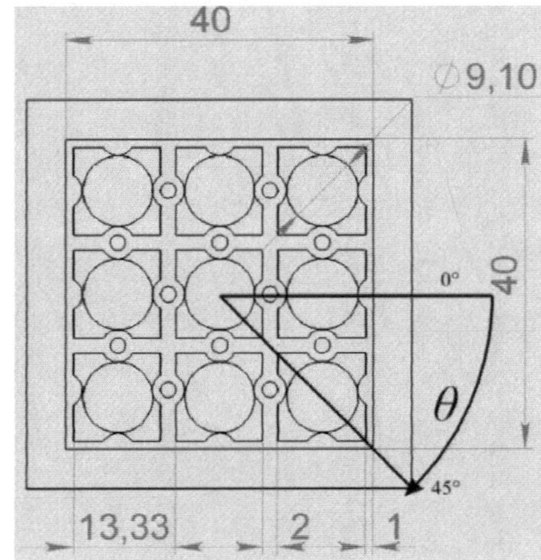


Figure.2. Scheme of channel cross-section. Dimensions are in millimeters.

carried out. The effect of spacer grid on the measured parameters was shown. In [15], was measured as averaged parameters using a gamma densitometer but also the local gas content using an optical sensor. Due to the method of introducing the probe into the test channel in the above works, the local void fraction was measured only on the line between the two rows of rods in the square layout assembly. In works [16, 17], more detailed studies of local gas content around a rod in an assembly are presented. But in these researches, the probe design did not allow measurements near the walls of the rods and in the narrow gaps between them.

In this work, due to the conductivity sensor introduction into the flow through one of the rods, it became possible to conduct a detailed measurement of local void fraction distribution near the walls of the rods and in narrow gaps between them.

## 2. Experimental apparatus and technique

The scheme of the experimental setup is presented on Fig. 1. The setup is a flow loop closed for liquid and open for gas. From the pump 1, the working liquid is fed into the channel 2 (Plexiglass tube of square section, 40 mm wide and 1800 mm long). Nine rods were mounted in the channel supported by four spacer grids. The cross section of the working channel is shown in Fig. 2. The bubbles generator 3 was installed at the bottom section of the channel. The liquid flow rate was defined by the pressure drop on flow orifice meter 4. The same way by the pressure drop on gas orifice meter 5 the gas volumetric flow rate was determined. The bubble mixture was fed into the tank 7 of the separator from the outlet of the working channel through a flexible plastic pipe 6. There the phases dividing were occurred. Then through the return pipe 9 the liquid returned to the main tank 8. The temperature of the liquid in the tank was kept constant and equal to 22 ± 0.1°C with the help of a system of thermal stabilization consisting of heater 10, valve 11, coil 12, thermocouple 13 and regulator 14. The resistance thermometer 15 was installed in the supply pipe to control the temperature of the fluid at the entrance to the test channel.

The test liquid was distilled water with the addition of potassium ferri- and ferrocyanide (0.01 N) and sodium carbonate (0.25 M) dissolved in it. The experiments were carried out at superficial liquid velocities $V_L$ in the range from 0.4 to 1 m/s, corresponding to Reynolds numbers Re = 4000 - 11000

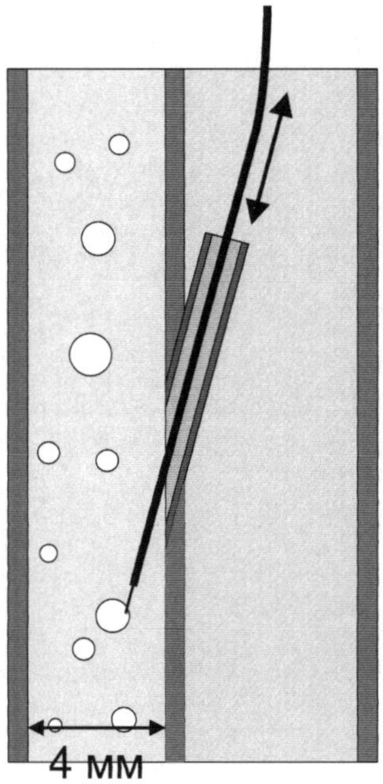


Figure.3 Scheme of void probe

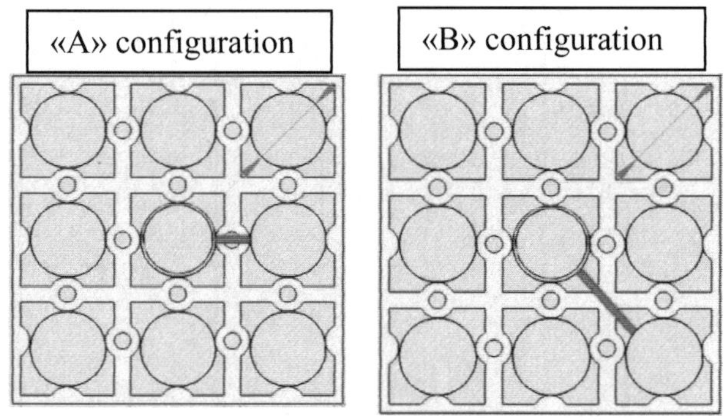


Figure.4. Measurement points

(Re = $\rho V_L D/\eta$). The hydraulic diameter was determined as the ratio of four areas of channel cross-sections to its perimeter D = 4S/P = 9.7 mm. Volumetric gas flow rate ratio $\beta$ was defined as $\beta = V_G/(V_G + V_L)$, where $V_G$ is superficial gas velocity. During the experiment, the gas content varied in the range from 3 to7%.

The conductivity probe was used to measure the local void fraction. The experimental technique was in detail described in [18]. The probe at an angle of 15° was inserted with the aid of a traversing device into the flow through a hole in the side or angular rod of the assembly as shown in Fig. 3. This method made it possible to measure local parameters near the walls of the rods and in the narrow gap between them.

## 3. Results and discussion

As a result of the experiments, local void fraction distributions were obtained on the lines connecting central and lateral, as well as central and angular rods of the assembly, as shown in Fig. 4, with different liquid and gas flow rate parameters. The distribution of local void fraction in a narrow gap between the rods shown on fig. 5. The distribution has a bell-shaped profile near the spacer grid, but at higher distance from the grid it becomes more filled. At large Reynolds numbers and low gas flow rate the distribution has minimum. Figure 6 shows the distribution of local gas content between central and angular rods. Near the spacer grid a bell-shaped profile is observed, as in the previous cases. But at some distance from the spacer grid, saddle-shaped void fraction profiles are realized. Besides the distribution maxima become more manifest with an increase of the Reynolds number and a decrease of gas flow rate. Probably, the distribution pattern far from the spacer grid is primarily related to the dispersion of the gas phase. If the bubble size is comparable to the size of the gap along which it moves, then the distribution will be bell-shaped. If the bubble is small enough compared to the size of the subchannel, the distribution will be saddle-shaped, which is typical for upward bubbly flows in pipes [1]. But this assumption must also be verified by optical methods.

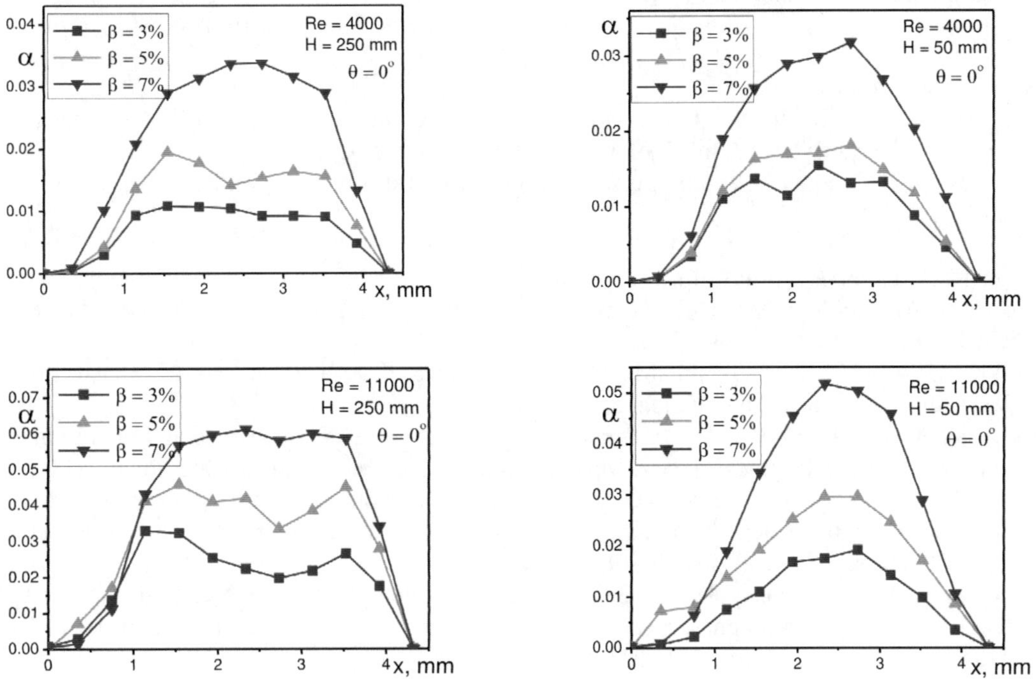


Figure.5. Distribution of local void fraction in «A» configuration.

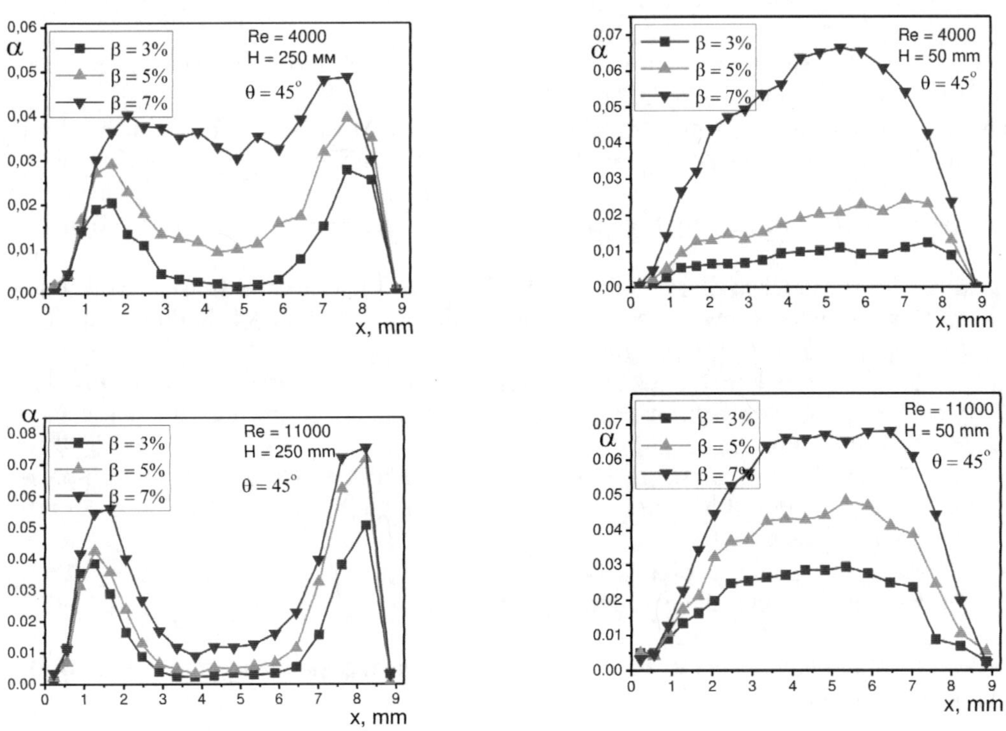


Figure.6. Distribution of local void fraction in «B» configuration.

The reported study was funded by RFBR and Government of Novosibirsk Region according to the research project. 19-48-543029 р_мол_а.

**References**

[1] Nakoryakov V.E., Kashinsky O.N., Burdukov A.P., Odnoral V.P. Local characteristics of upward gas-liquid flows // Int. J. of Multiphase Flow (1981)

[2] Kashinsky O.N., Randin V.V. Downward bubbly gas-liquid flow in a vertical pipe // // Int. J. of Multiphase Flow (1999)

[3] Jiacai Lu, Gretar Tryggvason. Effect of bubble deformability in turbulent bubbly upflow in a vertical channel. // Physics of Fluids 20. 040701 (2008).

[4] Prosperetti A. Bubble dynamics in a compressible liquid. Part 1. First-order theory // J.of Fluid Mechanics (1986)

[5] Hibiki T., Goda H., Kim S., Ishii M., Uhle J. Structure of vertical downward bubbly flow // 2004, International Journal of Heat and Mass Transfer, 47, pp. 1847-1862.

[6] Holloway M.V., Beasley D.E., Conner M.E. Single-phase convective heat transfer in rod bundles// Nuclear Engineering and Design Volume 238, Issue 4, April 2008, Pages 848-858

[7] Rehme K. The structure of turbulence in rod bundles and the implications on natural mixing between the subchannels // International Journal of Heat and Mass Transfer Volume 35, Issue 2, February 1992, Pages 567-581

[8] Holloway M.V., McClusky H.L., Beasley, D.E., Conner, M.E. The effect of support grid features on local, single-phase heat transfer measurements in rod bundles // Journal of Heat Transfer Volume 126, Issue 1, February 2004, Pages 43-53

[9] Dmitriev S.M., Dobrov A.A., Doronkov D.V., Pronin A.N., Solntsev D.N., Sorokin V.D., Khrobostov A.E. The study of hydrodynamic processes of a coolant flow in FA-KVADRAT PWR with various mixing spacer grids // Thermophysics and Aeromechanics, V. 25, I. 5, 2018, pp. 695-703.

[10] Anklam TM, Miller RF. Void fraction under high pressure, low flow conditions in rod bundle geometry. Nucl Eng Des. 1985;75:99–108.

[11] Kamei A, Hosokawa S, Tomiyama A, Kinoshita I, Murase M. Void fraction in a four by four rod bundle under a stagnant condition. J Power Energy Syst. 2010;4:315–326.

[12] Morooka S, Ishizuka T, Iizuka M, YoshimuraK. Experimental study on void fraction in a simulated BWR fuel assembly (evaluation of cross-sectional averaged void fraction). Nucl Eng Des. 1989;114:91–98.

[13] Clark C., Griffiths M., Chen Sh., Hibiki T., Ishii M., et.al. Experimental study of void fraction in an 8*8 rod bundle at low pressure and low liquid flow conditions // International Journal of Multiphase Flow 62 (2014) 87–100

[14] Rena Q., Pana L., Zhoua W., Dub S., Lib Z. Phase distribution characteristics of bubbly flow in 5×5 vertical rod bundles with mixing vane spacer grids // Experimental Thermal and Fluid Science 96 (2018) 451–459

[15] Kumamaru H., Kondo M., Murata H., Kukita Y. Void-fraction distribution under high-pressure boil-off conditions in rod bundle geometry // Nuclear Engineering and Design 150 (1994) 95-105

[16] Yun B., Park G., J. Julia E, Hibiki T. Flow Structure of Subcooled Boiling Water Flow in a Subchannel of 3*3 Rod Bundles // J. of Nuclear Science and Technology, Vol. 45, No. 5, p. 402–422 (2008)

[17] Hosokawa S., Hayashi K., Tomiyama A. Void distribution and bubble motion in bubbly flows in a 4×4 rod bundle. Part I: Experiments // J. of Nuclear Science and Technology, 51:2, 220-230 (2014).

[18] Kashinskii O.N., Kaipova E.V., Kurdyumov A.S. Application of the Electrochemical Method for Measuring the Fluid Velocity in a Two-Phase Bubble Flow // Journal of Engineering Physics and Thermophysics. 2003. T. 76. № 6. C. 1215-1220.

5th International Workshop on Heat/Mass Transfer Advances
for Energy Conservation and Pollution Control
August 13-16, 2019, Novosibirsk, Russia

# Effect of steam supply to the air-blown gasifier on hot syngas desulphurization

**A D Nikitin[1], N A Abaimov[1], E B Butakov[2], A P Burdukov[2] and A F Ryzhkov[1]**

[1] Department of Thermal Power Stations, Ural Federal University, Yekaterinburg, Mira Street, 19, Russia
[2] Laboratory of Ecological Problems of Thermal Power Engineering, Kutateladze Institute of Thermophysics SB RAS, Novosibirsk, 1 Academician Lavrentyev Avenue, Russia

e-mail: nikitin.a.d@yandex.ru, e_butakov@mail.ru

**Abstract**. The IGCC technology serves to efficiently produce thermal and electrical energy with minimal impact on the environment. In operating IGCC, wet desulphurization is used at temperatures below 200°C. The use of hot desulphurization at temperatures around 500°C will significantly improve IGCC efficiency. The preferred sorbent for hot gas cleaning is ZnO. At temperature of 450-500°C, ZnO begins decomposing because of reactions with syngas components (primarily hydrogen). Steam impedes reaction of ZnO with $H_2$ and increases ZnO thermal stability. Syngas $H_2/H_2O$ ratio is determined by gasifier operation mode. The purpose of this work is to determine maximum temperature of hot gas cleaning depending on condition of ZnO-sorbent thermal stability and steam-air-blown mechanically activated coal gasifier operation mode. To determine the effect of steam supply to syngas composition, experiments were performed on entrained-flow gasifier (1 MW). Experimental results were processed using thermodynamic analysis to determine idealized syngas composition and CFD-modeling to determine real experiment process parameters. Syngas $H_2O$ content was determined by CFD-modeling results. Study of ZnO-sorbent thermal stability depending on $H_2$ concentration and syngas $H_2/H_2O$ ratio was performed by TGA. As a result of experimentally confirmed thermodynamic calculations, ZnO-sorbent thermal stability was found to increase to 815°C due to steam dilution.

## 1. Introduction

The IGCC technology allows producing thermal and electrical energy with high efficiency and with minimal impact on the environment [1, 2]. The main IGCC unit is a solid fuel gasifier; and its operation mode affects the performance of all subsequent units, including gas cleaning, shift reactor (if pre-combustion scheme is used) and gas turbine unit (GTU). The composition of syngas depends on the type of gasifier: oxygen, steam-oxygen, air, etc. The steam-oxygen-blown and steam-air-blown syngases have increased humidity due to incomplete conversion of steam [3, 4]. An increase in syngas humidity improves a number of IGCC indicators: a growing rate of water shift reaction in the shift reactor increases its efficiency, and an expanding working fluid of the gas turbine improves its performance [5].

In operating IGCC, wet desulphurization is used at temperatures below 200°C. The use of hot desulphurization at temperatures around 500°C will significantly improve the efficiency of the IGCC cycle [6]. However, the problem of choosing a sorbent for hot gas cleaning has not been solved yet [7,

8]. The most preferred sorbent for hot desulphurization at 350-500°C is ZnO [9]. The upper limit of application is determined by the thermal stability of ZnO in the syngas. In the temperature range 450-550°C, a noticeable decomposition of ZnO begins as a result of interaction with $H_2$ [9]:

$$ZnO + H_2 \rightarrow Zn + H_2O$$

In addition to $H_2$, CO, $CO_2$ and $N_2$, the syngas supplied to desulphurization always contains $H_2O$ due to incomplete fuel gasification and excess steam supply to the gasifier. In accordance with Le Chatelier's principle, steam should prevent the ZnO decomposition reaction, increasing its thermal stability, respectively.

Studies of thermal stability of ZnO-sorbents were carried out either in pure gases [10, 11] or in model mixtures [12, 13], close to the design regime for a particular gasifier. At that, the composition of the syngas changes significantly depending on the operating mode and design of the gasifier. The aim of the work is to determine the thermal stability of the ZnO-sorbent to hydrogen destruction depending on the operation mode of the steam-air gasifier operating on mechanically activated fuel. Mechanoactivation of fuel is a promising method of improving the efficiency of the gasification process [14].

## 2. Materials and methods

### *2.1. Materials*

In experiments on steam-air-blown gasification, the bituminous Kuznetsk coal of D grade was used. Technical and elemental composition of coal is given in table 1.

**Table 1.** Technical and elemental composition of bituminous Kuznetsk coal.

| W, % | A, % | V, % | $C^{daf}$, % | $H^{daf}$, % | $O^{daf}$, % | $N^{daf}$, % | LHV, MJ/kg |
|---|---|---|---|---|---|---|---|
| 19.0 | 17.6 | 43.5 | 75.6 | 5.7 | 16.4 | 1.8 | 19.3 |

*daf – dry and ash free

Thermal stability was determined for the ground ZnO-sorbent, Katalco Johnson Matthew 32-4, fraction of 0-200 μm. Neither pre-drying nor calcination was realized. The composition of the ZnO-sorbent according to scanning electron microscopy is shown in table 2.

**Table 2.** Composition of ZnO-sorbent, Katalco Johnson Matthey 32-4.

| ZnO, % wt. | $Fe_2O_3$, % wt. | $Al_2O_3$, % wt. | CaO, % wt. |
|---|---|---|---|
| 94.42 | 0.63 | 3.55 | 1.41 |

### *2.2. Steam-air-blown gasification of mechanoactivated coal*

For the study of steam-air gasification of mechanoactivated coal, a steam-air-blowm entrained-flow gasifier of the Institute of Thermophysics of SB RAS with a capacity of 1 MW (for fuel) was used [14]. The first stage of the gasifier is a swirler, which is fed with mechanoactivated coal and blast air at room temperature (25°C). The second stage is the cylindrical reaction chamber, in which the steam superheated to 200°C is axially supplied through the nozzle. Mechanical activation of coal occurs immediately before the fuel supply to the gasifier in the disintegrator mill by grinding to a fraction of 0-70 microns. The average particle size is 18-30 μm. Due to mechanical activation, the ignition time is reduced and the combustion rate of coal increases, which eliminates additional heating of the air blast and the walls of the gasifier. The temperature regime is controlled by four thermocouples installed along the axis of the reaction chamber. The syngas composition is determined by a multicomponent gas analyzer TEST-1.

To determine the effect of steam supply on the syngas composition, two series of gasifier launch were realized. Before the gasification mode, the gasifier was heated in the mode of complete fuel combustion to a wall temperature of 1200°C. In the gasification modes, the fuel flow rate was 20 kg/h, the air flow rate was 40 $m^3$/h (the stoichiometric factor was 0.42), and the steam flow rate was 0, 0.235 and 0.5 kg of steam per kg of fuel.

### *2.3. Determining the thermal stability of ZnO-sorbent*

To study the effect of the syngas composition on the thermal stability of the ZnO sorbent, a series of non-isothermal experiments was carried out on the complex for thermogravimetric analysis (TGA), including the NETZSCH STA 449F3 device, the QMS 403C Aëolos quadrupole mass spectrometer and auxiliary equipment. The mass spectrometer was used to determine the composition and flow rate of gases at the outlet of the device. For this purpose, the mass spectrometer was in advance calibrated for mixtures of each measured gas ($H_2$, CO, $CO_2$, $CH_4$, $N_2$, $O_2$, $H_2O$) and argon. As a result, the correction factors were obtained for normalizing all ion currents of the mass spectrometer by argon. During the experiment, argon was fed with a constant flow rate to protect the block of scales; the flow rate and composition of the air blast were determined by the flow rate of argon and the ratio of normalized ion currents. Argon to protect the scales was supplied from below in counter flow to the main blast coming from the top. As the results of CFD modeling show, the shielding gas does not reach the sample and does not affect the conversion rate.

The weight of the sorbent sample was 25 mg. The heating was carried out to a temperature of 1000°C at the rate of 15°C/min. The flow rate of the mixture gas was 100 ml/min flow, and the flow rate of argon to protect the scales was 20 ml/min. Series of experiments were carried out with hydrogen dilution with inert gas (argon) and steam. The boundary of thermal stability (the temperature of the beginning of intense decomposition of ZnO) was determined by the dimensionless reaction rate:

$$R_\tau = \frac{1}{m_\tau - m_{end}} \cdot \frac{dm}{d\tau} > 0.0001,$$

where $m_\tau$ is the mass of the sample at the current time, and $m_{end}$ is the mass of the inert residue.

### *2.4. Thermodynamic modelling*

Thermodynamic modeling of thermal stability of ZnO and equilibrium composition of reaction products was performed using the entropy maximization method [15].

### *2.5. CFD-modelling*

The paper uses the CFD model of entrained-flow gasification of solid fuel proposed in [16, 17]. The modeling was carried out on an unstructured grid with 550 thousand of computational elements. The (standard) k–ε turbulence submodel, the verification of which is given in [18], is used. Discrete radiation submodel (32 rays) with the submodel of "gray" gases was applied. The heterogeneous response model is the diffusion-kinetic one with kinetic constants, obtained by the device for thermogravimetric analysis (TGA) using the method of [18]. The number of calculated trajectories of fuel particles is 2000. On the walls of the setup, the no-slip condition is fulfilled.

## 3. Results and discussion

### *3.1. The influence of steam supply on conversion of mechanically activated coal*

The composition of the syngas obtained by steam-air-blown gasification as a result of variation of the steam-fuel ratio is presented in table 3. Figure 1 shows the calculated dependence of the amount of steam coming out of the gasifier on the amount of steam supplied to the gasifier. From figure 1 it may be seen that the steam output from the gasifier in all modes is 95-99% of the steam at the inlet due to low initial temperature of the steam and short time of its residence in the gasifier. A serious contribution to the steam content in the syngas is due to the high fuel humidity and in the absence of steam supply; the humidity of the syngas is 10%.

**Table 3.** The effect of steam on the composition of products of mecanoactivated fuel conversion (data from CFD simulation are verified against the experimental data).

| Steam flow rate, kg/kg of coal | $H_2O$, % vol. | $CO_2$, % vol. | CO, % vol. | $H_2$, % vol. | $N_2$, % vol. | $Q_i^{raw}$, MJ/m$^3$ | $Q_i^{dry}$, MJ/m$^3$ |
|---|---|---|---|---|---|---|---|
| 0 | 10 | 13.5 | 5.6 | 1.9 | 69.0 | 0.91 | 1.01 |
| 0.235 | 20 | 11.2 | 7.0 | 2.5 | 59.3 | 1.15 | 1.44 |
| 0.5 | 26 | 9.2 | 7.1 | 3.1 | 54.7 | 1.23 | 1.66 |

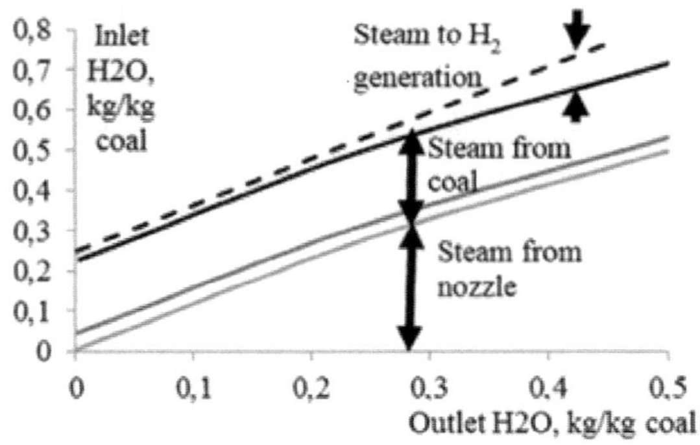


**Figure 1.** Dependence of the amount of steam from the gasifier on the amount of steam supplied to the gasifier.

### *3.2. Effect of hydrogen concentration and $H_2/H_2O$ ratio on thermal stability of ZnO sorbent*

The figure 2 demonstrates the curve of the sorbent mass loss during the experiment in pure hydrogen and the composition of gases at the outlet of the TGA device. There one may see a large excess of air blast. The figure 3 *a* shows curves of the sorbent mass loss when diluting pure hydrogen with an inert gas (argon), and the figure 3 *b* – when diluting with steam. With an increase in the proportion of argon, the rate of the sorbent decomposition and the temperature of the beginning of intense decomposition decrease due to reducing concentration of hydrogen reacting with ZnO. With an increase in the proportion of water, the temperature of the beginning of intense decomposition increases in comparison with dry hydrogen. Figure 4 demonstrates the comparison of the effect of argon and steam on the interaction of the sorbent with hydrogen according to TGA data and thermodynamic calculations. Thermodynamic estimates of thermal stability coincide with experimental data. The figure 5 shows the dependence of thermal stability of the sorbent in hydrogen on the $R_\tau$ value when it is diluted with argon and steam. It may be seen that the presence of a small amount of water significantly increases the thermal stability of the sorbent.

### *3.3. Determining thermal stability of ZnO-sorbent for investigated modes of steam-air-blown gasification of mechanoactivated coal*

Table 4 shows the maximum permissible temperature of hot gas cleaning using ZnO-sorbent for the obtained modes of steam-air-blown gasifier according to two criteria: the concentration of $H_2$ and the ratio $H_2/H_2O$ in the syngas. For the obtained modes, the presence of excess steam in the syngas allows increasing the thermal stability of the ZnO sorbent from 500 to 810-815°C.

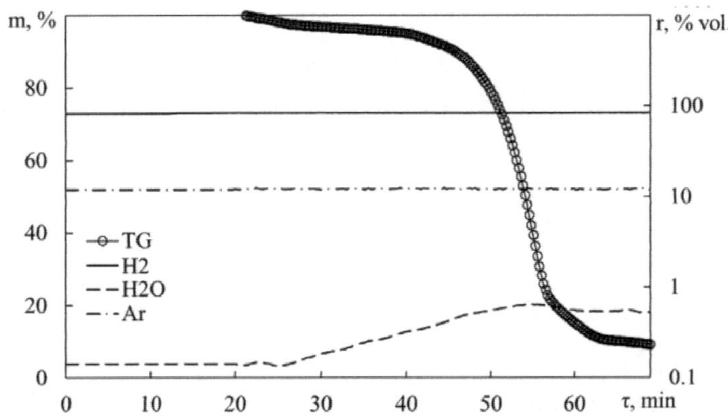


**Figure 2.** The loss of sorbent in pure hydrogen and the composition of gases at the outlet of the TGA.

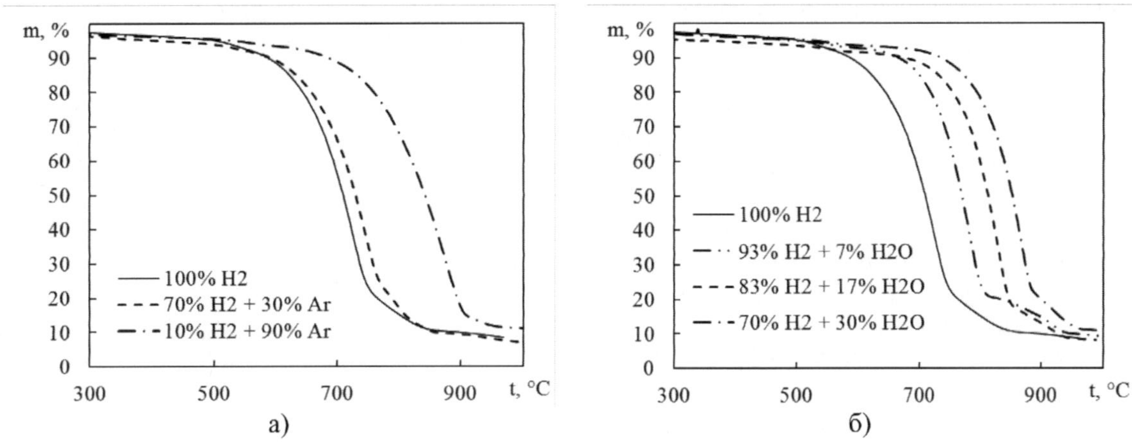


**Figure 3.** The curves of the sorbent mass loss at hydrogen dilution with argon (a) and steam (b).

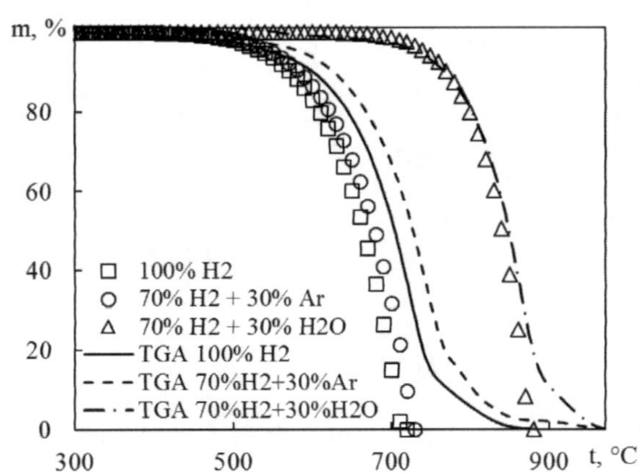


**Figure 4.** The effect of argon and steam on the interaction of sorbent and hydrogen: lines – experiment, dots – calculation.

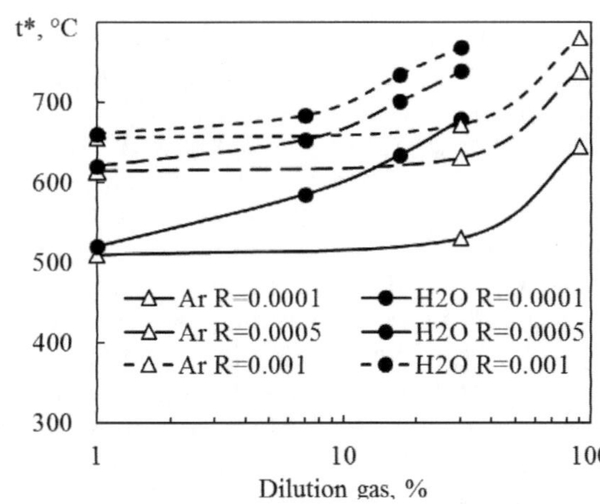


**Figure 5.** The dependence of thermal stability of the sorbent in hydrogen on the value of Rτ at dilution with argon and steam.

**Table 4.** The effect of steam supply at steam-air-blown conversion of mechnoactivated fuel on the temperature of gas cleaning using ZnO-sorbent.

| No. of the operating mode of the gasifier | Steam-fuel ratio, kg/kg | $H_2/H_2O$ ratio in syn-gas | Temperature limit depending on concentration of $H_2$, °C | Temperature limit on the ratio $H_2/H_2O$, °C |
|---|---|---|---|---|
| 1 | 0 | 0.19 | 720 | 790 |
| 2 | 0.235 | 0.13 | 715 | 810 |
| 3 | 0.5 | 0.12 | 710 | 815 |

## 4. Conclusion

As a result of experimental calculation (CFD) and studies of steam-air-blown conversion of mechanoactivated coal in a gasifier of a semi-industrial scale, the dependence of the conversion mode on the steam-fuel ratio has been obtained. With increasing steam flow rate, the concentration of combustible components and steam in the syngas increases. At that, a significant part of the supplied steam does not participate in chemical reactions and passes through the gasifier.

The dependence of thermal stability of the ZnO-sorbent on hydrogen concentration in the syngas and the ratio $H_2/H_2O$, obtained experimentally (TGA) and confirmed by thermodynamic calculations, has shown that the increase in the $H_2/H_2O$ ratio in the syngas allows increasing the thermal stability of the ZnO-sorbent to hydrogen destruction. At that, a low steam content in the syngas (10% vol.) increases the thermal stability by 100°C. For the investigated modes of operation of the steam-air-blown gasifier using mechanoactivated coal, the temperature of syngas desulfurization by ZnO-sorbent under the condition of thermal stability of the sorbent is 790-815°C.

## Acknowledgements

*The work was supported by Act 211 Government of the Russian Federation, contract № 02.A03.21.0006.*

## References

[1] Franco A, Diaz A 2009 *Energy* **34** 348
[2] Wang T, Stiegel G Integrated Gasification Combined Cycle (IGCC) Technologies 2017 (Elsevier)
[3] Giuffrida A, Romano M, Lozza G 2011 *Applied Energy* **88** 3949
[4] Qin S, Chang S, Yao Q 2018 *Applied Energy* **229** 413
[5] Komori T, Hara H, Arimura H, Kitauchi. Y 2003 Proc. Int. Gas Turbine Congress (Tokyo)
[6] Giuffrida A, Romano M, Lozza G 2010 *Applied Energy* **87** 3374
[7] Woolcock P, Brown R 2013 *Biomass&Bioenergy* **52** 54
[8] Meng X, Jong W, Pal R, Verkooijen A 2010 *Fuel Proc. Tech.* **91** 964
[9] Girard V, Baudot A, Chiche D, Bazer-Bachi D, Bounie C, Geantet C 2014 *Fuel* **128** 220
[10] Ko T, Chu H, Chaung L 2005 *Chemosphere* **58** 467
[11] Sun J, Modi S, Liu K, Lesieur R, Buglass J 2007 *Energy&Fuels* **21** 1863
[12] Jun H, Jung S, Lee T, Kim J 2004 *Korean J. Chem. Eng.* **21(2)** 425
[13] Fan H, Li C, Guo H, Xie K 2003 *J. Natural Gas Chem.* **12** 43
[14] Abaimov A, Butakov E, Burdukov A, Ryzhkov A 2018 *AIP Conf. Proc.* **2015** 020001
[15] Trusov B G 2005 III Int. simp. Combustion and Plasmochemistry (Almaty) 52
[16] Ryzhkov A, Abaimov N, Donskoi I, Svishchev D 2018 *Comb., Exp.&Shock Waves* **54(3)** 337
[17] Abaimov N, Ryzhkov A 2015 *Thermal Engineering* **62(11)** 767
[18] Dekterev A Osipov P Chernetskiy M Ryzhkov A 2017 *Solid Fuel Chem.* **51(1)** 17

5th International Workshop on Heat/Mass Transfer Advances
for Energy Conservation and Pollution Control
August 13-16, 2019, Novosibirsk, Russia

# Influence of the coal-derived synthesis gas composition on the thermal stability of ZnO sorbents in the desulfurization process

**Yu A Kagramanov, A F Ryzhkov, V G Tuponogov and A D Nikitin**

Department of Thermal Power Stations, Ural Federal University, Ekaterinburg, Mira Street, 19, Russia

e-mail: nikitin.a.d@yandex.ru, kagramanovya@yandex.ru

**Abstract.** The priority direction of thermal energy on solid fuels is to increase efficiency and environmental safety, approaching zero emissions of harmful substances. To achieve these goals, they are currently creating new schemes for combined cycle gas turbine units with intra-cycle gasification (IGCC). To improve the efficiency of the IGCC cycle, hot dry desulfurization (HGC) technologies are being developed. HGC uses regenerable sorbents that allow the repeated adsorption of hydrogen sulfide from synthesis gas, among which the most acceptable thermodynamic and operational characteristics are based on ZnO sorbents. The article uses the method of thermogravimetric analysis (TGA) to investigate the development of target reaction capture hydrogen sulfide ZnO-containing sorbent and sorbent reactions with components of synthesis gas in the temperature range up to 1000°C. The sorbent calcined at 900°C shows thermal stability in an inert atmosphere over the entire temperature range studied. Uncalcined sorbents at temperatures of 200 and 600°C emit carbon dioxide in an amount of 1.5-4.5% by weight of the sorbent. At temperatures up to 550°C, the target reaction is basic, and side reactions practically do not develop. The maximum rate of absorption of hydrogen sulfide by the sorbent is observed at a temperature of 600°C. At 500-550°C reactions of a sorbent begin with $H_2$ and CO, at 750-800°C – with carbon, at 850°C – with $CH_4$. The intensity of effect of reagents is defined on degradation of a sorbent by reactionary ability of gases more than temperature.

## 1. Introduction

The priority direction of thermal energy on solid fuels is to increase efficiency and environmental safety with the approach to near-zero emissions. To achieve these goals, they are currently creating new schemes for combined cycle gas turbine units with intra-cycle gasification (IGCC), as an alternative to direct fuel combustion. The synthesis gas produced in the IGCC gasifier must be purified from harmful impurities to the minimum concentrations that meet the requirements for the purity of the fuel gas when gas turbines are burned in the combustion chamber [1]. There is a tendency to switch from cold wet gas cleaning to hot dry HGC (hot gas cleanup) at intervals of 300-400°C to 1000°C and higher, since this increases the thermodynamic efficiency of the IGCC cycle by 1.8-2.5% [2,3]. In gas clean up studies in IGCC schemes, hot gas purification methods are of particular interest. The most costly gas cleaning processes include the removal of sulfur compounds contained in synthesis gas mainly in the form of $H_2S$ hydrogen sulfide with concentrations of 0.5 ml/l to 30 ml/l and more, depending on the grade of gasifiable fuel [4]. HGC uses regenerated sorbents that allow multiple reversible adsorption of sulfur from hydrogen sulfide synthesis gas, among which the sorbents based on ZnO have the most acceptable thermodynamic and operational characteristics, having a high affinity for $H_2S$ and having a working temperature range of 250-650°C, sulfur content at the level 200-300 g/kg and the degree of purification from $H_2S$ to <7 ppm [5-7]. Sulfur from a multicomponent synthesis gas is absorbed by several reactions, the main of which is the target reaction with $H_2S$:

$$ZnO + H_2S \rightleftharpoons ZnS + H_2O \tag{1}$$

In addition to the target reaction, non-sulfur-containing components of the synthesis gas ($H_2$, CO, $CH_4$) and condensed carbon can enter into complex interactions with the sorbent in the so-called side reactions (2) - (5), accompanied by release of metallic zinc, its evaporation and destruction or rebirth sorbent.

$$ZnO + H_2 \rightleftharpoons Zn + H_2O \tag{2}$$

$$ZnO + CO \rightleftharpoons Zn + CO_2 \tag{3}$$

$$4ZnO + CH_4 \rightleftharpoons 4Zn + CO_2 + 2H_2O \tag{4}$$

$$2ZnO + C \rightleftharpoons 2Zn + CO_2 \tag{5}$$

The currently existing calculation methods as a rule do not take into account the effect on the cleaning process of side reactions, as well as the reverse processes [8, 9], and the behavior of sorbents is usually studied on model gases [10, 11], or only the target reaction is considered [12-14]. In the article, the method of thermogravimetric analysis (TGA) traces the development of direct side reactions (2)-(5), compares with the target reaction (1), determines the beginning and rate of the thermochemical degradation of the ZnO-containing sorbent.

## 2. Materials and method

### *2.1. Materials*

As a sorbent, granulated (granule diameter 3 mm) and crushed (diameter less than 200 microns) ZnO-sorbent Katalco Johnson & Matthey 32-4 and zinc oxide powder (diameter less than 50 microns) were used. According to the data of scanning electron microscopy, in addition to Zn and O, Fe, Al and Ca are present in the range of 3-8% by weight of Katalco sorbent. Zinc oxide powder is 100% zinc and oxygen. The distribution of elements on the sample surface is uneven. The ratio of zinc atoms and oxygen in two measurements is 0.84-1.67 for Katalco and 1.29-1.92 for ZnO powder.

In experiments on TGA, gas mixture reagents hydrogen sulfide-argon (1% vol. $H_2S$), hydrogen argon (9.5% $H_2$), carbon monoxide-nitrogen (70% vol. CO), methane-argon (4.8% vol. $CH_4$), hydrogen. The solid reagent was BAU-A activated charcoal powder.

### *2.2. Method*

The thermostability experiments were carried out on a NETZSCH STA 449F3 instrument with a standard TG holder and a QMS 403C Aëolos quadrupole mass spectrometer. The sample mass of the sorbent was 20-25 mg, the flow rate of the gas mixture - 100-133 ml / min. The sample heating rate was 15 and 20 K/min. The temperature limit of thermal stability (the beginning of a significant decomposition of ZnO) was estimated by the relation:

$$R_\tau = \frac{1}{m_\tau - m_{end}} \cdot \frac{dm}{d\tau} > 0.0001$$

Where $m_\tau$ is the mass of the sample at the current time, $m_{end}$ is the mass of the inert residue.

The experiments were carried out in three stages. At the first stage, the effect of sorbent calcination (30 min, 900°C, air) on thermal stability in a neutral medium (argon) was evaluated. Previously, for calibration of the instrument, the baseline was removed without a sorbent, the resulting curve was subtracted from the subsequent experimental results.

At the second stage, the course of the direct target reaction (1) of the absorption of hydrogen sulfide by zinc oxide was recorded.

At the third stage, the flow of direct side reactions (2) - (5) of the synthesis gas purification process was studied: reduction of zinc oxide in hydrogen, carbon monoxide, methane and carbon separately.

## 3. Results and discussion

### *3.1. Behavior of the sorbent in a neutral environment*

The calcined sorbent throughout the entire temperature range (250-1000°C) shows thermal stability (curve 1 in figure 1). According to the Elingem diagram, zinc oxide dissociation occurs at temperatures above 2000 K. There is no gas output from the calcined sorbent. The uncalcined sorbent in the process of heating loses from 1.5 to 4.5% by weight, which is associated with local peak emissions of adsorbed gases and vapors ($CO_2$, $H_2O$). ZnO powder gives low-temperature (250°C) emissions of $CO_2$ (curve 5) and H2O. The crushed and granulated sorbent has a more pronounced medium-temperature (600°C) peak (curves 6 and 7). The release of $H_2O$ in non-powder sorbents was not observed.

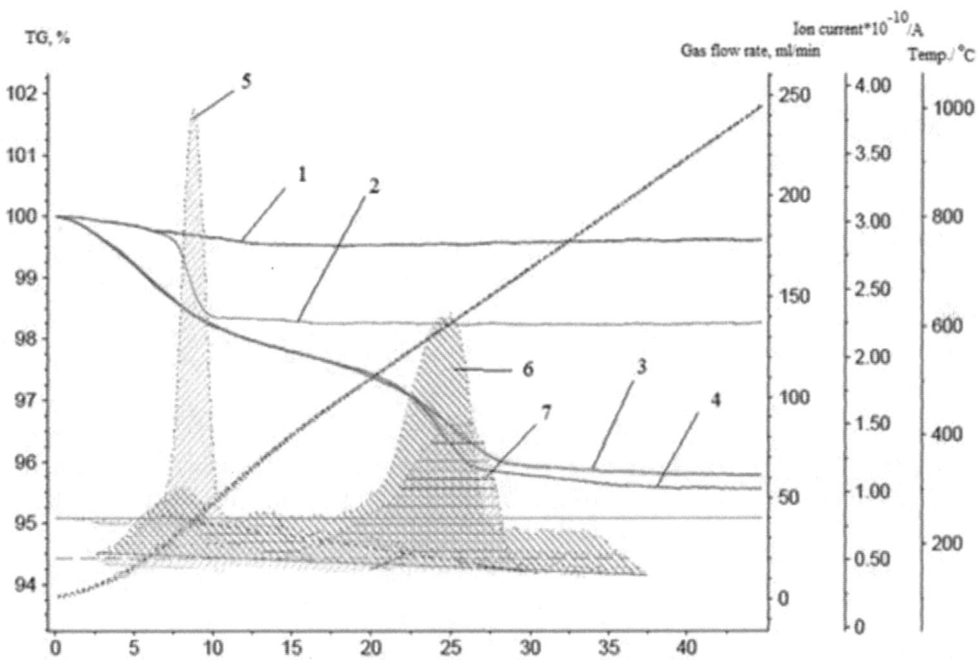


**Figure 1.** The results of TGA sorbents in argon: 1 - calcined crushed, 2 - uncalcined zinc oxide powder, 3 - uncalcined crushed sorbent, 4 - uncalcined granulated sorbent , 5 - carbon dioxide emission from the uncalcined powder, 6 - carbon dioxide emission from the uncalcined crushed sorbent, 7 - emission of carbon dioxide from uncalcined granulated sorbent.

### *3.2. Behavior of the sorbent in active media*

At a temperature of 350°C, the target reaction (1) begins. At 800°C, the target reaction is completed by saturation of the sorbent with an increase in mass of 16%. The maximum absorption rate at a temperature of 600°C.

In the temperature range of 500–1000°C, side reactions (2)-(5) develop, accompanied by significant mass loss. The sorbent exhibits the smallest temperature stability with respect to the main component of synthesis gas - hydrogen. A noticeable destruction of the sorbent in a mixture of hydrogen and argon begins at a temperature of 500-550°C. (figure 2). At close temperatures, the reaction of ZnO with carbon monoxide is activated (figure 2).

Carbon begins decomposition of ZnO at temperatures above 800°C (figure 2). The process is accompanied by the active release of $CO_2$ and the absorption of an oxygen impurity. At the end of the experiment, excess carbon remains in the crucible.

Methane, which is present in raw IGCC synthesis gases in insignificant quantities, acts on the sorbent in two ways. That part of methane, which is on the way from the gas generator to the gas clean up is subjected to thermal decomposition, is entered into the desulfurization reactor in the form of black carbon and hydrogen and acts according to reactions (5) and (2).

The other part of methane, directly entering the sorbent, at a temperature above 850°C, interacts with it according to the reaction (4) - curve 4 in figure 2 a. The interaction is accompanied by the release of $H_2$, $CO_2$, $H_2O$ and the binding of residual oxygen while maintaining the residual methane at the outlet of the installation (figure 3).

DTG curves (figure 2 b) show that the reactions with methane occur most intensively (each in its temperature range), then the reduction of zinc oxide with hydrogen, then - reactions with carbon and carbon monoxide, which have approximately the same rates, the target reaction is the least active. The limit of thermal stability of the sorbent in the components of the synthesis gas are shown in Table 1.

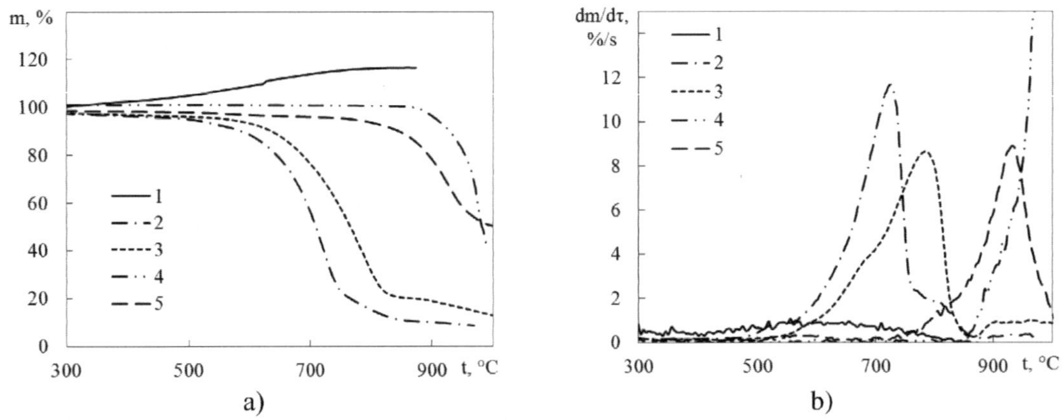


**Figure 2.** The results of the TGA sorbent based on ZnO in the environments of the components of the synthesis gas: TG-curves (a) and DTG-curves (b). The numbers correspond to the numbers of the reactions.

**Table 1.** The thermal stability limit of the sorbent based on ZnO.

| Reaction number | Reagent, % vol. | The temperature of the onset of significant decomposition of ZnO |
|---|---|---|
| 2 | 100% $H_2$ | 509 |
| 2 | 10% $H_2$ + 90% Ar | 644 |
| 3 | 70% CO + 30% $N_2$ | 561 |
| 2 | 70% $H_2$ + 30% Ar | 532 |
| 4 | 5% $CH_4$ + 95% Ar | 853 |
| 5 | 100% C | 772 |

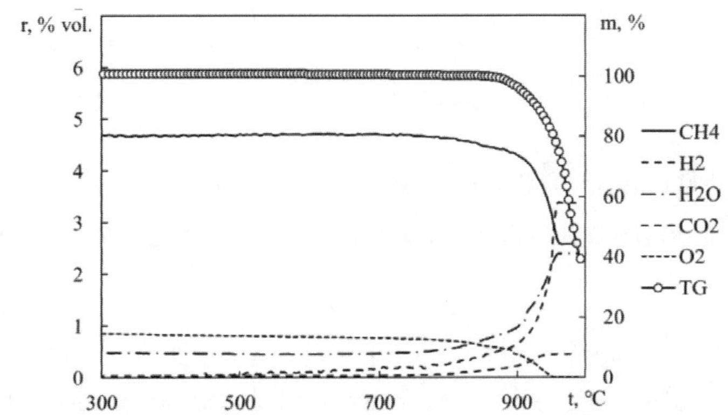


**Figure 3.** TG curve and gas composition in the interaction of the sorbent with methane.

## 4. Conclusion

The thermal stability of sorbents based on zinc oxide in a neutral medium was tested in the target temperature range of 400–1000 ° C. Sorbents showed stability over the entire temperature range. It has been found that at temperatures of 200 and 600 ° C, non-calcined sorbents emit carbon dioxide. Disposable emissions that do not affect the work of the sorbent.

Analysis of the interaction of zinc oxide with the components of the synthesis gas made it possible to study the intensity of each reaction separately in the target temperature range. At temperatures up to 550 ° C, the target reaction is basic, and side reactions practically do not develop. When the temperature rises above 550 ° C, the effect of direct side reactions is observed, leading to the destruction of the sorbent. At 500-550 ° C H2 and CO begin to act, at 750-800 ° C - carbon, at 850 ° C - methane. The intensity of the reagents action on the degradation of the sorbent is determined by the reactivity of gases to a greater extent than temperature. The most intense destruction produces methane, followed by hydrogen, the least intense destruction occurs as a result of the interaction of the sorbent with carbon monoxide and carbon.

## Acknowledgements

*The work was supported by Act 211 Government of the Russian Federation, contract № 02.A03.21.0006.*

## References

[1] Muller M 2013 2013 **108** 37
[2] Xiao Y 2012 *J. Ther. Sci.* **21** 391
[3] Giuffrida A 2011 *Applied Energy* **88** 3949
[4] Vamvuka D 2004 Environ. Eng. Sci. **21**(4) 525
[5] Gupta R 2007 High temperature sulfur removal in gasification applications. Res. Triangle Park
[6] Kohl A 1997 Gas Purification. Gulf Professional Publishing
[7] Gupta R 2018 RTI Warm Syngas Cleanup Operational Testing at Tampa Electric Company's Polk 1 IGCC Site. Research Triangle Park
[8] Konttinen J and Zevenhoven C 1997 *Ind. Eng. Chem. Res.***6** 2332
[9] Settino J 2015 *Energy Procedia* **82** 908
[10] Woods M 1989 Kinetics and Simulation Models for Novel High-Temperature Desulfurhtion Sorbents. Research Triangle Park
[11] Siriwardane R 1996 Patent Number 5 494 880
[12] Gibson J and Harrison D 1980 *Ind. Eng. Chem. Process Des. Dev.* **19** 231
[13] Atimtay A 1991 The use of zinc oxide sorbents to remove hydrogen sulfide from coal gases.
[14] Gangwal S and Roberts W 1998 Engineering Evaluation of Hot-Gas Desulfurization with Sulfur Recovery. Research Triangle Park

Journal of Physics: Conference Series **1369** (2019) 012031 doi:10.1088/1742-6596/1369/1/012031

# Investigating the gas-droplet flow generated by a pneumatic nozzle for a coal-water slurry

I S Anufriev[1,2], E P Kopyev[1], L I Maltsev[1], E Yu Shadrin[1] and O V Sharypov[1]

[1]Kutateladze Institute of Thermophysics SB RAS, Novosibirsk, Russia

[2]Novosibirsk State Technical University, Novosibirsk, Russia

E-mail: evgen_zavita@mail.ru

**Abstract.** The paper studies a promising method of spraying, based on the interaction of cumulative and near-wall jets with the impinging liquid and the Coanda effect. The measurements are based on shadow photography (SP) and interferometric method (IPI). In a wide range of performance parameters data have been obtained on the disperse composition of a gas-droplet flow at water spray by a special pneumatic atomizer intended for coal-water fuel. The dependences of droplet sizes on the flow rate of liquid and spraying air have been found.

## 1. Introduction

Currently, one of the most important areas in coal energy is the use of low-grade (substandard) coals and coal-enrichment wastes as fuel. A promising way to dispose of low-quality coal waste with high energy and environmental indicators is the combustion of coal in the form of coal-water slurry [1]. Combustion of such fuel requires spraying in the combustion chamber using pneumatic nozzles. Laboratory modeling to obtain detailed information about the structure and dispersed composition of the two-phase flow created by the nozzles is an important step in the creation and optimization of equipment for spraying and combustion of slurry fuel.

The perspective method of coal-water slurry spraying, developed by the researchers of IT SB RAS [2, 3], is based on the interaction of cumulative and near-wall jets created by the nozzle with the impinging liquid. In order to find the optimal operating parameters (atomizing air and liquid flow rates) for the creation and subsequent ignition of the gas-droplet flow, a detailed experimental study of the fuel atomization characteristics is necessary.

## 2. Experimental setup and technique

In this paper, the atomization by a pneumatic nozzle developed for coal-water fuel is studied using water as a test liquid. The specificity of liquid dispersion is as follows. A gas jet flowing out of a conical slit nozzle forms a converging jet gas stream outside the nozzle. As a result, flow along the nozzle axis in the forward direction and a high-speed return jet of the cumulative type are formed. The liquid is supplied through the central pipe under pressure. Due to the Coanda effect, the fluid flow is deflected to the walls of the diffuser nozzle and spreads along with them as a thin layer. High-speed return gas jet leads to dispersion of the liquid jet. The return flow of gas after collision with the liquid changes its direction to the opposite, spread along the walls of the diffuser and accelerates the flow of liquid. As a result of the interaction of gas and liquid flows at the outlet of the diffuser, a homogeneous fine gas-droplet flow is formed (Fig. 1). Previously [4], the characteristic angle of the jet opening varying within the range of 20÷25° was determined.

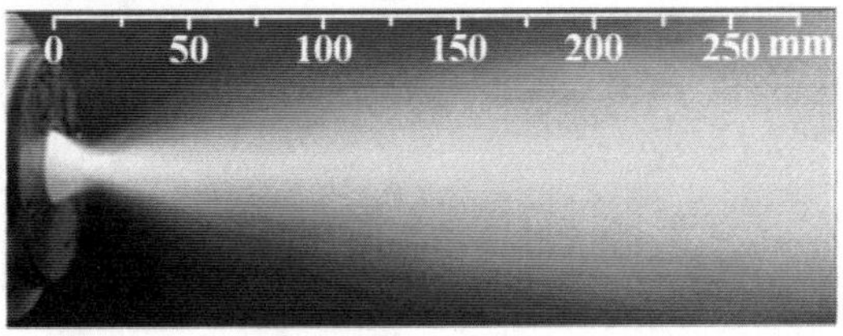


**Figure 1.**Gas-droplet flow generated by a pneumatic nozzle for coal-water slurry.

In the experiments, the regimes at the initial air overpressure in the annular gas chamber of the injector were studied within the range of $P_0$ = 1.0-4.0 bar (volume air flow $G$ = 80-100 m$^3$/h). The pressure was regulated by a valve and controlled by a standard pressure gauge. The liquid was supplied by a pump (flow rate $F_w$ = 20-80 g/s), and the flow control was carried out by means of a rotating vane flowmeter. The selected flow ranges of the carrier phase and liquid corresponded to the operation modes of the injector on a real boiler [3].

To study the characteristics of the liquid spray with a pneumatic nozzle, non-contact optical methods of flow diagnostics were used: the method of shadow photography (SP) [5] and the interferometric method for determining the droplet diameters (IPI) [6]. Figure 2-a shows the measuring areas (35x24 mm): on the axis of the jet and on the periphery.

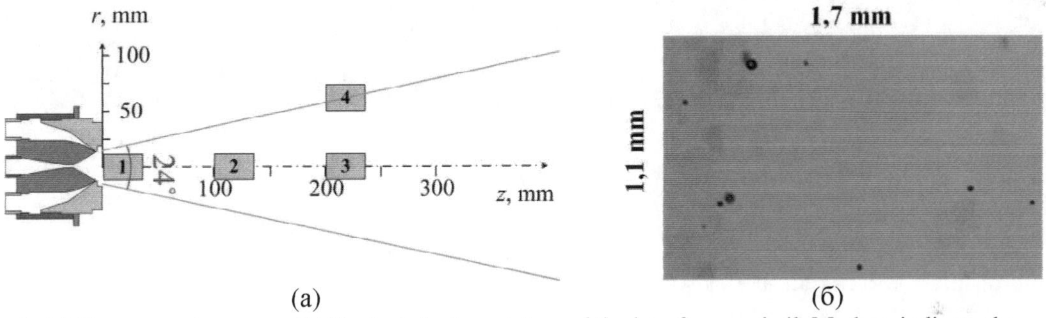


**Figure 2**. (a) The measuring areas; (b) Typical shadow picture of the jet of sprayed oil. Markers indicate the identified particles.

The SP method is based on the registration of a shadow photo of an object with a refractive index different from that of the environment. In this case, behind the object under study (relative to the camera), there is a diffuse light source with a uniform spatial distribution of intensity. Digital analysis of the shadow image allows determining the position and boundary of the object. Figure 2-b shows a characteristic shadow photo of the gas-droplet flow processed using special algorithms, where the identified particles are marked with markers. The IPI method is based on the registration of defocused droplet images, illuminated by a laser sheet. According to the Mie theory of scattering [6], the light reflected and two times refracted by the spherical surface of the droplet creates interference fringes in the images of droplets. Their frequency directly depends on the droplet diameter. Digital analysis of the obtained images allows determining the position and size of the droplets suspended in the stream. This method allows measuring particles with sizes from 10 microns.

To conduct experiments by the SP method, the "Polis" measuring complex was used. It includes the following components: CCD camera ImperX B4820-M (resolution 4904×3280 pixels, shooting frequency of 3.2 Hz, and minimum inter-frame delay of 200 ns) and a macro lens Tamron SP AF with a focal length of 180 mm, which ensured measurements with a good spatial resolution (magnification 1:1). As a light source, a background screen with a rhodamine-based fluorescent coating was used. It was pre-illuminated by a defocused beam of a pulsed Nd:YAG QuantelEVG laser (wavelength – 532 nm, pulse energy – up to 145 MJ, and pulse duration – 10 ns). To increase the contrast of shadow photography, a threshold light filter (560 nm) was used. Its bandwidth corresponded to the wavelength of light re-emitted by rhodamine.

The implementation of the IPI method also required the measuring complex "Polis", which includes the following parts: CCD camera ImperX B4820-M and a Nikon macro lens with a focal length of 105 mm. With the aim of increasing the method resolution and registration of individual particles in a dense gas-droplet stream, the images of drops decreased in one direction using an optical compression block specifically designed for the used lens. A pulsed laser Nd:YAG QuantelEVG was used as a light source.

The ActualFlow software with packages SP Kit and IPI Kit was used for digital processing.

## 3. Results

Shadow photos of the gas-droplet jet in different regions were obtained (Fig. 2-a). Figure 3 shows typical shadow images of the jet at different distances from the nozzle (in areas 1, 2 and 3, Fig. 2-a). It should be noted that near the axis the flow is dense and optically translucent, which prevents from using SP and IPI methods for the study. Therefore, the dispersed composition was investigated on the jet periphery (region 4, Fig. 2-a).

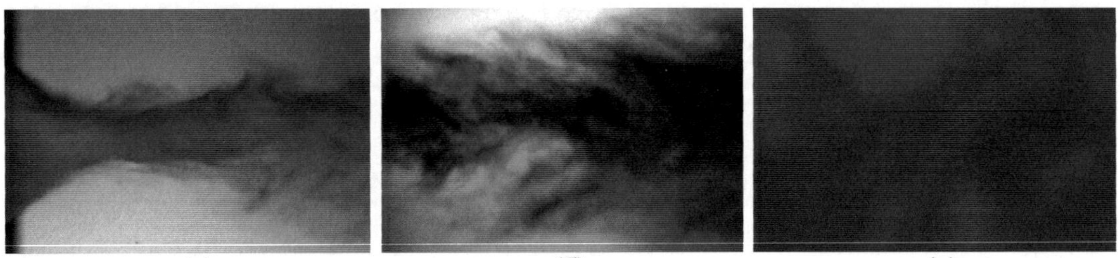

(a) (б) (в)

**Figure 3.** Characteristic shadow photos of the gas-droplet flow on the jet axis when spraying water through a pneumatic nozzle: (a) area 1 Fig. 2-a; (b) area 2; (b) area 3.

As a result of image processing in the measuring area, information on the dispersed composition of liquid droplets at different modes was obtained. Figure 4 presents distributions of the relative number of droplets by size depending on air overpressure $P_0$ in the nozzle at a constant flow rate $F_w$ (Fig. 4-a) and on flow rate at constant $P_0$ (Fig. 4-b,c). Data on fractions smaller than 10 μm are not available because the used systems do not allow recording particles with such dimensions.

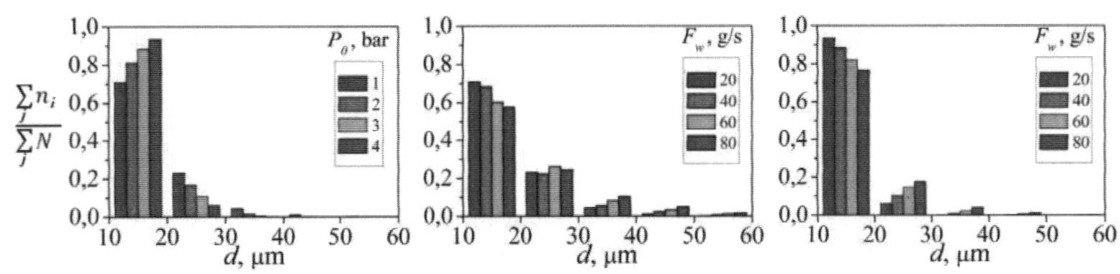


(a) (b) (c)

**Figure 4.** The dispersed composition of the gas-droplet flow when spraying water through a pneumatic nozzle (area 4, Fig. 2-a): (a) depending on $p_0$ at constant flow rate $F_w$ = 20 g/s; (b) depending on the flow rate $F_w$ at constant $p_0$ = 1 bar; (c) depending on the flow rate $F_w$ at constant $p_0$ = 4 bar
($n_i$ – the number of drops with sizes from the i-th range in the j-th image, $N$ – the total number of drops identified by the algorithm on the j-th image, j = 1 ... 200)

Analysis of the results shows that an increase in pressure increases the relative number of small droplets (10-20 microns), i.e., the increasing dynamic effects of the flow on the liquid increase the efficiency of fragmentation. At that, an increase in the flow rate leads to an increase in the relative number of larger drops.

## 4. Conclusion

With the use of modern optical methods, data have been obtained on the dispersed composition of the gas-droplet flow at water spraying through a pneumatic nozzle at different operating parameters, corresponding to the stable operation of the burner under study. The predominant droplet size at the periphery of the jet in the studied regimes was within the range of 10-20 μm. It is shown that the increase in air flow rate leads to a decrease in the proportion of large drops, i.e. an increase in the dynamic effect on the liquid increases the efficiency of fragmentation. At the same time, an increase in fluid flow rate leads to an increase in the proportion of large drops. The obtained results allow predicting the dispersion efficiency and optimizing the operating parameters.

The obtained experimental data may be used for numerical calculations of the combustion of pulverized fuel slurries.

**Acknowledgments**

The study was financially supported by a grant of the Russian Foundation for Basic Research (Project No. 18-38-00153 and Project No. 18-29-24007).

## References

1. Trubetskoi K N, Zaidenvarg V E, Kondrat'ev A S, Murko V I, Kassikhin G A, Nekhoroshii I Kh *Therm. Eng.* **55**(5). P.413-417.
2. Maltsev L I 2007 *RF Patent* 2346756 *(in Russian)*
3. Maltsev L I, Kravchenko I V, Lazarev S I and Lapin D A 2014 *Therm. Eng.* **61**(7) 486-490.
4. Anufriev I S, Maltsev L I, Shadrin E Yu and Sharypov O V 2018 *J. Phys.: Conf. Series.* **1128** 012045.
5. Watanabe H, Shoji Y, Yamagaki T, Hayashi J, Akamatsu F and Okazaki K 2016 *Fuel* **182** 259-265.
6. Lozhkin Yu A, Markovich D M, Pakhomov M A and Terekhov V I 2014 *Thermophys. Aeromech.* **21**(3) 294-307

# Investigating substandard liquid hydrocarbons spraying by a steam jet

I S Anufriev[1,2], E P Kopyev[1], M A Mukhina[1,3], Ya A Osintsev[1,2] and E Yu Shadrin[1]

[1]Kutateladze Institute of Thermophysics SB RAS, Novosibirsk, Russia

[2]Novosibirsk State Technical University, Novosibirsk, Russia

[3]Novosibirsk State University, Novosibirsk, Russia

E-mail: evgen_zavita@mail.ru

**Abstract.** Using the shadow photography (SP) method and the interferometric method (IPI), a two-phase gas-droplet flow was experimentally investigated when the spent transmission oil was sprayed by a jet of superheated steam. In a wide range of operating parameters, data on the dispersed composition of the steam-fuel jet have been obtained. The dependences of droplet sizes in the flow on the studied regime parameters have been established.

## 1. Introduction

High-quality fuel dispersion is an important technical task in the development of technologies for the combustion of substandard hydrocarbon fuels, such as waste oils, crude oil, etc. The use of such fuels in spray burners leads to the formation of a large amount of soot and harmful emissions, coking and clogging of spray nozzles, which cause failures in burners' operation. Therefore, when creating the burners for substandard fuel combustion, it is important to find the optimal solution to these problems by conducting laboratory simulation. Obtaining detailed information about the structure and dispersed composition of the gas-droplet fuel jet using modern optical methods (SP, IPI) [1-3] is important for optimizing the equipment and for the combustion process as a whole.

One of the promising methods of substandard liquid fuel spraying was developed by the staff of IT SB RAS [4] on the basis of interaction of liquid and high-speed gas jet. Its distinctive feature is the lack of contact between the fuel and the surface of the spray nozzle, which prevents from possible coking of surfaces and subsequent failures in operation. The use of superheated steam as a spraying medium also allows achieving high thermal and environmental performance. In order to find the optimal operating parameters (temperature and steam flow rate, fuel flow rate) for the creation and subsequent ignition of the gas-droplet flow, a detailed experimental study of the fuel atomization characteristics by the proposed method is necessary.

## 2. Experimental setup and technique

In this paper, the atomization of substandard liquid hydrocarbons by a jet of superheated steam in a direct-flow burner (without combustion) is studied by the example of the spent transmission oil. This fuel, as an example of low-grade liquid hydrocarbons with relatively high viscosity ($\approx$ 69 cSt at 40 °C) and high content of impurities, is chosen to investigate the performance of spray unit in application to such fuels.

The process of dispersion and combustion of liquid fuels is as follows. The steam flows out of the nozzle (diameter of 0.5 mm) in the form of a jet. Liquid fuel flows freely into its base where it is atomized by the jet, resulting in a finely dispersed gas-droplet flow. At the same time, superheated steam increases the temperature of the fuel droplets, thereby improving the process of mixture formation and furthering stable ignition. Steam gasification of products of thermal decomposition of fuel also takes place in the combustion zone, which improves the combustion characteristics of liquid hydrocarbons similar to the case of evaporative burners [6]. To reduce toxic emissions and increase the completeness of fuel combustion, it is important to ensure high uniformity of the gas-droplet flow, the smallest possible fragmentation of fuel, and high stability of the torch.

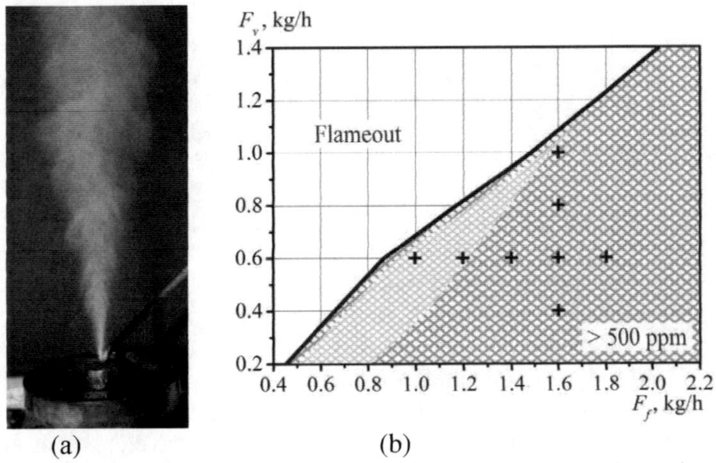


**Figure 1**. (a) Photo of fuel jet at spraying spent transmission oil by superheated steam jet; (b) Regime map of the burner ($T_s$=250°C): I – flameout; II – zone of "ecological" stable combustion; III – flame with a high content of CO in combustion products (> 500 ppm); "+" – studied regimes.

Preliminary studies served to obtain a range of stable combustion modes typical for this design of the burner device (Fig. 1-b). The range of steam flow rates $F_v$ corresponds to the working range of the dosing water pump. Limits of fuel flow rates $F_f$ correspond to the permissible power of the burner in laboratory measurements. The temperature of superheated steam in the experiments varied in the range $T_s$= 150..350 °C. The steam pressure depending on the values of $F_v$ and $T_s$ ranged from 3 to 11 bar, and steam superheating ($T_s - T_b$) reached 100 °C ($T_b$ – temperature of saturated steam). In the experiments, the waste oil was heated to a temperature of 40 °C to reduce the viscosity and ensure a stable supply. For the study by the methods of SP and IPI in this work, characteristic modes marked on the map by symbols "+" were selected.

To study the characteristics of liquid fuel spray, non-contact optical methods of flow diagnostics were used: the method of shadow photography (SP) [1-2] and the interferometric method for determining droplet diameters (IPI) [1, 3]. The SP method is based on a shadow photo of an object with a refractive index different from its environment. In this case, behind the object under study (relative to the camera) there is a diffuse light source with a uniform spatial distribution of intensity. Digital analysis of the shadow image allows determining the position and boundary of the object. Figure 2 shows a typical shadow photograph of a jet of sprayed oil, processed using special algorithms. The IPI method is based on the registration of defocused droplet images, illuminated by a laser sheet. According to the Mie theory of scattering [3], the light reflected and two times refracted by the spherical surface of the droplet creates interference fringes on the images of droplets. Their frequency directly depends on the diameter of the droplet. Digital analysis of the obtained images allows determining the position and size of the droplets suspended in the stream. This method allows measuring particles with sizes from 10 microns.

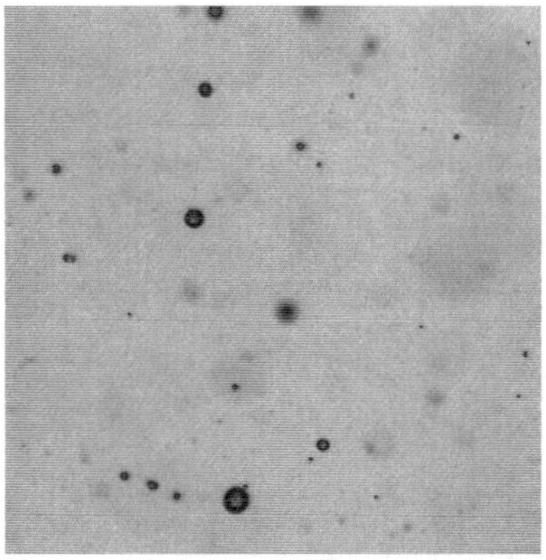

**Figure 2**. Typical shadow picture of the sprayed oil jet. The markers indicate the identified particles.

To conduct experiments by the SP method, the "Polis" measuring complex was used. It included: CCD camera ImperX B4820-M (resolution 4904×3280 pixels, shooting frequency of 3.2 Hz, and minimum inter-frame delay of 200 ns) and a macro lens Tamron SP AF with a focal length of 180 mm, which allowed measurements with good spatial resolution (magnification 1:1). As a light source, a background screen with a rhodamine-based fluorescent coating was used. It was pre-illuminated by a defocused beam of a pulsed Nd:YAG QuantelEVG laser (wavelength – 532 nm, pulse energy – up to 145 MJ, and pulse duration – 10 ns). To increase the contrast of shadow photographs, a threshold light filter (560 nm) was applied. Its bandwidth corresponded to the wavelength of light re-emitted by rhodamine.

For experiments with the use of the IPI method the measuring complex "Polis" was used also. It included: CCD camera ImperX B4820-M and Nikon macro lens with a focal length of 105 mm. To compress the image in one direction, an optical compression unit was used. It was specially designed for the used lens. Pulsed Nd:YAG laser QuantelEVG was applied as a light source.

For digital processing, ActualFlow software with SP Kit and IPI Kit packages was used.

## 3. Results

As a result of processing images obtained by oil spraying with a jet of superheated steam, the following features have been revealed (Fig. 3). The predominant particle size in the studied regimes is 10-20 μm, which is a sufficient condition for efficient fuel combustion [6] and indicates the advantages of the method used for dispersing liquid fuel.

With changes in fuel flow rate, the dispersed composition remains unchanged (Fig. 3-a). It is an obvious fact, since the liquid spraying under constant influence is determined by the parameters of this liquid and is independent of the flow rate. The steam temperature above 250 °C has no effect on the droplet size (Fig. 3-c): with adiabatic expansion of the steam jet flowing out of the nozzle, its temperature drops sharply and in the measuring area (where the fuel ignites in the burner) it reaches approximately the same value for different initial steam temperatures. Downstream the number of small drops increases (Fig. 3-d), which is associated with the evaporation of fuel.

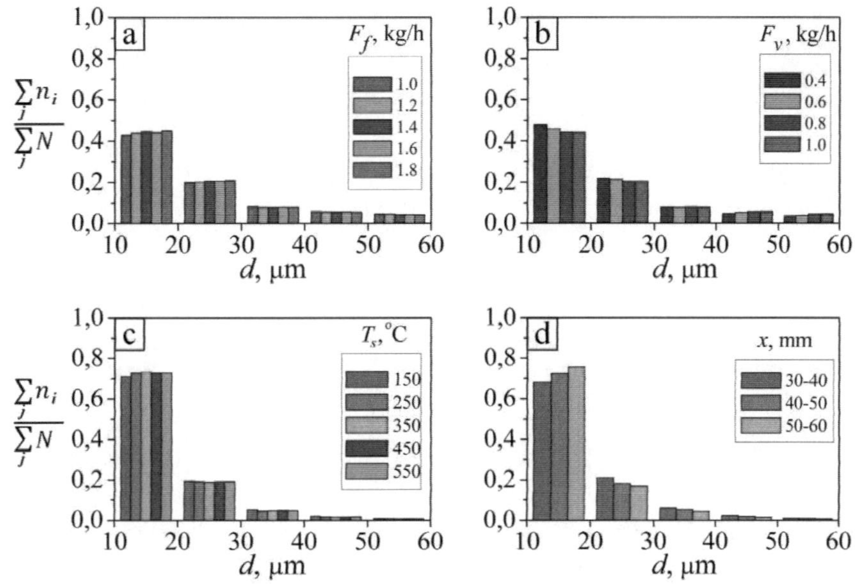


**Figure 3.** The dispersed composition of the gas-droplet flow formed by spraying the exhaust transmission motor oil with a jet of superheated steam:

(a) at constant fuel flow rate $F_f$ =1.6 kg/h ($T_s$=(260±10) and oil temperature of 40 °C (IPI-measurements);
(b) at constant steam flow rate $F_v$=0.6 kg/h ($T_s$=(260±10) °C) and oil temperature of 40 °C (IPI-measurements);
(c) at constant steam flow rate $F_v$=0.6 kg/h, fuel flow rate $F_f$=1.0 kg/h and oil temperature of 40 °C (SP-measurements);
(d) at constant steam flow rate $F_v$=0.6 kg/h, fuel flow rate $F_f$=1.0 kg/h and oil temperature of 40 °C ($T_s$=(260±10) °C) at different distances from the nozzle (SP- measurements).
($n_i$ is the number of drops with dimensions from the i-th range in the j-th image, and $N$ is the total number of droplets identified by the algorithm in the j-th image, j = 1…200)

## 4. Conclusion

Using the method of shadow photography and interferometric method, the dispersed composition of the gas-droplet flow has been studied at spraying the spent transmission oil by a jet of superheated steam at different operating parameters corresponding to the stable operation of the burner under study. It is shown that the characteristic size of the identified droplets is about 10-20 microns in the whole range of the studied regimes, which is sufficient for efficient combustion of liquid fuel. Steam temperatures above 250 °C do not affect spray performance. The obtained experimental data are in demand for numerical calculations of the combustion processes of liquid hydrocarbons with atomization.

**Acknowledgments**
The study was financially supported by grant of Russian Science Foundation (Project No. 18-79-10134).

## References

1. Kuznetsov G V, Strizhak P A, Volkov R S and Vysokomornaya O V 2016 *Int. J. Therm. Sci.* **108** 218-234

2. Watanabe H, Shoji Y, Yamagaki T, Hayashi J, Akamatsu F and Okazaki K 2016 *Fuel* **182** 259-265
3. Lozhkin Yu A, Markovich D M, Pakhomov M A and Terekhov V I 2014 *Thermophys. and Aeromech.* **21**(3) 294-307
4. Vigriyanov M S, Alekseenko S V, Sharypov O V and Anufriev I S 2019 *RF Patent* 2678150 *(in Russian)*
5. Alekseenko S V, Anufriev I S, Vigriyanov M S, Kopyev E P and Sharypov O V 2016 *Combust. Expl. Shock Waves* **52**(3) 286-293
6. Danis A M, Namer I and Cernansky N P 1988 *Combust. Flame* **74**(3) 285-294

# Investigating characteristics of diesel fuel spray in a perspective burner

I S Anufriev, E P Kopyev, M A Mukhina, Ya A Osintsev and E Yu Shadrin

Kutateladze Institute of Thermophysics SB RAS, Novosibirsk, Russia

E-mail: evgen_zavita@mail.ru

**Abstract.** Gas-droplet flow at diesel fuel spraying by superheated steam jet has been experimentally investigated in this work with the use of interferometric method (IPI). Data on the dispersed composition of the steam-fuel jet have been obtained in a wide range of operating parameters (steam and fuel flow rates) corresponding to the area of stable operation of the burner. Dependences of droplet sizes in the flow on the regime parameters have been established.

## Introduction

Developing methods for creating dispersed gas-liquid flows with specified parameters is an important scientific and technical task. Gas-droplet flows are widely used in various fields: medicine, energy, etc. One of the important trends in the use of such flows is the combustion of liquid hydrocarbon fuels. The efficiency of combustion and emission of harmful combustion products largely depends on the characteristics of the liquid fuel spray, which determines the intensity of the interphase exchange and the composition of the fuel mixture. Understanding the patterns of fuel atomization is necessary to control the combustion process, used in a wide range of industrial technologies. The development of new efficient and environmentally safe methods of fuel combustion with spraying requires a detailed study of spray devices, i.e. nozzles, using laboratory modeling. Obtaining detailed information about the structure and dispersed composition of the gas-droplet fuel jet using modern optical methods (SP, IPI) [1-2] is important for optimizing the equipment and the combustion process as a whole.

One of the promising devices for fuel combustion is a burner developed by the staff of IT SB RAS [3]. This burner uses superheated steam for fuel atomization. Its distinctive feature is the lack of contact between the fuel and the surface of the spray nozzle, which prevents from possible coking of surfaces and subsequent failures in operation. In order to find the optimal operating parameters for the creation and subsequent ignition of the gas-droplet flow, there is a need in a detailed experimental study of the characteristics of fuel atomization by the nozzle used in this burner.

## Experimental setup and technique

In this paper, the atomization of liquid hydrocarbons by a jet of superheated steam in a direct-flow burner (without combustion) is studied by the example of diesel fuel. This type of fuel, as the most common and widely used liquid fuel, is chosen to investigate the base characteristics of the spray unit.

The process of atomization and combustion of liquid fuels is as follows (Fig. 1-a). The steam flows out of the nozzle (diameter of 0.5 mm) in the form of a jet. Liquid fuel flows freely into its base and interacts with the steam jet to form a finely dispersed gas-droplet flow. At the same time, superheated water vapor increases the temperature of fuel droplets, thereby improving the mixture formation and fostering its stable ignition. Along with this, in the combustion zone there is a steam gasification of the thermal decomposition products of the fuel, which, as in the case of evaporative burners [4], improves the combustion characteristics of liquid hydrocarbons. To reduce toxic emissions and increase the completeness of fuel combustion, it is important to ensure high uniformity of the gas flow, the smallest possible fragmentation of fuel, and high stability of the torch.

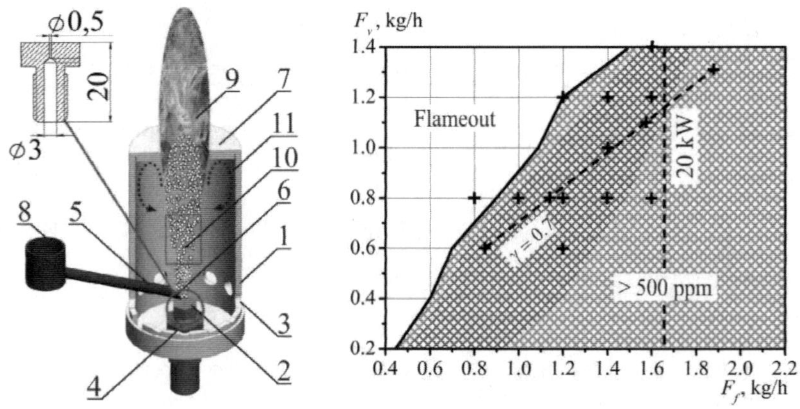


(a) (b)

**Figure 1**. (a) Scheme of the experimental burner: cylindrical housing – 1, steam atomizer – 2, air supply holes – 3, steam line – 4, fuel feed tube – 5, chamfer – 6, nozzle – 7, fuel receiver – 8, torch – 9, steam-fuel jet – 10, recirculation zone – 11; (b) Regime map of the burner ($T_s$=250°C): I – flame blowout; II – zone of "ecological" stable combustion; III – flame with a high content of CO in combustion products (> 500 ppm); "+" – regimes studied with IPI.

Preliminary studies [5] of this burner design (Fig. 1-b) allowed determining the conditions for stable combustion of diesel fuel. The investigation concerned the regimes in the range of steam flow rate $F_v$ corresponding to the working range of the metering water pump and to the performance of laboratory electric steam generator, required to superheat the steam to a predetermined temperature ($T_s$ 550 °C). The limits of variation of fuel flow rate $F_f$ corresponded to the permissible power of the burner in laboratory measurements ($\gamma = F_v/F_f$ is the relative mass flow rate of steam). The temperature of superheated steam in the experiments was set constant $T_s$=(250±10) °C, since it had been previously found that a further increase in the degree of steam superheating (at $T_s$>250 °C) did not affect the combustion process. Depending on $F_v$ and $T_s$ values, steam pressure ranged from 3 to 11 bar, and steam overheating $T_s - T_b$ reached 100 °C ($T_b$ – saturated steam temperature).

In accordance with the data obtained for combustion modes, this paper studies characteristics of fuel dispersion without combustion in the same range of fuel and steam flow rates (their values are shown in Fig. 1-b with symbol "+"). The measurements were carried out in the flow region 10, shown in Fig. 1-a, where the dispersed fuel evaporated intensively and mixed with water vapor.

The characteristics of liquid fuel spraying by a steam jet were studied by a non-contact optical method for diagnosing flows, namely an interferometric method for determining droplet diameters (IPI) [1, 6-7]. The method is based on the registration of defocused droplet images, illuminated by a laser sheet. According to the scattering theory [6-7], the light reflected and two times refracted by the spherical surface of the droplet creates interference fringes in the droplet images. Their frequency directly depends on the droplet diameter. Fig. 2 shows an example of an image obtained using the IPI method. Digital analysis of the obtained images allows determining the position and size of the droplets suspended in the stream. This method serves to measure particles with sizes from 10 microns.

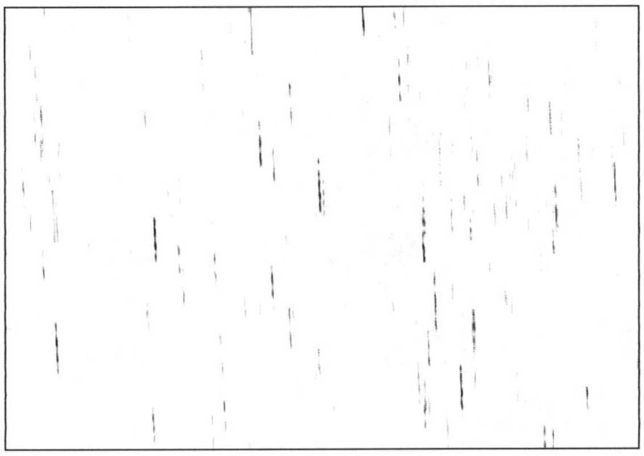

**Figure 2**. The inverted images of the investigated droplets obtained by the IPI method.

To conduct the experiments, the "Polis" measuring complex was used. It includes: CCD camera ImperX B4820-M (resolution 4904×3280 pixels, shooting frequency of 3.2 Hz, and minimum inter-frame delay of 200 ns) and a Nikon macro lens with a focal length of 105 mm. To compress the image in one direction, an optical compression unit specially designed for the lens was used. A pulsed laser Nd:YAG QuantelEVG (wavelength – 532 nm, pulse energy – up to 145 MJ, and pulse duration – 10 ns) was used as a light source.

In digital processing, the ActualFlow software with IPI Kit package was used to implement the following algorithms:

- Image search algorithm. When searching for images, the scales corresponding to the size of the droplet image are selected by applying a continuous wavelet transform to the image, and the intensity maxima on the resulting image correspond to the position of the images. To eliminate failures, the algorithm includes a step of the found image filtering by the following parameters: the average brightness of the image; the brightness of the image relative to the background; and the distance between the maximums (elimination of multiple operations of the algorithm in one image).
- Frequency calculation algorithm. The frequency of interference fringes in the image is determined by the maximum in the amplitude Fourier spectrum, built over the contour along the image of the drop.
- Algorithm for screening images. In this algorithm, the following filters are used: screening of images located on the image border; screening by the signal/noise ratio; screening by the minimum and maximum distance between the interference fringes.
- Calibration algorithm. This algorithm is used to convert the frequency of interference fringes in identified images into the droplet size. Input parameters are refractive indices of environment (air) and atomized liquid (fuel); laser radiation wavelength; the size of one pixel of the camera matrix; as well as the defocus distance and optical magnification obtained from preliminary calibrations of the optical system.

**Results**

For each mode under study, a series of 100 images was obtained. The processing took into account the total distribution of particles in all images, normalized by the total number of identified particles. As a result of IPI-image processing, detailed information has been obtained on the disperse composition in the measuring area of the gas-droplet flow when diesel fuel is sprayed by a high-speed flow of superheated steam depending on the operating parameters (fuel flow rate and steam flow rate and temperature).

It is shown that the predominant particle size in all studied regimes is in the range of 10-20 μm (Fig. 3), which is a sufficient condition for efficient fuel combustion [7] and indicates the advantages of the method used for dispersing liquid fuel.

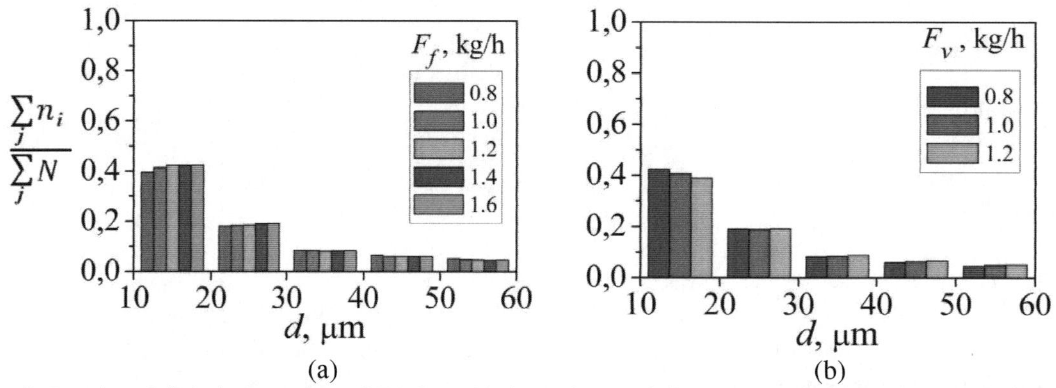


**Figure 3.** Results of digital processing of IPI data: (a) dependence of disperse composition of diesel fuel spray on fuel flow rate ($F_f$), $F_v$ = 0.8 kg/h; (b) dependence of disperse composition of diesel fuel spray on vapour flow rate ($F_v$), $F_f$ = 1.4 kg/h, $T_s$ = 250 °C.
($n_i$ – number of droplets with sizes of the $i$-th range in the $j$-th image, $N$ – total number of droplets identified by the algorithm in the $j$-th image, $j$ = 1…100)

With changes in fuel flow rate in the studied range (at a fixed steam flow rate) droplet sizes do not change (Fig. 3-a). Therefore, flameout and high CO content in combustion products observed in certain modes (Fig. 1-b) are not related to the dispersed composition of the spray.

With an increase in steam flow rate from 0.8 to 1.2 kg/h at constant fuel flow rate (1.4 kg/h), there is a slight decrease in the share of identified droplets of predominant size (10-20 μm) and an increase in the number of larger droplets (Fig. 3-b). Due to a high flow rate the time of phase interaction decreases and less fuel evaporates, respectively.

## Conclusion

In this work, using the interferometric method for determining the droplet diameter, the gas-droplet flow has been experimentally studied at diesel fuel spraying by a jet of superheated water vapor in a straight-flow burner (without combustion). In a wide range of operating parameters corresponding to different combustion modes, size distributions of fuel droplets have been obtained. The characteristic size of the identified droplets in all studied regimes is shown to be 10-20 μm, which is sufficient for efficient combustion of liquid fuel. It has been revealed that the modes of flameout with high content of CO in the combustion products are not associated with the dispersed composition of the spray, but are caused by fuel to oxidizer ratio that is non-optimal for combustion. The obtained experimental data are in demand for numerical calculations of the combustion processes of dispersed liquid hydrocarbons.

## Acknowledgments

This work was carried out within the framework of the state contract with the IT SB RAS.

## References

1. Kuznetsov G V, Strizhak P A, Volkov R S and Vysokomornaya O V 2016 *Int. J. Therm. Sci.* **108** 218-234
2. Watanabe H, Shoji Y, Yamagaki T, Hayashi J, Akamatsu F and Okazaki K 2016 *Fuel* **182** 259-265
3. Vigriyanov M S, Alekseenko S V, Sharypov O V and Anufriev I S 2019 *RF Patent* 2678150 *(in Russian)*
4. Alekseenko S V, Anufriev I S, Vigriyanov M S, Kopyev E P and Sharypov O V 2016 *Combus. Expl. Shock Waves* **52**(3) 286-293

5. I S Anufriev, Kopyev E P, Sharypov O V, Arsent'ev S S, Vigriyanov M S and Osintsev Ya A 2018 *J. Phys.: Conf. Series* **1105** 012036
6. Lozhkin Yu A, Markovich D M, Pakhomov M A and Terekhov V I 2014 *Thermophys. Aeromech.* **21**(3) 294-307
7. Danis A M, Namer I and Cernansky N P 1988 *Combust. Flame* **74**(3) 285-294

5th International Workshop on Heat/Mass Transfer Advances
for Energy Conservation and Pollution Control
August 13-16, 2019, Novosibirsk, Russia

# Investigation of applicability area for turbulence models in the problems of mass transfer intensification by the control of a rotary divergent flow

**I.K. Kabardin[1], M. Kh. Pravdina[1], N.I. Yavorsky[1], V.I. Polyakova[1], V.G. Meledin[1], D. P. Ezendeeva[1], S.V. Kakaulin[1], M.R. Gordienko[1], D.V. Kulikov[1], O. Yu. Sadbakov[1], I.A. Klimonov[2] and E.V. Usov[2]**

[1] Kutateladze Institute of Thermophysics SB RAS, Lavrentyev ave. 1,, Novosibirsk, 630090, Russia

[2] Nuclear Safety Institute RAS, 52, Bolshaya Tulskaya Street, Moscow, 115191, Russia

E-mail: kabardin@itp.nsc.ru

**Abstract.** The fundamental scientific problem of this kind arises in mechanics, chemistry and catalysis in the intensification of mass transfer. The turbulent mass transfer characteristics were analyzed experimentally in a controlled rotational-divergent flow. The laser Doppler anemometry was used for turbulent mass transfer diagnostic. The verification of numerical calculations by computational hydrodynamics has been carried out according to the results of experimental studies. Calculation of turbulent kinematic is build on solving of a system of continuity and the Navier-Stokes equations. The fluid is incompressible and isothermal. For averaged equations closure, semi-empirical turbulence models were used: the k-ε model of turbulence and the Reynolds stresses transfer model. The calculation has showed the existanceof a near-wall jet after a rotating device and the generation of reverse flow zones. Comparison with experimental data has shown that all models of turbulence adequately model the rotational-divergent flow only until the formation of flow gaps and the formation of reverse flow zones. The discrepancy with the experiment occurs while controlling the flow to equalize the velocity profiles.

## 1. Introduction

The fundamental scientific problem of this kind arises in mechanics, chemistry and catalysis in the intensification of mass transfer [1-3]. The efficiency of mass transfer is influenced by the velocity field and small-scale turbulence in the reacting flow. Reagent streams in real installations and reactors interact under conditions of complex geometry with turning and expanding sections in which wall jets are formed, flow reconnections occur, reverse flow zones are present. Such effects usually increase the hydrodynamic resistance of the flow and disturb the turbulent mass transfer uniformity. For the numerical simulation of such phenomena, verification and modernization of turbulence models are required for an adequate description of mass transfer [4]. The verification of numerical calculations by computational hydrodynamics has been carried out according to the results of experimental studies. Numerical simulation of turbulent gas flow is based on solving of a system of continuity, the Navier-Stokes and energy conservation equations. The medium is considered incompressible and isothermal.

For averaged equations closure, semi-empirical models of turbulence were used: the k-ε model [5] of turbulence (Launder 1984) and the Reynolds stresses transfer model [6-7]. Velocity fields measurements in the control sections has carried out compare the experimental data with computational one. A noncontact optical laser Doppler anemometry method for achieving the kinematic flow properties is necessary. TheLAD-05 on the basis of this method system was made at the IT SB RAS.

## 2. Experimental setup and technique

For investigation of applicability area for turbulence models in the problems of mass transfer intensification by the control of a rotary divergent flow the experimental model was made (Fig. 1) and an experimental setup for flow characteristics investigation in this model. The test section is made from several functional blocks: the inlet; section before the bend, the bend, the control section and the section before the second bend. Fluid was through the fan. Then fluid pass through the flow meter. The temperature in the flow is measured by thermal transducer, and the overpressure by pressure sensors. The hydraulic resistance of the flow is measured on the differential pressure on a bend and a section before the second bend using the differential pressure transmitters. The working fluid was air. The flow rate was equal 50,100,150,200 and 250 n.m3/h. The maximum overpressure was 0.03 MPa. The temperature varied from 23 to 26 ° C. The Reynolds varied from 10000 to 50000.

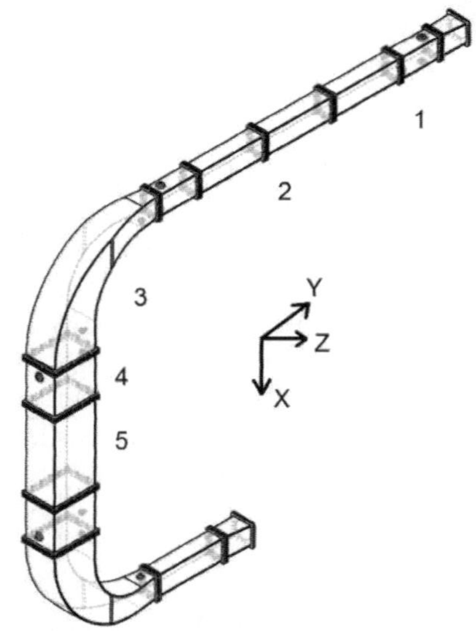


**Figure 1.** 3D geometry of the investigated model (1- the inlet; 2- section before the bend, 3 - bend, 4- control section, 5 - section before the second bend)

### *2.1. Velocity measuring in the experiment*

For numerical calculations verification, velocity fields measurements in the model cross-sections were carried out. Measurement of the two-dimensional kinematic characteristics of the flow required a modern invasive optical methodic for aerodynamic flow diagnostics. As methodic laser Doppler anemometry was used (LDA). The LAD-05 system based on LDA method was designed and manufactured at the Institute of Thermophysics SB RAS (Fig. 2). The measurement system was mounted on the actuator.

Journal of Physics: Conference Series **1369** (2019) 012034 doi:10.1088/1742-6596/1369/1/012034

**Figure 2.** Measuring system LAD-05

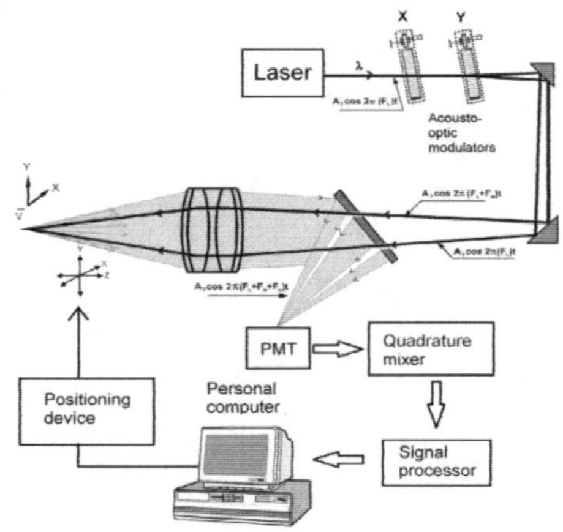


**Figure 3.** The functional scheme of the measurement system

The functional scheme of the measurument system conclude a laser, two orthogonal acoustic-optical modulators, a receiving-transmitting optical system, and a photoelectron multiplier. A two-channel Doppler signal processor was connected to the photodetector and transmits the processed data to the computer. Scanning of the measuring volume is carried out by moving the optical unit by the coordinate-positioning device. To measure the velocity, the laser beam after passing the elements is sent to the acousto-optic switch-modulator (Fig. 3). The action of the modulating voltage at the output of the modulator serves to form two light beams, diffracted in zero and minus the first order. The split beams pass sequentially through the system of optical elements and are directed by the lens to the studied flow region. Intersecting in the flow, laser beams form an interference pattern with a known spatial-temporal periodic structure. When the scattering particle crosses the probing optical field at the output of the photodetector the radio pulse of photoelectric current appears. Its Doppler frequency is a known linear velocity function, and the duration is equal to the time of passage of the measuring volume by the particle. As a result, measurement system can measure two velocity vector projections with value from 0.001 to 400 m/s with relative error less than $10^{-1}$ percent. The measurument volume is 0.1x0.1x0.5 mm. The actuator can move the measuring system in the range of 250 x 250 x 250 mm with step of 0.1 mm (Fig. 4).

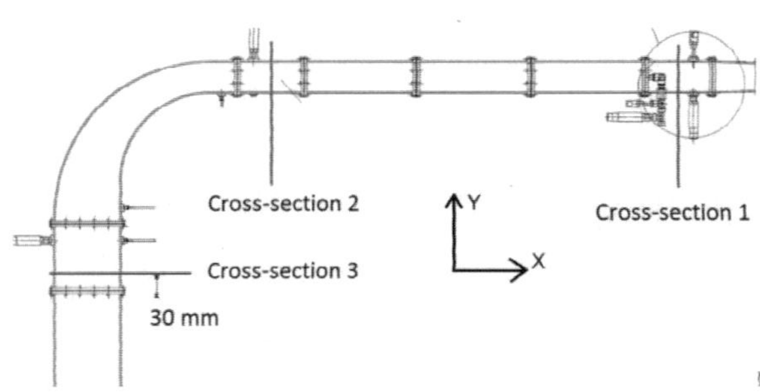


**Figure 4.** Three measurements cross-section in the model

## 3. Numerical simulation methods

The validation of CFD-calculation was carried out. The velocity and pressure distribution in the test pipe were simulated. CFD-calculation of turbulent gas flow was based on the solution of the system of continuity equations and Navier-Stokes equations [4]. The medium was considered incompressible and isothermal. The semiempirical turbulence models is used for closing the averaged equations. These models are the k-ε turbulence model, [5] and the model of Reynolds stress transfer [6-7]. The nonuniform grid with 1 500 000 cells was used in computations. The computation grid was condensed near the walls to get resolution of laminar sublayer near the wall. Near the walls the computational grid was rectangular and in the flow core the computational grid was triangular. The cell size near the walls was about 0.2 mm and in the flow core it was about 8 mm. In the inlet the experimental measured velocity distribution was used to get better accuracy of computation. The condition on the constant pressure of 1 bar was set at exit of the gas.

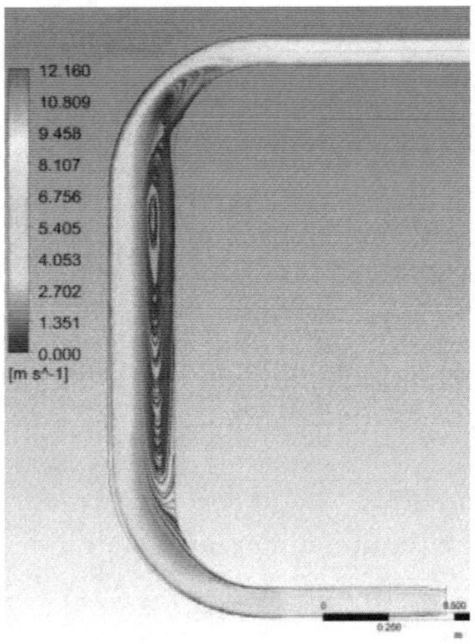


**Figure 5.** The picture of the streamlines after the rotator, obtained using the Reynolds stress transfer model. The colour bar represent the velocity magnitude in the flow

## 4. Results

Vx components values were measured in the middle plane (Fig. 5). The results are shown in Figure 6. The CFD-results for the longitudinal velocity component in middle cross sections line are shown in Figure 7. Comparison of experimental and calculated data shows that for the flow after the rotation, there are noticeable differences for all models. The best comparison is for the Reynolds-Stress transfer model.

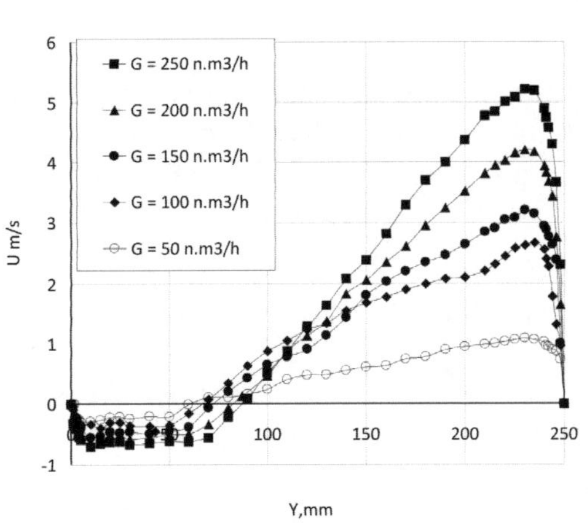


**Figure 6.** Velocity profiles after the bend at different values of mass flow rate

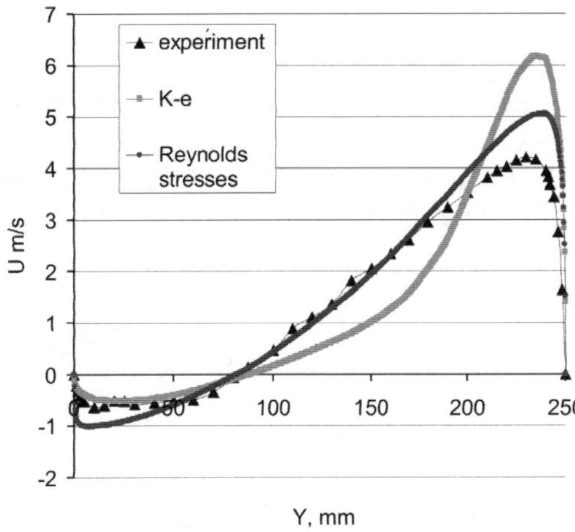


**Figure 7.** Comparison of experiment and calculation for flow rate G = 200 n.m3/h

As characteristics of applicability area for turbulence models the width of recirculation zone in the channel after bend was chosen. The recirculation zone width in cross-section 3 after the bend for different values of flow rate are shown in figure 8. From experimental data it can be seen that the recirculation zone width takes values X/Xmax = 0.24÷ 0.36, where Xmax is 250 mm. For Reynolds-Stresses model the recirculation zone width take close values to the experimental one till flow rate of 200 n.m3/h. For k-ε turbulence model the values of recirculation zone differ from experimental widths. The model of Reynolds stress transfer adequately describes the flow in the range of flow rates from 50 to 250 n.m3/h. And k-ε turbulence adequately describes the flow in the range 200-250 n.m$^3$/h.

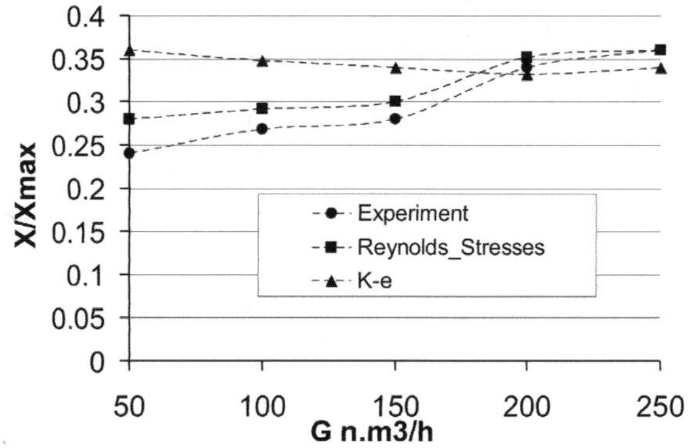


**Figure 8.** The recirculation zone width in cross-section 3 after the bend for different flow rate values

## 5. Conclusion

The turbulent mass transfer characteristics were analyzed experimentally in a controlled rotational-divergent flow. The laser Doppler anemometry was used for turbulent mass transfer diagnostic. The verification of numerical calculations by computational hydrodynamics has been carried out according to the results of experimental studies. Numerical simulation of turbulent gas flow is based on solving of a system of continuity, the Navier-Stokes and energy conservation equations. The medium is considered incompressible and isothermal. For averaged equations closure, semi-empirical models of turbulence were used: the k-ε model of turbulence and the Reynolds stresses transfer model. The model of Reynolds stress transfer adequately describes the flow in range of flow rates from 50 to 250 n.m3/h. And k-ε turbulence adequately describes the flow in the range 200-250 n.m$^3$/h.

**Acknowledgements**

The experimental study of flow kinematic properties was carried out under state contract with IT SB RAS (AAAA-A18-118051690120-2), the study of applicability area for turbulence models in the problems of mass transfer intensification by the control of a rotary divergent flow was financially supported by RFBR (Project No. 18-31-20036 ).

## References

[1] D. Jecha, J.Martinec, V. Brummer , P. Stehlík, P. Leštinský Metal catalysts for VOC oxidation, Chemical engineering transaction, 3, 1-6.( 2013)

[2] B.S. Bal'zhinimaev, I. V. Kovalyov., V.V. Kaichev , A.P. Suknev , V.I. Zaikovskii Catalytic abatement of VOC over novel Pt fiberglass catalysts, Topics in Catalysis, 60,73-82 (2017).

[3] B.S. Bal'zhinimaev, V.N. Parmon The Innovative Russian Approaches to Catalystss Design: New Generation of Fiberglass Catalysts, Topics in Catalysis, 55, 1289-1296 (2017).

[4] H.K. Versteeg, W. Malalasekera An Introduction to Computational Fluid Dynamics. The Finite Volume Method. ( Longman Scientific and Technical). 257 (1995).

[5] B.E. Launder, D.B. Spalding The Numerical Computation of Turbulent Flows, Computer methods in applied mechanics and engineering, 3, 269–289 (1974).

[6] D. C. Wilcox Turbulence modeling for CFD ( DCW Industries. Inc. La Canada. California), (1998).

[7] D.K. Walters, D. A. Cokljat Three-Equation Eddy-Viscosity Model for Reynolds-Averaged Navier-Stokes Simulations of Transitional Flow, Journal of Fluids Engineering, 130, 121401-1 – 121401-14(2008).

5th International Workshop on Heat/Mass Transfer Advances
for Energy Conservation and Pollution Control
August 13-16, 2019, Novosibirsk, Russia

# Investigation of high speed steam jet effect on combustion of substandard liquid hydrocarbons

**E.P. Kopyev [1], I.S. Anufriev [1,2], Ya.A. Osintsev[1,2] and M.A. Mukhina[1,3]**

1 Kutateladze Institute of Thermophysics SB RAS, Novosibirsk, 630090, Russia

2 Novosibirsk State Technical University, Novosibirsk, 630073, Russia

3 Novosibirsk State University, Novosibirsk, 630090, Russia

E-mail: kopyeve@itp.nsc.ru

**Abstract**

Combustion of liquid hydrocarbon fuel dispersed by superheated steam jet was experimentally studied. Characteristics of the combustion process (heat release, emission of toxic products, flame temperature) were investigated in a laboratory direct-flow atmospheric burner (capacity up to 20 kW) at different operating systems with waste transmission automotive oil as fuel example. Regimes were found which provoke stable combustion of non-quality liquid hydrocarbons. It was shown dependences of heat, flue gas parameters on superheated steam flow rates, fuel flow rates, flame temperatures.

## Introduction

It is known that adding water to the process of burning organic fuel can be used in different ways. One of them is to disperse heavy fuels, suppress detonation in combustion chambers and reduce heat load [1]. Also, it can be applied at water-fuel emulsion creation [2-4]. In addition, to reduce the concentration of $NO_x$ [5] by lowering the flame temperature, water is also used. This is due to the fact that in the presence of water vapor there is an increase in the heat capacity of the combustible mixture. At considering the use of the technology in gas turbines some researchers note that water vapor has the greatest effect on suppressing the formation of thermal $NO_x$ compared to $CO_2$ and $N_2$ when using the technology of recirculation of waste gases [6-9]. But when considering works that investigate water influence on liquid fuel combustion, for example, articles with study with kerosene spray [10, 11], it can be noted that this point is not fully understood today. Many important questions are still open and require further study.

A method of combustion liquid hydrocarbons in a superheated steam presence [12], based on gasification of liquid hydrocarbons incomplete combustion carbon-containing particles, was proposed and investigated on evaporator-type burners in works [13, 14]. It was shown that superheated steam positively effects on the main combustion characteristics: stabilization of the ignition, increase combustion efficiency, decrease the content of flue gas toxic components. This method can be used for environmentally friendly disposal of low and non-quality liquid hydrocarbons and combustible industrial waste with heat produce. Despite the positively steam effect, utilization of high viscosity fuel in evaporative burners is difficult due to relatively fast coking of burner surfaces. Therefore, to burn off-grade liquid hydrocarbons in the steam gasification mode, it is necessary to develop and research fundamentally different ways of supplying fuel and mixing it with water vapor.

One of the promising methods of spraying substandard liquid hydrocarbons is a technical solution based on the interaction of a liquid with a high-speed steam flow without contact of the fuel with a nozzle [15]. This method of dispersion has significant technological advantages when using substandard liquid fuels associated with preventing coking of the nozzle and clogging of the fuel supply channels, which improves the performance and reliability of the burner.

In the present work, combustion of liquid hydrocarbon fuel dispersed by superheated steam jet was experimentally studied in a laboratory burner device (capacity up to 20 kW).

In this paper, the process of burning liquid hydrocarbons sprayed by a superheated steam jet is experimentally studied using the waste transmission automotive oil as an example. The main goal of the research is to study the effect of regime parameters on the characteristics of the combustion process.

**Experimental equipment and methods**

The investigation was conducted on an experimental setup, which consists of a steam and fuel supply systems. The steam with necessary parameters (flow rate up to 1.4 kg / h, temperature up to 550 ° C) is produced by an electric steam generator (there are three heated elements with a summary power consumption of 2.2 kW). The water, which is required for steam generating, is provided by a plunger dosing pump ND 0.5R 1.6/100 K14A (accuracy class - 0.5). The fuel (up to 2 kg / h) is supplied by a fuel pump thought prefilter and dosed by a fuel injector. In case of high fuel viscosity, a heating system is applied (up to 110 ° C). To control the steam and fuel flow rates, Acom PC-100W-5 and Acom PC-100W-10H electronic scales (limit of permissible errors are 0.5 g and 1 g) are used.

Fig. 1 shows the photograph and the scheme of work of the atmospheric burner laboratory sample. The burner device (Fig. 1) includes several main components. They are: a gas generation chamber, consisting from a body and an output nozzle; a base with installed coaxially in the center and oriented vertically upwards a steam nozzle (diameter 0.6 mm, the opening angle of 17°), connected to a holder with a steam line; a fuel pipeline with a receiver. The device size is 50 mm in external diameter, 140 mm in the height. The outlet nozzle has a diameter of 25 mm. The fuel pipeline is a tube with a beveled end. It is placed at an acute angle to the burner base that this end is placed close to the steam jet base. This design allows supplying a steady fuel flow rate for further formation of a quality homogeneous gas-droplet flow. Also, the gas generation chamber lower part has holes providing an air supply from the atmosphere into the reaction zone for fuel ignition in the device. The steam nozzle connected through the holder with the steam line to the electric steam generator. Thus, the laboratory direct-flow atmospheric burner has the following scheme of liquid fuel spraying and burning. The water pump supplies water with a given flow rate to an electric steam generator heated to the desired temperature. The produced steam flows from the nozzle in the burner and heats it. The power control system of the steam generator ensures desired superheated steam temperature. A liquid fuel with a given flow rate is supplied through the fuel tube into the steam jet base, as a result, a uniform fine gas-droplet flow is formed. In addition to spraying the fuel, superheated water vapor raises the temperature of the fuel droplets, which intensifies the mass exchange and the mixture formation, contributing to steady ignition. Also, a recirculation area is formed in the peripheral zone near the inner plane of the outlet nozzle due to jet structure. In this way here ignition is initiated during start-up and the ignition of the steam-oil jet during device operation stabilizes. At the same time, in the combustion zone, steam gasification of the products of thermal decomposition of the fuel ($H_2O + C \rightarrow CO + H_2$) occurs, which also influences the combustion process. The resulting combustible mixture of CO and $H_2$ mixing with oxygen from the external atmosphere burns in the flame. The burner design is protected by RF patent [15].

Parameters, which were experimentally studied and used to analyze the characteristics of the combustion method of liquid fuels, were: heat release, flue gas composition and flame temperature.

Heat ($q$) and the thermal power $W$ of the combustion of the waste transmission automotive oil sprayed by superheated steam was measured by a flow calorimeter [13] at different burner regimes. The burner flame was inserted into the internal channel of the calorimeter after establishing a constant difference in water temperature at the inlet and the outlet. The temperature of the coolant (water) was measured at steady state thermal conditions by using chromel-alumel thermocouples at the inlet and outlet of the calorimeter. All results were recalculated as the difference between the thermal energy received by the calorimeter (calculated by heat equation) and the energy spent on maintaining the process (per unit time) [13].

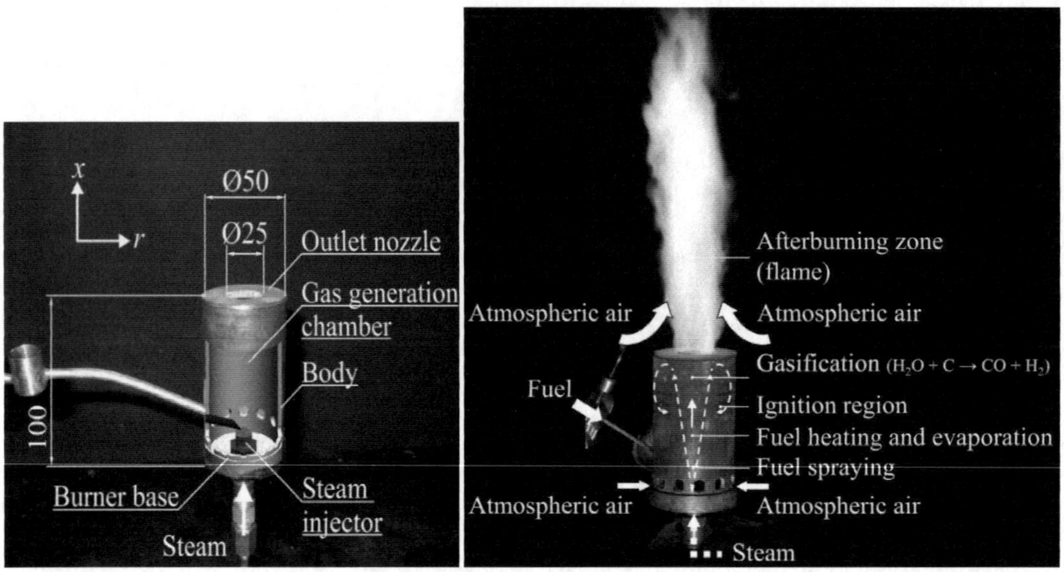


Fig.1. The burner device scheme

The flue gas composition ($O_2$, CO, NO, $NO_2$, $SO_2$, $CO_2$) was controlled by a TESTO 350 gas analyzer (5% error). A sampling of the reaction products was carried out at the exit of the calorimeter.

To control the burner operation modes, the flame temperature was measured using a platinum-rhodium-platinum-rhodium thermocouple (0.3 mm diameter). Temperature profiles were measured in an external plume in the plane of symmetry of the burner. The thermocouple was moved using an automated coordinate-moving device (spatial step 10 mm along the torch and 3 mm in the radial direction, the measurement time at the point was at least 10 s, the delay time before measurements at the point was 7 s).

**Results**

Experiments on the measurement of heat generation, emission of toxic products and flame temperature were carried out at various operating burner regimes: at the steam mass flow rate $F_v$ in the range of 0.2–1.4 kg/h and fuel flow rate $F_f$ in the range of 0.4–2.2 kg/h (deviations when adjusting from the given average values of rates - within 5% for steam and 2.5% for fuel). The range of $F_v$ values corresponds to the working range of the water pump, as well as the performance of the laboratory steam generator, necessary for overheating the steam to a predetermined temperature. The limits of fuel flow rate $F_f$ correspond to the allowable power of the burner during laboratory measurements.

It was previously established [16] that the temperature of superheated water vapor has a weak effect on the main indicators of liquid fuel combustion, so the measurements were carried out at a constant steam temperature $T_s$ = (260±10)°C, which ensures sufficient steam overheating for the reaction and – minimum energy consumption for steam overheating.

At the measurements waste transmission oil was used (density - 863.25 kg/m3, viscosity - 41.8 mm2/s, the highest calorific value - 45.318 MJ/kg, C - 85.6%, H - 13.2%, S - 0.6% (by weight)). Measurements were carried out for characteristic regimes in the region of stable combustion along the line $\gamma = 0.6$ (Fig. 2), where $\gamma = F_v / F_f$ is the relative mass flow rate of steam. The oil was preheated to a temperature of 60°C to reduce the viscosity.

Fig. 2 also shows a map of the CO concentration in various modes of operation of the burner. The boundaries of the flameout area are based on the results of visual observations of flameout, which depends on the oxidizer/fuel ratio. An approximating straight line was also constructed for the minimum values of CO: $F_v = 1.2F_f - 0.7$ (relationship of steam supply for various power modes of the burner operation).

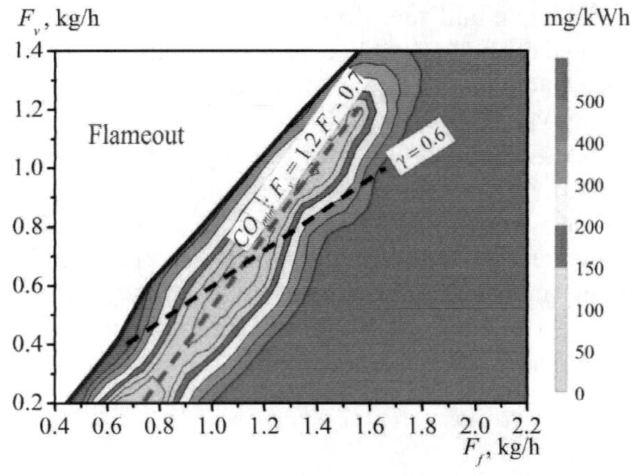


Fig.2. The map of CO concentration at the waste automotive oil combustion in a jet of superheated steam at $T_s = (260\pm10)$°C.

Fig. 3 shows the dependences of heat generation $q$, useful heat power $W$, the amount of CO and $NO_x$ per 1 kWh of released thermal energy and the temperature of the external flame T from the steam flow rate at a constant parameter $\gamma = 0.6$. The maximum value $q \sim 44$ MJ/kg is achieved with a burner power $W = 14$ kW and steam flow rate $F_v = 0.7$ kg/h (Fig. 3-a); fuel combustion efficiency ratio = 97%. At the same time, the minimum amount of CO and $NO_x$ ({CO} = 20 mg / kWh, {$NO_x$} = 60 mg/kWh) is achieved, Fig.3-b. In the area of increasing the {CO} value, the q values decrease sharply, which indicates incomplete combustion of the fuel. {$NO_x$} values decrease with an increase in steam flow rate (Fig. 3-b), because the combustion occurs at low air consumption. The maximum temperature in the torch reaches ~ 1330°C (Fig.3-c).

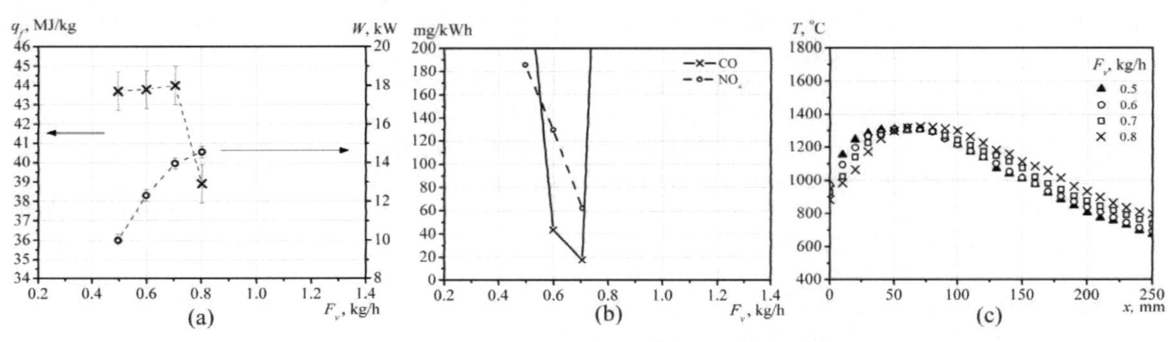


Fig.3. Dependencies of $q$, $W$, contents of {CO}, {$NO_x$} and the temperature level of the external flame $T$ ($T_s = (260\pm10)$°C) from steam flow rate at a constant parameter $\gamma = 0.6$.

## Conclusions

Using the example of waste transmission automotive oil, the process of burning liquid hydrocarbons sprayed by superheated water vapor was experimentally investigated. The experiments were carried out on a laboratory sample of a burner device (up to 20 kW), which implements a promising method of dispersing substandard liquid hydrocarbons due to the interaction of a liquid with a high-speed steam flow, without contacting the fuel with a nozzle. This method has significant technological advantages associated with the prevention of coking of the nozzle and clogging of the fuel supply channels, which improves the performance and reliability of the burner.

The effect of operating parameters (steam and fuel flow rate) on the characteristics of the combustion process is studied: heat generation and composition of combustion products, flame temperature. Depending on the steam and fuel flow rates, a regime map was obtained, including: the flame breaking boundary was established, a stable combustion area and a zone with a high CO content in the combustion products were found, and steam supply dependency was obtained for various power modes, which ensured minimum values CO.

The maximum value of 44 MJ / kg is achieved with a burner power of 14 kW and steam flow rate of 0.7 kg/h, the coefficient of completeness of combustion of fuel is 97%. In this case, the minimum amount (per the released thermal energy) of carbon monoxide 20 mg/kWh and nitrogen oxides 60 mg/kWh is produced. The maximum temperature in the flame reaches 1330°C.

Thus, the possibility of substandard liquid hydrocarbons combustion in a direct-flow atmospheric burner is shown. The advantages of combustion with steam jet atomization allow using the burner for heating low-rise industrial and residential facilities.

## Acknowledgements

The studies were carried out with the financial support of the RF Ministry of Education and Science, RFMEFI60417X0185.

## References

[1] Ballester JM Fueyo N and Dopazo C 1996 *Fuel* **75** 695-705
[2] Nadeem M Rangkuti C Anuar K Haq MRU Tan IB and Shah S S 2006 *Fuel* **85** 2111-2119
[3] Samec N Kegl B and Dibble RW 2002 *Fuel* **81** 2035-2044
[4] Armas O Ballesteros R Martos FJ and Agudelo JR *Fuel* **84** 1011-1018
[5] Dryer F L 1977 *Concepts and applications, Symposium (International) on Combustion* **16** 279-295
[6] Farokhipour A Hamidpour E and Amani E 2018 *Fuel* **212** 173-176
[7] Le Cong T and Dagaut P 2009 *Energy & Fuels* **23** 725-734
[8] Lee M C Seo SB Yoon J Kim M and Yoon Y *Fuel* **102** 431-438
[9] Pugh D G Bowen PJ March R Crayford AP Runyon J Morris S Valera-Medina A and Giles A 2017*Combustion and Flame* **177** 37-38
[10] Furuhata T Kawata T Mizukoshi N and Arai M 2010 *Fuel* **89** 3119-3126
[11] Jozsa V and Sztanko K 2017 *Thermal Science* **21** 1021-1030
[12] Vigriyanov MS Salomatov VV Alekseenko SV 2003 *Russian Federation Patent No.2219435. Method of soot-free burning of fuels* (in Russian)
[13] Alekseenko SV Anufriev IS Vigriyanov MS Kopyev EP and Sharypov OV 2016 *Combust Expl Shock Waves* **52** 286-293
[14] Anufriev I Arsentyev S Vigriyanov M Kopyev E and Sharypov O 2017 *MATEC Web of Conferences* **115** 03013
[15] *Patent RF 2018105056 Steam oil burner*(in Russian)
[16] Alekseenko SV Anufriev IS Sharypov OV Kopyev EP Arsentyev SS 2018 *Proceedings of the Seventh Russian National Conference on Heat Exchange* **1** 359-363

5th International Workshop on Heat/Mass Transfer Advances
for Energy Conservation and Pollution Control
August 13-16, 2019, Novosibirsk, Russia

# Effect of the angle of rotation of structured packing layers on separation efficiency in distillation column

**A N Pavlenko[1], N I Pecherkin[1], V E Zhukov[1], A D Nazarov[1], G Meski[2], and P Houghton[2]**

[1]Kutateladze Institute of Thermophysics SB RAS, 1 Lavrentiev Ave., Novosibirsk, 630090, Russia
[2]Air Products, Inc., 7201 Hamilton Boulevard, Allentown, PA 18195-1501, USA

pecherkin@itp.nsc.ru

Abstract. It is well known that a good quality of liquid distribution over the surface of structured packings is crucial for high performance of large diameter columns. The angle of rotation of the packing layers and their number are among the main parameters characterizing the distribution of liquid in a structured packing over the column cross-section and height. This paper presents investigation results on the influence of the rotation angle of structured packing layers on separation efficiency, pressure drop, distribution of local composition of mixture and local liquid flow rate under the packing. The experiments were carried out in distillation column with the internal diameter of 0.9 m on a structured packing with a specific area of 500 $m^2/m^3$, and the corrugation angle of packing equal to 45°. The mixture of refrigerants R114 and R21 at the pressure of 3 bar and total reflux was used as the working fluid. The column was packed with 11 packing layers with rotation angles of 20, 45, 70, and 90 degrees. According to experimental results, a decrease in the angle of packing layer rotation from 90 to 20 degrees leads to a decrease in separation efficiency, while the HETP value increases by 10–12%.

## 1. Introduction

Irrigation of structured packing in distillation columns is characterized by the fact that the number of drip points in the liquid distributor is approximately an order of magnitude smaller than the number of packing channels. When single liquid jets flow onto the upper layer of the packing, liquid flows, oriented in the direction of irrigated packing sheets, are formed at the layer exit since liquid spreading along the sheets is significantly higher than in the transverse direction. In practice, to level liquid distribution over the cross-section, each subsequent layer is turned at a certain angle when assembling the column. After each rotation of the layer, uniformity of underlying layer irrigation increases. The rotation angle of the packing layers and their number are the main parameters characterizing liquid distribution in a structured packing over the cross-section and height of the column. To achieve uniform distribution of liquid over the column cross-section, several layers of packing are usually required [1].

It is well known that achieving a good quality of liquid distribution over the surface of structured packing is crucial for high efficiency of the large-diameter columns [2, 3]. The relationship of the efficiency of mixture separation with maldistribution of local flow characteristics over the column cross-section was shown in [4]. A new approach to develop the method for dynamically controlled packing irrigation by the liquid distributor is suggested in [5]. This method is aimed at the destruction of large-scale stationary structures of maldistribution of the local flow rates and mixture composition over the packing cross-section and, as a result, at the achievement of high separation efficiency.

In [6], the authors studied the effect of the angle of layer rotation on the efficiency of mixture separation at different methods of structured packing irrigation. In the experiments, the rotation angle of the structured packing layers was varied taking the values of 20° and 70°. The initial maldistribution of liquid irrigation at the packing inlet was generated by closing the holes in the bottom of the liquid distributor. The results of measuring the distribution of the local liquid flow under the packing showed that under the closed row of holes in the liquid distributor at the outlet of packing with a layer rotation angle of 20°, a zone with significantly reduced flow rate was observed. At that, the efficiency of mixture separation was greatly decreased.

This paper presents the results of continuing research on the effect of the rotation angle of the structured packing layers on separation efficiency, pressure drop, distribution of local mixture composition, and local flow rate under the packing.

## 2. Experimental setup

The experiments were carried out in a column with a diameter of 0.9 m using a packing with a specific surface of 500 $m^2/m^3$. The description of the experimental setup and measurement methods are given in [7]. A mixture of refrigerants R114 and R21 at a pressure of 3 bar and total reflux was used as the working fluid. The packing was irrigated by a liquid distributor with a rectangular drip point pattern [6]. The density of drip points was 450 $1/m^2$, approximately. There were 11 packing layers in the column, each 0.2 m high. Packing layers were assembled of the corrugated aluminum sheets with the inclination angles of corrugations 45° relative to the horizon. The channels in the adjacent packing sheets were directed in opposite directions. Each subsequent layer, starting from the top, was rotated clockwise at a certain angle. The angles of layer rotation are given in Table 1. This Table also shows the data on a change in the orientation of the packing layer sheets from the top to the bottom of the column. Presentation of data on the rotation angles in radians allows us to estimate the number of a complete rotation of the packing layers in the package by 180° ($\pi$ radians) relative to the liquid distributor (column body). In the experiments, the total angle of a change in orientation of the packing sheets in the layers from the top to the bottom of the column was from 1.1 to 5 $\pi$ radians. To compare the results obtained in the packing of various heights, the Table also presents the data on a change in the position of layers per a unit of the packing height.

Table 1

| The angle of layer rotation, degrees | 20 | 45 | 70 | 90 |
|---|---|---|---|---|
| Total angle of layer rotation, $\pi$ rad | 1.11 | 2.5 | 3.9 | 5.0 |
| The total angle of layer rotation per a unit of height, $\pi$ rad/m | 0.49 | 1.1 | 1.71 | 2.2 |

In the experiments, the mixture composition was measured in the vapor and liquid phases at the inlet and outlet of the column. Using these data, the separation efficiency was determined as a height equivalent to a theoretical plate (HETP). The pressure drop across the structured packing, distribution of the local flow rate of flowing from the bottom end of the packing, and concentration of the volatile component in the flowing liquid were also measured. It should be emphasized that the data on the distribution of the local characteristics of the liquid flow, presented in the work, were obtained in a distillation column in the process of mixture separation. The local flow rates under the packing were measured by the time-of-flight method using a low-pressure flow meter with a collector diameter of 28 mm. The collector was moved according to a given program along two coordinates (radius and angle), measuring the flow rate at 420 positions over the column cross-section. In these positions, the temperature of the liquid in the collector was measured; then, it was used to determine the mixture composition in the liquid phase in the cross-section of the column. The method for calculating the composition of mixture by measuring the temperature based on relationship between the concentration of a volatile component in the mixture and its temperature is given in [8]. Data on the distribution of local characteristics are presented in the form of topograms, data on the pressure drop and separation efficiency are presented in the form of dependence of corresponding parameters on the vapor load of the column $K_v = \sqrt{\rho_v} U_v / \sqrt{\rho_l - \rho_v}$, m/s, where $\rho$ is the liquid or vapor density, kg/m$^3$; $U_v$ is superficial vapor velocity, m/s.

## 3. Distribution of the local liquid flow rate over the column cross-section

Data on the distribution of the local liquid flow under the packing in the column cross-section is shown in Fig. 1. The orientation of corrugated sheets in the upper and lower packing layers are shown in the topograms as the rectangles. The position of the lower column layer is the same in all experiments, while the position of the upper layer is varied relative to the column body. The orientation of the rows of holes in the liquid distributor relative to the sheets of the top irrigated packing layer was the same in all experiments. Data are presented as relative flow rates, i.e., the values of local liquid flow rates at a given point, reduced to the average liquid flow rate at the packing outlet under the conditions of experiments. The scale corresponding to the range of changes in the local flow rates is shown next to the topograms.

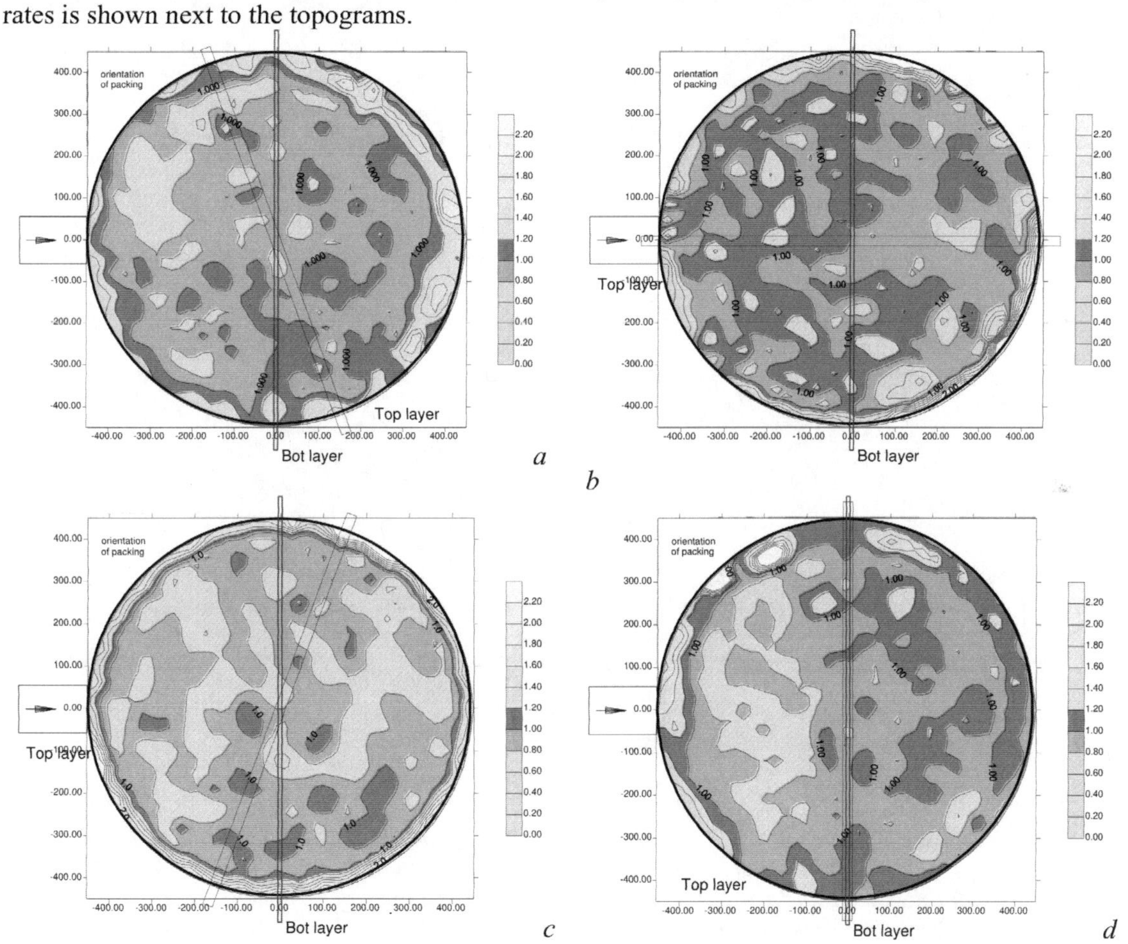


Fig. 1. Distribution of local liquid flow rates at the packing outlet. $F_v = 1.4$ Pa$^{1/2}$.
The angle of layers rotation: *a* – 20°; *b* – 45°; *c* – 70°; *d* – 90°.

In all cases, the zones in the form of rings 50–100 mm wide with an increased liquid flow are observed at the packing periphery. In topograms, for the angles of 20° and 70°, these zones are located almost along the entire perimeter, and for the angles of 45° and 90° they occupy half of the perimeter. In most part of the center area of the column cross-section, the local flow rates of liquid correspond to the average flow rate. For the packing with the angle of layer rotation of 90°, distribution of the local liquid flow in the central part is symmetrical relative to the diametrical line coinciding with the orientation of the corrugated sheets of the packing in the top and bottom layers. In that part of column cross-section, where the flow at the packing periphery is maximal, in the center of packing cross-section, the flow rate of liquid is minimal. The most uniform distributions of local liquid flow in the center section of the column cross-section are observed for the angles of layers rotation of 45° and 70°.

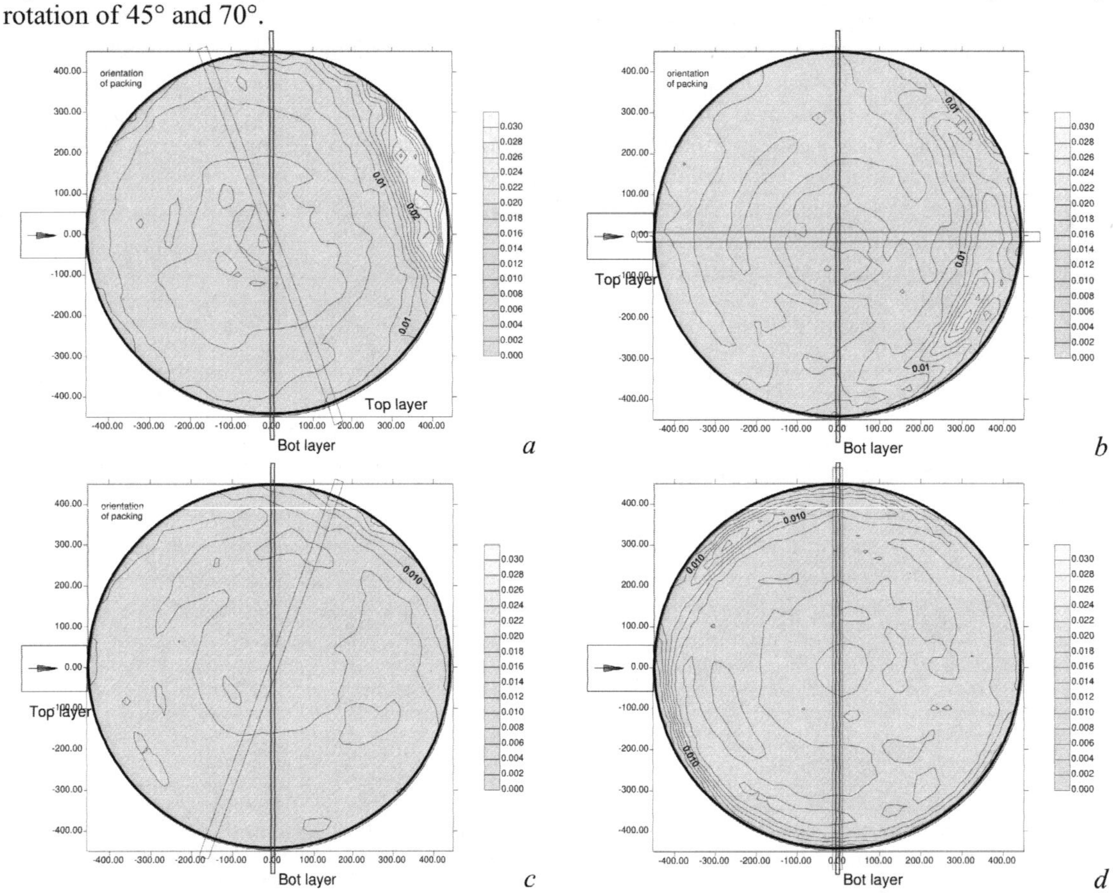


Fig. 2. Distribution of liquid phase composition at the packing outlet. $F_v = 1.4$ Pa$^{1/2}$.
The angle of layers rotation: *a* – 20°; *b* – 45°; *c* – 70°; *d* – 90°.

Topograms of distribution of volatile component concentration (R114) in the liquid phase at the packing outlet are shown in Fig. 2 for four values of the layers rotation angle. Orientation of corrugated sheets in the upper and lower packing layers is also shown there. Concentrations shown in this figure are expressed in the mole fractions of a volatile component in the mixture. They are obtained by measuring the temperatures of the liquid in the collector under the package, which are then converted to concentrations. These results on the distribution of mixture composition in the column cross-section can be used to compare qualitatively the data obtained under different conditions. To calculate the parameters characterizing separation efficiency, it is necessary to use the data on mixture composition obtained by the chromatograph. Significant maldistribution is typical of all cases and maximal maldistribution is observed at the rotation angle of 20° (Fig. 2 *a*). Also, a rather large area with an increased concentration of volatile component was obtained in experiments with a layer rotation angle of 45° (Fig. 2 *b*). Maximal concentration in these experiments is much less than for the rotation angle of 20°. On the packing with rotation angles of 70° the mixture composition over column cross-section is more uniform (Fig. 2 *c*). On the packing with a rotation angle of 90°, an annular region with increased concentration of a volatile component is distinguished at the packing periphery (Fig. 2 *d*). Locations of concentration maximum for almost all cases correspond to the zones with increased flow rates in Fig. 1.

These results show significant dependence of the distribution of the local liquid flow and mixture composition over the column cross-section on the rotation angle of the layers of structured packing. The most uniform distribution of local parameters over the column cross-section was obtained in experiments with the angles of layers rotation of 45° and 70°.

## 4. Effect of the angle of packing layers rotation on separation efficiency

Data on the effect of vapor flow rate on the efficiency of mixture separation for different angles of packing layers rotation are shown in Fig. 3. In the studied range of changes in the vapor flow rate, a decrease in separation efficiency is observed for all angles of packing layer rotation with a decrease in vapor flow rate. The highest HETP values are obtained for the angle of layers rotation of 20°.

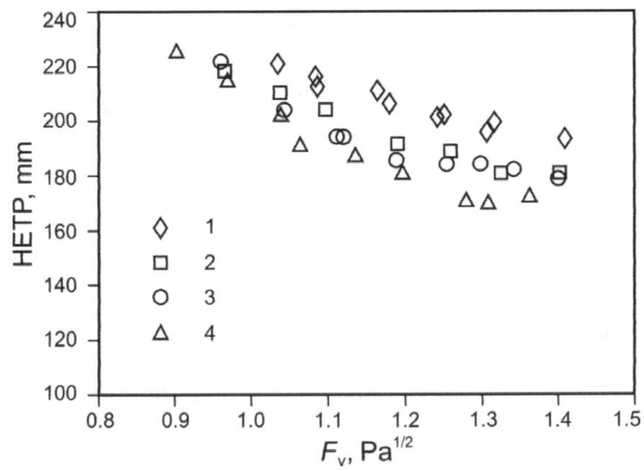


Fig. 3. Dependence of HETP on *F*-factor for different turn angles of the packing layers.

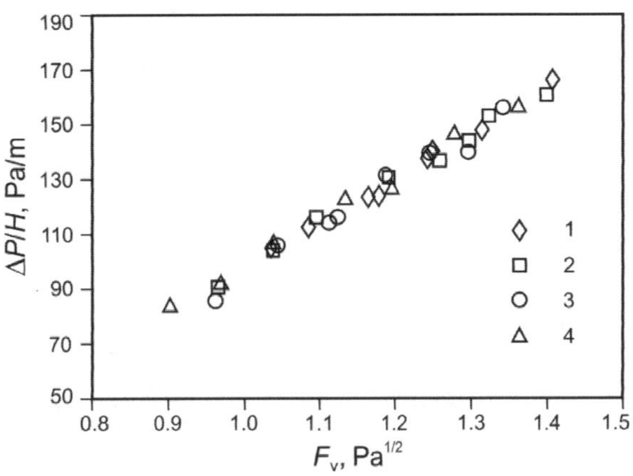


Fig. 4. Dependence of the relative pressure drop on *F*-factor for different turn angles of the packing layers.

With an increase in the rotation angle, the separation efficiency increases. For rotation angles of 45° and 70°, the difference in HETP is small. The relative pressure drop is almost independent of the angle of packing layer rotation (Fig. 4).

In experiments with a rotation angle of 20°, there is a minimal change in the nature of irrigation of the underlying packing layers along the package height. For example, incomplete wetting of the upper layer surface due to the relatively low drip point density may be preserved on more layers than in the case of intensive redistribution of liquid on the packing with a large angle of layers rotation. As it was already noted in Introduction, the defect of irrigation at the column top can be spread for these conditions to the entire height of the structured packing. Improving the liquid mixing along the packing height with an increase in the angle of layers rotation led to an increase in the efficiency of mixture separation, Fig. 5. According to the results of this study, increasing the angle of packing layers rotation from 20 to 90 degrees leads to an increase in separation efficiency, while the HETP value decreases by 10–12%.

## 5. Conclusion

In this paper, we studied the effect of the rotation angle of structured packing layers on the separation efficiency of a binary mixture in a distillation column. It is shown that a decrease in the layer rotation angle leads to a significant decrease in the efficiency of mixture separation. A decrease in separation efficiency is associated with an increase in maldistribution of the local liquid flow rate and mixture composition over the cross-section and height of the column.

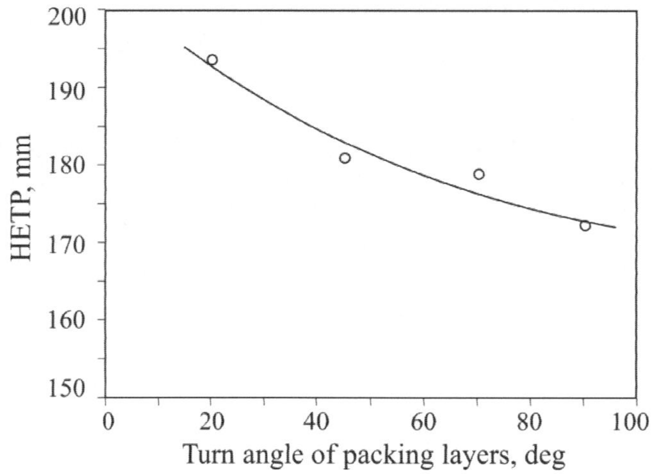


Fig. 5. The efficiency of mixture separation vs. rotation angle of structured packing layers, $F_v = 1.4$ Pa$^{1/2}$.

**Acknowledgments**

The efficiency of mixture separation was studied within the joint project of Kutateladze Institute of Thermophysics SB RAS and Air Products Inc., USA. Distribution of mixture composition was studied under the state contract with IT SB RAS (AAAA-A17-117030310025-3).

## References

[1] Hoek P J, Wesseleingh J A and Zuiderweg F J 1986 *Chem Eng Res Des* 64 431
[2] Olujić Ž and Jansen H 2015 *Chem Eng Res Des* **99** 2
[3] Spiegel L 2018 *Chemical Engineering Transactions* **69** 138
[4] Pavlenko A, Zhukov V, Pecherkin N, Chekhovich V, Volodin O, Shilkin A and Grossmann C 2014 *AIChE J.* **60** 690
[5] Pavlenko A N, Zhukov V E, Pecherkin N I, Slesareva E Yu, Nazarov A D, Li X, Sui X 2018 *Chemical Engineering Transactions* **69** 26
[6] Pavlenko A N, Pecherkin N I, Chekhovich V Yu, Zhukov V E, Sunder S and Houghton P 2009 *Theor. Found. of Chem. Eng.* **43** 1
[7] Pavlenko A, Pecherkin N, Chekhovich V, Zhukov V, Sunder S, Houghton P, Serov A and Nazarov A 2005 *Journal of Engineering Thermophysics* **13** 1
[8] Pavlenko A N, Zhukov V E, Pecherkin N I, Chekhovich V Y, Sunder S and Houghton P 2010 *Theor. Found. of Chem. Eng.* **44** 869

5th International Workshop on Heat/Mass Transfer Advances
for Energy Conservation and Pollution Control
August 13-16, 2019, Novosibirsk, Russia

# Effect of the wall properties on the shock wave / laminar boundary layer interaction

**A I Kutepova[1], P A Polivanov[1] and A A Sidorenko[1]**

[1]Khristianovich Institute of Theoretical and Applied Mechanics SB RAS, 630090, Institutskaya, 4/1, Novosibirsk, Russia

E-mail: a.kutepova@g.nsu.ru

**Abstract.** The CFD calculations have been performed based on RANS approach with the Menter transition model [1]. The boundary layer developing on a flat plate was considered. To generate the incident shock wave, a wedge was installed over the plate. The wall was represented by one cm thick solid flat plate made from aluminum and polyacetal. Flow parameters were $M_\infty$=1.41, $P_0$=0.7 bar, $T_0$=293K. As a result of solving the conjugate heat transfer problem, the SWBLI flow was studied. The temperature and pressure distributions on the wall for different surface materials were obtained. It was found that the streamwise heat fluxes inside the wall can lead to a change of the separation zone length.

## 1. Introduction

Transonic aircraft of the next generation should have significantly lower $CO_2$ emission and have higher fuel efficiency [2]. These can be achieved by improving the engines, aerodynamics of the aircraft and using lighter materials. One of the means of drag reduction of the aircraft is to provide the laminar flow in the boundary layer on the wing and engine nacelle. The adverse pressure gradient generated by a shock wave can lead to separation of the boundary layer [3]. The separation of the laminar boundary layer results in significant instability leading to earlier laminar-turbulent transition. For correct calculations of the aircraft aerodynamic characteristics, an exact definition of the transition position is mandatory. The wall properties, particularly temperature distribution can significantly affect the flow characteristics at Shock Wave/Boundary layer interaction (SWBLI). In CFD calculations it is commonly assumed that the wall boundary conditions are either adiabatic or isothermal. In reality, heat fluxes inside a solid wall can lead to a redistribution of wall temperature.

In the case of a cold wall (the temperature of the wall is less than the recovery temperature), the separation zone decreases, while for the hot wall the opposite effect is observed. An example of the influence of wall temperature can be found in [4], in which the SWBLI zone was studied by direct numerical simulation for the case of an incoming turbulent boundary layer. This work demonstrates a significant change not only of a stationary flow in the interaction zone but also a significant change in non-stationary characteristics in the shear layer. Therefore, when modeling SWBLI, it is important to set the temperature of the model wall correctly.

Adiabatic conditions on the model wall are often assumed when simulating experimental test cases performed in blow-down wind tunnels. But along the SWBLI zone, the recovery temperature changes significantly. This is due to a change in the Mach number at the boundary of the shear layer and heat fluxes in the shear layer. The emergence of the longitudinal temperature gradient on the surface of the model should lead to the appearance of heat fluxes within the experimental model, which will lead to a redistribution of temperature on the surface of the model. The influence of this phenomenon can be estimated by solving the conjugate problem.

An example of such calculations for high-speed flows can be found in papers [5, 6]. In the work [5] it was shown that even for unseparated flows in high Mach number flight, for a good comparison with the experimental results it is necessary to calculate the heat fluxes in the model. A good agreement between the numerical results and the experiments was achieved in the work [6], in which the cooling efficiency of the turbine blades was evaluated. However, this phenomenon is often not taken into account when modeling SWBLI.

Heat flows generated inside the model can be used to control the flow. In the work [7] a method of flow control based on the creation of transverse jets is proposed. The pressure for generating a jet is created by the evaporation of a liquid, which is heated by taking heat from an external stream with a high stagnation temperature.

The temperature effect is important not only for improving the accuracy of the calculation. Another important point is to improve the accuracy of the experimental methods. For example, when using in the experiment the method of measuring the surface temperature with a thermal imaging camera, incorrect interpretation of the data is possible since the heat fluxes inside the model can lead to a significant change in temperature fields. In the work [7] a complex combination of materials was used, which significantly reduced the effect of heat fluxes from the lower and fore part of the model on the flow field under study.

The purpose of this study is to estimate the influence of the wall material on the development of separated flow and the position of the laminar-turbulent transition in the shock wave/boundary layer interaction.

## 2. Problem formulation

The flow parameters and geometry of the computational domain correspond to the experiment described in [3]. The computational domain was as close as possible to the real geometry, which consisted of a plate, a wedge and two bypass channels located in the lower and upper parts of the domain. The presence of sharp steps and caverns in the bypass channels was due to the design considerations. In the experiment, the region of interaction of the shock wave generated by the wedge with the boundary layer developing on a flat plate was studied. Therefore, when constructing a computational grid in a given area, the grid refinement was used. All calculations were performed at a wedge angle of attack of 4°. The main material of the plate was polyacetal. To ensure the requirements of rigidity the nose section was made of steel. The heat fluxes inside the solid body were calculated only for the plate. The boundary conditions for the remaining walls were the condition of zero heat flux.

The computational domain was covered by a regular hexahedral computational grid. The total number of the computational domain cells finally amounted to 1.1 mil. The problem was simulated numerically with the aid of the software package ANSYS Fluent. The steady Navier–Stokes equations were solved with the aid of a density-based solver. An implicit second-order scheme in time and space

Journal of Physics: Conference Series **1369** (2019) 012037 doi:10.1088/1742-6596/1369/1/012037

was used together with the AUSM method of splitting the convective fluxes. The calculations were performed using the Menter transition model *κ-ω-γReθ* and the turbulent model *κ-ω SST*. The freestream parameters were specified at the inlet boundary of the computational domain. At the outlet boundary, the flow expiration condition was set. The capability of conjugate heat transfer calculation is accomplished by integrating a heat conduction procedure in a solid model (plate).

In total, three calculation cases with different plate material were considered. The base case was the calculation, where the main material was polyacetal, and the nose part was steel. In the second case, the entire plate was steel. In the third configuration, adiabatic boundary conditions were set on the plate wall.

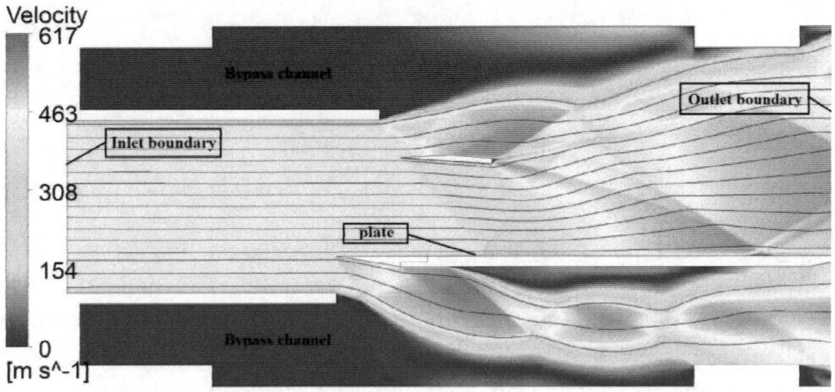


**Figure1.** Velocity distribution in the computational domain

## 3. Calculation results

Figure 1 shows the velocity field for the base case. On the left input boundary, the flow parameters were specified, which corresponded to the values calculated for the nozzle under the required conditions. It is seen that when the incident flow interacts with the wedge, a shock wave is formed, which interacts with the boundary layer developing on the upper part of the plate. The deflection angle of the lower nose of the plate is 9°, which leads to the formation of a powerful shock wave. In the absence of the bypass channel, the interaction of this shock wave with the lower wall could lead to the choking of the wind tunnel. The presence of a sudden expansion formed by the bypass channel leads to the formation of rarefaction waves weakening the shock wave formed by the nose of the plate. Further downstream, the flow accelerates in the fan of the rarefaction waves generated in the zone of the end of the wedge-shaped part of the plate. This leads to a significant decrease in pressure, which becomes less than the static pressure of the bypass channel. As a result, an oblique shock wave is formed with the formation of the developed separation zone in the lower part of the plate. The main difference between the flow formed in the upper bypass channel is the small geometrical dimensions of the wedge. This leads to the formation of a weaker shock wave and rarefaction waves on its upper part. The absence of a significant circulation zone in the upper bypass channel is due to the small geometrical dimensions of the wedge.

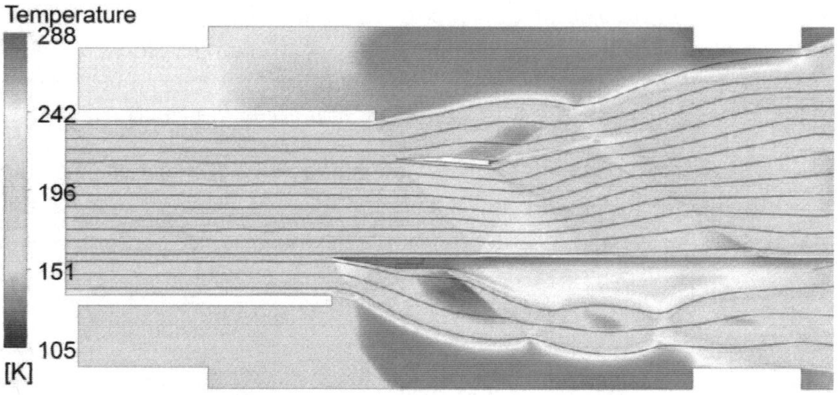


**Figure 2.** Temperature distribution for the base case in the computational domain

Figure 2 shows the temperature field, from which the formation of the region with the minimum temperature in the acceleration section under the plate is seen. This is accompanied by an increase in the Mach number, which should lead to a decrease in the recovery temperature on the bottom wall of the plate. Since the difference in the Mach number on the lower and upper walls of the nose part of the plate is not large (at the top M = 1.45, at the bottom M = 3), there is no significant difference in the recovery temperature. Therefore, the nose of the plate has a uniform temperature distribution throughout the area. Since the flow velocity is subsonic in the separation zone under the plate, the temperature of the bottom wall coincides with the flow temperature in the separation zone. It can be seen in Fig. 3 that in the separation zone the stagnation temperature is about 230K, which is significantly lower than the total temperature of the free stream 293K. As a result, heat flux is generated, directed from the upper wall to the lower one, which reduces the temperature on the upper wall of the plate in the SWBLI zone.

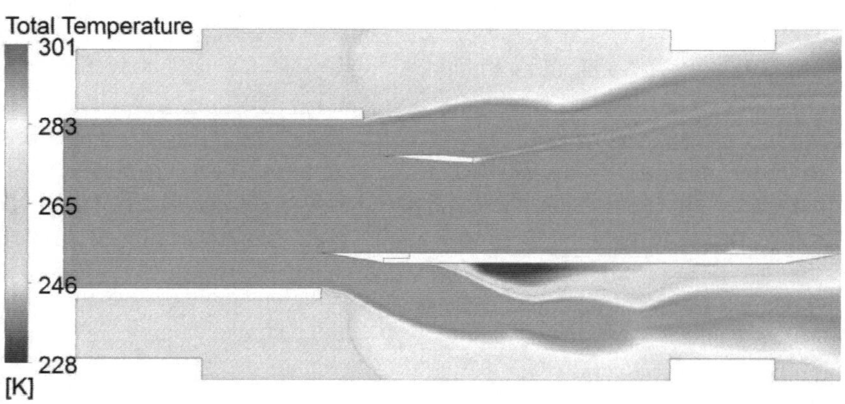


**Figure 3.** The distribution of the total temperature for the base case in the computational domain

a)

б)

**Figure 4.** The temperature distribution in the plate for the calculated cases of base (a); second configuration (b)

Figure 4 presents the temperature distribution inside the solid body for the first and second configurations. It can be seen, that for the case of an all-steel plate (Figure 4b), due to the high value of the thermal conductivity coefficient, the plate temperature is equalized along the normal coordinate. Before the SWBLI zone, the plate temperature rises, and in the wake zone, it falls. The reason for this is the redistribution of the stagnation temperature in the separation zone under the plate.

In the case where the main material of the plate was polyacetal, one can see a significant temperature non-uniformity for the plastic part of the model along the normal coordinate. This is due to the low thermal conductivity of the plastic, which makes the temperature distribution closer to the adiabatic one, so the temperature in the SWBLI zone increases in the model due to the rise in the recovery temperature (the Mach number decreases).

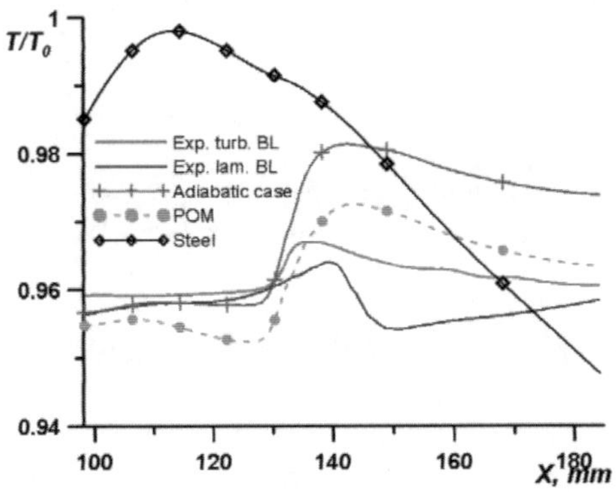


a)

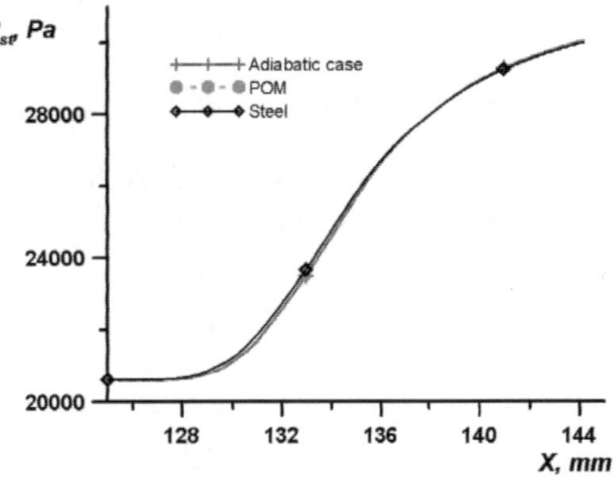


b)

**Figure 5.** Comparison of experimental results with calculated temperatures for the upper part of the plate (a); pressure distribution in the stream (b)

The coordinate x=130 corresponds to the intersection of the incident shock wave generated by the wedge with the plate. In this region, an increase in temperature occurs for the adiabatic case. It was noted above that the reason for this is the drop in the Mach number. In the case of polyacetal, the temperature distribution is qualitatively similar to the adiabatic one. However, there is a decrease in temperature gain and overall cooling of the upper wall. The most interesting case is the steel plate. The obtained temperature distribution (Figure 5a) does not allow determining the flow structure on the upper part of the plate. The red curve corresponds to the experimental data obtained on a plastic model for a turbulent boundary layer. Temperature measurement was carried out by thermal imaging method. One can see a good qualitative agreement of the results. The difference obtained may be attributed to the absence of three-dimensional effects in a two-dimensional calculation, as well as the neglect of nonstationary phenomena in separation zones of RANS methods. Despite the significant effect of the plate material on the temperature of the upper wall in the SWBLI zone, this does not have a significant effect on the flow structure, as can be seen from the pressure distribution.

## 4. Conclusion

As a result of the calculations, it was shown that the material of the experimental model has a significant effect on the temperature distribution on the surface of the model. A significant effect of the flow formed in the lower part of the model on the temperature distribution in the zone of experimental measurements was found. Most likely, the reason for the discrepancy between the calculated and experimental temperature distributions is the limitations of flow modeling under the plate. RANS modeling methods do not allow conducting simulation with sufficient accuracy of the circulation zone found under the bottom surface of the plate.

## Acknowledgments

The work was supported by the grant of RSF 18-19-00547.

## References

[1] Menter F R, Langtry R B, Likki S R, Suzen Y B, Huang P G and Völker S 2004 *J. Turbomach*, **128**(3), 413-422.

[2] Budd T, Suau-Sanchez P 2016 *Research in Transportation Business & Management*, **21**, 68-75.

[3] Polivanov P A, Sidorenko A A, Maslov A A 2015. 53rd AIAA Aerospace Sciences Meeting 2015 (Kissimmee, Florida, USA, 5-9 Jan., 2015) : Conference Paper. **18** No.AIAA 2015-1974.

[4] Bernardini M, Asproulias I, Larsson J, Pirozzoli S and Grasso F 2016 *Physical Review Fluids* **1**(8):1-18.

[5] Chandra Murty, Manna P and Debasis Chakraborty 2012 *Proceedings of the Institution of Mechanical Engineers, Part G: Journal of Aerospace Engineering* **227** 1672-1681

[6] Peng Wang, Yu Li, Zhengping Zou, Weihao Zhang 2013 *Propulsion and Power Research* **2**(1):56–69.

[7] Liang Zhu, Xiong Chen, Yingkun Li, Omer Musa, Changsheng Zhou 2018 *Acta Astronautica* **142**, 300-313.

5th International Workshop on Heat/Mass Transfer Advances
for Energy Conservation and Pollution Control
August 13-16, 2019, Novosibirsk, Russia

# PIV measurement of shock wave / laminar boundary layer interaction

**O I Vishnyakov[1], P A Polivanov[1], A D Budovsky[1] and A A Sidorenko[1]**

[1]Khristianovich Institute of Theoretical and Applied Mechanics SB RAS, 630090, Institutskaya, 4/1, Novosibirsk, Russia

E-mail: vis_ol@itam.nsc.ru

**Abstract**. The paper deals with study the area of interaction incident shock wave with a laminar boundary layer by PIV technique. The several data processing techniques to determine PIV measurements usability in transonic flow with small typical scales are compared in the paper. The results show that the major limitation of the PIV technique are associated with the seeding particles inertia.

## Introduction

Resource limits and the necessity to reduce harmful emissions in areas of human living require permanent improvements in the efficiency of aircraft [1]. The problem can be solved by several methods: improvement of engines, aerodynamics and decreasing the weight of aircraft by the implementation of new material. One of the ways to aerodynamic improvement is providing laminar boundary layer at wings and engine manacles to reduce the drag. At the same time aircrafts have to be able to flight maximally fast at transonic and supersonic speeds which is accompanied by shock wave existence on the surfaces of aircrafts. An adverse pressure gradient generated by a shock wave can lead to separation of the boundary layer[2]. Laminar boundary layer separation can lead to significant instability, which provokes upstream shifting of laminar-turbulent transition [3]. The accurate prediction of the position of laminar-turbulent transition in the zone of Shock Wave / Boundary layer Interaction (SWBLI) is important for correct calculation of aerodynamic characteristics.

For engineering RANS calculations the transition position is mainly determined using $e^N$ method approach [4] or the four-parametric Menter's model [5]. These models use empirical data which depend on flow parameters. Therefore, the experimental data are necessary for calibration and testing the approaches for using in transonic flow calculations. The feature of transonic researches is the considerable difficulty of using probes, such as a hot-wire anemometer or Pitot tube, because of probable flow blockage. Therefore, non-invasive methods, such as PIV, are more preferable for use in transonic flows [6]. In addition, the velocity fields obtained by PIV measurements can be used to calculate the pressure field in the measurement area [7].

Zone of laminar flow separation has a large dimension in the streamwise direction, but small size in the normal direction. The spatial resolution of the PIV method in the vertical and horizontal directions is usually approximately the same. Therefore, the measurements in such flows can be accompanied by the problems of measuring the flow parameters in the boundary layer. In addition, in high-speed flows, the inertia of seeding particles leads to additional inaccuracies in the measurements, which has to be taken into account in the data processing. For example, in paper [8], the PIV method was used to study the interaction of an oblique shock wave with a laminar and turbulent boundary layer with a Mach number $M_\infty$ = 1.7. During the data processing, it was found that the concentration of particles in the laminar boundary layer is significantly lower than in the turbulent one. It occurs because the inertia of the particles prevents them from penetrating into the laminar boundary layer. The data processing algorithm with averaging of cross-correlation field allows decreasing the influence of incorrect (artificial) vectors in the area with low seeding particle density. The data processing algorithm with averaging of cross-correlation field allows decreasing the influence of incorrect crosscorrelation peaks in the area with low particle density.

The paper is devoted to testing various PIV data processing techniques by the example of the problem of the interaction of a shock wave with a laminar boundary layer at the condition of flow with small supersonic Mach number.

## 1. Experimental setup

The experiments similar to described in the paper [3] were chosen as the test task. The experiments were carried out in T-325 wind tunnel at $M_\infty = 1.46$; $T_0 = 291$ K; and $P_0 = 0.7 \times 10^5$ Pa, unit Reynolds number ($Re_1 = 11.5\times 10^6$ m$^{-1}$). The cross section of the wind tunnel working part (160 × 200 mm) was rectangular. The scheme of the experimental equipment consisting of a flat plate with a sharp leading edge and wedge (at the top) that generates the oblique shock wave is presented in Fig. 1. The incidence angle of the wedge was set to 4° to correspond to the regular mode of the shock reflection from the model for a nonviscous flow. The position of the shock wave generator was chosen so as to provide the location of the incident point of the shock wave on the plate at the center of the optical window (134 mm from the leading edge). For these free-stream parameters, the boundary layer remained laminar. In addition, the measurements for the turbulent boundary layer were also performed. For turbulization of the oncoming boundary layer, the roughness was placed in spanwise direction at the distance of 40 mm from the leading edge. The roughness has a saw-like shape in plan and thickness (height) of 0.2 mm.

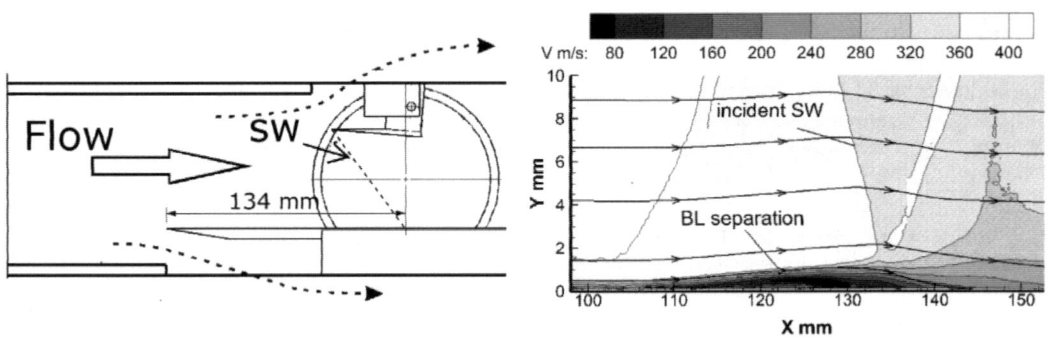


Fig.1 Experiment scheme (at left) and flow velocity field (at right) obtained by PIV

The study of the velocity field was carried out with a DantecDynamics measuring complex, with a LitronNano 135-15 laser. The measurements were carried out on two scales to study the possibility of increasing the spatial resolution of the method. In the one case, the Camera Phantom 310m (1280x800 pix) was used with a Nikkor Lens f105 / 2.8 lenses; the measurement area was 54.51x34.14 mm covering the entire SWBLI area. The laser frequency was 15 Hz, and the delay between frames was 0.9 μs. In the second case, a Hamamatsu camera with a resolution (1344x1024 pix) was used, with a Nikkor Lens f105 / 2.8 lenses, which gave a measurement area of 16.75x12.78 mm. During a series of experiments with a large scale, the camera moved in streamwise direction (along X coordinate), to obtain velocity fields in the entire SWBLI zone. Measurements with a Hamamatsu camera were carried out with two-time delays of 0.9 μs - the minimum delay at which it was possible to obtain data for cross-correlation analysis and with a delay between laser pulses of 0.3 μs. In the last case, both laser pulses occurred during the exposure of one frame. Therefore, to analyze these data, autocorrelation algorithms were used.

## 2. Results and discussion

At the image processing routine, the necessary operation to obtain high-quality velocity fields was the elimination of the vertical plate movements caused by wind tunnel vibration, which led to image oscillations up to 10 pixels. These oscillations were eliminated by searching the Y position of the plate on each frame and equating it to zero. The position of the plate was determined based on the positions of the flare from particles sticking to the plate during the start of the wind tunnel run.

As noted above, the lack of convective flows in the laminar flow prevents the penetration of particles into the laminar boundary layer. The turbulization of the boundary layer leads to the formation of strong vortices, which quickly increase the concentration of particles in the shear layer. Fig.3 shows the distribution of the smoothed intensity of the averaged images of particles (averaged over 2000 frames) at Y=0.3 mm and normalized to the maximum value that was in the area of inviscid flow behind the shock wave.

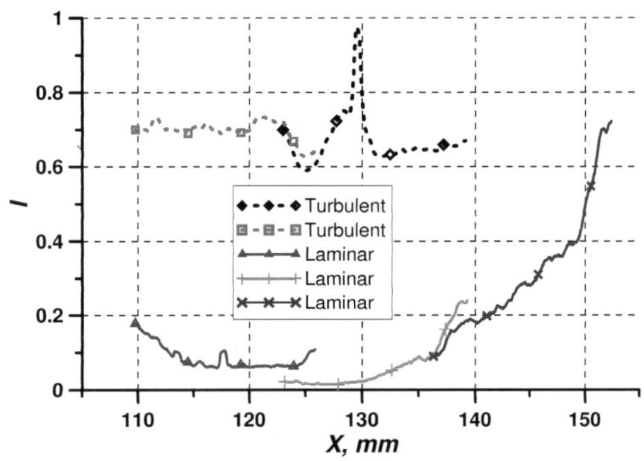


Fig.2 Intensity of particles seeding along the zone of SWBLI at Y = 0.3 mm

The data presented in the figure can be interpreted as the concentration of particles distribution in the measurement area. Different curves correspond to measurements performed during different wind tunnel runs. It is clearly seen that for a turbulent boundary layer the concentration of particles is much higher than in that the laminar case. Moreover, with passing through the region of the laminar separation bubble, the concentration of seeding particles drops to almost zero. Turbulization of the shear flow after the separation zone leads to a rapid increase in the concentration of particles. This means that in order to obtain correct data using standard PIV processing methods in the laminar separation region, it is very important to use strict conditions for filtering false (incorrectly determined) velocity vectors. An alternative way is to use the averaged correlation or autocorrelation algorithm. In this algorithm, when analyzing the data, the averaging of correlation distributions, rather than velocity values, occurs. This allows for minimizing errors due to the absence of particles or the disappearance of particles in the measurement area. The disadvantage of the algorithm is the lack of data on the non-stationary characteristics of the velocity fields.

For the averaged autocorrelation algorithm, the calculation was performed on 2000 images using a 32x8 pixel interrogation area. Data normalization was not performed in the course of autocorrelation searching; this made it possible to minimize the influence of regions without particles on the resulting averaged autocorrelation distribution. An example of such distribution with and without low-frequency filtering (near zero speeds) is shown in Fig. 3. It should be noted that when using autocorrelation processing methods it is impossible to determine the direction of velocity. Therefore, it must be determined on the basis of physical considerations.

For a large measuring area, the recovery of the velocity field was made only by using the cross-correlation algorithm implemented in DantecDynamics software using interrogation area 16x8 pixels and with implemented overlapping of 50x50%. The resulting velocity field was obtained by averaging 2700 instantaneous velocity fields.

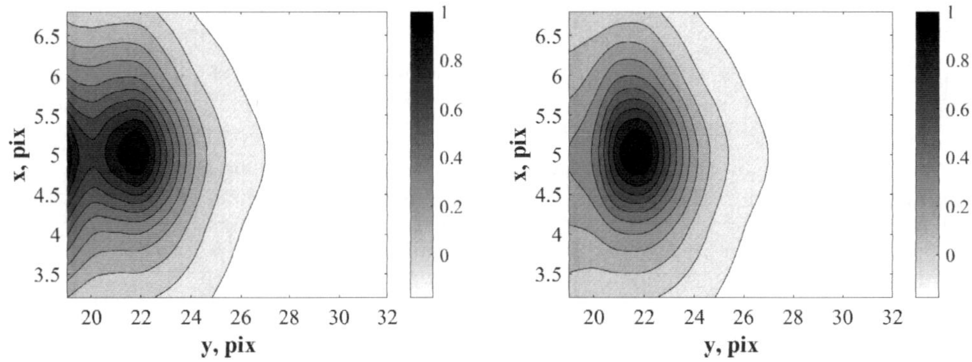


Fig.3 Example of autocorrelation peak with zero peak filtering (right) and without filtering (left)

For a small measurement area, the analysis was performed using three algorithms. For images obtained with a larger time delay of 0.9 μs, the processing was performed using DantecDynamics software. An iterative cross-correlation algorithm was used with a 64x8 pixel interrogation window with overlapping of 75x25%. The resulting velocity field was obtained by averaging 2000 instantaneous velocity fields.

Testing of autocorrelation algorithms was performed on the example of experiments carried out with a small time delay between laser pulses - 0.3 μs. Two techniques were investigated. The first technique was based on using a standard cross-correlation algorithm to search for the movement of particles implemented in ActualFlow software, with an interrogation area size of 32x8 pixels without overlapping. This software allows setting a limit of the minimum and maximum particles displacement (in pixels), which allows discarding the correlation peaks associated with multiplying the intensity of the particles on itself and giving a zero speed. However, as can be seen in Fig. 4, the results of this processing give a serious discrepancy with cross-correlation and averaged auto-correlation algorithms. This is caused by the effect of losing the particles pair. As a result, the algorithm gives a large number of false vectors that ultimately reduces the average value of the velocity outside the boundary layer. At the same time, in the boundary layer, the situation is aggravated by a small number of particles, as a result, there are significant numbers of vectors with a higher velocity value.

Comparing the velocity profiles shown in Fig.4 for large and small scales and obtained both at a time delay of 0.9 μs and at 0.3 μs shows that an increase in the spatial resolution of the image does not allow better resolving the velocity in the boundary layer due to particle inertia. This conclusion was obtained based on the calculations the time constant of the used seeding particles for the case of a laminar boundary layer (Fig. 5). The time constant was calculated based on the Stokes formula for a particle with a diameter of 0.8 μm and a density of 1200 kg / m$^3$ (boundary layer thickness $\delta$ = 0.55 mm). The obtained values of the time constant were multiplied by the local values of the velocities, which made it possible to obtain the typical length of free path of the particle. One can see that the length of the free path of a particle near the scope of the boundary layer is approximately 3.8 of the thickness of the boundary layer (about 2 mm). This means that although in this study, the PIV method has provided a large frequency range (up to 200 kHz), large values of speed will lead to significant smoothing of the velocity fields. Thus, decreasing the displacement of particles between frames up to the characteristic typical lengths of the particles runs not lead to a significant improvement in the results of data processing.

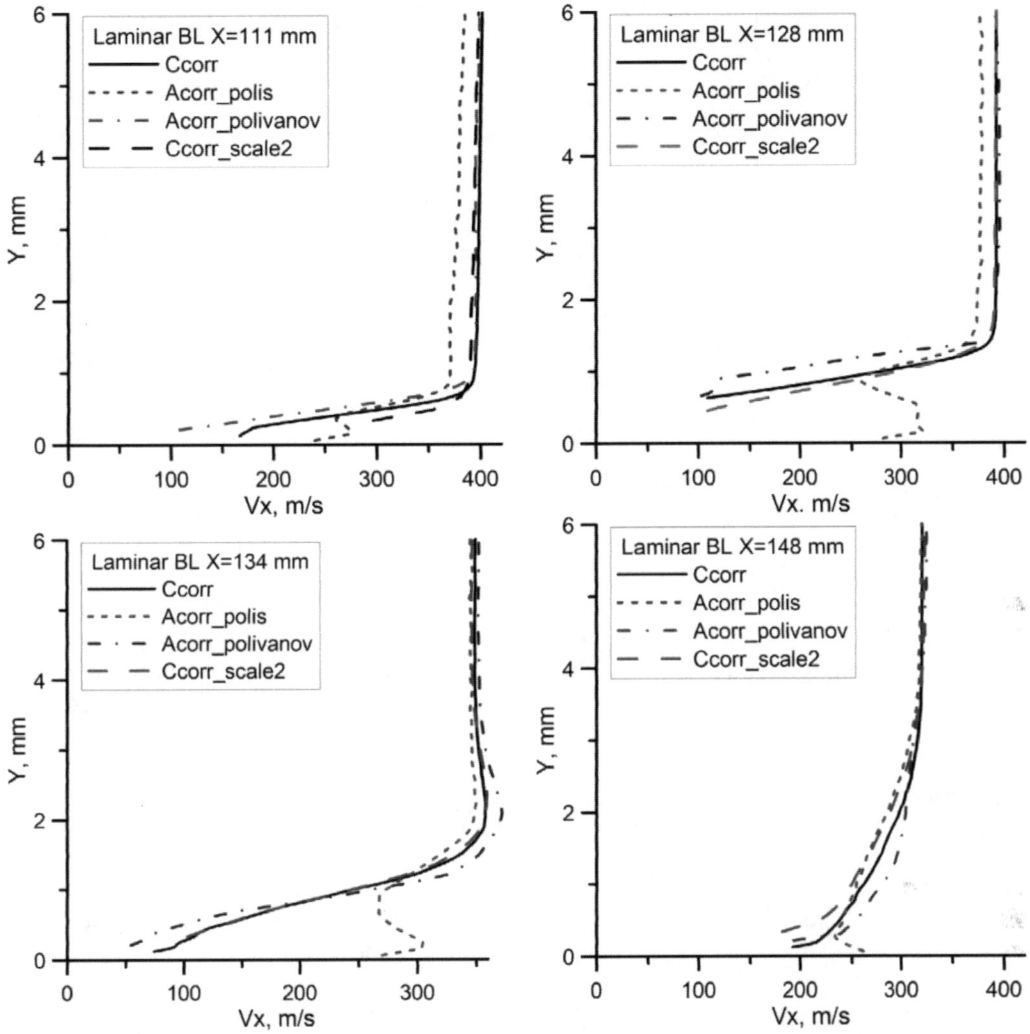


Fig. 4 Profiles of the longitudinal component of velocity obtained by PIV

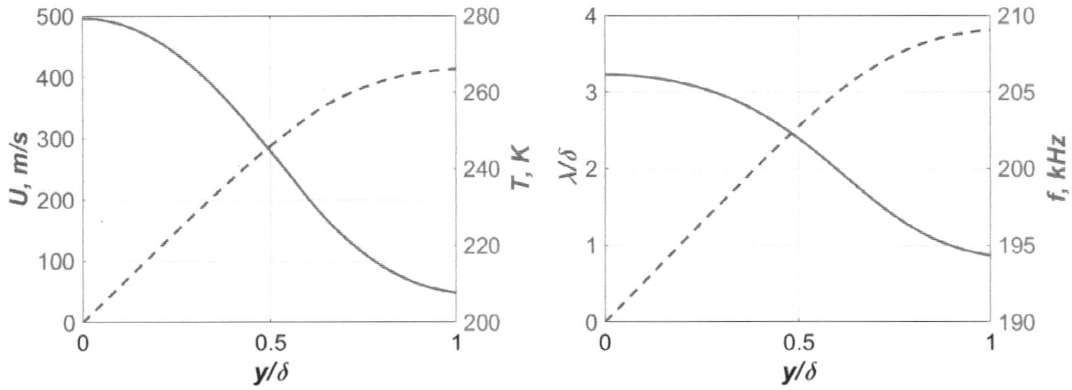


Fig.5 On the left – the velocity and temperature profile in the boundary layer, on the right – the characteristic length of the particles in the boundary layer

**Conclusions**

Comparison of the fields obtained for the small size of the measurement area, with the normal movement of particles (about 8 pixels, autocorrelation analysis) and significant movement of particles (about 30 pixels, cross-correlation analysis) did not reveal a significant difference. The absence of a significant difference was also shown when comparing the velocity fields obtained at different measurement scales, but with the same delay between frames. The reason for this is the significant typical length of trackers (particles) runs for the problem under study, which leads to the smoothing of the velocity fields. The results of the study show that increasing the spatial resolution more than the characteristic length of the free path of a particle does not improve the quality of the velocity field measurements.

**Acknowledgments**

The work was supported by the grant of RSF 18-19-00547.

**References**

[1] Budd T and Suau-Sanchez P 2016 Assessing the fuel burn and CO2 impacts of the introduction of next-generation aircraft: A study of a major European low-cost carrier *Res. Transp. Bus. Manag.* 21 68–75

[2] Polivanov P A, Sidorenko A A and Maslov A A 2017 Effective plasma buffet and drag control for laminar transonic aerofoil *Proc. Inst. Mech. Eng. Part G J. Aerosp. Eng.* 0954410018795542

[3] Polivanov P A, Sidorenko A A and Maslov A A 2015 The influence of the laminar-turbulent transition on the interaction between the shock wave and boundary layer at a low supersonic Mach number *Tech. Phys. Lett.* 41 933–7

[4] Sidorenko A, Kirilovskiy S, Boiko A, Demyanko K and Nechepurenko Y 2018 Comparison of two methods of laminar-turbulent transition prediction for transonic speeds *AIP Conf. Proc.* 2027 40082

[5] Menter F R, Langtry R and Völker S 2006 Transition Modelling for General Purpose CFD Codes Flow, *Turbul. Combust.* 77 277–303

[6] Stryczniewicz W, Placek R and Szczepaniak R 2016 PIV Measurements of Flow Separation Over Laminar Airfoil At Transonic Speeds *J. KONES. Powertrain Transp.* 23 329–36

[7] Polivanov P A 2018 Calculating pressure fields on the basis of PIV-measurements for supersonic flows *Thermophys. Aeromechanics* 25 789–92

[8] Giepman R H M, Schrijer F F J and van Oudheusden B W 2015 High-resolution PIV measurements of a transitional shock wave–boundary layer interaction *Exp. Fluids* 56 1–20

5th International Workshop on Heat/Mass Transfer Advances
for Energy Conservation and Pollution Control
August 13-16, 2019, Novosibirsk, Russia

# Stretched flame behavior in a planar meso-scale channel with heat conducting walls

**S Mokrin[1,2], D Sharaborin[3], E Odintsov[4,1], G Uriupin[1] and S Minaev[1,2]**

[1] Far Eastern Federal University, Sukhanova Str. 8, Vladivostok 690950, Russia

[2] Institute for Applied Mathematics FEB RAS, Radio Str. 7, Vladivostok 690041, Russia

[3] Kutateladze Institute of Thermophysics SB RAS, Academician Lavrentyev Ave. 1, Novosibirsk, 630090, Russia

[4] Khristianovich Institute of Theoretical and Applied Mechanics SB RAS, Institutskaya str., 4/1, Novosibirsk, 630090, Russia

msn_primat@mail.ru

Abstract. The influence of the wall on flow field, combustion regimes and flame extinction limit of premixed methane/air flames stabilized in the planar channel with heat conducting walls was experimentally investigated. Experimental data were compared with the results obtained in previous theoretical works. The influence of channel walls on gas flow was estimated by using PIV experiments with non-reacting gas flow and compared with the results of numerical simulation.

## 1. Introduction

Understanding of structure and extinction limit of premixed flames is important fundamental problem in combustion science and it is relevant in the field of development of clean combustion technologies. Counterflow flames are the basic objects for detailed investigations of combustion wave characteristics. This one-dimensional configuration is very convenient for both experimental and numerical investigations. The structure and flammability limits of counterflow flames at high and moderate flow rates under conditions of normal gravity were studied in early experimental and theoretical works [1–3]. The experimental investigations of low-stretched flames in normal gravity conditions are complicated because of the buoyancy effect related with natural convection. Microgravity experiments [4–6] yield valuable information about flame structure but they are expensive and need special facilities like drop tower, airplane, etc.

One of the ways to overcome the buoyancy effect is the stabilization of flame in a narrow channel. In papers [7–8] it was shown that such system allows the investigation of low-stretched flames in the normal gravity conditions. Proposed experimental approach is also very convenient to apply various experimental technics for flame diagnostics such as contact methods, optical methods, gas chromatography, spectroscopy, and etc. In our recent work [9] we obtained the flammability limit and combustion regimes of premixed $CH_4$/Air flames stabilized in 5 mm channel. It was found that two stationary combustion modes exist: distant flame (DF) and near-stagnation plane flame (NSF) or "flame tube". Besides, the shape of extinction limits curve differs from C-shaped typical for conventional counterflow flames. In a narrow channel it has ε-shaped form. One-dimensional numerical simulation with the thermal-diffusion model using an overall one-step reaction conducted in paper [10] allowed us to reveal prime physical mechanisms of flame quenching on different branches of the extinction limit curve. These mechanisms, beginning with the upper limit and ending with the lower one on ε- shaped curve, are the following: stretch; radiative and convective heat losses; weakening of heat recirculation; heat dissipation to the burner. First two are direct analogs of the upper and lower limits on conventional C-shaped curve, while two latter are related with the effects of channel walls.

The main objective of this paper is to study the influence of the wall on flow field, combustion regimes, and extinction limit of premixed stretched flames stabilized in the planar channel with various gap size.

## 2. Experimental Setup

Experimental setup consists of counterflow slot-jet burner, mass flow controllers and AD/DA converter connected to PC. Counterflow burner consisted of two opposite directed slot-jet burners similar to that used in the work [11] mounted horizontally at the distance 2L=50 mm. Nozzle size was achieved by using the flanges with rectangular hole. We used the flanges with nozzles 40x5 mm and 40x3 mm to construct 5 mm and 3 mm channel respectively. Two rectangular grooves were made in the flanges under and above the nozzle to install the quartz plates. We used 50x50 mm and 1.3 mm thickness quartz plates as channel walls. To uniform flow, we filled each burner's body by 1 mm ceramic balls and installed two layers of stainless steel mesh in front of the flange. To maintain the constant temperature of inlet gas we mounted the water-cooling system on the burner's body. For flame visualization two digital HD photo cameras Nikon D7200 were used. The scheme of counterflow burner is shown in Fig. 1.

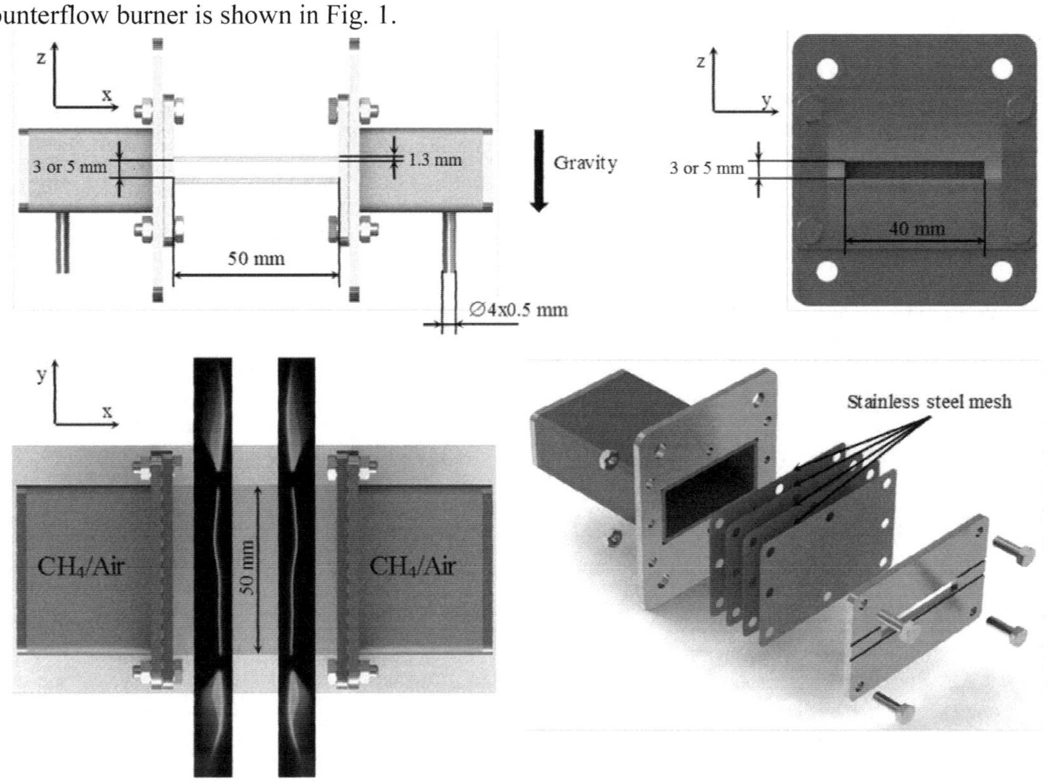


Fig. 1. Scheme of counterflow slot-jet burner.

Lean methane/air mixture was used in experiment. Mixture with equivalence ratio $\phi = 0.8$ for 5 mm channel and with $\phi = 0.9$ for 3 mm channel was ignited by external heater at chosen value of stretch rate. Furthermore, the stationary state was determined by indications of K-type thermocouples mounted on the outer surface of the quartz plate and at the nozzles between the layers of stainless steel mesh. After that, the equivalence ratio was reduced on the value $D\_\phi = 0.01$ to establish new stationary state. We considered that the stationary state was established when the temperature change was less than +/- 0.1 C/min. After measurement of flame characteristics, we reduced equivalence ratio to study next stationary state. When the flame extinguished, we fixed the extinction limit, and reduce the set value of stretch rate, and repeat the procedure from the beginning.

To examine the effect of gravity field on flame stabilization we rotated the counterflow burner from horizontal position to vertical one. It was found that influence of gravity field on flame location is insignificant even at low values of stretch rate. To study the influence of the wall on the flow field we applied particle image velocimetry (PIV) method. This method allows the investigation of the velocity field, vorticity, flow rate pulsations, shear stresses in the region of interest during one measurement. PIV method is also applied for reacting systems.

The scheme of PIV experimental setup is shown in Fig. 2. The PIV system consists of a high-resolution CCD camera (16 Mpx Bobcat ImperX), a double-pulsed Nd:YAG Laser (Beamtech Vlite 200, 200 mJ/pulse), a host computer, and synchronizer. Titanium dioxide particles (D = 1 μm) were used as tracers. The double-pulsed laser sheets were directed to the space between quartz plates parallel to them. The delay between laser flashes determining the displacement of the particles between two frames was in the range from 50 to 500 μs for different flow rates. As can be seen in Fig. 2 CCD camera was mounted perpendicularly to the quartz plates. Control of PIV system and data processing were implemented by "Actual Flow" program package developed in Kutateladze Institute of Thermophysics.

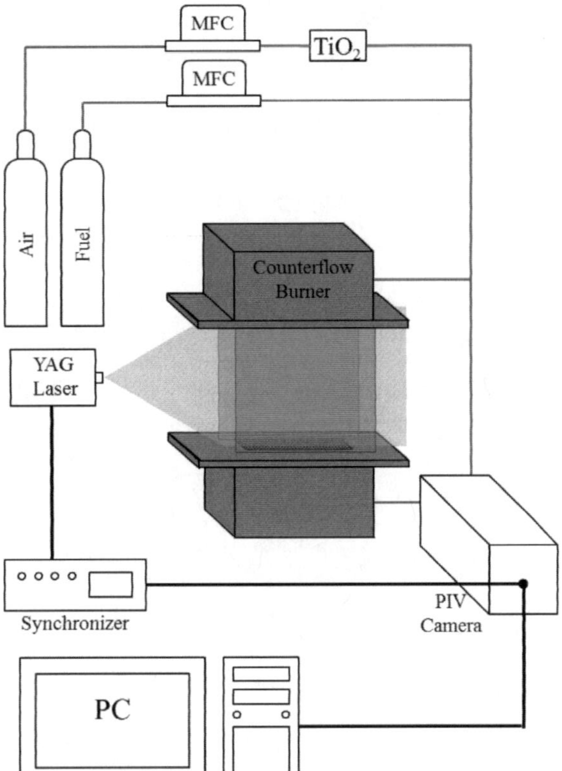


Fig. 2. Scheme of PIV experimental setup.

## 3. Results and Discussion

Flame extinction limits and combustion regimes were experimentally obtained for premixed methane/air flames stabilized in 3 mm and 5 mm planar channels in the range of stretch rates from 44 $s^{-1}$ to 10 $s^{-1}$. The stretch rate is determined as $a=V_0/L$, where $V_0$ is inlet gas velocity and the L is the half distance between burners.

In Fig. 3a the experimental dependence of flame position on stretch rate for 5 mm channel is shown. As can be seen from the figure the distant flame shifts to near-stagnation plane flame at stretch rate value more than 30 s-1. On the other hand, results of numerical simulation (see Fig. 3b) obtained in paper [10] demonstrate that both stationary regimes exit in region of non-dimensional stretch rate from $10^{-2}$ to $10^{-1}$. This disagreement between experiment and simulation can be explained by features of experimental method. During the experiments we fixed the stretch rate and reduced the equivalence ratio until the flame quenching. Fixing the equivalence ratio with increasing the stretch rate will reveal two stationary combustion regimes at same stretch rates predicted in paper [10].

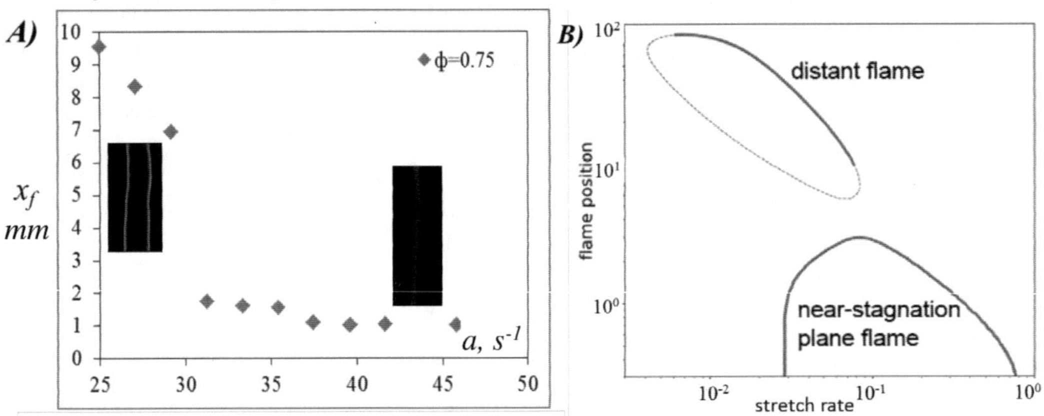


Fig. 3. Dependence of the flame front position on stretch rate. A – experiment, B – non-dimensional flame position vs non-dimensional stretch rate obtained numerically in [10].

It is necessary to note that only normal flame regime was observed in experiments with 3 mm channel. We suggest that it is related with high convective heat losses from outside surface of channel walls.

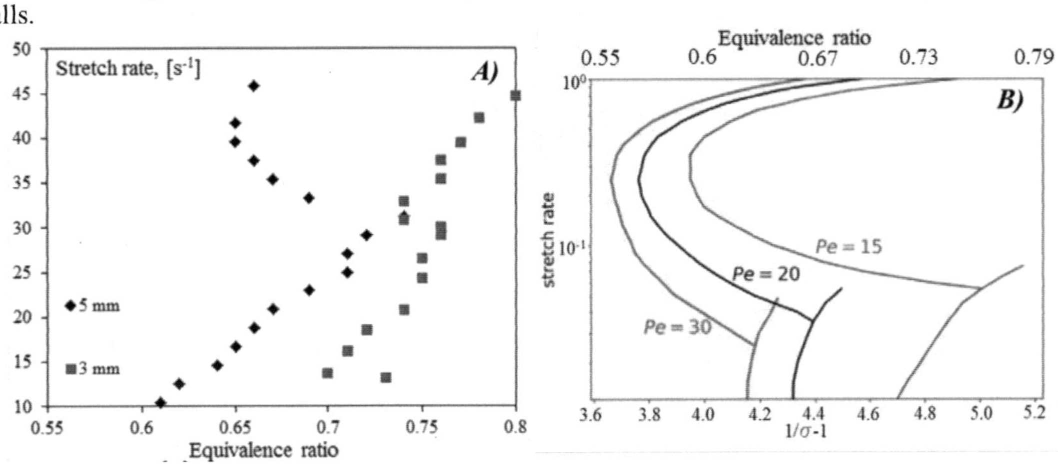

Fig. 4. A) Experimental flame extinction limit curves obtained for 3 mm (green squares) and 5 mm (black rhombs) channels. B) Non-dimensional extinction limit curves calculated for different values of Peclet numbers.

In paper [10] it was shown that decreasing of the channel size leads to the intensification of interphase heat exchange and results to the expansion of flame extinction limit at low values of stretch rate. But as can be seen from Fig. 4a in experiments decreasing of channel gap resulted to the narrowing of flame extinction limit not only in the region of moderate stretch rates but in low one too. This disagreement with the theory is due to the fact that convective heat losses from channel walls were not included in the model described in paper [10]. When we modify the model, we obtained the calculated flame extinction limits for different Peclet numbers (see Fig 4b) which were in good agreement with experimental one.

To estimate the influence of the wall on flow field we conduct the series of PIV experiments with cold gas. Non-reacting flow field was calculated using the model including continuity and momentum equation with assumption of ideal gas. Numerical simulation was conducted using OpenFOAM CFD simulations software with icoFoam solver for incompressible laminar Navier-Stokes equations. All presented results were obtained with uniform 100x100 grid.

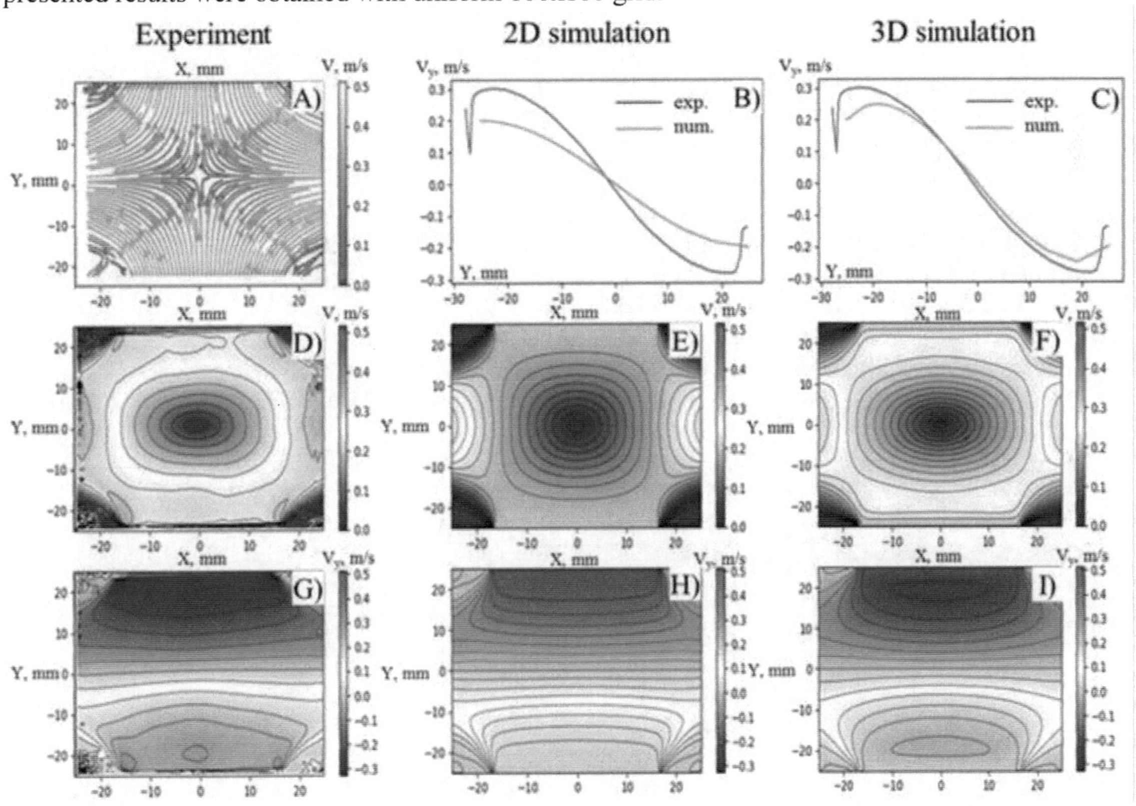


Fig. 5. Comparison between PIV results and numerical simulation.

Results of PIV experiments and numerical simulation for non- reacting flow in 3 mm channel are shown in Fig. 5. Streamlines corresponds the instant flow field is shown in Fig 5a. Figure 5b illustrates the comparison between velocity component $V_y$ along the axis X=0 obtained experimentally and via 2D numerical simulation. In Fig. 5c the numerical simulation was conducted for 3D model and velocity component $V_y$ calculated along the axis X=0 on the plane Z=0. Line DEF in Fig. 5 corresponds to velocity module on the plane Z=0 and line GHI in Fig. 5 corresponds to the velocity component $V_y$ plotted on the plane Z=0 obtained in the experiment, 2D simulation, and 3D simulation respectively. Fig. 5 clearly shows that results of 3D numerical simulation are in better quantitative agreement with experimental data than two-dimensional one. It is explained by the fact that the 3D calculations considered the friction between the gas and the channel walls.

## 4. Conclusions

Reduced mathematical model proposed in paper [10] allowed us to explain the physical mechanisms determining extinction limits of counterflow premixed flames stabilized in a narrow channel with heat conducting walls. However, experiments showed that external heat losses should be considered during numerical simulation. Comparison of results of PIV experiments and numerical simulation shows that friction between gas flow and channel walls has a big influence and also must be taking into account.

## 5. Acknowledgment

The research was carried out within the state assignment of FASO of Russia No. 075-00400-19-01. Authors would like to thank Dr. Vladimir Dulin from Kutateladze Institute of Thermophysics SB RAS and Dr. Anatoly Chernov from Voevodsky Institute of Chemical Kinetics and Combustion SB RAS for their help in conducting of PIV experiments.

## References

[1] Law C K, Ishizuka S and Mizomoto M 1981 *Proc. Combust. Inst.* **18** 1791
[2] Sato J 1982 *Proc. Combust. Inst.* **19** 1541
[3] Ishizuka S and Law C K 1982 *Proc. Combust. Inst.* **19** 327
[4] Maruta K, Yoshida M, Ju Y and Niioka T 1996 *Proc. Combust. Inst.* **26** 1283
[5] Fursenko R, Minaev S, Nakamura H, Tezuka T, Hasegawa S, Takase K, Li X, Katsuta M, Kikuchi M and Maruta K 2013 *Proc. Combust. Inst.* **34** 981
[6] Okuno T, Akiba T, Nakamura H, Fursenko R, Minaev S, Tezuka T, Hasegawa S, Kikuchi K and Maruta K 2018 *Combust. Flame* **194** 343
[7] Lee M J and Kim N I 2014 *Combust. Flame* **161** 2361
[8] Lee M J, Cho M S and Kim N I 2015 *Proc. Combust. Inst.* **35** 3439
[9] Mokrin S, Odintsov E, Uriupin G, Tezuka T, Minaev S and Maruta K 2017 *Combust. Flame*, **185** 261
[10] Fursenko R, Mokrin S and Minaev S 2019 *Proc. Combust. Inst.* **37** 1655
[11] Kaiser C, Liu J B and Ronney P D 2000 *38th AIAA Aerosp. Sci. Meet. Exhib.* 1

5th International Workshop on Heat/Mass Transfer Advances
for Energy Conservation and Pollution Control
August 13-16, 2019, Novosibirsk, Russia

# Experimental research of supersonic flow heat transfer in the wake of a backward-facing step

**A I Leontiev[1,2], S S Popovich[1], Y A Vinogradov[1] and M M Strongin[1]**

[1]Institute of Mechanics, Lomonosov Moscow State University, 1 Michurinsky Prospect, Moscow, 119192, Russia

[2]National Research University of technology "Bauman Moscow State Technical University", 5 2nd Baumanskaya Street, Moscow, 105005, Russia

pss@imec.msu.ru

**Abstract.** The results of an experimental study of heat transfer for a supersonic flow behind a backward-facing step are presented. Input flow Mach number was 2.2, Reynolds number was over 20 million when calculated with the use of the dynamic boundary layer length from the critical section to the exit section of the nozzle. The height of the step ranged from 8 to 14 mm. The thickness of dynamic boundary layer for plane model flow was about 6 mm in the region of nozzle exit section. The heat transfer coefficient and adiabatic wall temperature for the wake flow behind the step have been analysed in comparison with the plane model flow. The experiment was made with the use of chromel-kopel thermocouples with thermal compensation of cold junction, stagnation and static pressure meters, NI LabView powered programs of experimental automation. The graphs of the wall temperature, stagnation flow temperature, adiabatic wall temperature, the stagnation and static pressure, the distribution of the temperature recovery coefficient and the relative Stanton number along the model length are presented.

## 1. Inroduction

Due to viscous dissipation in the boundary layer of a supersonic compressible gas flow the total temperature redistribution takes place. As a result, the temperature in the inner part of the boundary layer is lower, while temperature in the outer part is higher than the rate in the main flow [1, 2]. If there is no heat transfer through the wall, the total temperature of the near-wall gas layer takes the value of the adiabatic wall temperature.

Heat transfer data, as a rule, are presented in the form of a heat transfer coefficient, which in the general case is determined by the ratio of the specific heat flux into the wall to the temperature difference between the instantaneous temperature of the streamlined wall and a certain determining temperature in the flow [3, 4]:

$$h = \frac{q_w}{T_{aw} - T_w}$$

The determining temperature may be the thermodynamic temperature in the flow, the stagnation temperature, the initial temperature (at the entrance to the channel with heat sources or sinks), or the adiabatic wall temperature. For high-speed flows the adiabatic wall temperature is applied as determining one which guarantees the accuracy of transferring the data obtained on wind tunnels to the actual operating conditions of the designed machine.

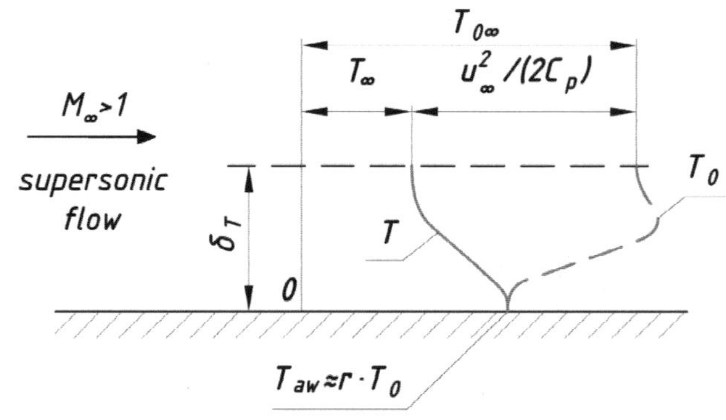


**Figure 1.** Temperature profile in the thermal boundary layer for a supersonic compressible gas flow around a plane surface.

The ratio of the heat transfer intensity to the total potential heat transfer of the gas flow is determined by the Stanton number [4, 5]. The heat transfer data presented in the form of this dimensionless heat-transfer criterion are less sensitive to changes in the flow velocity as well as to changes in the wall temperature than the common heat transfer coefficient *h*:

$$\mathrm{St} = \frac{q_w}{\rho_\infty u_\infty C_p \left(T_{aw} - T_w\right)}$$

Determination of the Stanton number using the above relation faces the problem of the adiabatic wall temperature calculation. As usual for engineering applications this value is calculated with the use of the temperature recovery factor *r*, the stagnation temperature of the flow $T_0$ and the input flow Mach number M:

$$T_{aw} = T_0 \cdot \frac{1 + r\dfrac{k-1}{2}\mathrm{M}^2}{1 + \dfrac{k-1}{2}\mathrm{M}^2}$$

Different experimental researches for gas flows [4, 5] have demonstrated that for a developed turbulent supersonic flow around a plane surface the temperature recovery factor appears to lie in the range flor 0.875 to 0.895. There is an interesting Eckert-Weise effect of temperature recovery factor decreasing and the corresponding adiabatic wall temperature reducing down to the meanings lower than the static temperature for flow with subsonic velocity across the cylinder [6-8]. It appears that the effect of temperature reducing propagates downstream in the direction of the cylinder wake. In the first papers on this topic authors suggested that the vortex wake flow formed by a streamlined body causes this effect. For flows with supersonic velocities adiabatic wall temperature decreasing was registered for flows around a cylinder behind circular ribs, on a plane surface behind obstacles in the form of a rib or a backward-facing step, on cones behind a sphere, cylinder, disk and cones of different sizes [9-11]. Cold regions in the wake flow behind the cut of turbine blade are also registered in the distribution of the stagnation temperature [13].

The research continues the experimental study of the previously recorded [14-18] effect of "aerodynamic cooling" of the wall (reduction of the temperature recovery factor) in the wake of a streamlined body in a supersonic flow. The aim of this work is to study the heat transfer parameters in the supersonic wake flow of backward-facing step.

## 2. Experimental apparatus, instrumentation and technique

The studies were carried out at the AR-2 supersonic wind tunnel (Figure 2) with input flow Mach number 2.2 and a stagnation temperature of 294 K with a turbulent flow regime (Reynolds number over 20 million). The objective of the study was to determine the adiabatic wall temperature and heat transfer coefficient for supersonic flow around a smooth wall and in the wake behind the backward-facing step.

The supersonic wind tunnel working channel has a rectangular cross-section 70×98 mm (Figure 2). On the side walls of the working channel optical windows were mounted for visualization by an optical method using the shadow photographs (Figure 3). A flat model made of non-heat-conducting plexiglass (thermal conductivity coefficient was 0.19 W/(m·K)) to eliminate heat leakage in the longitudinal and transverse directions in order to increase the accuracy of measuring heat transfer parameters [19]. The model was installed on the lower wall parallel to the main flow. The heat transfer parameters were determined by varying the height of the step from 8 to 14 mm. The thickness of the unperturbed boundary layer during flow around a flat wall was about 6 mm.

The adiabatic wall temperature in the experiment is determined using the processing technique [20-22], which allows one to determine the heat flux into the wall already in the process of starting the wind tunnel and then extrapolate it to zero. The wall temperature in this case will be the adiabatic wall temperature. Information from all thermocouples and pressure meters of the experimental facility was collected in the temperature and pressure sensor blocks, then through amplifiers was directed toward the analog-digital converter. The automation program for experimental data obtaining and further processing was programmed in LabVIEW software and could be seen in the form of virtual instruments on the experimental monitor during each run of the wind tunnel.

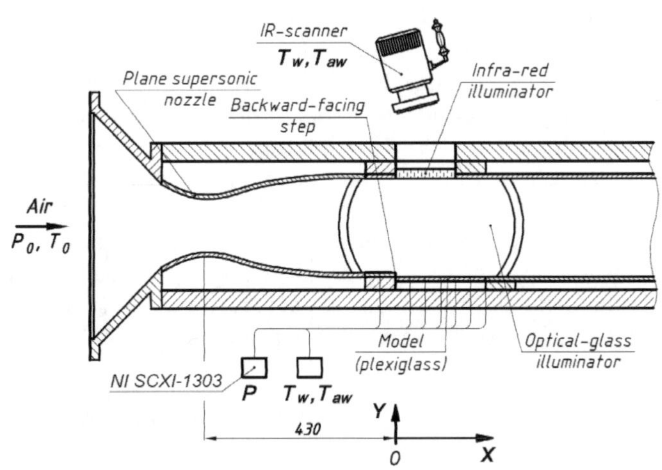


**Figure 2.** Experimental equipment and the scheme of the supersonic wind tunnel.

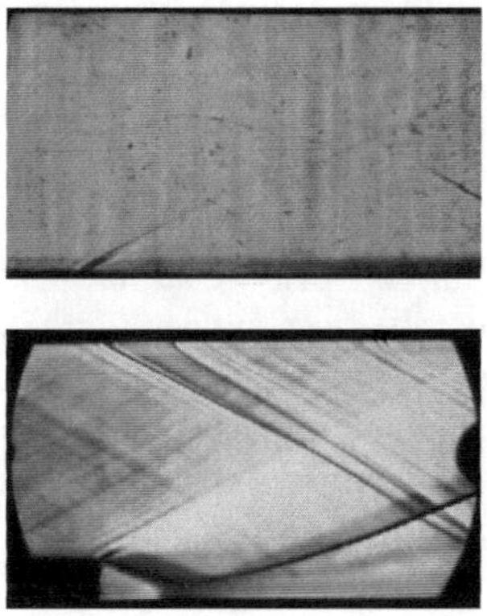

**Figure 3.** Shadow photos for supersonic plane model flow and for wake flow behind a step.

The uncertainties of the main parameters were defined for 95% confidence interval [23]: ±0.5K for wall temperature, ±0.3K for flow stagnation temperature, ±6kPa for stagnation pressure, ±0.5kPa for static pressure, ±1.2% for Mach number, ±2.7% for Reynolds number, ±1% for temperature recovery factor, ±9.5% for Stanton number.

## 3. Results and discussion

As a result of the experimental data processing, the Mach number in the flow, the adiabatic wall temperature, the heat flux, the temperature recovery factor and the Stanton number were calculated. Data in Figures 4 and 5 is presented for the case of continuous supersonic flow around plane model without the step and reflects the typical changing of the thermal gas dynamics parameters during the experiment.

Figure 4 shows a graph of change of total pressure in the pre-chamber of supersonic wind tunnel and static pressure on the model surface during the experiment. After opening the valve, launching the wind tunnel takes about 4 seconds, after which the pressure in the flow is set up and practically does not change during the experiment. The ratio of total and static pressure allows us to determine the Mach number of the input flow that is 2.2. All further experiments were carried out at the same input flow Mach number.

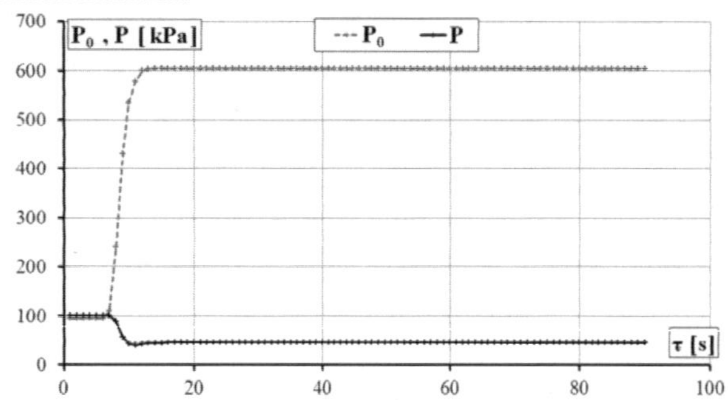


**Figure 4.** Graph of the change of total pressure in the pre-chamber of supersonic wind tunnel and static pressure on the plane model surface during the experiment.

Figure 5 shows a graph of change of the total temperature, the model wall temperature and adiabatic wall temperature (obtained as a result of data processing) during the launching and operation of the wind tunnel. There was no possibility of heating the flow or the model during the experiment, so the initial model temperature was determined by the environmental conditions. The increase in the total temperature at the time of launching is due to the reaction of the thermocouple located in the pre-chamber to a sudden increase in pressure. As it can be seen, the input flow total temperature is getting set up more slowly than the pressure (Figure 4), but also remains almost constant in the course of the experiment. The model is cooled down by the flow and in the limit it can reach the adiabatic wall temperature, but in this experiment the state of thermal equilibrium was not achieved purposely during the study. Thus the heat transfer parameters were obtained by data processing technique [21, 22].

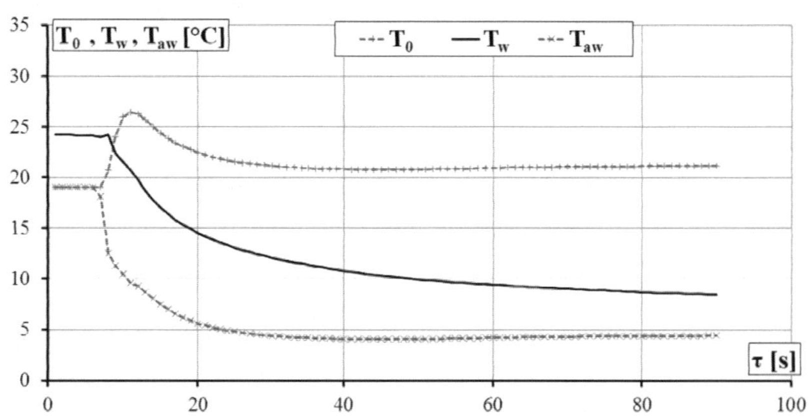


**Figure 5.** Graph of the change of total temperature in the pre-chamber, the model wall temperature and the adiabatic wall temperature (obtained as a result of data processing) during the experiment.

Figure 6 shows the temperature recovery factor calculations in the places of thermocouples mounting. The non-dimensional length of the model – the abscissa axis – was calculated as the ratio of the coordinate from the beginning of the model at the nozzle exit section to the corresponding height of the step (8-14 mm). An increase in the temperature recovery factor in comparison with the flow around the smooth model directly behind the step and a decrease in the wake flow were obtained. The maximum value of the temperature recovery factor reaches in the region of the boundary layer reattachment *x/H*=2. This effect corresponds to the results of a previous research on the falling shock wave influence on the adiabatic wall temperature [14, 15]. At the same time, a decrease in the temperature recovery factor along the entire length of the model at a distance of up to 20 calibres downstream of the step is observed. As it can be seen the value of temperature recovery factor in the wake is the lower, the higher the height of the step. This effect correlates with the results of the research of the supersonic flow behind the rib [16-18], but the decrease in the temperature recovery factor is less pronounced.

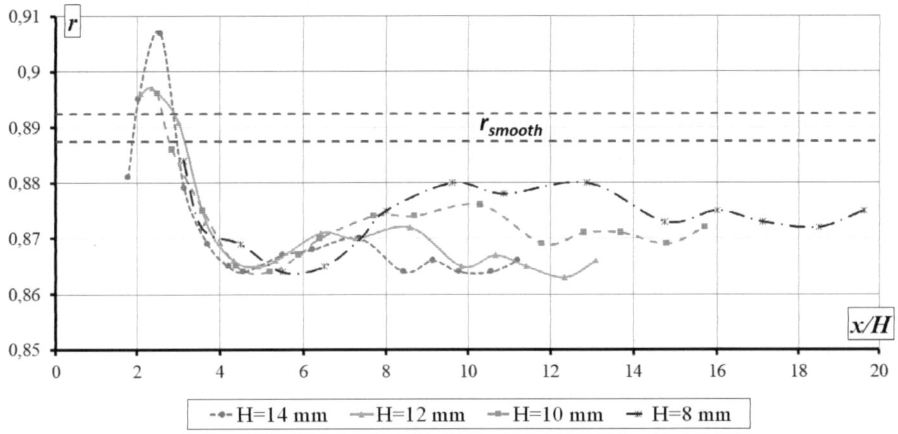


**Figure 6.** Field distribution of temperature recovery factor along the dimensionless length of the plane model for supersonic wake flow behind a backward-facing step of varied height.

The heat transfer coefficient (Stanton number) behind the step, related to the heat transfer coefficient in the smooth model flow $St/St_0$, was less than 1 directly behind the step and increased as the distance from the step (Figure 7). Starting from the coordinate *x/H*=3÷4, heat transfer enhancement is observed, remaining at the level of 20% over the entire measurement area along the length of the model – at a distance of up to 20 calibres downstream of the step.

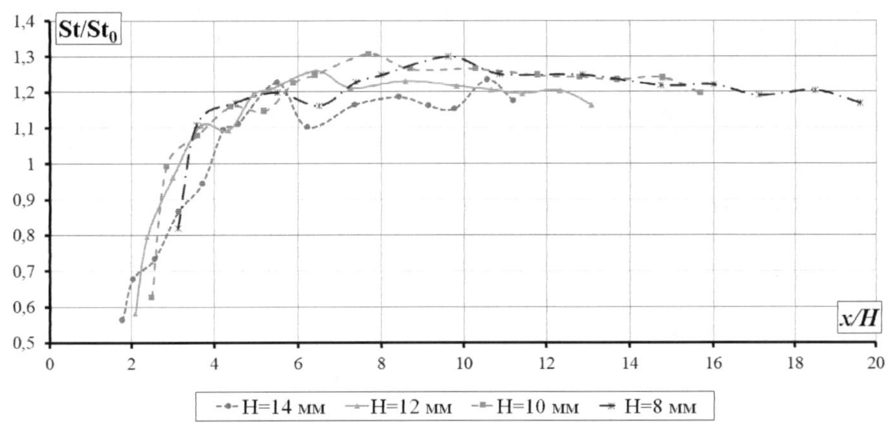


**Figure 7.** Field distribution of Stanton number along the dimensionless length of the plane model for supersonic wake flow behind a backward-facing step of varied height.

Thus, in the area of the boundary layer reattachment behind the step, there is a maximum adiabatic wall temperature (temperature recovery factor) and a rather sharp increase in the heat transfer coefficient. At the same time, at a considerable distance (up to 20 calibers) downstream, the cold wake flow region and the heat transfer enhancement are fixed. Perhaps this effect is associated with the entry into the boundary layer of high-speed low-temperature particles of the main flow, carried away by the recirculated flow behind the step. However, to confirm this hypothesis it is necessary to conduct additional research.

## 4. Conclusions

The results of an experimental study of heat transfer parameters for a supersonic flow in the wake of a backward-facing step are presented. The maximum temperature recovery coefficient and a sharp increase in the heat transfer coefficient in the region of the boundary layer reattachment at a distance of about 2 heights of the step were recorded. At a distance of up to 20 heights of the step downstream, there is a decrease in the temperature recovery coefficient down to 0.865 and the heat transfer enhancement at the level of 20% compared to the smooth model flow. The value of temperature recovery factor in the wake is the lower, the higher the height of the step.

**Acknowledgements**

The research was supported by The Council for grants of President of Russian Federation (Project SP-631.2018.1).

## References

[1] Leontiev A I 2002 *Fluid Dynamics* **37** pp 512-529
[2] Hayes J R and Neumann R D 1992 *AIAA PrAA* **142** pp 63-114
[3] Moffat R J 1998 *Int J Heat Fluid Flow* **19** pp 90–101
[4] Kutateladze S S and Leontiev A I 1990 *Heat Transfer, Mass Transfer, and Friction in Turbulent Boundary Layers* (Taylor and Francis)
[5] Schlichting H 1968 *Boundary layer theory* (McGraw-Hill)
[6] Eckert E R G 1986 *Int Commun Heat Mass Transf* **13** pp 127-43
[7] Goldstein R J and Kulkarni K S 2008 *J Heat Transf* **130** pp 1-9
[8] Aleksyuk A I and Osiptsov A N 2018 *Int J Heat Mass Transf* **119** pp 665–677
[9] Brinich P F 1958 *NACA TN* 4213
[10] Gadd G E, Cope W F and Attridge J L 1960 *R&M* 3148 (London)
[11] Chang P K 1970 *Separation of flow* (Pergamon Press, Oxford)
[12] Vinogradov Y A, Zditovets A G, Leontiev A I, Popovich S S and Strongin M M 2017 *JPCS* **891**
[13] Carscallen W E, Currie T C, Hogg S I and Gostelow J P 1999 *J Turbomach* **121** 4 pp 703–708
[14] Popovich S S, Egorov K S and Vinogradov Y A 2014 *Proc 15th Int Heat Transf Conf* (Kyoto)
[15] Vinogradov Y A, Zditovets A G, Leontiev A I, Popovich S S and Strongin M M 2017 *JPCS* **891**
[16] Popovich S, Strongin M, Vinogradov Y and Zditovets A 2015 *Proc 8th Int Symp Turb, Heat Mass Transf* (Sarajevo) pp 865-868
[17] Leontiev A I, Popovich S S, Vinogradov U A and Strongin M M 2018 *JPCS* **1129**
[18] Leontiev A I, Popovich S S, Vinogradov U A and Strongin M M 2018 *Proc 16th Int Heat Transf Conf* (Beijing)
[19] Schultz D L and Jones T V 1973 *AGARDograph* **165** 157 p
[20] Mee D J, Chiu H S and Ireland P T 2002 *Int J Heat Mass Transf* **45** pp 3287–3297
[21] Leontiev A, Popovich S, Strongin M and Vinogradov Y 2017 *EPJ Web Conf* **159**
[22] Popovich S S 2018 Software Engineering (New technologies, Moscow) **9** pp 35-45
[23] Moffat R J 1988 *Exp Therm Fluid Sci* **1** pp 3-17

5th International Workshop on Heat/Mass Transfer Advances
for Energy Conservation and Pollution Control
August 13-16, 2019, Novosibirsk, Russia

# A laboratory setup based on porous burner for coal particles combustion investigation

**A A Ponomareva[1,2], K A Tcoi[1], S N Mokrin[1,2] & K A Shtym[1]**

[1] Far Eastern Federal University, Suhanova St. 8, Vladivostok, Primorsky Krai, 690950, Russia
[2] Institute for Applied Mathematics FEB RAS, Radio Str. 7, Vladivostok, Primorsky Krai, 690041, Russia

ponomareva.aa@dvfu.ru

In this work, physicochemical features, thermal behaviour, and combustion properties of a coal dust of Pavlovskiy brown coal were studied. For investigations of kinetic properties of fine coal particles combustion, the laboratory-made setup was developed and constructed. An experimental research on the properties of fine coal particles was carried out. Additionally, thermophysical characteristics of coal dust, compositions of thermal decomposition products and particle size distribution were investigated.

## 1. Introduction

The study of effective combustion of fossil fuels is important for finding optimal regimes for burning coal particles in various domestic and industrial boilers and power plants [1, 2].

The investigation of the complex physicochemical process of coal dust burning in the combustion chamber is difficult. Experimental studies of the combustion of a single particle as an elementary component of a coal dust torch have found wide application [3-5].

According to the current models [6-8], it is known that the combustion of coal particles is a complex physicochemical process in which several stages of organic matter conversion during the heating of a coal particle are occurred such as an evaporation of moisture from the surface of a coal particle, a multi-stage yield and combustion of volatile components at thermal decomposition of organic mass, heterogeneous reactions of the char and oxidizer with the formation of complete and incompletely combustion products, and others.

The literature analysis [9, 10] revealed that the kinetic features of the thermal transformation and combustion of coal dust with particle sizes less than 150 μm have not been fully studied. Further investigations require special experiments and a more complex theoretical analysis using improved numerical models.

In this work, physicochemical features, thermal behavior, and combustion properties of a coal dust of Pavlovskiy brown coal were studied. For investigations of kinetic properties of fine coal particles combustion, the laboratory-made setup was constructed.

## 2. Experimental

A coal dust of Pavlovskiy brown coal was studied in present work. The particle sizes of coal dust were analyzed by Analysette 22 Nano Tec (Fritsch, Germany). Ash content was determined using standard method (GOST 11022-95). The lowest calorific value was measured by Calorimeter IKA C 6000

(IKA -Werke GmbH & Co, Germany) as an average value of three measurements. Thermogravimetric analysis was performed using TGA-HP150s equipment (TA Instruments, USA). IR specters of thermal decomposition gas products were recorded using the Nicolet iS10 FT-IR Spectrometer (Thermo Electron Corporation, USA). The experiments were carried out in a chamber at atmospheric pressure in the air. Samples were heated up to 800°C with a heating step of 5°C/min. Outlet gases came out into separate gas line heated up to 180°C connected to the spectrometer.

### *2.1. Laboratory-made setup*

The scheme consists of several units (fig.1): a coal dust supply unit, a coal particles transporting line, a gas-air mixture feed unit, a porous burner, a cooling zone, a wet ash collector, and a decompression section. In the laboratory-made setup the thermal radiation used for coal particles decomposition and ignition was emitted by porous burner during propane/butane combustion. The porous cylindrical Ni-Al alloy burner prepared by self-propagating high-temperature synthesis was used as external heater [11]. Such type of heater was chosen because of its high radiative and heat flux density [12].

a

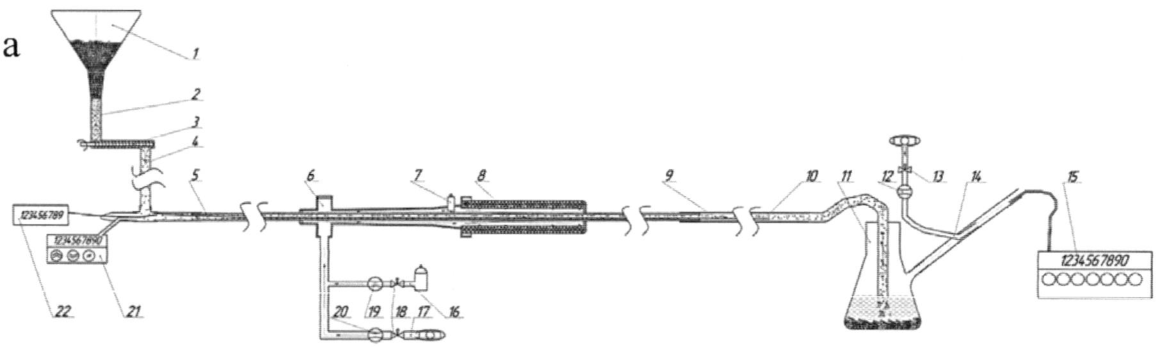


1- coal dust batcher; 2, 4 - plastic tube with d=11 mm; 3 - screw coal feeding unit; 4 - quartz tube with d(inner) = 6 mm; 6 - gas and air mixer; 7 - spark plug; 8 - porous burner; 9 - cooling zone (metal tube with d=8 mm); 10 - thermocouple, 11 - wet ash collector; 12,19, 20 - flow meter; 13,18 - flow valve; 14 – jet pump; 15 - gas analyzer; 16 - gas supply tube; 17 - air supply tube; 21 - draft gauge, 22 - thermocouple controller.

b

Fig. 1. The laboratory-made setup: a – scheme, b - photo.

Coal dust is poured into the "coal feeder". Under gravity and decompression the dust moves through a tube of 11 mm in diameter to the screw feeding unit that regulates the metering of coal dust. Then the dust gets into a transporting line that is a quartz tube with an internal diameter of 6 mm and a wall thickness of 2 mm. When the coal particles enter the zone of burner, they is heated up and ignited. The burning dust particles come out from the zone of the porous burner and move along the quartz tube into the cooling zone. This zone is performed as a stainless steel tube with an internal diameter of 8 mm and a length of 150 mm. Then the residues of particles fall into the wet ash collector. The pressure in the particles transporting line is controlled by a draft gauge.

Thermocouples (K types), a gas analyzer, flowmeters (Bronkhorst) with a flow regulators, a thermal imager (NEC TH 9100), a video camera (Nicon), a high-speed camera (Photron), and a thermometer are used in the setup for research purposes. The recording of coal dust particles movements with a high-speed camera and/or a video camera can be carried out at two points: in front of the burner and after it.

## 3. Results and Discussion

The distribution of measured particle sizes is illustrated on figure 1.

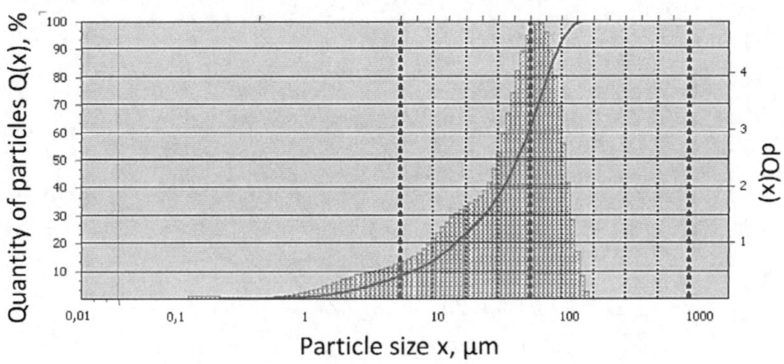


Fig. 2. The particles size distribution of coal dust.

The results of particle size measurements show that the sizes of coal dust particles are less than 120 µm. Ash content measured is about 24.5 %. It was found that the lowest calorific value is higher for fine coal particles (3914 kcal/kg) than for a raw coal of the same rank (3250 kcal/kg) in according to our previous investigations [13].

FTIR spectroscopy of fine coal particles during thermal transformation in air show that the trace of CO has already detected at 200°C. On the spectra recorded at 300°C the trace of methane was appeared. Absorbance peak maximums of CO and $CH_4$ were detected at about 400°C and 450°C, respectively. At the same time TG curve demonstrates two well-defined step-like zones that are from 35 to 110°C and from 190 to 550°C. In the first zone the rapid decrease in weight with rate of about 4.8 mg/°C is observed. The rate of weight decreasing in the second zone is about 1 mg/°C. The rest of diagram (from 550 to 800°C) is characterized by the weight-loss rate less than 0.3 mg/°C [14].

After the laboratory setup (fig. 1) was constructed the combustion of coal particles under different conditions were investigated.

Table 1. Temperature conditions initiated by the porous burner (the pressure in transporting line is -2 hPa)

| Flow gas rate, l/min | Flow air rate, ml/s | Fuel–air ratio, φ | Heat release*, kW/m$^2$ | Outer temperature, °C | Inner temperature, °C |
|---|---|---|---|---|---|
| 1.5 | 750 | 0.8 | 395 | 906 | 915 |
| 1.7 | 850 | 0.8 | 448 | 915 | 955 |
| 1.8 | 900 | 0.8 | 475 | 924 | 1000 |
| 2 | 1000 | 0.8 | 528 | 965 | 1058 |
| 2.5 | 1250 | 0.8 | 660 | 1066 | 1129 |
| 4 | 2000 | 0.8 | 1052 | 1211 | 1283 |

* - the heat release of the internal surface of the burner

Temperatures on the surface of porous burner and in the middle of quartz tube (coal particles transporting line) were measured using the thermal imager and the thermocouple. The obtained results in dependence of burner power are presented in table 1. In the table 1 maximum temperatures are pointed out. Additionally, temperatures of zones near burner were investigated. It was determined that the temperature before and after the burner decreases slightly during the transition from a decompression pressure P = -2 hPa to P = -5 hPa. The temperature is reduced by 10-15% but cooling in the cooling zone is slower and temperatures are higher at 2- 4%.

The mass flow rate of coal dust was determined at different level of decompression, for instance at -2.5 hPa the mass flow rate is 205 mg/min.

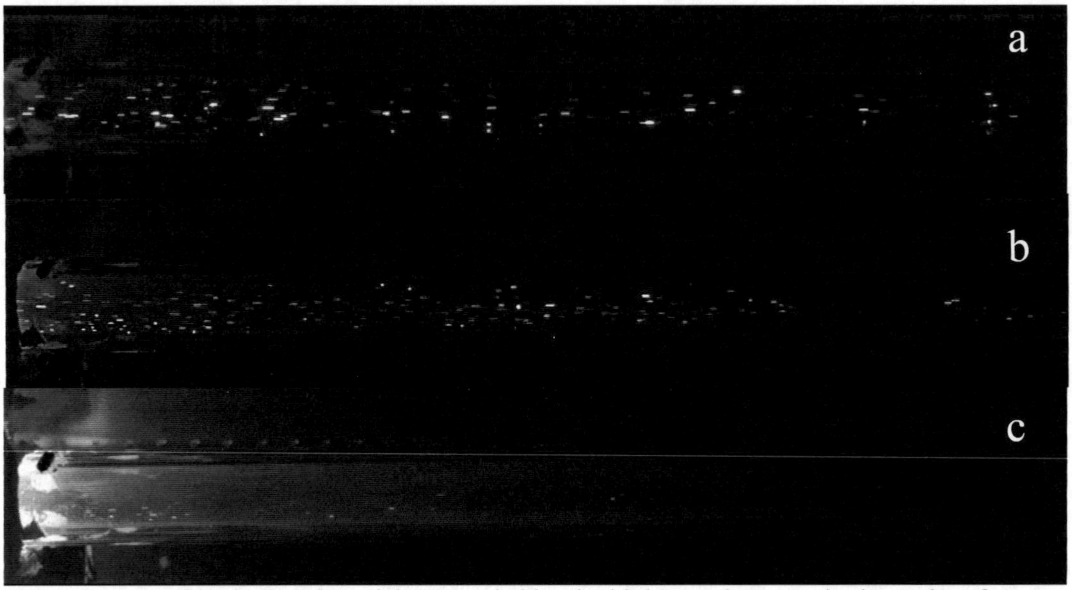


Fig. 3. Images of ignited coal particles recorded by the high-speed camera in the outlet of porous burner in dependence of radiant flux density: a – 395 kW/m$^2$, b – 475 kW/m$^2$, c – 1052 kW/m$^2$. (the decompression in transporting line is -2 hPa)

To study the flow rate of coal particles, a high-speed camera was used. During experiments the camera mode was selected which was 20,000 frames per second. Laser illumination was also used to better display the coal particles. The measured speed of coal particles at -2 hPa was 3 m/s before the burner and 11.6 m/s after it. Figure 3 shows the ignited coal particles recorded in the outlet of porous burner. As can be seen in the images, the linear dimensions of the particles decrease with increasing of burner power. At the same time, it was found that the number of outlet particles are initially increased and then rapidly decreased. Such behaviour could be explained by the fragmentation process [15, 16].

## 4. Concluding Remarks

Experimental results such as coal properties, decomposition rates (temperature dependencies of mass loss), and product distributions, coal particle size analysis were obtained in our work.

The scheme of the laboratory setup was developed and constructed. When developing the scheme of the laboratory setup, the concept of practicality and multifunctionality was adhered to. Combustion of coal dust of Pavlovskiy brown coal was investigated using the laboratory-made setup.

## 5. Acknowledgments

The research was carried out within the state assignment of FASO of Russia No. 075-00400-19-01.

Journal of Physics: Conference Series **1369** (2019) 012041 doi:10.1088/1742-6596/1369/1/012041

## References

[1] Beer J M 2000 *Progress in Energy and Combustion Science* **26** 301
[2] Tcoi K, Babenko G and Lesnykh A 2017 *Scientific Rev.* **2** 30
[3] Lee H and Choi S 2015 *Combustion and Flame* **162** 2610
[4] Bai X, Lu G, Bennet T, Sarroza A, Eastwick C, Liu H and Yan Y 2017 Thermal and Fluid Science **85** 322
[5] Riaza J, Khatami R, Levendis Y A, Álvarez L, Gil M V, Pevida C, Rubiera F and Pis J J 2014 *Combustion and Flame* **161** 1096
[6] Saloojee F 2011 *Kinetics of pyrolysis and combustion of a South African coal using the distributed activation energy model* Master's dissertation Johannesburg 155 p
[7] Boiko E A, Pachkovskii S V and Didichin D G 2005 *Combust. Explo. Shock+* **41** 47
[8] Ranzi E, Pierucci S, Aliprandi P C and Stringa S 2011 *Energy and Fuel* **25** 4195
[9] Wang F Y and Bhatia S K 2001 *Chemical Engineering Science* **56** 3683
[10] Babiy V I 1986 *Combustion of coal dust and calculation of a pulverized torch.* (Moscow: Energoatomizdat)
[11] *Patent RU 2640305* - Minaev S S, Gushchin A N, Tcoi K A, Gushchin D A, Kirdyashkin A I, Shtym K A, Maznoi A S and Filippov D A
[12] Fursenko R, Maznoy A, Odintsov E, Kirdyashkin A, Minaev S and Sudarshan K 2016 *Int. J. Heat and Mass Transfer* **98** 277
[13] Ponomareva A, Lesnykh A, Tcoi K, Urjupin G, Babushok V. and Shtym K. *Proc. Fourteenth ICFD* 2017 Sendai, Japan 208
[14] Ponomareva A A, Grebenyuk I V, Tcoi K A, Lesnykh A V, Babushok V I and Shtym K A 2018 *IOP Conf. Series: J. Phys.: Conf. Series* **1115** 042031
[15] Bar-Ziv E and Kantorovich I I 2001 *Progress in Energy and Comb.* **27** 667
[16] Suzuki A, Yamamoto T, Aoki H and Miura T 2002 Proc. of the Combustion Institute 29 459

# Explosive vaporization of water-based nanofluid on a flat microheater

**I A Kozulin**[1,2], **V V Kuznetsov**[1,2]

[1] SB RAS, Institute of Thermophysics, Lavrentieva av. 1, Novosibirsk, 630090, Russia
[2] Novosibirsk State University, Pirogova str. 1, Novosibirsk, 630090, Russia

E-mail: vladkuz@itp.nsc.ru

**Abstract**. The explosive vaporization of the nanofluid formed by water and $SiO_2$ nanoparticles on a heater with the size of 100×110 μm was studied. In experiments, the reflected optical signal produced by heater lighting using laser beam was measured. The patterns of the initial stage of the nanofluid explosive vaporization were obtained and the effect of nanoparticles on the nucleation temperature was discussed. The variation of the nucleation temperature and vaporization time depending on the temperature increase rate was defined.

## 1. Introduction

Investigations of the explosive vaporization of liquid on microscale are widespread recently due to needing in the design of MEMS systems for high technologies. The MEMS allows for obtaining high temperature increase rate and homogeneous nucleation using microheaters. The regimes of explosive vaporization of water were obtained in [1-3]. The initial stage of explosive vaporization of water on a rectangular heater with a small size was studied in [4]. The bubble formation around gold nanoparticles during pulsed heating was studied in [5]. It is well known that the explosive vaporization occurs when the superheat limit of the liquid is achieved. This work is aimed on the experimental investigation of the explosive vaporization of the nanofluid formed by water and $SiO_2$ nanoparticles on a heater with the size of 100×110 μm. To investigate the initial stage of the explosive vaporization, the reflected optical signal produced by heater lighting using the laser beam was measured that shows the microbubbles appearance.

## 2. Experimental equipment and methods

The experimental setup for investigation of the nanofluid explosive vaporization produced by the heater with a small size is shown in figure 1 (a). The multilayer thin-film heater with the size of 100x110 μm is shown in figure (b) was used in the experiments. This heater has external layer manufactured from silicon carbide. To supply stepwise current for a conductive layer of the heater, the pulse generator was used. The duration of the current pulse was selected so that it is sufficient to produce the explosive vaporization. The supplied heat divided by resistor surface area was used as the "effective" heat flux. The heater (1) was placed in the water or nanofluid (2), and the heater surface was lighting by a laser beam (3). The reflected optical signal was measured using the microscope (4), the photodiode (6) and high-speed ADC (7). The diaphragm (5) selects the area of the heater surface where the nucleation occurs and the reflected optical signal begins to go down due to microbubbles appearance. Therefore, the variation of the reflected optical signal corresponds to the dynamics of the

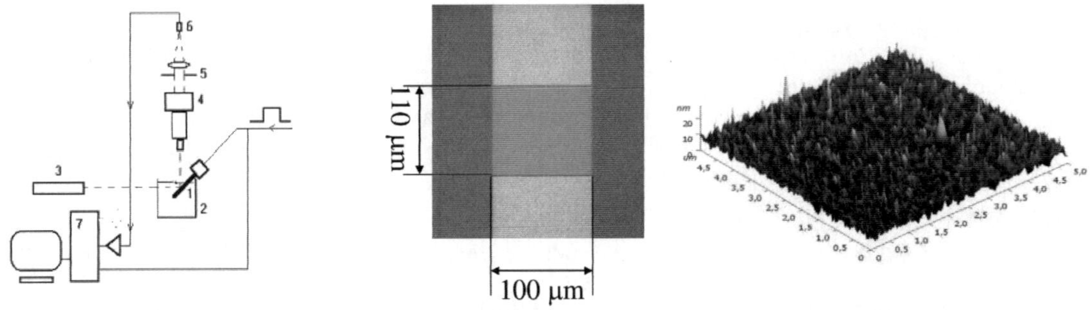


(a) (b) (c)

**Figure 1.** Experimental setup (a), photo of the microheater with size 100x110 μm (b), the morphology of the heater surface (c).

vapor bubbles appearance. This method was used for investigation of the patterns of initial stage of the vaporization depending on the effective heat flux $q_{eff}$ and liquid temperature increase rate. The initial liquid temperature of the liquid was supported in the range from 18 to 21 °C and the static pressure equals atmospheric pressure.

When preparing the nanofluid, the required amount of silica nanoparticles Aerosil 200 is added to the deionized water and vibrated in the ultrasonic bath for a long time. The volume concentration of nanoparticles in a liquid $c_0$ is determined using the density of water and amorphous silicon.

To find the flux transferred to liquid for a heater consists of four layers produced from different materials, the Fourier equation was solved numerically. The liquid temperature, liquid temperature increase rate and heat flux were obtained depending on time for given heat released in the conductive layer. The literature data were used to select the thermal conductivity and heat capacity of the heater layers.

## 3. Experimental results

The variation of the dimensionless reflected optical signal in time during explosive vaporization of deionized water for (a) $q_{eff}$ = 37.6 kW/cm$^2$, $dT_{liq}/dt$ = 26.3 MK/s, (b) $q_{eff}$ = 136 kW/cm$^2$, $dT_{liq}/dt$ = 176 MK/s is shown in figure 2 (a, b) respectively. The dimensionless reflected optical signal was defined as measured value divided on reflected optical signal at the beginning of liquid heating. Here, thin and bold lines show the time dependence of the normalized current on the heater resistor and the normalized intensity of the reflected light beam, respectively, the main stages of the explosive phase

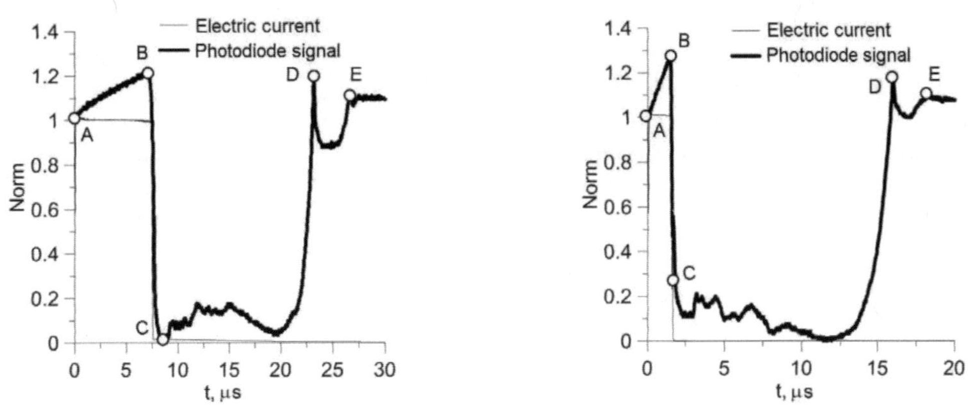


(a) (b)

**Figure 2.** The dimensionless reflected optical signal depending on the time for explosive vaporization of water: (a) $q_{eff}$=37.6 kW/cm$^2$, (b) $q_{eff}$=136 kW/cm$^2$.

change are indicated by letters. Zero time in figure 2 shows the beginning of the current pulse. Stage A-B corresponds to liquid heating, stage B-C corresponds to the time of explosive vaporization, when the microbubbles are appeared and grow up. Stage C-D corresponds to the combining microbubbles into the vapor cavity, the cavity grows and collapses. After cavity collapses, the existence of a satellite bubble is observed if the thermal energy accumulated during liquid heating is sufficient (stage D-E). The time interval B-E is defined as the duration of the explosive phase change $t_{boil}$, and the interval AB is defined as the heating time before the nucleation starts $t_{ini}$.

The variation of the dimensionless reflected optical signal in the time during explosive vaporization of water containing nanoparticles of Aerosil 200 with the volume content of 0.001 is shown in figure 3. For presented data, the effective heat flux $q_{eff}$ and the temperature growth rate $dT_{liq}/dt$ correspond to: (a) $q_{eff}$=32.7 kW/cm$^2$, $dT_{liq}/dt$=20.2 MK/s, (b) $q_{eff}$ =130 kW/cm$^2$, $dT_{liq}/dt$=156 MK/s. Comparing the reflected optical signal depending on the time during explosive vaporization of pure water and water with Aerosil 200 one can see that the nanoparticles lead to the decrease in the time of nucleation and increase the flux of nuclei (decrease the time of vaporization). The existence of the threshold for explosive phase change is seen from the increase in the duration of stage B-C in figure 3 (a).

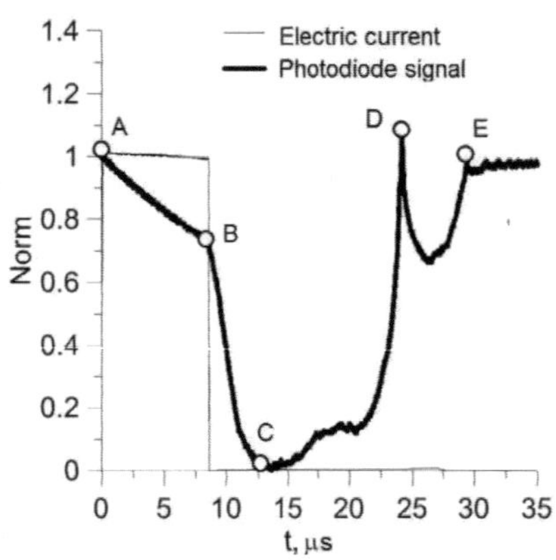


(a)

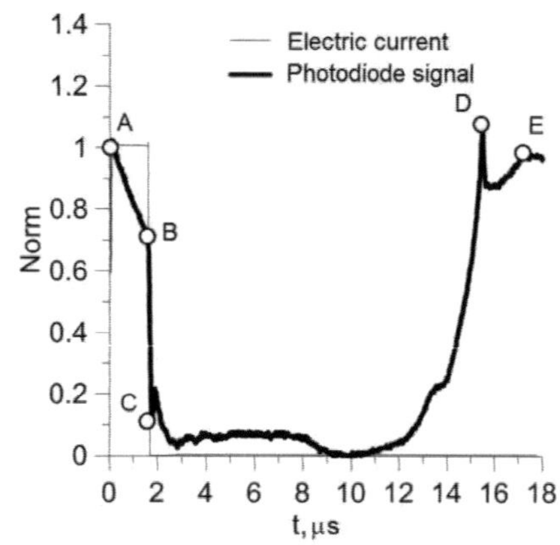


(b)

**Figure 3.** The dimensionless reflected optical signal depending on time for explosive vaporization of water with Aerosil 200 nanoparticles with volume content of 0.001: (a) $q_{eff}$ =32.7 kW/cm$^2$, (b) $q_{eff}$ =130 kW/cm$^2$.

Using the results of numerical calculations of the thermal problem for a multilayer heater and measurements of the time of the nucleation, the dependences of the temperature corresponding to the appearance of vapor bubbles on the rate of liquid temperature growth were determined for pure water and water with silicon oxide nanoparticles. The dependence of the nucleation temperature on the temperature increase rate is shown in figure 4 for pure water and water with Aerosil 200 at particle content of 0.001. The line shows the temperature of superheat limit for water, determined by the equation

$$T_{lim}/T_{cr} = 0.905+0.095\cdot(T_{sat}/T_{cr})^8 \quad (1)$$

that obtained from experimental data [7]. It is known that for water, at relatively low rates of temperature growth, the explosive vaporization occurs at the temperature lower than that given by equation (1) due to decrease in the work of nucleation near the metal surface. As can be seen from figure 4, this is also observed for silicon carbide surfaces, which is characterized by the presence of areas with low surface energy and degraded wetting for water. With the increase in the rate of liquid temperature growth, the temperature of the nucleation for pure water increases and reaches the value of the temperature of the superheat limit given by equation (1). The data for water with silicon

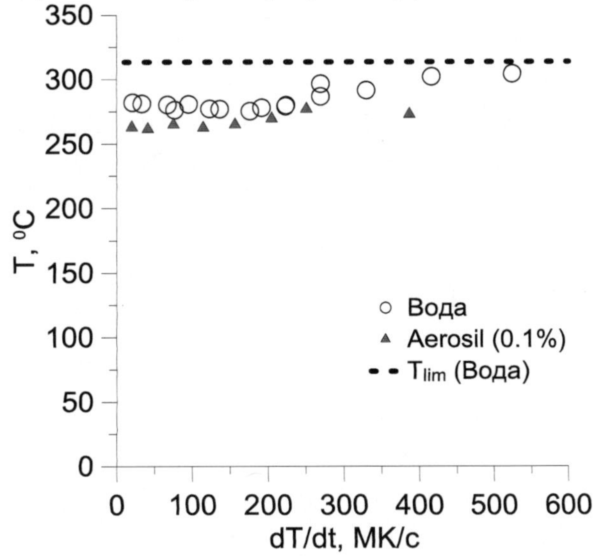


**Figure 4.** The temperature of explosive vaporization depending on temperature increase rate for pure water and water containing nanoparticles.

nanoparticles shown in figure 4 reveal a decrease in the temperature of nucleation due to the reduction of the work of nucleation in the presence of nanoparticles. This result is important for the development of the methods to control the temperature of explosive vaporization in metastable systems.

## 4. Conclusions

Results of the present study were obtained using the original optical technique, which allows resolving processes at times of the order of ten nanoseconds during rapid phase change near the microheater surface. It allowed us to find new patterns of the explosive vaporization of the nanofluid formed by water and $SiO_2$ nanoparticles and show that the explosive vaporization in the liquid containing inhomogeneities at the nanoscale occurs in the region of temperatures between the binodal and the spinodal by the mechanism of heterogeneous nucleation. This statement is confirmed by a decrease in the temperature of vaporization and increase in the flux of bubbles stimulated by the nanoparticles.

## Acknowledgment

This work was carried out under a state contract with IT SB RAS (project 5.3 KPFI SB RAS, AAAA-A18-118112690094-4).

## References

[1] Hong Y, Ashgriz N, Andrews J 2004 *J. Heat Transfer* **126**(2) 259
[2] Zhao Z, Glod S, Poulikakos D. 2000 *Int. J. Heat Mass Transfe*r **43** 281
[3] Avedisian C T, Osborne W S, McLeod F D, Curley C M 1999 *Proc. Royal Soc. London A: Mathematical, Physical, Engineering Sci.* **455** 3875
[4] Kuznetsov V V, Kozulin I A 2010 *J. Eng. Thermophys.* **19**(2) 102
[5] Lombard J, Biben T, Merabia S 2014 *Phys. Rev. Letters* **112** 105701.
[6] Bhaskar E V, Aden J S 1985 *Hewlett-Packard Journal* **36**(5) 27
[7] Skripov V P 1974 *Metastable Liquids* John Wiley & Sons
[8] Skripov P V, Skripov A P 2010. *J. Thermophys.* **31**(4–5) 816

# Fundamental issues related to flow boiling and condensation in microchannels – experimental challenges and opportunities

**V V Kuznetsov, A S Shamirzaev**
[1]Institute of Thermophysics Siberian Branch of Russian Academy of Sciences
1, Academician Lavrentyev Av., Novosibirsk, 630090, Russian Federation

E-mail: vladkuz@itp.nsc.ru

**Abstract**. In recent years, application of the microchannels has received considerable attention due to their capability for thermal management of microelectronic equipment, high power devices and development of compact evaporators. The present paper focuses at a discussion of the heat transfer during flow boiling and condensation in microchannels of the heat sink to identify the methods to improve heat transfer performance and select the physically based models for heat transfer prediction. The main purpose of this study is to discuss the mechanisms of heat transfer in microchannels at different aspect ratio for the three-sided heating case using the experimental data were obtained for refrigerant R134a and dielectric fluid.

## 1. Introduction

The ability of cooling system to remove sufficient heat is important for the design of the advanced electronic components, high-performance computer chips and high heat-load optical systems. Nowadays, the advanced electronic components generate heat fluxes higher then 100 W/cm$^2$, while some future electronic components, such as high-power laser and electronic radar systems, have been projected to generate heat fluxes over 1000 W/cm$^2$ [1]. To dissipate large amount of heat, flow boiling can be organized inside the multiple parallel microchannels named as microchannel heat sink [2]. Microchannel heat sink that uses refrigerants or water as a coolant allows absorbing high heat fluxes due to the latent heat of vaporization and maintains the desired temperature control. Considerable limitation of two-phase cooling is an existence of critical heat flux (*CHF*) refers to the maximum heat flux just before the boiling crisis occurs. The CHF can be increased in case of surface (subcooled) boiling [3] when a sudden increase of surface temperature will occur at higher heat flux than for saturated boiling. The use of water for electronic equipment cooling is limited by the high boiling point and more promising is the use of refrigerants or dielectric fluids, such as FC-72, that have a lower boiling point.

It is well known that the flow boiling heat transfer coefficient increases for refrigerants with increasing heat flux [4, 5], but in some experiments it was shown that the heat transfer coefficient depends on the vapor quality, showing growth [6] or reduction [7] when the quality is increased. The difficulties in selection of the reliable correlations for predicting the flow boiling heat transfer constrain the development of two-phase microchannel heat sinks and micro tube evaporators. This problem was discussed in many studies, e.g. [5], however, the flow boiling heat transfer in microchannels remains understood incompletely.

It is obvious that the design of a small scale two-phase cooling systems needs to development of the microchannel condensers with high efficiency [8, 9]. Microchannels increase the shear stress exerted upon the film interface, that greatly decreases the film thickness and resulting in very high condensation heat transfer coefficients [8]. The mechanism of condensation in microchannels depends on the relative importance of the surface tension, gravity and shear forces, which, in turn, depend on several parameters, such as vapor quality, mass flux, fluid properties and channel geometry [9].

The objective of this paper is the experimental consideration of the mechanisms of flow boiling and condensation heat transfer in microchannels with different aspect ratio for the three-sided heating case. Existing models for predicting the heat transfer coefficient inside rectangular microchannels were verified using the experimental data obtained for refrigerant R134a and dielectric fluid perfluorohexane.

## 2. Experimental equipment and methods

Experiments were performed in the closed loop presented in detail in [10]. The working liquid is supplied from the condenser through the filter and damper of pulsation to the flow controller via the plunger pump. The desired inlet vapor quality is obtained when the fluid goes through the pre-evaporator placed before the test section. Two experimental test sections were used to investigate the flow boiling and condensation heat transfer phenomena in the microchannel heat sink. The heat sink with the channels having a low aspect ratio is shown in figure 1a. The oxygen-free copper microchannel plate and heating block are soldered by tin to the stainless steel shell. There are twenty-one microchannels on the plate. The cross-section of each microchannel is 0.335 μm × 0.930 μm and the spacing between the channels is 0.650 μm, the aspect ratio $\alpha=a/b$ is 0.36, where $a$ is the width and $b$ is the height of the channel. The size of the microchannel plate is 20 × 40 × 2.5 mm. Two layers of the thermocouples are placed along the microchannels into the solder on both sides of the partition sheet made from stainless steel. Their locations, measured from the inlet of the microchannels, are 5 mm, 15 mm, 25 mm, and 35 mm. After mounting into the stainless steel shell, the microchannel plate is covered by thin sheet of copper and stainless steel plate. The inlet and outlet temperature and pressure are measured inside inlet and outlet chambers.

Heat is supplied to thermo-insulated test section from the heat cartridges mounted in the copper block as shown in figure 1a. In the case of flow condensation, heat cartridge is replaced by the Peltier module and external water heat exchanger. External local heat flux that passes into microchannel plate is determined from temperature difference $\Delta T_{w,i}$ across the partition sheet. The heat flux on the wall of the microchannel $q_{w,i}$ (internal heat flux) is calculated accounting the difference in the internal and base surface area of the microchannel plate. The internal wall temperatures $T_{w,i}$ is determined accounting the local heat flux and measured temperature $T_{m,i}$. The saturation temperature related to $i$-th thermocouple is calculated using linear approximation of the measured input and output pressure, and accounting the dependence of the saturation temperature on pressure for working liquid.

The heat sink with two microchannels having a high aspect ratio is shown in figure 1b. The

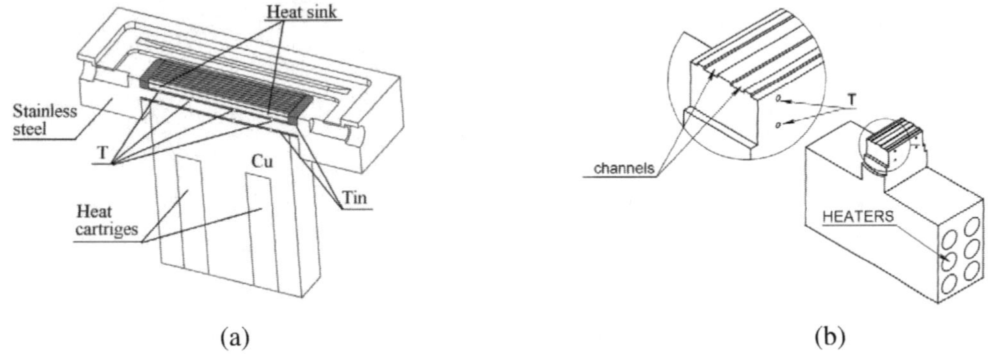


**Figure 1.** Schematic of the test sections with low (a) and high (b) channel aspect ratio.

microchannels with cross section 2 mm×0.36 mm were milled on the top of the copper block made from oxygen free copper with channel-to-channel spacing equal to 2 mm and covered by a thin nickel layer using electrochemical treatment. The length of micro-channels is 16 mm, the surface roughness Ra is 0.67 μm, the aspect ratio of the channels is 5.56. Heat cartridges are mounted into the copper block as is shown in figure 1b. Two layers of four thermocouples are placed into the copper block along the length of microchannels. The thermocouples locations, measured from the inlet of the microchannels, are 3 mm and 13 mm. The thermocouple layers are placed at distance 4.5 mm from each other and the top layer is placed at 1.5 mm below the surface of the microchannels. The microchannels are covered by a polished stainless steel plate. The inlet and outlet chambers have a cylindrical shape with a diameter of 8 mm and a length of 6 mm. The external local heat flux to the microchannel plate is determined from the measured temperature gradient in the copper. The inner wall temperature is calculated using a linear approximation of the temperature gradient in the copper block. The vapor quality at the inlet was determined considering heat production in the pre-evaporator $Q_{coil}$.

## 3. Flow boiling heat transfer

Experimental data on local heat transfer coefficients for R-134a flow boiling in a microchannel heat sink with microchannels having a low aspect ratio (figure 1a) were obtained in the range of internal heat flux from 20 to 500 kW/m$^2$ and mass flux $G$ from 300 to 600 kg/m$^2$s. The local thermodynamic vapor quality was varied from -0.02 to 1.2 and static pressure from 6 to 16 bars. The heat transfer coefficient was determined accounting the fin efficiency that exceeded 0.98. The experimental data show a considerable effect of heat flux on the magnitude of the heat transfer coefficient. It shows that nucleate boiling is the dominant mechanism for heat transfer. Nevertheless, it was obtained that the pool boiling equation cannot be used for predicting flow boiling heat transfer. The comparison of experimental data for saturated flow boiling of R-134a in copper microchannel heat sink with prediction according to the equation for pool boiling on cooper surface [11] is shown in figure 2a. The effect of vapor quality $x$ on the heat transfer coefficient is presented for the average mass flux of 380 kg/m$^2$s, saturation temperature 24.2 °C and heat flux on internal channel surface of 22.3 kW/m$^2$. Points show experimental data, the calculation results according to equation [11]

$$h_{boil} = 100 q_w^m p_r^{0.45} \left(-\log(p_r)\right)^{-0.8} R_a^{0.2} M^{-0.5} \quad (1)$$

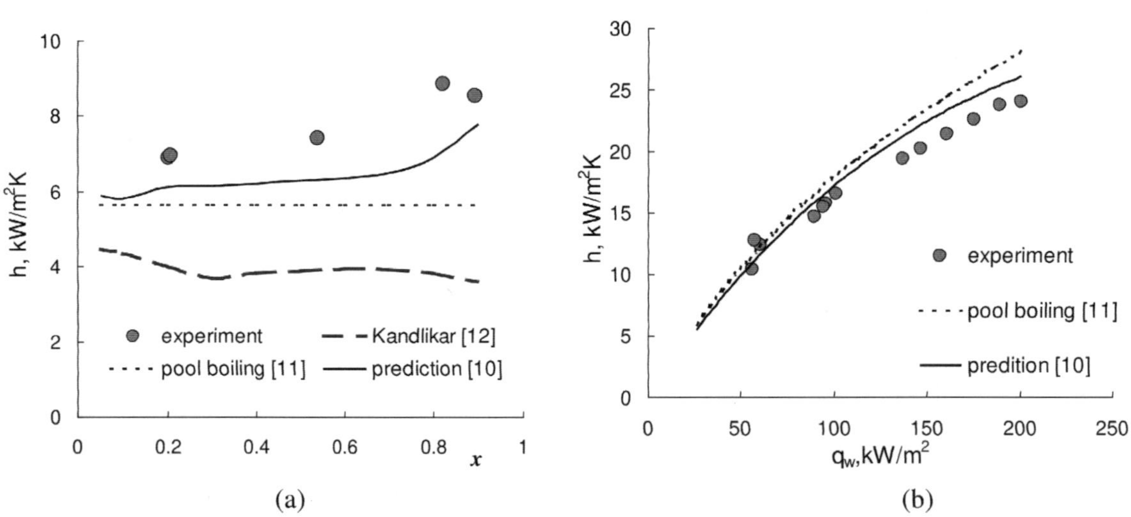


**Figure 2.** Heat transfer coefficient vs. vapor quality for heat flux of 22.3 kW/m$^2$ (a), and vs. heat flux for x < 0.3 and G = 550 kg/m$^2$s (b). Points show experimental data, solid lines show the calculation results according to [10], dotted and dashed lines according to [11] and [12].

are shown as a dotted line, were $R_a$ is roughness, $m = 0.9 - 0.3 p_r^{0.2}$ and $p_r = p / p_{cr}$ is the reduced pressure, the dashed line corresponds to the calculations according to [12]. As is seen, the pool boiling equation considerably underpredicts the experimental data, especially at high vapor quality. The reason for the increase of heat transfer coefficient in compare with pool boiling model is the importance of the evaporation mechanism of heat transfer because in microchannels the liquid film becomes very thin due to high interfacial shear stress. It should be noted that increasing the heat flux for the same mass flux causes the reduction of the heat transfer due to nucleate boiling suppression.

The data were obtained show that reliable approach for predicting the flow boiling heat transfer in microchannels should account for the determining mechanisms such as nucleate boiling suppression occurs when the liquid film thickness becomes sufficiently small, forced convection and evaporation of a thin liquid film. The physical basis of the approach proposed in [10] consists in accounting all these mechanisms in one general equation. The proposed method has the privilege over simple summing of corresponding terms because it allows clearly distinguishing the contribution of each term under the conditions when it is predominant. To determine the heat transfer coefficient for the evaporating wavy-turbulent liquid film, the model of [13] was used in this paper with the calculation of the shear stress according to [14]. The comparison of the calculation results on dependence of heat transfer coefficient from heat flux with experimental data obtained for G = 550 kg/m$^2$s, $p_r$ = 0.22 and vapor quality less then 0.3 is presented on figure 2b. One can see that model [10] better suits the experimental data both for low and high heat fluxes in compare with pool boiling model. It happens so because the proposed model takes into account both the nucleate boiling suppression and thin-film evaporation.

Measurements of the dependence of internal averaged heat flux on wall superheat during R-134a flow boiling in the microchannels having a high aspect ratio (figure 1b) are presented in figure 3a. The data were obtained for saturated and subcooled conditions at the inlet in the range of internal heat flux from 50 to 500 kW/m$^2$ and mass flux of 840 kg/m$^2$s at static pressure 7.3 and 7.35 bar accordingly. For flow boiling of refrigerant R-134a, the visible impact of heat flux on the magnitude of wall superheat was observed. It was obtained that the equation [11] for pool boiling heat transfer cannot be used for prediction of the flow boiling heat transfer in saturated conditions. Points in figure 3a shows the experimental data, dotted line show the result of calculations according to [11]. As is seen, good agreement with experimental data has the model [10] accounted for both the nucleate boiling suppression and evaporation of liquid film. It is also seen that in a subcooled region, the experimental data are above the calculation results that shows enhancement of heat transfer due to liquid film

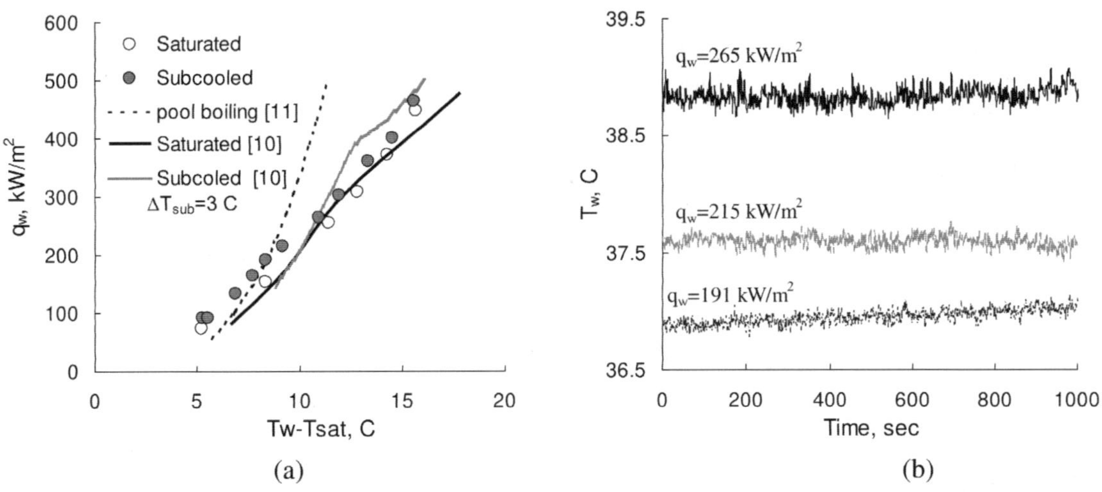


**Figure 3.** The dependence of heat flux from wall superheat (a) and the local wall temperature from the heat flux for G=840 kg/m$^2$ (b).

evaporation. It happens because after nucleation the produced vapor remains as the bubbles attached to the surface while growing and collapsing. In the region of saturated boiling, the experimental data are below the calculations due to increasing the level of wall temperature fluctuations that produced unsteady conditions inside the microchannels, see figure 3b.

## 4. Flow condensation heat transfer

Measurements of the local heat transfer coefficients during flow condensation of dielectric fluid perfluorohexane in the microchannels having a low aspect ratio (figure 1a) are shown in figure 4a. During condensation, the heat was removed from the wall of the test section using Peltier modules that allow varying local wall subcooling. The data were obtained in the range of wall subcooling from 1 to 8 °C and mass flux from 100 to 200 kg/m$^2$s. The local vapor quality was varied from 0.1 to 0.92 and static pressure from 1 to 1.2 bars. The heat transfer coefficient was determined accounting the fin efficiency that exceeded 0.98. As is seen, the condensation heat transfer coefficient is highest near the channel

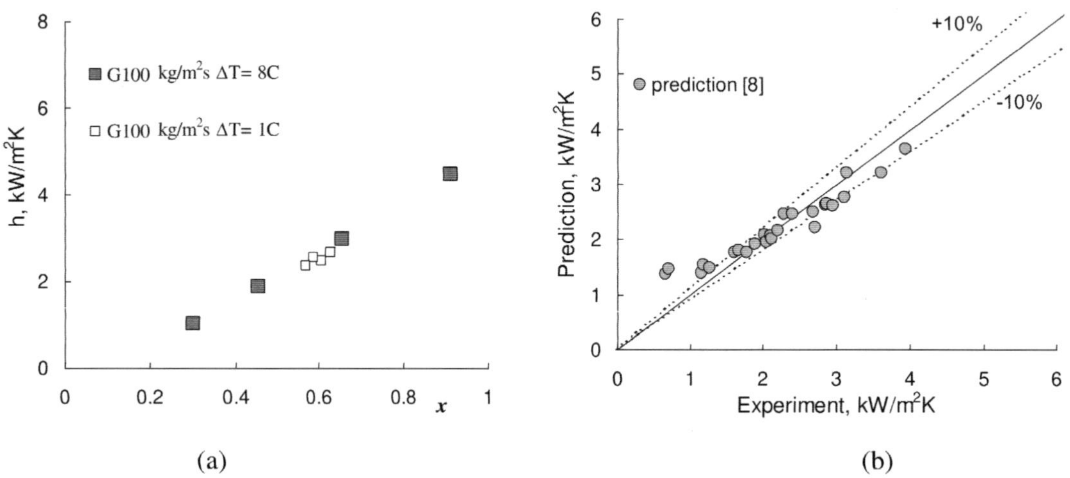


**Figure 4.** The dependence of heat transfer coefficient on vapor quality at G = 100 kg/m$^2$s (a), comparison of measured and predicted heat flux for flow condensation in microchannels (b).

inlet and decreases along the microchannels. Increasing the mass velocity increases the condensation heat transfer coefficient due to thinning the liquid film when the vapor interfacial shear stress is increased. For flow condensation of perfluorohexane, the impact of wall subcooling on the magnitude of the heat transfer coefficient was not observed. It shows a considerable influence of the capillary forces on local film thickness that was discussed in [15]. The comparison of experimental data with predictions according to the model [8] for the studied range of mass flux and local vapour quality is shown in figure 4b. One can see that calculation results are in good agreement with the experimental data. The deviation of experimental data from calculations is caused by the transition to elongated bubble flow.

## 5. Conclusions

The presented results show that saturated and subcooled flow boiling of refrigerant R134a in microchannels with low and high aspect ratio for the three-sided heating case is characterized by the visual impact of heat flux on the magnitude of heat transfer coefficient. Nevertheless, the pool boiling equation cannot be used for prediction of the flow boiling heat transfer. The reason for the increase of heat transfer coefficient in comparison with the pool boiling model is the importance of the evaporation mechanism of heat transfer because the liquid film becomes very thin at high interfacial shear stress. Therefore, the reliable approach for predicting the flow boiling heat transfer should account for the determining mechanisms such as nucleate boiling suppression occurs when the liquid film thickness becomes sufficiently small, forced convection and evaporation of a thin liquid film. Comparison of the experimental data with the calculations according to Kuznetsov and Shamirzaev model [10] that accounts these mechanisms shows better suits the experimental data both for low and high heat fluxes.

The experimental data on flow condensation of dielectric fluid perfluorohexane in microchannels at low aspect ratio for the three-sided cooling case shows that the impact of wall subcooling on the magnitude of the heat transfer coefficient is not observed. It was obtained than increasing the mass velocity enhances the condensation heat transfer due to thinning the liquid film when the vapor interfacial shear stress is increased. The comparison of the experimental data with prediction according to the Kim and Mudawar [8] model shows that the calculation results are in good agreement with the experimental data. The deviation of experimental data from calculations occurs at the transition to elongated bubble flow.

**Acknowledgments**
This work was performed in the Kutateladze Institute of Thermophysics SB RAS by a state contract with IT SB RAS, AAAA-A17-117022850026-8 (section 4) and by Russian Science Foundation, Project No 16-19-10519 (section 3).

**References**
[1] Kandlikar S G, Bapat A V 2007 *Heat Transfer Eng.* **28** 911
[2] Tuckerman D B, Pease R F W 1981 *IEEE Electron Device Letters ELD.* **2(5)** 126
[3] Zeigarnik Y A, Privalov N P, Klimov A I 1981 *Thermal Eng.* **28(1)** 40
[4] Qu W, Mudawar I 2004 *Int. J. Heat Mass Transf.* **47** 2045
[5] Bertsch S S, Groll E A, Garimella S V 2009 *Int. J. Multiphase Flow.* **35** 142
[6] Kuznetsov V V, Shamirzaev A S 2007 *Heat Transfer Eng.* **28** 738
[7] Karayiannis T G, Shiferaw D, Kenning D B R, Wadekar V V 2010 *Heat Transfer Eng.* **31** 257
[8] Kim S M Mudawar I 2012 *Int. J. Heat Mass Transfer.* **55** 958
[9] Al-Zaidi A H, Mahmoud M M, Karayiannis T G 2018 *Exp. Therm. Fluid Sci.* **90** 153
[10] Kuznetsov V V, Shamirzaev A S 2016 *Heat Transf. Eng.* **37** 1105
[11] Ribatski G, Saiz Jabardo J M 2003 *Int. J. Heat Mass Transfer.* **46** 4439
[12] Kandlikar S G 2010 *Heat Transfer Eng.* **31** 159
[13] Butterworth D 1974 *Int. J. Multiphas. Flow.* **1** 671
[14] Kim S M, Mudawar I 2013 *Int. J. Heat Mass Transfer.* **58** 718
[15] Safonov S A, Kuznetsov V V 2018 J. Phys. Conf. Ser. **1105** 012062

# Two-phase compact heat exchangers: a review and future applications for high heat flux removal

**V V Kuznetsov, A S Shamirzaev**
Institute of Thermophysics Siberian Branch of Russian Academy of Sciences
1, Academician Lavrentyev Av., Novosibirsk, 630090, Russian Federation

E-mail: vladkuz@itp.nsc.ru

**Abstract**. The compact flow heat exchangers are important components of the equipment for power generation, chemical processing, and thermal environment control. This paper presents the flow pattern characterization and heat transfer performance for two-phase compact and ultra-compact heat exchangers. The scaling analysis is presented in this paper to identify the effect of various forces. The heat transfer enhancement mechanisms during flow boiling and condensation in compact heat exchangers are discussed using experimental data for refrigerant R-21. The mechanisms of subcooled and saturated flow boiling in ultra-compact heat exchangers based on microchannels are discussed using experimental data obtained for water.

## 1. Introduction

Two-phase compact heat exchangers are widely used in the power engineering, chemical, and food industries [1]. These heat exchangers with preferably rectangular channels are characterized by the specific surface area above 1000 $m^2/m^3$ and can be used as evaporators or condensers, to control the temperature of the process fluid and solid surface, recover or reject heat and so on. For example, cryogenic plate-fin exchanger has about 10% of the volume of an equivalent shell-and-tube exchanger and they can be plain, perforated or interrupted, such as offset strip fins that have the largest heat transfer efficiency. The important application has ultra-compact two-phase heat exchangers based on microchannels for electronic cooling because vapor/liquid phase change allows us to optimize the cooling strategy for the next generation of heat load systems. These heat exchangers have the potential to meet the large heat dissipation demands of high power electronics and computing systems. Currently, most of the advanced electronic components already generate heat fluxes higher than 100 W/$cm^2$, while future power electronic components, such as high-power laser and electronic radar systems, have been projected to generate heat fluxes over 1000 W/$cm^2$ [2]. To dissipate heat from planner surfaces, flow boiling can be achieved inside multiple parallel channels that are formed in a high conductivity substrate, named as a microchannel heat sink [3]. The two-phase microchannel heat exchangers also can be used for many small head load applications including condensers of heat pumps and air-conditioning systems [4].

This paper considers the flow pattern characterization and heat transfer performance in two-phase compact and ultra-compact heat exchangers based on experimental observations and numerical studies. To further explore the scale effects on flow boiling, the scaling analysis is presented in this paper to identify the effect of various forces. The heat transfer enhancement mechanisms in compact heat exchangers, as well as their advantages and limitations, are discussed using the data for refrigerant R-21. The mechanisms of subcooled and saturated flow boiling in ultra-compact heat exchangers based on microchannels are discussed using experimental data obtained for water.

## 2. Channel size characterization for compact heat exchangers with rectangular channels

Existing definitions of the criterion for the transition to microscale [5] give only a rough estimation of the channel dimension when this transition occurs. In order to show the influence of the channel size on flow patterns, the elongated bubble flow in a capillary is considered. The Navier-Stokes equations for the steady-state laminar liquid flow in the system of coordinates associated with the head of a bubble

$$\rho_{liq}\mathbf{u}\cdot\nabla\mathbf{u} = -\nabla p_{liq} + \mu_{liq}\nabla^2\mathbf{u} + \rho_{liq}\mathbf{g}, \quad \nabla\cdot\mathbf{u} = 0. \tag{1}$$

and conjugation conditions at the gas-liquid interface

$$p_{\rm liq} = p_{\rm gas} - \sigma\kappa \;,\; \kappa = R_1^{-1} + R_2^{-1},\; e_{ij}\tau_i n_j = 0 \tag{2}$$

are reduced to dimensionless equation assuming the constant gas pressure in the bubble as follows

$$We\tilde{\mathbf{u}}\cdot\nabla\tilde{\mathbf{u}} = -\nabla\tilde{\kappa} + Ca\nabla^2\tilde{\mathbf{u}} + Bo(\mathbf{g}/g). \tag{3}$$

Here $\mathbf{u}$ is the liquid velocity, $p$ is the pressure, $\mu_{\rm liq}$ is the viscosity, $\mathbf{e}$ is the stress tensor in the liquid, $\mathbf{n}$ is unit normal vector directed to the gas phase, $\boldsymbol{\tau}$ is a unit tangent vector, $\kappa$ is interface curvature and $R_i$ is principal radii of curvature. The dimensionless variables are scaled by bubble velocity $u_b$, the longitudinal, transverse coordinates and curvature are scaled by radius of the channel $D_h/2$. The dimensionless numbers are defined as follows: $We = \rho_{\rm liq}u_b^2 D_h/(2\sigma)$, $Ca = u_b\mu_{\rm liq}/\sigma$, $Bo = \rho_{\rm liq}gD_h^2/(4\sigma)$. When reducing the diameter of the channel, the order of magnitude of the Bond number becomes much less than unity and gravity forces are negligible. For flow with $Bo \ll 1$ and $We \geq 1$ the inertial forces remain significant and flow characteristics are different from that for the conventional channel, this case can be considered as the flow in minichannel. At further reducing diameter of the channel, the Weber number becomes much less than unity $We \ll 1$, this case can be considered as the flow in microchannel. The dependence of the Weber number and Bond number on bubble velocity for refrigerant R-21 flow is in the channel with $D_h = 200$ μm is shown in figure 1. As is seen, the same channel can be considered as minichannel of microchannel depending on flow conditions.

For studying the nitrogen-water flow in a microchannel with cross-section of 269x362 μm the setup described in [6] was used. It was obtained that the typical flow patterns are periodic flow with elongated bubbles, transition flow, and annular flow. The boundaries between the flow patterns are blurred and can be characterized by a change in the statistical characteristics of the flow. Figure 2 shows the shape of the elongated bubble for a flow with $J_{\rm lig} = 0.064$ m/s, $J_{\rm gas} = 0.068$ m/s (We=0.058).

## 3. Heat transfer during flow boiling and condensation in plate-fin heat exchangers

The problem of prediction of the heat transfer in two-phase compact heat exchangers with the low-temperature difference between the streams has acquired recently special attention in connection with the development of advanced systems for liquefying natural gas, cryogenic separation, and heat pumps. These heat exchangers are manufactured on the basis of minichannels, that allows considerably enhanced the heat and mass transfer due to capillary forces action in small size non-circular channels.

Compact evaporators and condensers are able to transfer the heat between the streams at a very low temperature difference. Experimental study of heat transfer during flow boiling in plate fin evaporators was performed in [7, 8]. The model based on conduction and convection through the film flowing down

Journal of Physics: Conference Series **1369** (2019) 012044 doi:10.1088/1742-6596/1369/1/012044

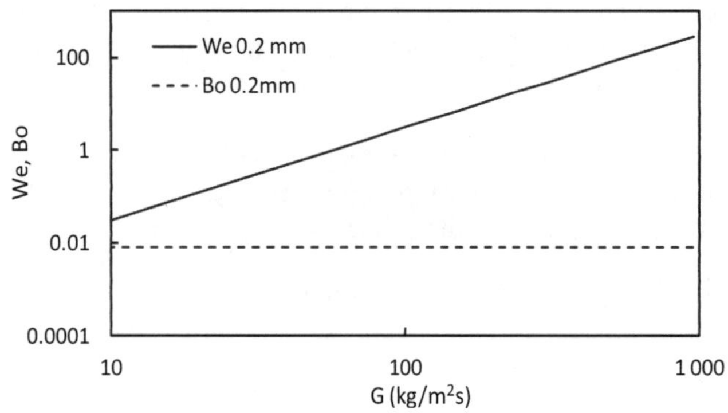


(a)

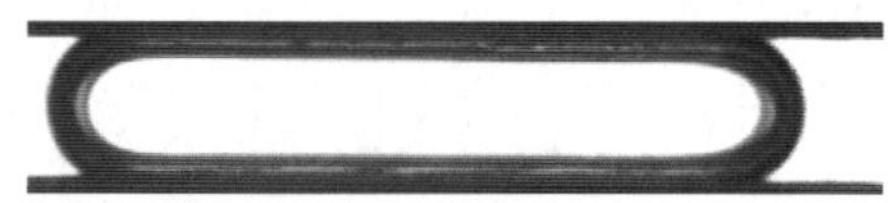

(б)

**Figure 1.** Dependence of the Weber number and Bond number on bubble velocity for $D_h$ = 200 μm, $p_s$= 0.236 MPa, vapor quality $x$ = 0.05.

**Figure 2.** Characteristic bubble shape for nitrogen-water flow in microchannel with $J_{liq}$ = 0.064 m/s, $J_{gas}$ = 0.068 m/s.

along the walls of the channels was proposed in [7] to calculate heat transfer with evaporation.

The experimental test section for studying heat transfer during boiling and condensation of R-21 in a plate fin heat exchanger with 800 fins per meter and fins height of 6.1 mm is shown in figure 3a. This refrigerant was selected because its thermal properties at room temperature are very similar to nitrogen properties at cryogenic temperatures. The experiments were performed using closed loop that re-circulate refrigerant presented in [9]. The experiments showed a significant difference in the heat transfer coefficient behavior during boiling and condensation for temperature difference near 1 °C. For flow condensation, the heat transfer coefficient growth with increasing of vapor quality. In contrary, for flow boiling, the deterioration of heat transfer was observed for quality higher than 0.8. Figure 3b shows the dependence of the heat transfer coefficient on vapor quality during flow boiling R-21 with wall superheat near 1.5 °C, mass flux G=43 kg/m$^2$s and static pressure of 0.236 MPa.

To explain this result, the mathematical model of the heat transfer during flow evaporation and condensation in rectangular microchannel proposed in [10] was used. The model is based on the coupling of liquid flow near the channel corner, limited by the interfacial meniscus, and in the film on a

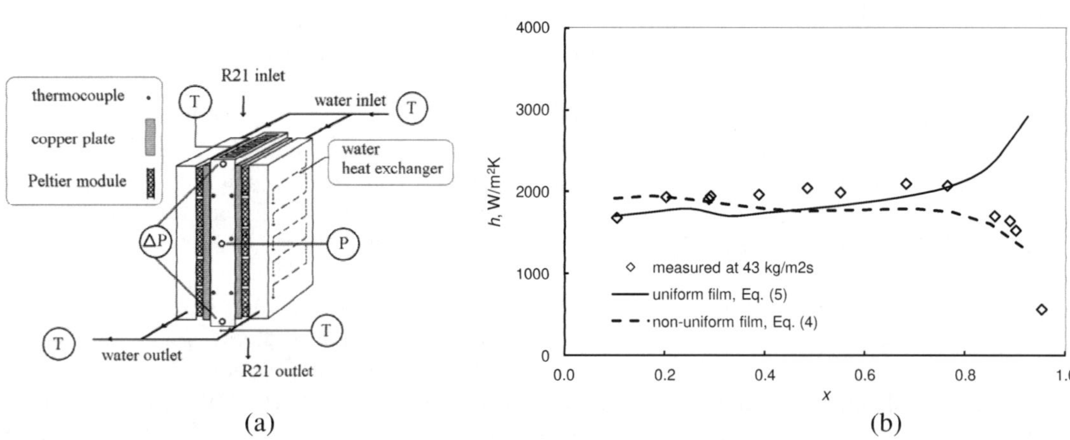


(a) (b)

**Figure 3.** Scheme of test section (a) and dependence of heat transfer coefficient on vapor quality for wall superheat near 1.5 °C, mass flux of 43 kg/m$^2$s and pressure of 0.236 MPa (b). The lines show the calculations for uniform and non-uniform film.

wall. The governing equations were supplemented by the kinematic condition, boundary conditions, and the integral equation of mass conservation for the film-meniscus system. For down flow with $\varepsilon = \delta_0/a \ll 1$, where $\delta_0$ is initial thickness of the liquid film and $a$ is half the width of the long side of the channel, the Navier-Stokes equations are reduced to the equation

$$(\gamma m^3 + 1.5\frac{\kappa}{\varepsilon}m^2)_x + (m^3 m_{yyy})_y = \frac{3}{\varepsilon^4}(Ga(\frac{m_y}{m})_y - \frac{G_0\Theta_{w,i}}{m\varepsilon}) \quad (4)$$

where $\gamma = 1 - (dp_g/dx)/\rho_f g$, $\kappa = \tau/\rho_f g a$, $Bo = \rho_f g a^2/\sigma$, $Ga = A_0/(6\pi a^2\sigma)$, $m = \delta/\delta_0$ – film thickness, longitudinal $x$ and transverse $y$ coordinates are dimensionless on $a\mathrm{Bo}/\varepsilon$ and $a$ accordingly, $\tau$ is shear stress on the surface, $\Theta_{w,i} = (T_{w,i}-T_s)/T^*$, $A_0$ is the Hamaker constant, $h_{fg}$ is the latent heat, $T_{w,i}$ is the internal wall temperature, $T_s$ is the saturation temperature and $T^* = T_{w,e}-T_s$ is defined by external wall temperature. A special feature of this equation is the consideration of the disjoining pressure in ultrathin liquid films, which becomes especially important for flow in nanochannels. The mass flow at the interphase surface is determined by the model of conductive heat transfer in a film. In the case of evaporation, the conditions for film rupture [10] were applied to predict the dry spot formation between rivulet and meniscus. After obtaining the configuration of the liquid surface, the Fourier equation for liquid and wall are solved together. Figure 4 shows the calculated interface shape during condensation at mass velocity of 49.3 kg/m$^2$s and wall subcooling of 2 K in minichannel of 1.6x6.3 mm: (a) corresponds to refrigerant R-21 at $T_s$ = 293 K, (b) corresponds to nitrogen at 95 K. The calculation shows the formation of thin film and an increase in the heat transfer coefficient in the vicinity of the channel corner during condensation and formation of the dry spot in this area during evaporation. The calculated dependence of the heat transfer coefficient averaged along the channel perimeter on vapor quality is shown in figure 3b as the dotted line. The evaporation model predicts well the deterioration of heat transfer coefficient at vapor quality higher than 0.8 as it was observed in experiments.

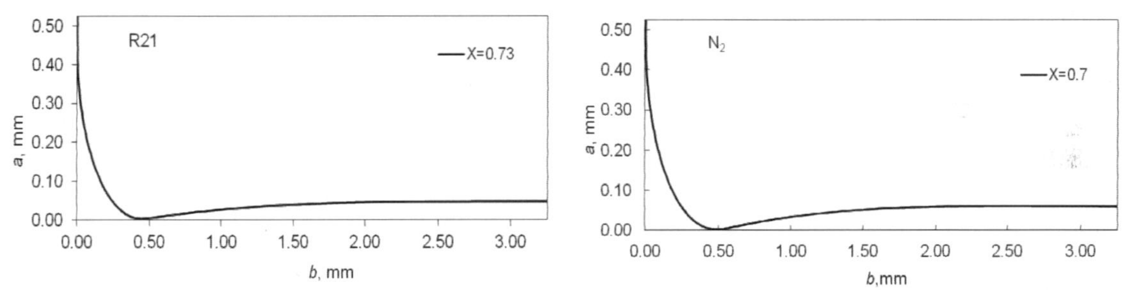


(a) (b)

**Figure 4.** The interface shape during condensation at the mass velocity of 49.3 kg/m$^2$s and wall subcooling of 2 K: (a) corresponds to R-21 at $T_s$ =293 K, (b) corresponds to nitrogen at $T_s$ = 95 K.

The alternative model of the heat transfer during flow boiling was proposed in [11]. This model takes into account the contribution of nucleate boiling $h_{boil}$, two-phase forced convection $h_{con}$ and evaporation of the liquid film $h_{ev}$. The proposed model has the privilege over simple summing of corresponding terms because it allows clearly distinguishing the contribution of each term under the conditions when it is predominant. A feature of this model is account for the nucleate boiling suppression in thin liquid film ($\Psi_{sup}$ factor), nucleate boiling near the channel corners, and evaporation in liquid film as follows

$$h^2 = (h_{con}F)^2 + (h_{boil}\Psi_{sup}S)^2 + (h_{ev}E)^2, \quad (5)$$

where F, S and E are the enhancement factors for corresponding terms. The heat transfer coefficient for forced convection was determined from [4] for three-side heating at T-constant. To determine the heat transfer coefficient for evaporating wavy-turbulent liquid film, the model [12] was used in this paper with the calculation of the shear stress according to [13]. The result of the calculation of the dependence of the heat transfer coefficient on vapor quality is shown in figure 3b as a solid line. As is seen this model reasonably predicts the heat transfer for quality less than 0.8 when liquid film rupture occurs.

## 4. Heat transfer during flow boiling in ultra-compact heat exchangers based on microchannels

A feature of the heat transfer in two-phase microchannel systems used to cool heat-stressed equipment is extremely high heat fluxes before the crisis arises. Despite the considerable efforts in this area, the data of experimental studies are quite contradictory. In a number of papers, it was found that heat transfer coefficients depend mainly on heat flux and pressure, and weakly depend on the flow rate and mass vapor content. In other works, it was noted that the contribution of nucleate boiling to heat transfer is not decisive and heat transfer coefficients increase with increasing mass flux and vapour quality. There is a general agreement that the heat transfer coefficient during flow boiling in microchannels increases for refrigerants with increasing heat flux [14], but in some experiments it was shown that the heat transfer coefficient depends on the vapor quality, showing growth [15] or reduction [16] when the quality increases. The difficulties in the selection of reliable equations for prediction of flow boiling heat transfer constrain the development of ultra-compact two-phase microchannel heat exchangers. Despite the fact that this problem was addressed in several studies, e.g. [14], the flow boiling heat transfer in microchannels is still understood incompletely.

For studying heat transfer during subcooled and saturated flow boiling of water in microchannel heat sink the experimental setup described in [17] was used. During the tests, water is supplied from the condenser through the filter and damper of pulsation to the flow controller via the plunger pump. Then it goes through the pre-evaporator to attain the flow with desired vapor quality or subcooling. The copper block has two horizontal microchannels with cross section 2 mm × 0.36 mm and channel-to-channel spacing of 2 mm. The length of micro-channels is 16 mm, surface roughness Rp is 1.5 μm and static pressure at the channel inlet is near 0.1 MPa. Heat cartridges and set of thermocouples are mounted into the copper block to supply heat and measure averaged heat flux and wall temperature.

The dependence of the heat flux on wall temperature for saturated flow boiling of water ($x_{in} \cong 0.1$) is presented in figure 5a for G = 450 kg/m$^2$s. The dotted line shows the calculation for pool boiling according to Yagov equation [18]

$$q_{boil} = A\frac{k_{liq}{}^2 \Delta T_s^3}{\nu_{liq}\sigma T_s}\left(1+\sqrt{1+800B}+400B\right) \tag{6}$$

Here $A = 3.43\cdot 10^{-4}$, $B = h_{fg}\left(\nu_{liq}\rho_{gas}\right)^{1.5}/\left(\sigma\cdot(\lambda T_s)^{0.5}\right)$, $k_{liq}$, $\nu_{liq}$ are the heat conductivity and kinematic viscosity of the liquid, $h_{fg}$ is the specific heat of vaporization, $\rho_{gas}$ is the vapor density. As is seen, the experimental data are considerably below then the calculations. The solid line shows the calculations according to equation (5) when nucleate boiling suppression in the area of the liquid film was taken into account and $A = 0.47\cdot 10^{-4}$ as recommended in [19]. The heat transfer coefficient for forced convection was determined from [4] for three-side heating (microchannel aspect ratio is five). For this case, the calculations correspond well to the experimental data.

The dependence of the heat flux on wall temperature for subcooled flow boiling ($T_{in}$ = 50 °C) is presented in figure 5b for G = 1170 kg/m$^2$s. The dotted line shows the calculation for pool boiling according to [18]. At subcooling flow boiling, the wall superheat for forced convection, nucleate boiling and evaporation is different [20]. For this case the equation 5 can be rewritten as follows

$$q_w{}^2 = \left(h_{con}F\left(T_w - T_{liq}\right)\right)^2 + \left(h_{boil}\Psi_{sup}S\left(T_w - T_{sat}\right)\right)^2 + \left(E\varphi h_{ev}\left(T_w - T_{sat}\right)\right)^2 \tag{7}$$

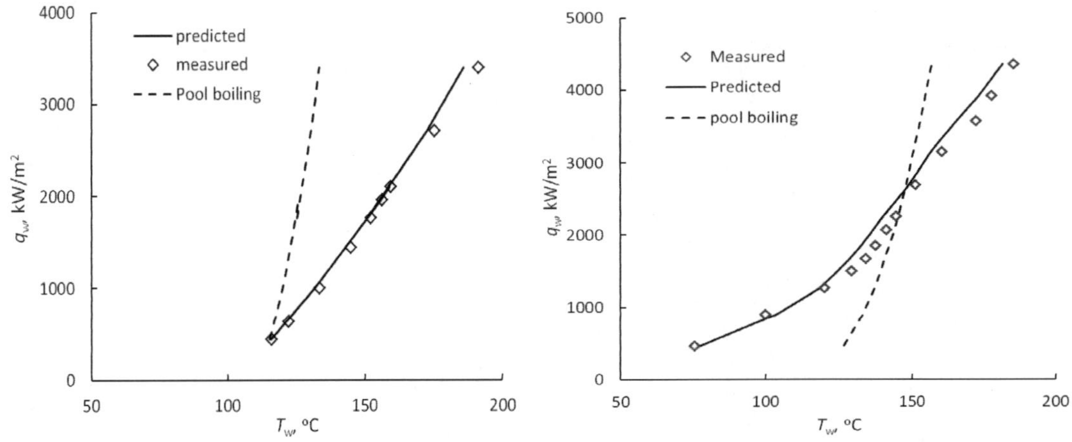


(a) (b)

**Figure 5.** The dependence of heat flux on wall temperature for saturated boiling of water at G = 450 kg/m$^2$s (a) and subcooled boiling at G = 1170 kg/m$^2$s (b). The dotted lines show the calculation for pool boiling equation (6), solid lines show the calculations for saturated boiling equation (5), and subcooled boiling equation (7) accordingly.

It is assumed that all functions in this equation are the same as for saturated boiling and correspond to vapor quality equals to zero. Therefore the factors in equation (12) are: $F = 1.0$, $E = 0$ and $S < 1$. Solving this equation for given wall heat flux one can obtain the wall temperature depending on heat flux in the subcooled boiling region. The solid line in figure 5b shows the calculations according to equation (7) when nucleate boiling suppression was taken into account in case of local saturated boiling and $A=0.47 \cdot 10^{-4}$ in equation (6) as recommended in [19]. For this case, the calculations correspond well to the experimental data.

## 5. Conclusions

The results are presented show the high efficiency of the compact heat exchangers for operation with a low-temperature difference and ultra-compact heat exchangers for high heat flux removal which allows for the implementation of previously unavailable technologies. In minichannels, the high intensity of heat transfer is observed during flow condensation due to deformation of the liquid film surface by capillary forces. In contrary, during flow boiling, the deterioration of heat transfer was observed at high vapor quality due to the formation of the dry spots. To realize these possibilities, the methods for predicting phase change in mini- and microchannels was proposed that show not only the advantages of using the channels of small size but also the problems associated, for example, with the suppression of nucleate boiling in a thin liquid film.

## Acknowledgments

This work was performed in the Kutateladze Institute of Thermophysics SB RAS by a state contract AAAA-A17-117022850026-8 (sections 2, 3) and by Russian Science Foundation, Project No 16-19-10519-C (section 4).

## References

[1] *Two-Phase Flow Heat Exchangers: Thermal-Hydraulic fundamentals and Design* 1988 ed. Kakaç S, Bergles A E, Fernandes E O Kluwer Publishers
[2] Kandlikar S G, Bapat A V 2007 *Heat Transfer Eng.* **28** 911
[3] Tuckerman D B, Pease R F W 1981 *IEEE Electron Device Letters ELD* **2**(5) 126

[4] Kandlikar S, Garimella S, Li D, Colin S, King M R 2005 *Heat transfer and fluid flow in minichannels and microchannels* Elsevier
[5] Mehendal S S, Jacobi A M, Shah R K 2000 *Appl. Mech. Rev.* **53**(7) 175
[6] Kuznetsov V V, Shamirzaev A S, Kozulin I A, Kozlov S P 2013 *Heat Transfer Eng.* **34**(2-3) 235
[7] Robertson J M 1982 *Proc. 7th Int. Heat Transfer Conference*, Munich 341
[8] Robertson J M, Lovergrove P C 1983 *J. Heat Transfer* **105**(3) 605-610
[9] Kuznetsov V V, Shamirzaev A S 2017 *J. Eng. Thermophys.* **26**(3) 353
[10] Kuznetsov V V, Safonov S A 2013 *Minichannels and Microchannels, Heat Pipes and Solid Sorption Transformations: Fundamentals and Practical Applications* ed. Vasiliev L L, Kakac S CRC Press. 465–496
[11] Kuznetsov V V 2019 *Heat Transf. Eng.* **40** 711
[12] Butterworth D 1974 *Int. J. Multiphas. Flow.* **1** 671
[13] Kim S M, Mudawar I 2013 *Int. J. Heat Mass Transf.* 58 718
[14] Bertsch S S, Groll E A, Garimella S V 2009 *Int. J. Multiphase Flow* **35** 142
[15] Kuznetsov V V, Shamirzaev A S 2007 *Heat Transf. Eng.* **28** 738
[16] Karayiannis T G, Shiferaw D, Kenning D B R, Wadekar V V 2010 *Heat Transfer Eng.* **31** 257
[17] Kuznetsov V V, Shamirzaev A S 2018 *Tech. Phys. Lett.* **44** 938
[18] Yagov V V 2008 *Thermal Engineering* **55** 245
[19] Dedov A V, Komov A T, Varava A N, Yagov V V 2002 *Proc. 12th Int. Heat Trans. Conf.* 731
[20] Liu Z, Winterton R H S 1991 *Int. J. Heat Mass Transf.* **34** 2759

5th International Workshop on Heat/Mass Transfer Advances
for Energy Conservation and Pollution Control
August 13-16, 2019, Novosibirsk, Russia

# Direct numerical simulation of turbulent flow and heat transfer in a hexagonal rod bundle

**K M Sergeenko, A Yu Chulyunin and B I Krasnopolsky**

Institute of Mechanics, Lomonosov Moscow State University, Moscow, Michurinsky ave. 1, 119192, Russia

E-mail: km.sergeenko@gmail.com

**Abstract.** The aspects of numerical simulation of heat transfer in a flow along the hexagonal rod bundle are discussed in the present work. The direct numerical simulation is used to investigate the flow with Reynolds and Prandtl numbers varying in the range from 2500 to 10000 and 0.0043 to 7 correspondingly. The obtained simulation results demonstrate good agreement with available experimental data and semi-empirical correlations for the thermo-hydraulic characteristics of the channel. The empirical correlations for the Nusselt number distributions summarizing the results of the numerical simulations are proposed. The preliminary calculations have shown that the modelled time averaging interval is comparable with the initial transition time interval. This observation allows applying the approach, proposed in [7], to speed up the simulations.

## 1. Introduction

Heat transfer devices used in practical applications usually have a complex geometry. This fact makes almost impossible to obtain analytical solutions to predict their hydraulic and heat transfer characteristics. Rod bundle, the essential element of the nuclear reactors, is the typical example of such heat exchangers. It consists of a packed array of fuel rods, surrounded by the liquid coolant. Flow and heat structures observed in such configuration have been investigated over many years and remain an object of interest up to the present time.

Despite the intensive development of numerical methods and corresponding simulation software, a physical experiment remains the primary tool to analyze the thermo-hydraulic characteristics of rod bundles. Usually, results of the experiments are presented in the form of semi-empirical expressions, reproducing the functional dependency of the characteristics of interest on the problem parameters. Several examples of the corresponding expressions for the Prandtl numbers Pr<<1 are presented below. The correlation for the Nusselt number, valid for the range of Peclet numbers Pe<2200 and the pitch-to-diameter ratio $1.1 \leq \chi \leq 1.5$ was proposed in [1]:

$$\mathrm{Nu} = \begin{cases} 24.15\,\mathrm{lg}(-8.12 + 12.76\chi - 3.65\chi^2), \ \mathrm{Pe} < 200, \\ 24.15\,\mathrm{lg}(-8.12 + 12.756\chi - 3.65\chi^2) + \\ 0.0174\{1 - \exp\left[-6(\chi - 1)\right]\}(\mathrm{Pe} - 200)^{0.9}, \ 200 < Pe < 2200 \end{cases} \quad (1)$$

where Nu is the Nusselt number, non-dimensional heat transfer coefficient, defined as a ratio of convective to conductive heat transfer rates, Pe is the Peclet number, non-dimensional coefficient, which puts convective and diffusive transport phenomena in correlation, and $\chi$ is the pitch-to-diameter ratio, characterizing the rod geometry. The similar correlation, covering another set of experimental data for the range 150<Pe<4000 and $1.2\leq\chi\leq2.0$ was presented in [2]:

$$\mathrm{Nu} = 0.25 + 6.2\chi + (-0.007 + 0.032\chi)\mathrm{Pe}^{0.8-0.24\chi}. \quad (2)$$

Additional correlation for Pe<4000 and $1.3\leq\chi\leq2.0$ was formulated in [3]:

$$\mathrm{Nu} = 7.55\chi - 20\chi^{-13} + \frac{3.67}{90\chi^2}\mathrm{Pe}^{0.56+0.19\chi}. \quad (3)$$

Recently, the authors compiled in [4] several sets of experimental data and proposed the relation

$$\mathrm{Nu} = 0.047\{1 - \exp[-3.8(\chi - 1)]\}(\mathrm{Pe}^{0.77} + 250) \quad (4)$$

as a fit for the range $1.1\leq\chi\leq1.95$ and 30<Pe<5000.

The presented semi-empirical correlations do not exhaust the list of expressions available in the literature but show the limited range of applicability for each of them. These correlations do not cover the range of low Peclet numbers (Pe~10) or provide observable deviation in the corresponding predictions. To all other, the range of low Peclet numbers is hardly achievable in the experiments due to increasing role of the sensitivity of measuring equipment. This, however, is a good example when the high-fidelity numerical methods can provide the missing data and supplement the physical experiments. The present study deals with direct numerical simulation of turbulent flow and heat transfer in a hexagonal rod bundle [5] for several liquids and liquid flow rates.

## 2. Computational setup

In the present paper the flow in a periodic segment of the rod bundle is investigated (figure 1). The ratio of the geometric parameters, $\chi=S/(2R)$, in accordance with [5] is set to 1.4. The modelling liquid is considered as Newtonian and incompressible fluid with constant properties. The flow of the corresponding liquid is described by the following system of governing equations, written in the non-dimensional form:

$$\nabla\cdot\mathbf{u} = 0, \quad (5)$$

$$\frac{\partial\mathbf{u}}{\partial t} + \mathbf{u}\cdot\nabla\mathbf{u} = -\nabla p + \frac{1}{\mathrm{Re}}\Delta\mathbf{u}, \quad (6)$$

$$\frac{\partial T_i}{\partial t} + \mathbf{u}\cdot\nabla T_i = \frac{1}{\mathrm{Pr_i Re}}\Delta T_i, \quad (7)$$

where $p$ is the pressure, $T_i$ is the temperature of i-th liquid, $\mathbf{u}$ is the velocity vector, $\mathrm{Pr_i}$ is the Prandtl number of i-th liquid, and $\mathrm{Re} = UD/\nu$ is the Reynolds number ($D$ is the hydraulic diameter, $\nu$ is the kinematic viscosity, and $U$ is the bulk velocity). The corresponding calculations are performed for three Reynolds numbers 2500, 5000, and 10000, and five Prandtl numbers 0.0043, 0.0154, 0.0324, 0.71, and 7.0. With selected assumptions, heat transfer characteristics for multiple Prandtl numbers can be calculated at once, thus leading to only three different simulations with varying Reynolds numbers.

The model is supplemented with no slip boundary condition with constant heat flux on the heated walls. The periodic boundary condition with constant mass flux is preserved in the streamwise direction. The periodic conditions are also specified on the other boundaries.

The structured hexahedral grid is constructed to perform the simulations (figure 2). A grid stretching normal to the wall is used to provide the sufficient near-wall grid resolution with $y^+<1$. A stretching factor normal to the wall is chosen to place several grid points into the viscous sublayer. In the third direction along the rod bundle, the homogenous grid is used. The parameters of the grids are summarized in table 1. Here, parameter $L$ denotes the length of the computational domain, $N_{\mathrm{cell}}$ is total grid size, $N_{\mathrm{cyl}}$ is the number of grid points along the cylinder arc, $N_z$ is the number of cells in the longitudinal direction. The $\Delta y^+$, $\Delta\varphi^+$, $\Delta z^+$ values characterize the near-wall grid step size in the directions normal to wall, along the cylinder arc and along the rod bundle correspondingly. The $\Delta^+_{\text{center}}$ characterizes the maximal grid size in the center of the channel.

The range of Reynolds numbers covered in the paper includes both transient and developed turbulent flows. To model flow with the intermittency (alternation of laminar and turbulent flow region along the pipe), concerned for the lowest considered Reynolds number, the length of the computational domain is increased up to 50 hydraulic diameters.

**Table 1.** Parameters of the computational models.

| Re | $L$ | $N_{\text{cell}}$ | $\Delta y^+$ | $\Delta\varphi^+$ | $\Delta z^+$ | $\Delta^+_{\text{center}}$ | $N_z$ | $N_{\text{cyl}}$ |
|---|---|---|---|---|---|---|---|---|
| 2500 | 50 | $9.45\times10^6$ | 0.95 | 2.02 | 3.57 | 2.02 | 1400 | 45 |
| 5000 | 20 | $11.02\times10^6$ | 0.92 | 3.07 | 7.40 | 3.37 | 1000 | 54 |
| 10000 | 20 | $31.5\times10^6$ | 0.95 | 4.07 | 9.05 | 4.95 | 1500 | 75 |

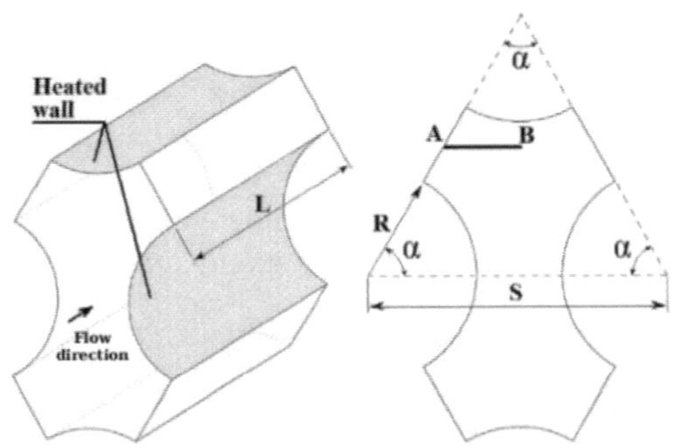


**Figure 1.** Geometry of the computational domain (α = 60°).

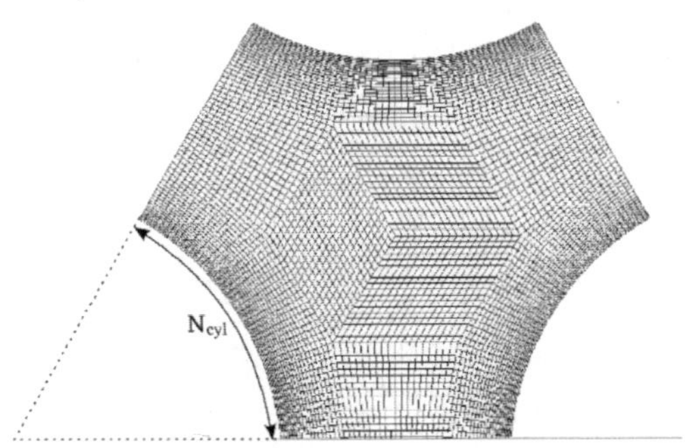


**Figure 2.** Fragment of the computational grid.

## 3. Results

The current section presents results of preliminary investigations and validation of the simulation results with the experimental data. Despite the high impact of corresponding data in the engineering practice, the only several experiments for low Peclet numbers in the hexagonal rod geometry can be found it the literature, e.g. [5,6]. Comparison of the mean streamwise velocity profiles and profiles of the velocity pulsations with the data, presented in [6] is shown in figure 3. Results are plotted along the A-B segment, marked on figure 1. The experimental data is available in only several points and the measured and simulated data differs in the Reynolds numbers, which complicates the comparison.

However, distributions of the mean velocity and the velocity intensity generally agrees with the experimental data.

The experimental data, published in [5] includes the integral hydraulic and heat transfer characteristics, measured in the range of Reynolds numbers $10^4 - 1.3 \cdot 10^5$. Comparison of the simulation results with the series of experiments for the lowest Reynolds number is presented in table 2. The hydraulic characteristics (friction factor coefficient) demonstrate about 10% variation and the Nusselt number for the liquid with Pr=0.0324 shows 4% error.

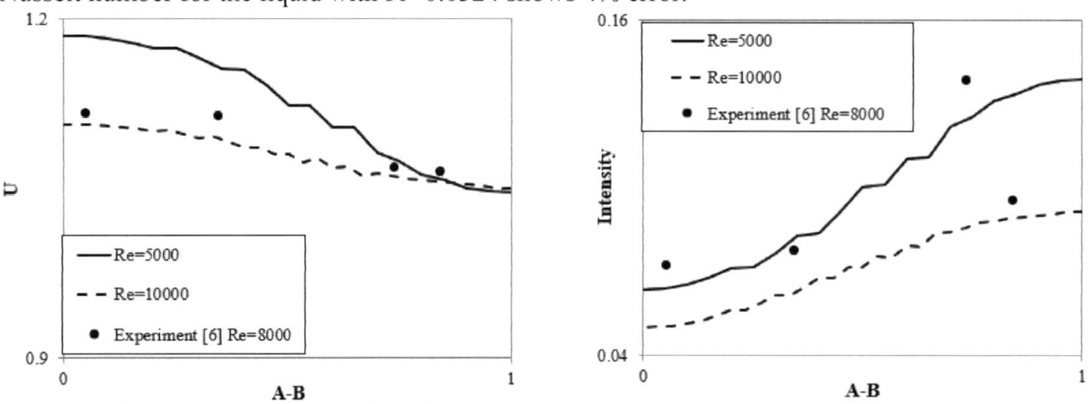


**Figure 3.** Comparison of the numerical results with the experimental data [6].
Left - mean velocity profiles, right - velocity intensity profiles.

**Table 2.** Comparison of the simulation results with the experimental data [5].

| | Re | FF | Nu |
|---|---|---|---|
| Simulation | 10000 | 0.0342 | 14.51 |
| Experiment | 10200 | 0.031 | 14 |

It should be noted the semi-empirical correlations (1)-(4) underpredict the Nusselt number for the chosen model parameters. The observed variation of the experimental data with the empirical model predictions is within 9-32%. Thus, it can be concluded the results of DNS provided better correlation with the experimental data compared the widely used semi-empirical correlations.

The normalized instantaneous distributions of the streamwise velocity (figure 4), $u^* = u/U_{max}$, where $U_{max}$ is the maximum velocity value, illustrate the evolution of the spatial scales of the vortex structures with increasing the Reynolds number. It can be seen that for the lower Reynolds number the large scale vortices dominate in the flow and the flow is close to the laminar regime. For the higher Reynolds number, the scale of the vortices decreases, and the developed turbulent flow is observed.

The normalized temperature distributions, presented in the form

$$T_i^* = \frac{(T_i - T_{imin})}{(T_{imax} - T_{imin})}, \tag{8}$$

where $T_{imax}$ - the maximum value of a temperature of i-th liquid, $T_{imin}$ - the minimum value of a temperature of i-th liquid, allow to compare the correlation of velocity pulsations with the corresponding pulsations of temperature fields. Indices follow the increase of Prandtl number: 0.0043 (i=1), 0.0154 (i=2), 0.0324 (i=3), 0.71 (i=4), and 7.0 (i=5).

The temperature distributions vary dramatically with increasing the liquid's Prandtl number. The distributions for the low Prandtl numbers demonstrate weak sensibility to the flow perturbations: the dominating conductive heat transfer mechanism smooths out perturbations occurring as a result of convective transport. The growth of the Prandtl number increases the role of convective heat transfer. For the Pr=0.7 the thermal structures generally reproduce the scales of the flow structures. The further increase in the Prandtl number decreases the scale of the thermal structures and strengthens the role of the convective heat transfer mechanism.

The averaged integral characteristics including the friction factor coefficients and the Nusselt numbers for all performed simulations are summarized in table 3. The presented results clearly show the difference for the heat transfer processes in the liquids with low and high Prandtl numbers: the fourfold increase of the Reynolds for the Pr~0.01 leads to only 7% growth of the Nusselt number, while the same situation for the liquid with Pr~10 leads to a fivefold heat exchange intensification. The presented results are compared with the ones predicted by the semi-empirical expressions (1)-(4). In the range of applicability of these correlations, i.e. Pr << 1, the variation of the results of numerical simulations and the corresponding predictions is within 20%. The use of these formulas for the higher range of Pr numbers, expectedly, leads to significant growth of the prediction error. Using the obtained numerical results, the following approximation can be suggested:

$$\mathrm{Nu} = -3.55 + 0.081 \cdot \mathrm{Pr}^{\left(-0.88+0.164\cdot \mathrm{Re}^{0.157}\right)} \cdot \mathrm{Re}^{\left(0.21+0.47\cdot \mathrm{Pr}^{0.114}\right)} \quad (9)$$

In the considered range of parameters, the present correlation allows to predict the Nusselt numbers within 11 %.

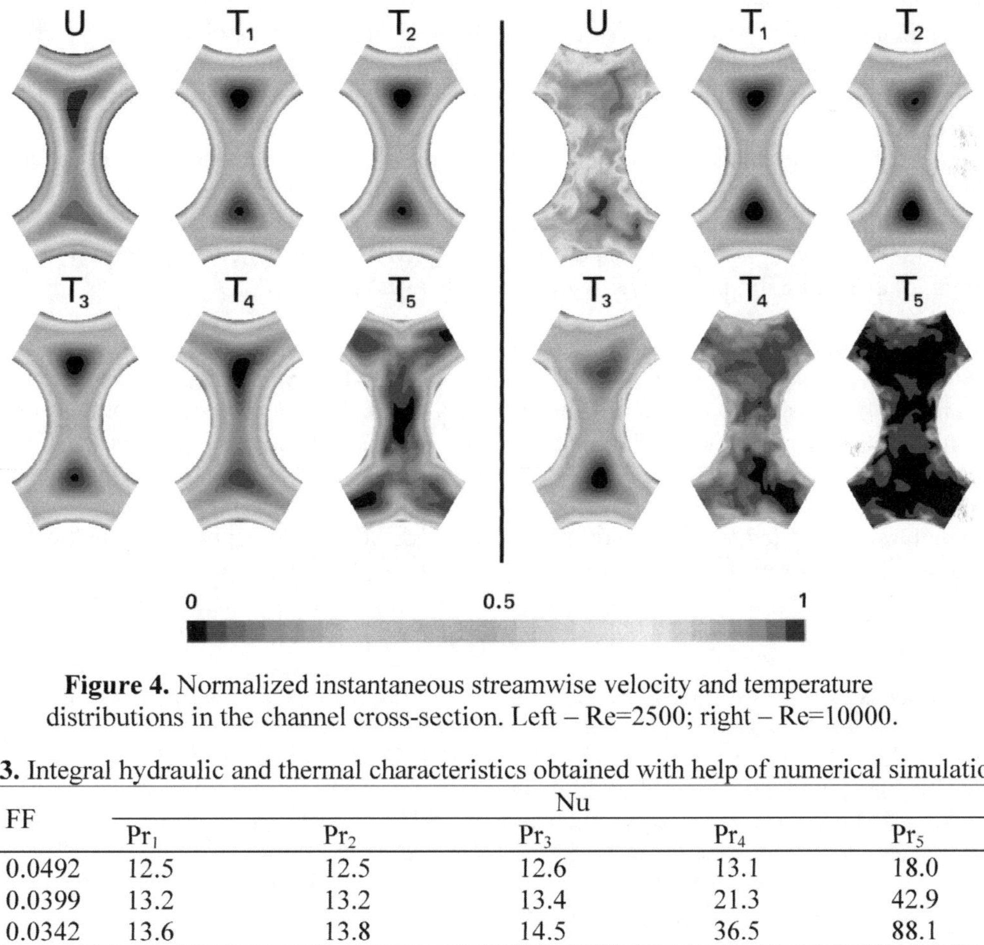


**Figure 4.** Normalized instantaneous streamwise velocity and temperature distributions in the channel cross-section. Left – Re=2500; right – Re=10000.

**Table 3.** Integral hydraulic and thermal characteristics obtained with help of numerical simulation.

| Re | FF | Nu | | | | |
|---|---|---|---|---|---|---|
| | | $Pr_1$ | $Pr_2$ | $Pr_3$ | $Pr_4$ | $Pr_5$ |
| 2500 | 0.0492 | 12.5 | 12.5 | 12.6 | 13.1 | 18.0 |
| 5000 | 0.0399 | 13.2 | 13.2 | 13.4 | 21.3 | 42.9 |
| 10000 | 0.0342 | 13.6 | 13.8 | 14.5 | 36.5 | 88.1 |

The presented above characteristics of turbulent flows are obtained as a result of averaging in time together with spatial averaging in the homogeneous directions. The need to perform the statistical averaging of the thermal characteristics significantly increases the corresponding time averaging interval. This aspect further increases the computational costs to perform these simulations, and the corresponding simulations take several days using tens of compute nodes.

The preliminary calculations have shown that despite the symmetry in the computational domain, the modelled time averaging interval, $T_A$, is comparable with the initial transition time interval, $T_T$, and the ratio of the intervals, $\beta = T_A/T_T$, varies in the range 1.7-12.3. This observation allows to apply the approach, proposed in [7], to speed up the simulations. The corresponding approach is based on the idea of simultaneous modelling of multiple uncorrelated turbulent flow states and combining of averaging in time together with ensemble averaging. The present problem formulation is planned to be used for further efficiency investigation of the proposed approach in the scientific and engineering applications.

## 4. Conclusion

Comparison of numerical results obtained from DNS for the turbulent flow in a hexagonal rod bundle for a variation of the Reynolds numbers (2500, 5000, and 10000) and Prandtl numbers (0.0043, 0.0154, 0.0324, 0.71, and 7.0) with the experimental data and semi-empirical correlations shows, that the DNS approach is capable for predicting of thermo-hydraulic characteristics of such channels. The difference for the heat transfer process in the liquids with low and high Prandtl numbers is indicated: the fourfold increase of the Reynolds for the Pr~0.01 leads to only 7% growth of the Nusselt number, while the same situation for the liquid with Pr~10 leads to a fivefold heat exchange intensification. This fact is a result of switching the dominating heat transfer mechanism and increasing the role of convective heat transfer.

The presented simulations are characterized by long time integration intervals to obtain reliable turbulent statistics, and the averaging time interval dominates over the initial transition interval performed to obtain statistically steady turbulent flow. This observation allows to apply the approach based on the idea of simultaneous modelling of multiple uncorrelated turbulent flow states and combining of averaging in time together with ensemble averaging.

## Acknowledgements

The presented work is supported by the RSF grant No. 18-71-10075. The research is carried out using the equipment of the shared research facilities of HPC computing resources at Lomonosov Moscow State University and computing resources of the federal collective usage center Complex for Simulation and Data Processing for Mega-science Facilities at NRC "Kurchatov Institute" (ministry subvention under agreement RFMEFI62117X0016), http://ckp.nrcki.ru/.

The authors acknowledge P. Lobanov for providing details of the experiment used to validate the presented simulation results.

## References

[1] Borishanskii V, Gotovskii M and Firsova E V 1969 *Sov. Atom. Energy* **27 (6)** 1347–50
[2] Gräber H and Rieger M 1972 *Atomkernenergie* **19 (1)** 23–40
[3] Ushakov P A, Zhukov A V and Matyukhin N M 1977 *High Temp.* **15** 868–73
[4] Mikityuk K 2009 *Nucl. Eng. Des.* **239 (4)** 680–7
[5] Pacio J, Litfin K, Batta A, Viellieber M, Class A, Doolaard H, Roelofs F, Manservisi S, Menghini F and Böttcher M 2015 *Nucl. Eng. Des.* **290** 27–39
[6] Kashinsky O N, Kulikov D V, Lobanov P D, Pribaturin N A and Svetonosov A I 2018 *Thermophys. Aeromech.* **25(2)** 219–23
[7] Krasnopolsky B 2018 *Comput. Phys. Commun.* **229** 8–19

5th International Workshop on Heat/Mass Transfer Advances
for Energy Conservation and Pollution Control
August 13-16, 2019, Novosibirsk, Russia

# The influence of the surface structuring type on heat transfer in falling films of the refrigerant mixture

**O A Volodin[1], N I Pecherkin[1], A N Pavlenko[1] and N N Zubkov[2]**

[1] Kutateladze Institute of Thermophysics, Acad. Lavrentiev Ave., 1, Novosibirsk, Russia

[2] Bauman Moscow State Technical University, Baumanskaya 2-ya, 5, Moscow, Russia

E-mail: volodin_o@mail.ru

**Abstract.** The paper considers the influence of enhanced surfaces on heat transfer during evaporation and boiling of a falling laminar-wave liquid film. A mixture of R114/R21 refrigerants was used as a test fluid. The liquid film flowed on the outer surface of vertically-oriented cylinders. To intensify the evaporation process, the authors used the mesh coatings with variable mesh sizes, whereas the surfaces with variable characteristics of microstructure obtained by deformational cutting were used to enhance boiling. The report makes a conclusion about the studied types of structured surfaces, most effective for enhancement of heat transfer in different regimes. It is shown that the mesh coatings intensify heat transfer at evaporation twice as compared to a smooth surface; the microstructured coatings with partly-closed micropores obtained by deformational cutting allow fourfold heat transfer intensification at boiling.

## 1. Introduction

In connection with the development of processing techniques and ever-increasing technological requirements for the amount of heat flux taken away (for example, there is already a need to remove up to 100 W/cm$^2$ from the surface of computing modules), it is necessary to continue research on the development and improvement of enhanced surfaces, efficient under different conditions, and expand fundamental knowledge about the processes of boiling and evaporation.

The first experiments on intensifying the process of water boiling were carried out back in 1931 by Jakob and Fritz, allowing an increase in the heat transfer coefficient up to three times in comparison with a smooth surface (the structured surface has a square mesh of machine grooves), however, it turned out that for the "aging effect", resulting intensification disappeared after a few days [1]. So the result was considered useless for practical application.

Interest in the research of intensification with the help of developed surfaces returned in the 50s of the last century. Already in the first such works [2], it was shown that boiling enhancement occurs with an increase in the density of active nucleation sites due to the artificially created stable (and long-living) centers. It was also shown that the creation of artificial nucleation sites leads to a decrease in the temperature drop.

In the 1970s, researchers developed and registered patents for a number of boiling surfaces, such as Thermoexcel-E [3], GEWA-T [4], etc. (see [5]). Processing methods and the enhanced surfaces obtained with their help continue to be improved.

In 1997, Thors et al. registered a patent for the Turbo-BIII surface, which turns out to be 70% more efficient than the earlier version of this type of surface, Turbo-B [6]. Approximately at the same time (in 1996), the first patent of Zubkov and Ovchinnikov [7] appeared, and in 2007 and 2013, there were the joint patents of Thors and Zubkov [8, 9], which described the technology of surface microprocessing using the so-called deformational cutting method (MDC). Boiling surfaces used in this work were created using this method.

Starting from [10, 11], attempts have been made to create the models that describe boiling of liquid on artificial sites of structured coatings, but at the moment, this area is still being developed. Despite a large number of thematic works, there is still a need for further research on boiling enhancement using microstructured surfaces, especially for falling liquid films.

This paper presents also the data obtained during the study of intensification of falling film evaporation with the help of mesh coatings. Such a fairly simple method of surface development can also give positive results and help to shed light on the physical mechanisms of intensification. Not many works are devoted to the study of heat transfer intensification using mesh coverings. In [12], the possibility of intensifying the process of heat and mass transfer during the evaporation of falling films of water was shown. The results of test apparatus performance indicated evaporation rates up to 3-fold higher using the mesh promoter than without it (at water inflow rates of 0.72-1.00 g/s and fixed carrier gas inflow rate of $2.44 \cdot 10^{-3}$ $m^3/s$).

In [13], the dielectric fluorocarbon PF 5060 was used. The experiments were carried out at pool boiling, as well as under geometrically limited conditions when the mesh was covered with a plate of glass or polycarbonate. The largest increase in heat transfer coefficient at boiling reached 4 times as compared with a smooth surface for a mesh containing 20 wires/cm.

In [14], heat transfer intensification was investigated using the thin metal meshes with different numbers of layers (from 1 to 4) at pool boiling. Alcohol was used as a working fluid. The authors achieved a significant enhancement of boiling heat transfer (about 8 times) and a significant decrease in the temperature drop. It is shown that the use of the 4th layer of the mesh affects heat transfer. Some models presented in the paper do not describe adequately the experimental data obtained by the authors.

Thus, it can be noted that at present there are not so many results on the study of the mesh coating effect on heat transfer, and available research is mainly devoted to pool boiling. That is also true for the studies of the effectiveness of microstructured boiling surfaces. At the same time, such studies are necessary, since the falling films have a number of well-known advantages in comparison with heat transfer at pool boiling, as well as specific features (bubbles driftage, drop entrainment, influence of film velocity, and contribution of evaporation from the free surface even in the boiling regime) that distinguish them from pool boiling.

The purpose of this work was an experimental study of the effect of different types of coatings on heat transfer enhancement at evaporation and boiling. For each of the coating types (MDC-microstructures, mesh coatings), characteristic parameters were varied. It is shown that external mini-scaled mesh structures with respect to a smooth heat releasing surface contribute to the heat transfer enhancement in evaporation regime, while microstructures with partially covered micropores (being rather subsurface) obtained by MDC, enhance boiling greatly.

The results will be useful for the further study of heat transfer enhancement during evaporation and nucleate boiling of liquids (primarily for the case of falling films) using surfaces structured on different scales, for developing an understanding of the physical mechanisms of intensification, as well as for designing new high-performance types of enhanced surfaces.

## 2. Experimental setup and techniques

The scheme and description of setup for studying heat transfer at film flow of refrigerant binary mixtures are given in [15, 16]. The R114/R21 refrigerant mixture film flowed on vertically oriented cylinders with a diameter of 50 mm. The concentration of volatile component (R114) of the mixture at the inlet of the test section was 15% and it did not change. For the mixture used, the effect of a change in its concentration during evaporation on the heat transfer coefficient was not noticeable (this issue is discussed in [15]). The cylinder wall thickness was 1.5 mm. The length of the structuring area was 80 mm, and the length of the heat releasing zone was 70 mm. The experiments were carried out as on a smooth cylindrical surface with roughness $R_a$ = 2.5 μm, so on the MDC-structured surfaces and the sections with mesh coatings. When calculating the heat flux density, an increase in the area of the structured heat transfer surface due to the formation of (micro-) relief was not taken into account.

The surfaces obtained by the method of deformational cutting were used as boiling surfaces. Photographs of one of the surfaces, microstructured by MDC, in various sections are shown in figure 1. Table 1 shows the characteristic parameters of the studied MDC-surfaces.

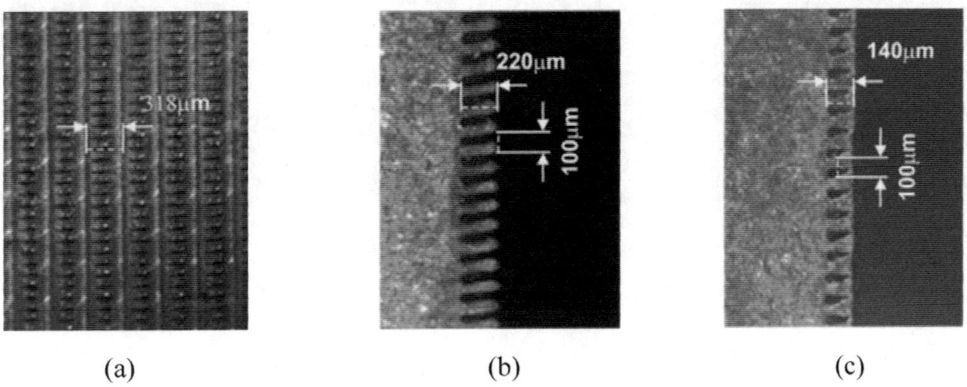


**Figure 1.** Surface No. 1: a – frontal view; b – section along the fins; c – section along with the knurling.

**Table 1.** Parameters of microstructured boiling surfaces.

| No. | Fin pitch, μm | Fin height, μm | Knurling pitch, μm |
|---|---|---|---|
| 1 | 100 | 220 | 318 |
| 2 | 200 | 440 | 318 |
| 3 | 60 | 120 | 318 |

Figure 2 (a-d) shows the used test sections with the wire mesh coatings, the parameters of meshes are given in Table 2.

**Figure 2.** Test sections with mesh coatings Nos. 1-4 (a-d).

**Table 2.** Parameters of the studied mesh coatings.

| No. | Material | Cell size, mm | Wire diameter, mm | Mesh disposition |
|---|---|---|---|---|
| 1 | Brass | 1.6×1.6 | 0.4 | Vertical |
| 2 | Brass | 1.6×1.6 | 0.4 | Diagonal |
| 3 | Stainless steel | 3.0×3.3 | 0.5 | Vertical |
| 4 | Stainless steel | 6.0×6.0 | 0.7 | Vertical |

## 3. Experimental results and discussion

Figure 3 in *q-dT* coordinates shows a comparison of the data obtained by authors for microstructured surfaces Nos. 1-3 (from Table 1) and sample smooth surface [16] with experimental data of [17] for R113. The data of [17] were obtained for the smooth and structured surfaces (Gewa-K19, Gewa-T19D) at pool boiling of R113 and water. The surface structure of Gewa-K19 is the plain transverse finning of the tube, while the tops of the fins of the Gewa-T19D have the flattened T-shaped form, close to the form of the flattened fins of microstructures Nos. 1-3 (see figure 1c).

It can be seen in figure 3 that the values of heat flux for microstructures Nos. 1 and 2 in the boiling region lie above the values obtained for Gewa-T19D, and the boiling curves have the sharper growth than the dependences shown in the figure for all other surfaces. The heat flux values for surface No. 3, which does not give intensification as compared to a smooth surface at film flow [16], nevertheless, lie above the heat flux values corresponding to pool boiling on a smooth surface [17] (this is due to the characteristic features of evaporation and boiling in thin films of liquid that intensify heat transfer in comparison with pool boiling) and practically coincide with the data for Gewa-K19. The absence of enhancement for surface No. 3 can be related to the absence of notable effect of small micropores on "suction and evaporation" process (in terms of Nakayama [10]), so characteristics of this structure could be considered as limiting (the structure dimensions) from below.

The observed enhancement of heat transfer for microstructured surfaces Nos. 1, 2 in comparison with Gewa-T19D and Gewa-K19 (in the boiling region), besides the differences caused by the specificity of heat transfer in falling films, apparently, is directly related to the smaller characteristic sizes of microtextures Nos. 1, 2 and, as a result, a large number of nucleate sites per unit area. The finning pitch for surfaces 1, 2 was 100 and 200 μm, respectively (see Table 1), while the pitch for Gewa-K19 and Gewa-T19D surfaces was 1.35 mm [17].

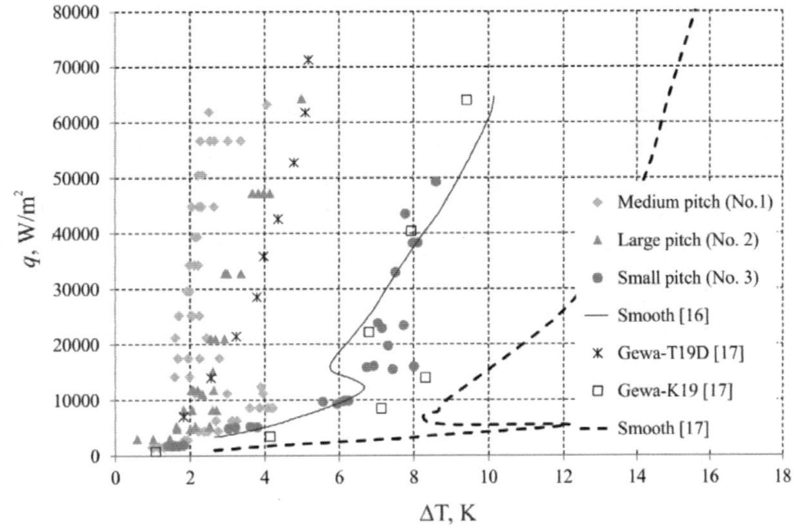


**Figure 3.** Comparison of boiling data for MDC-surfaces and Gewa-series surfaces [17].

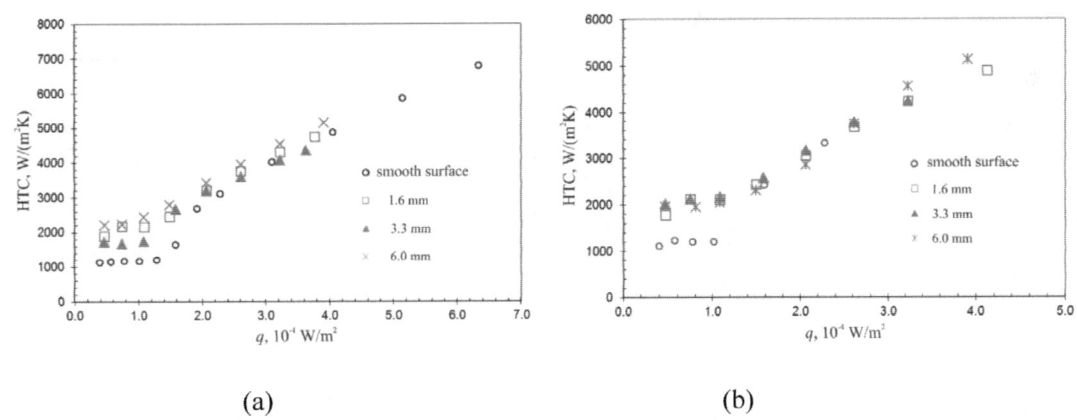


**Figure 4.** The dependence of the heat transfer coefficient on the heat flux density for a smooth surface and mesh coated surfaces (Nos. 1, 3, 4 from Table 2): a – Re = 640, b – Re = 1040.

Dependences of the length-averaged heat transfer coefficient on the heat flux density for different mesh coatings is shown in figure 4a, b in comparison with the results for a smooth surface for two different film Reynolds numbers (see definition of film Re number in [15, 16]).

Experiments showed that the data for the evaporation heat transfer coefficient for a large mesh No. 4 (Table 2), slightly exceeding the values for other meshes in the case of Reynolds number 640, are close in magnitude to the heat transfer values for meshes with a smaller cell size for other Reynolds numbers (208-1040) and they are approximately two times higher than the heat transfer values obtained for a smooth surface (see figure 4).

The increase in the heat transfer coefficient in evaporation regime on the meshed structures, apparently, is due to the change in hydrodynamics of the wave flow of the film caused by capillary forces. It was shown in [18] that intensification of heat transfer using wire coatings occurs due to the appearance of recirculation zones arising because of the pressure gradient behind the transverse rings of the wire.

In the regime of nucleate boiling ($q > 1 \cdot 10^4$ W/m$^2$), a noticeable increase in heat transfer coefficients in comparison with a smooth surface is not observed for the studied mesh coatings (see figure 4a, b).

## 4. Conclusions

The presented results demonstrate the effect on heat transfer in falling films with the formation of external or internal (with respect to a smooth surface) ordered structural disturbances.

At the time of the research, the minimum knurling pitch was limited by 318 μm, and the knurling roller had the sharpened teeth, which made it impossible to create full-fledged semi-closed pores during knurling. These factors could limit the magnitude of intensification obtained for the MDC-surfaces.

Sufficiently high enhancement results obtained with the use of MDC-structures [19] for pool boiling demonstrate also that the possibilities of heat transfer intensification at nucleate boiling using deformational cutting are not exhausted yet and require additional studies.

## Acknowledgments

The study was performed at Kutateladze Institute of Thermophysics (IT SB RAS) with the support of the BSI SAS Program for 2017-2020 (project III.18.2.3, reg. no. AAAA-17-117030310025-3)

## References

[1] Jakob M 1949 *Heat Transfer*, John Wiley & Sons, New York 636-638
[2] Kurihari H M and Myers J E 1960 *AIChE J.* **6** 83
[3] Fujie K, Nakayama W, Kuwahara H and Kakizakci K 1977 U.S. Patent 4060125
[4] Saier M, Kastner H W and Klockler R 1979 U.S. Patent 4179911
[5] Webb R L 2004 *J. Heat Transfer* **126** 1051
[6] Thors P, Clevinger N R, Campbell B and Tyler J T 1997 U.S. Patent 5697430
[7] Zubkov N N, Ovchinnikov A I 1996 European Patent EP 0727269
[8] Thors P and Zoubkov N 2007 U.S. Patent 7311137
[9] Thors P and Zoubkov N 2013 U.S. Patent 8573022
[10] Nakayama W, Daikoku T, Kuwahara H and Nakajima T 1980 *J. of Heat Transfer* **102** 451
[11] Chien L-H and Webb R L 1988 *Int. J. Heat and Mass Transfer* **41** 2183
[12] Salvagnini W and Taqueda M A 2004 *Ind. Eng. Chem. Res.* **43** (21) 6832
[13] Gerlach D W and Joshi Y K 2005 *In Proc. IMECE2005* 82595
[14] Dąbek L, Kapjor A and Orman Ł J 2016 *In Proc. AIP Conf.* **1745** 020005
[15] Pecherkin N I, Pavlenko A N and Volodin O A 2015 *Int. J. Heat and Mass Transfer* **90** (11) 149
[16] Volodin O A, Pecherkin N I, Pavlenko A N and Zubkov N N 2017 *Interf. Phenomena and Heat Transf.* **5** 215
[17] Ayub Z H 1986 Retrospective Theses and Dissertations Paper 7979
[18] Åkesjö A, Gourdon M, Vamling L, Innings F and Sasic S 2019 *Int. J. Heat and Mass Transf.* **131** 237
[19] Shchelchkov A V, Popov I A and Zubkov N N 2016 *J. Eng. Physics and Thermophysics* **89** (5) 1152

5th International Workshop on Heat/Mass Transfer Advances
for Energy Conservation and Pollution Control
August 13-16, 2019, Novosibirsk, Russia

# Results of Thermal Tests of Recirculation Cooling Installations

**Boris Balunov[1], Vitalii Lychakov[1], Mikle Egorov*[1,2], Andrey Sheglov[1] and Yurii Suhorukov[1]**

[1]Joint-Stock Company "I.I. Polzunov Scientific and Development Association on Research and Design of Power Equipment", St. Petersburg, 191167, Russian Federation

[2]Peter the Great St.Petersburg Polytechnic University, Polytechnicheskaya 29, St. Petersburg, 195251, Russian Federation

Abstract. The article presents the results from thermal tests of some recirculation installations for cooling air in nuclear power plant premises, including the volume under the containment. The cooling effect in such installations is produced by pumping water through its heat-transfer tubes. Air from the cooled room is blown by a fan through a bundle of transversely finned tubes and is removed to the same room after having been cooled. The finning of tubes used in the tested installations was made of Grade 08Kh18N10T and Grade 08Kh18N10 stainless steels or Grade AD1 aluminum. Steel fins were attached to the tube over their entire length by means of high-frequency welding. Aluminum fins were extruded on a lathe from the external tube sheath into which a steel tube had preliminarily been placed. Although the fin extrusion operation was accompanied by pressing the sheath inner part to the steel tube, tight contact between them over the entire surface was not fully achieved. In view of this, the air gap's thermal resistance coefficient was introduced in calculating the heat transfer between the heat-transferring media. The air gap average thickness was determined from the test results taking into account the gap variation with temperature due to different linear expansion coefficients of steel and aluminum. These tests were mainly aimed at checking if the obtained thermal characteristics were consistent with the values calculated according to the standard recommendations with introduction, if necessary, of modifications to those recommendations.

## 1. Introduction

Nuclear power plant (NPP) premises are equipped with recirculation cooling installations (RCIs), also known as recirculation ventilation installations (RVIs). Air from the room being ventilated is pumped (by means of a fan) through the heat exchanger available in such an installation and returned to the same room. Such installations are of particular relevance for being placed in rarely attended premises whose air may contain harmful substances.

The NPO CKTI carried out thermalhydraulic tests of the Type ROU-6.3 recirculation cooling installation designed for the rated air flowrate $G_{air}$=6300 m$^3$/h and intended to be used in the system for cooling air under the containment of the Rostov NPP units № 3, 4 [1]. The installation was manufactured by the Volgodonsk Works of Ventilation Articles (VZVI). Similar installations produced for the Rostov and Leningrad NPPs were tested. The latter installation used flashing freon for air cooling purposes; the freon itself was cooled by service cooling water circulating in the compressor-type chilling installation (air conditioner). The RUV-60 recirculation ventilation installation with $G_{air}$=6×10$^4$ m$^3$/h for the Novovoronezh NPP was the largest tested installation.

The tested installations, which had the same staggered arrangement of finned tubes (referred to hereinafter as tubes) in the heat exchanger and the same annular shape of fins, differed in the used materials, geometrical characteristics, and manner in which the tube was engaged in contact with the fin: by welding a fin to the tube over the entire contact length by means of high-frequency welding or by crimping an aluminum tube around a steel tube with simultaneously extruding helical finning from the aluminum tube (Figure 1). The results from a numerical analysis of heat transfer intensity from the tube carried out according to the standard recommendations [2–4] have demonstrated excellent results. The aim of this study is to check the validity of the calculations against the results obtained from the tests of a few large heat-transfer installations carried out with the actually existing air velocities (Re).

## 2. Description of the tested installations, test facility, and measurement procedure

The tested installations comprised a heat exchanger and a centrifugal or an axial fan interconnected by an air duct (Figure 2). A louver-type moisture separator was placed inside the air duct for removing the moisture generated during steam condensation from humid air in the course of its cooling in the heat exchanger. The technical data of the finned tubes and tested ROU-6.3 and RUV-60 installations are given in [5]. The RUV-60 installation contained two heat exchangers connected in a series by air and having different numbers of identical tubes—540 pcs. in the large heat exchanger and 216 pcs. in the small heat exchanger—and with the corresponding number of rows along the bundle depth (10 and 4). Water was pumped through one heat exchanger during the tests. The tubes externally cooled by air were arranged horizontally in all installations. Water flowed sequentially through the row of tube passing over the height with the cross flow of heat-transferring media.

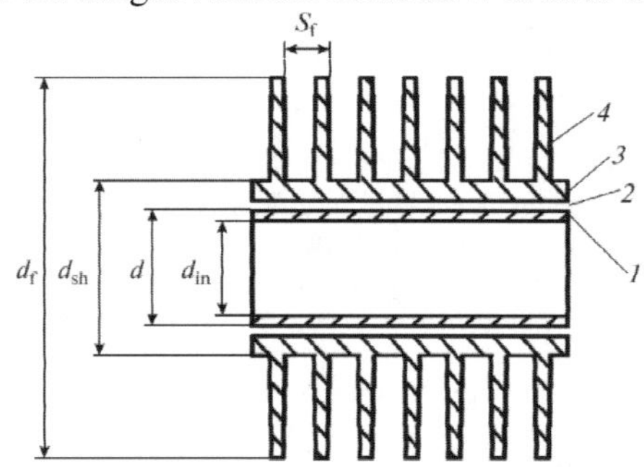


**Figure 1.** Finned tube manufactured using the knurling method. (1) Tube, (2) air gap between the tube and sheath, (3) tube finning sheath, and (4) fin.

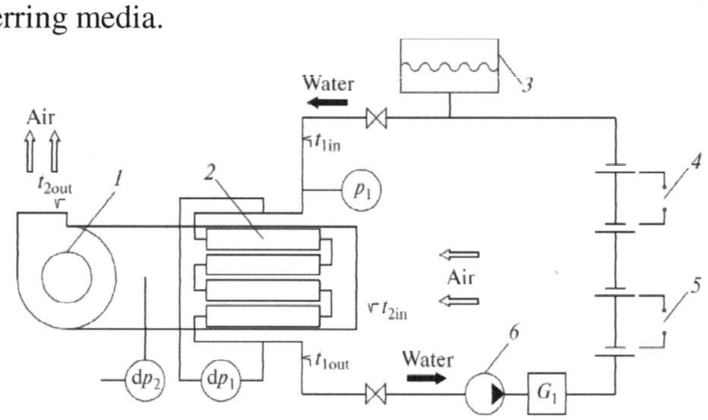


**Figure 2.** Test facility circuit arrangement: (1) fan, (2) heat exchanger, (3) pressurizer, (4) and (5) electric heaters, and (6) circulation pump driven by an adjustable speed motor.

In this article considers operation of the installations in the air-cooling mode without condensing the steam contained in the air. By changing the water temperature and flowrate at the inlet, as well as the air flowrate from one experiment to another, we succeeded in obtaining a certain amount of experimental data on the coefficients of heat transfer from the finned tube surface to air for comparison with the calculated results. The Type KSk3-6 air heater was in its delivery state just a heat exchanger, which was additionally fitted for the tests with a fan driven by an adjustable-speed electric motor. For testing the RUV-60 installation, a petal-type shutter was installed downstream of the axial fan outlet pipe, and flow limiting orifices were installed at the edge of the fan outlet pipe during some tests of the ROU-6.3 and ROU-35 installations for the possibility to adjust the air flowrate of the fan.

The water circuit instrumentation (see Figure 2) included means for measuring the pressure at the inlet $p_1$ and pressure drop in the heat exchanger $dp_1$ and the flowrate $G_1$ and temperature at the heat exchanger inlet $t_{1in}$ and at its outlet $t_{1out}$. The water flowrate was determined by measuring the pressure difference across the Venturi flow tube. The air circuit instrumentation included means for measuring underpressure downstream of the heat exchanger d$p_2$ and the temperature at its inlet $t_{2in}$ and outlet $t_{2out}$.

All thermocouples used to measure water temperature were duplicated. Three to eight thermocouples were installed for measuring the field of air-flow temperatures over the air duct cross sections. A uniform field of temperatures at the fan outlet was obtained owing to air being agitated in the fan. Apart from subjecting all instruments to qualification at the Test-SPb Institution, the thermocouples passed additional calibrations before each series of tests, which included the following operations.

Owing to these measures, we succeeded in keeping the maximum absolute error (MAE) or the media temperature determination uncertainty according to [6] to be as low as 0.5°C. The same value of MAE was estimated for the measured changes in the temperature of each medium $\delta t_1$ and the average difference of temperatures between the heat-transferring media $\Delta t_{av}$.

The rms error (RMSE) evaluated taking into account the errors introduced by the sensor reading transmission and automatic recording circuits did not exceed 0.212% for pressure measurements and 0.224% for pressure difference measurements. Computer-aided recording of instrument readings was performed at a 1 Hz frequency. The RMSE of determining the water flowrate $\delta G_1$ in the course of data processing did not exceed $5.5\times10^{-3}$. During the tests, the combination of media flowrates and temperatures at the inlet was selected so that the change in the temperature of each medium and the average difference of temperatures between the media were no less than 10°C. As a result, the RMSE of determining the heat exchanger power $\delta N$ did not exceed 0.093, and that of air flowrate $\delta G_2$ calculated based on the data from determining the heat exchanger power in its water circuit did not exceed 0.118. Our attempts to determine the air flowrate using the data from measurements of its velocity field in the fan outlet pipe, as well as attempts to set up the air flow stabilization section (during other tests) of the required length (more than 5 m) between the heat exchanger and fan in measuring the air velocity field by means of a heat-loss anemometer, gave essentially higher RMSE values as compared with that of the balance method used in the present study.

## 3. Results and Discussion

The further processing of the investigation data included calculations of the heat transfer coefficient $k$ related to the total outer surface area $F_s$ of the tested heat exchanger: $k = N/(F_s\Delta t_{av})$, where $F_s = \pi d_{sh}Ln_t\varphi$ and $L$ is the length of the tube working (finned) part.

The value of $k$ can also be represented as follows

$$k=1/\varphi\{d_{sh}/(\alpha_1 d_{in})+d_{sh}\ln(d/d_{in})/(2\lambda_w)+d_{sh}\ln[(d+2\delta_g)/d]/(2\lambda_g)+d_{sh}\ln[d_{sh}/(d+2\delta_g)]/(2\lambda_f)+1/(\alpha_2\varphi)\}^{-1}, \quad (1)$$

where $\delta_g$ is the mean gap between the tube and sheath, m; $\lambda_w$, $\lambda_f$, and $\lambda_g$ are the heat conductivities of the tube wall, fin, and air in the gap (the latter is determined at the air temperature in the gap), W/(m K); $\alpha_1$ is the coefficient of heat transfer from water to the tube inner surface, W/(m$^2$ K); and $\alpha_2$ is the coefficient of heat transfer to air related to the total outer surface area of tubes, W/(m$^2$ K).

For the installations with steel finning, the fins in which were welded over the entire length to the tube outer surface, the third and fourth terms in the denominator of (1) are equal to zero, and the sheath outer diameter is replaced by the tube outer diameter. The same applies to the expression for calculating the surface area. Therefore, by using the $\alpha_1$ value calculated from the Krausold–Nusselt relationship [2–4] for turbulent flow of water inside the tubes, we determined the sought value of $\alpha_2$ from expression (1). The fraction of thermal resistance $(\alpha_2)^{-1}$ in the performed tests was 44–88% of its total value $k^{-1}$, and, with a small error of calculating the other terms in (1), $\delta(\alpha_2) = 0.101$–$0.134$.

The regulatory documents [2, 3] use the same expression for calculating the heat transfer coefficient related to the total outer surface of tubes

$$\alpha_{calc}=(E\mu_f\psi F_f/F_s+F_t/F_s)\alpha, \quad (2)$$

where $E = f(\beta h_f)$ is the finning efficiency coefficient, the value of which for annular fins is calculated analytically using the Bessel functions; $\mu_f$ is the coefficient taking into account the effect of fins becoming wider toward their base; $\psi = (1-0.058\beta h_f)$ is the coefficient that takes into account heat transfer nonuniformity over the fin surface; $\beta=(2\alpha/(\lambda_f\delta_{av})^{0.5})$; $h_f$ is the fin height, m; $F_f$ and $F_t$ are the fin and tube surface areas, m$^2$; and $\alpha = \mathrm{Nu}\lambda_2/l_0$.

The difference between the methods described in [2] and [3] lies in the quantity taken as the characteristic size $l_0$: this is the tube outer diameter $d$ (or the sheath outer diameter $d_{sh}$ if there is a sheath) in [3], and this is a more intricate complex in [2] calculated from the formula

$$l_0=(1-F_f/F_s)d+(F_f/F_s)(0.25\pi(d_f^2-d^2))^{0.5}. \quad (3)$$

For the case of staggered tube bundles placed in a cross flow, the Nusselt criterion related to the total outer surface area is given, according to [2], by

$$\mathrm{Nu} = 0.36\mathrm{Re}^n\mathrm{Pr}^{0.33}\varphi^{-0.5}C_zC_s, \quad (4)$$

where $n = 0.6\varphi^{0.07}$; $C_z$ is the correction factor for the number of tube rows $z$ in the bundle along the flow ($C_z = 1.0$ at $z>4$); and $C_s=[(\sigma_1-1)/(\sigma_2^*-1)]^{0.1}$ is the bundle form factor. Formula (4) is recommended to be used at $\mathrm{Re} = w_2l_0/\nu_2 = 5\times10^3$–$3.7\times10^5$, $\varphi=1.0$–$21.2$, $l_0=0.012$–$0.178$ m, and $(\sigma_1-1)/(\sigma_2^*-1)=0.46$–$2.2$, where $\sigma_2^*=(\sigma_1^2/4+\sigma_2^2)^{0.5}$; $w_2$ is the air velocity in the narrow section; $\nu_2$ is the kinematic viscosity coefficient; and $\sigma_1$ and $\sigma_2$ are the transverse and longitudinal relative pitches of tubes in the bundle.

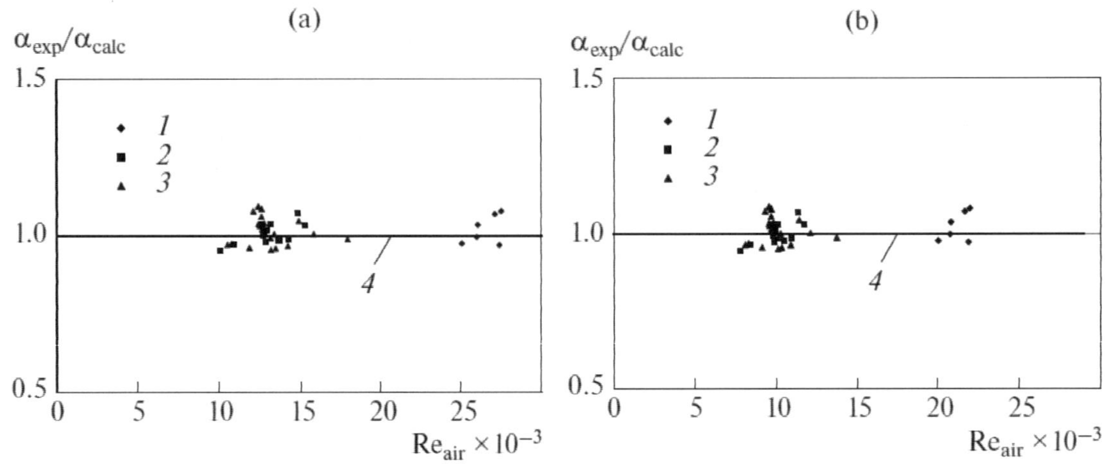


**Figure 3.** Dependence $\alpha_{exp}/\alpha_{calc}$ vs. the Re number for installations with welded steel finning. (1) OU-6.3 installation (experiment), (2) and (3) large and small RUV-60 heat exchangers (experiment), and (4) $\alpha_{calc}$ calculated according to (a) the recommendations of [2] and (b) according to [3].

In [3], another correlation was proposed:

$$\mathrm{Nu} = 0.113\mathrm{Re}^n\mathrm{Pr}^{0.33}C_zC_s, \quad (5)$$

where $n = 0.7 + 0.08\xi + 0.005\varphi$; $\xi = \tanh(x)$; for staggered tube bundles, the following holds: $x=\sigma_1/\sigma_2-1.26/\varphi-2$; $C_s=(1.36-\xi)[11/(\varphi+8)-0.14]$; and $C_z = 1.0$ at $z>7$.

The authors of [3] did not specify the application boundaries for expression (5). However, the authors of [7], the recommendations of which are similar to those used in [3], indicated them: $\mathrm{Re} = w_2d/\nu_2 = 5\times10^3$–$2\times10^5$; $\varphi=1.2$–$39$; $\sigma_1=1.7$–$6.5$; $\sigma_2=1.3$–$9.5$; and $\sigma_1/\sigma_2=0.3$–$5.2$. It should also be pointed out that, according to [7], the coefficient that considers the heat transfer nonuniformity over the fin surface is given by $\psi=1-0.016(d_f/d-1)[1+\tanh(2\beta h_f-1)]$.

Figure 3 shows a comparison between the experimental coefficients of heat transfer to air $\alpha_{exp}=\alpha_2$ for installations with welded steel finning and their calculated values $\alpha_{calc}$ evaluated according to the recommendations given in [2, 3].

The relative deviations $\delta\alpha = |\alpha_{exp}/\alpha_{calc}-1|$ obtained in the calculations carried out according to [2] are as follows for their maximal, algebraic mean, and rms values, respectively: $\delta\alpha_{max}=0.09$, $\delta\alpha_{a.m}=0.03$, and $\sigma=0.04$; in the calculations carried out according to [3], $\delta\alpha_{max}=0.11$, $\delta\alpha_{a.m}=0.04$, and $\sigma=0.04$. All the above-mentioned values of the quantities are within the $\alpha_{exp}$ determination error and within the permissible deviation from the calculation according to [2, 3].

The regulatory document [4], which is the last one in the chronological order, contains an outdated procedure for calculating the Nu criterion [8] without separating the contribution the heat transfer from the nonfinned part of the tube surface and that from the surface of fins introduce in the average value of $\alpha$ with taking into account the heat conductivity of their material, whereas expression (2) does take this separation into account. The calculation procedures developed by foreign authors [9, 10] are also simplified in nature and are inferior to the procedures outlined in [2, 3].

Both installations the test results of which are analyzed in Fig. 3 have geometrical characteristics that are often used in real heat exchangers, including $\varphi$=7.0 and 7.5 as well as the air flow velocities $w_{air}$=5.6–13.7 m/s. In all likelihood, the researchers who developed formulas (3)–(5) used the combination of these parameters. The tested assemblies fitted with aluminum fins had a higher value of the finning ratio $\varphi$, and, as the results of experiments reported in [11] have shown—which were carried out on small tube assemblies with welded fins at an increased value of $\varphi$—the value of the Re number exponent $n$ in (5) is overestimated, and the ratio $\alpha_{exp}/\alpha_{calc}$ decreases monotonically as Re increases. The exponent $n$ in (4) under the same conditions has a smaller value and satisfactorily describes the function $\alpha_{exp}=f(\mathrm{Re})$. Therefore, those who analyze the results of tests on installations with aluminum finning should give preference to the recommendations outlined in [2]. It should be pointed out as an argument in favor of this choice that both the recommendations also consider the case of a nonfinned tube ($\varphi$=1), for which $n$=0.6 in the considered range of Re (at $l_0=d$) according to the recommendations of [2–4], which is not the case in [3]. It can also be conjectured that with $\varphi\to\infty$ (a flat slit channel), $n\to0.8$. These limiting conditions are fulfilled in [2] and give higher values of $n$ according to [3].

The recommendations mentioned above were used for calculating the values of $\alpha_1$ and $\alpha_2$, and the results from testing the ROU-35 installation (an installation with knurled aluminum finning that does not have positively tight contact between the aluminum sheath inner surface and the steel tube outer surface) were generalized. It was found from that analysis that the third term in (1) accounts for 4–16% of the total thermal resistance to heat transfer between the heat-transferring media. Its fraction increases with increasing the calculated air temperature in the considered gap $t_g$, which was attributed to the difference between the linear expansion coefficients of Grade AD1 aluminum $\alpha_{al}$ and Grade 08Kh18N10T steel $\alpha_{st}$. For the considered temperature range of $t_g$, we have $\alpha_{st}=16.1\times10^{-6}$ K$^{-1}$ and $\alpha_{al}=23.5\times10^{-6}$ K$^{-1}$ [12, 13].

With the aluminum tube wall heat conductivity $\lambda_{al}$=200 W/(m K) adopted according to [12], the solution of (1) with respect to the average air gap enabled us to obtain the dependences $\delta_g=f(t_g)$ (Fig. 4): experimental (1) and calculated (2) derived from the equation

$$\delta_g=\delta_0+(\alpha_{al}-\alpha_{st})t_g d/2, \qquad (6)$$

where $\delta_0$ is the conditional average gap, μm at 0°C, which was determined proceeding from the condition of minimizing the average difference between the values of $\delta_g$ obtained from the experiment and calculated from (6) (for the considered tests, $\delta_0$=8.5 μm).

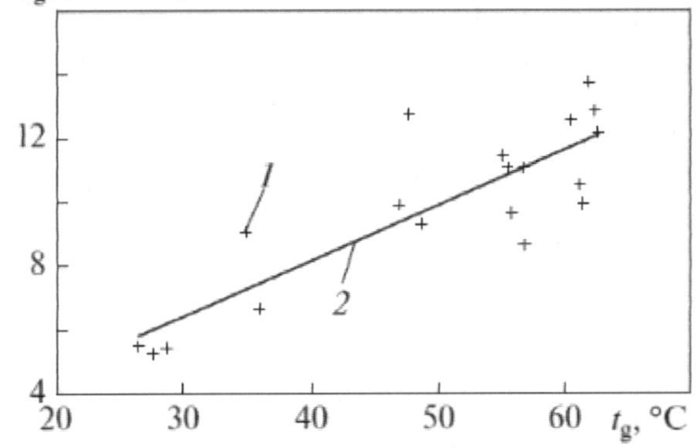


**Figure 4.** Dependence of $\delta_g$ vs. $t_g$ for the ROU-35 installation for $\delta_0$ = 8.5 μm.

Expression (6) is simplified in nature and does not take into account the difference between the tube wall and finning sheath temperatures nor does it take into account elastic deformation of mating materials when they come into force contact with each other on a certain part of the surface.

The results obtained from the tests of tubes with aluminum finning we carried out at JSC Kalorifernyi Zavod (Air Heater Works) in the city of Kostroma in different years have shown that the technologies for manufacturing them have been improved, as a result of which smaller values of $\delta_{g.in}$ have been obtained. Thus, the previous tests of small tube assemblies manufactured at the Works in 2000 gave $\delta_0$=22.5 μm, whereas the gap measured in the tubes of the Type KSk3-6 air heater manufactured in 2014 was found to be $\delta_g$=4–8 μm at $t_g$=53–116°C. This testifies that the roughness protrusions on the contacting surfaces experience elastic deformation and that its value decreases (without loss of contact between the mating surfaces) as $t_g$ increases.

## 4. Conclusions

(1) The results obtained from the tests of real recirculation air cooling installations for NPPs have corroborated the validity of the recommendations set out in the *RD* (Guiding Document) 24.035.05-89 for calculating heat transfer at the external finned surface of heat-transfer tubes.
(2) Some mistakes in RD 24.035.05-89 were corrected and its application area was extended.
(3) The thermal resistance of the air gap between the tube and finning must be taken into account if positive contact between their surfaces (like that secured by welding, soldering, etc.) cannot be guaranteed. The procedure proposed for calculating the air gap thermal resistance considers the difference between the linear expansion coefficients of the tube and finning materials.

## References

[1] Balunov B, Balashov V, Il'in V, Krayushnikov V, Lychakov V, Meshalkin V, Ustinov A, Shcheglov A 2013 *Thermal Engineering* **60** 653–659
[2] 1991 *Thermal-Hydraulic Calculation of Heat Transfer Equipment for NPPs* (Leningrad: TsKTI) p 214
[3] 1998 *Thermal Calculation of Boiler Units* (St. Petersburg: NPO TsKTI) p 256
[4] 2009 *Calculation Ratios and Methods for Calculation of Hydrodynamic and Thermal Characteristics of Elements and Equipment of Water-Cooled Nuclear Power Plants* (Moscow: NTTs YaRB) p 254
[5] Balunov B, Lychakov V, Il'in V, Shcheglov A, Maslov O, Rasskazova N, Rakhimov R, Boyarov R 2017 *Thermal Engineering* **64** 856–863
[6] 2009 *Recommendations on Interstate Standardization. Joint Use of the Concepts of "Error" and "Measurement Uncertainty"* (Moscow: Standartinform) p 8
[7] Pis'mennyi E 2004 *Heat Transfer and Aerodynamics of Packages of Cross-Sectioned Pipes* (Kyiv: Al'terpress) p 244
[8] Zhukauskas A 1982 *Convective Transfer in Heat Exchangers* (Moscow: Nauka) p 472
[9] Rouillard E 1993 *N&O JOERNALL* **9(2)**
[10] Feten T 2013 *Master Thesis* (Trondheim: Norwegian Univ. Science and Technology)
[11] Lychakov V 2016 *Candidate Dissertation in Engineering* (St. Petersburg: Central Boiler and Turbine Inst.) p 149
[12] Grigor'ev I, Meilikhov E 1991 *Physical Quantities* (Moscow: Energoatomizdat) p 1232
[13] 1989 *Standards for Strength Calculation of Equipment and Pipelines of Nuclear Power Units* (Moscow: Energoatomizdat) P 525

5th International Workshop on Heat/Mass Transfer Advances
for Energy Conservation and Pollution Control
August 13-16, 2019, Novosibirsk, Russia

# Heating and cooling of a battery pack of a solar electric car

**Ivan Kasatkin, Mikle Egorov*, Evgeny Zakhlebaev, Evgeny Kotov**

Peter the Great St.Petersburg Polytechnic University, Polytechnicheskaya 29, St. Petersburg, 195251, Russian Federation

* Corresponding author: mikhail.yu.egorov@gmail.com

Abstract. The purpose of the research is to analyse forced air cooling of a solar electric vehicle battery. Due to the conditions of safe battery performance, it is obligatory to keep its temperature under 45°C, so cooling should be provided. The battery pack is assembled of 420 battery cells with square pitch. At normal operating conditions of the solar car a forced air cooling thermal calculation is performed. Hydraulic calculation is carried out. The air pressure loss magnitude is calculated. A comparison of the geometric parameters of the corridor beam is presented. Based on a steady-state heat transfer analysis and hydraulic calculation, suitable airflow parameters and battery pack layout are selected.

## 1. Introduction

Solar energy appliance has advanced considerably in recent years. [1]. Moreover, the concept of creating a vehicle that is capable of travelling over long distances via only solar energy is developing [2, 3].

Every year competitions named Solar Challenge are held, which goal is to pass the distance of about 3000 kilometers using the solar energy. The battery pack of a solar car can be fully charged before the race, but then it receives energy only from the solar array.

A student team «Polytech Solar» designs a solar car [6], Figure 1. The car is powered by a lithium-ion battery pack that is charged by an array of photovoltaic cells.

The design [7], and the technical performance of the vehicle are specified by the competition rules, as well as the desire for aerodynamic drag reduction [8–10].

The battery pack represents 420 battery cells of the 18650 size, see Figure 2. It is assembled in a square pitch layout, see Figure 3.

The battery is heated during operation conditions. The battery performance degradation takes place at a high temperature magnitude. It can cause its capacity decline [11, 12] and even thermal runaway [13], which clarifies the necessity for cooling.

**Figure 1.** «Polytech Solar» car.

**Figure 2.** 18650 size battery cell.

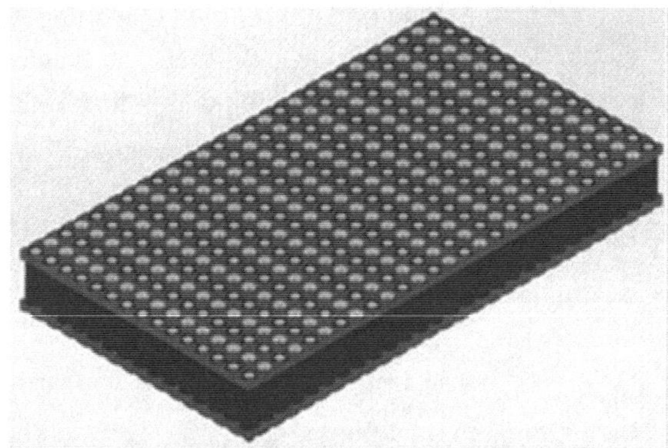

**Figure 3.** The Battery assembly.

## 2. Methods

### *2.1. Heat dissipation analysis*
The maximum continuous current passing through a cell is 1.5 A. So, the maximum cell heat dissipation power is 0.27 W, and for all 420 cells – 110 W. Heat dissipation calculations are based on a battery model, described in [14].

### *2.2. Initial data*
The battery assembly includes n = 15 cells in width and m = 28 cells in length. The length of an 18650 cell is l = 0.065 m, diameter d = 0.018 m. The transverse $s_1$ and longitudinal $s_2$ pitches are $s_1 = 0.03$ m and $s_2 = 0.02$ m respectively. The total surface area of heat exchange (the side walls are not counted) is:

$$F = \pi \cdot l \cdot d \cdot (n–1) \cdot m = 1.44 \text{ m}^2.$$

The ambient temperature of $t_{amb} = 38°C$ was used for the thermal calculation. The key condition is that the battery temperature at steady state cooling does not exceed 45°C.

### *2.3. Forced air cooling calculation*
Fan forced air cooling is the most appropriate in these conditions [15]. Airflow should pass along the narrowest cross section of the battery assembly [16].

The border, top and bottom sides are wrapped to be isolated from the ambience. Thereby, the airflow goes directly through the cells.

Several assumptions were proposed:

- The internal heat dissipation and rate of heat flow remains constant across the whole heat-exchange surface.
- Specific heat is independent of air temperature.
- Airflow enthalpy is independent of air pressure, and equals to:

$$\int dh = c_p \cdot \Delta t_{air},$$

where $h$ is specific air enthalpy, $c_p$ – specific heat and $\Delta t_{air}$ – air temperature difference between inlet and outlet.

Rate of heat flow between the airflow and the battery is conditioned by the internal heat at steady state and equals to:

$$Q_v = Q_\alpha = G \cdot c_p \cdot \Delta t_{air} , \quad (1)$$

where $Q_\alpha$ is heat flow rate between solid and liquid, $Q_v$ is the internal heat power in the battery and $G$ is mass airflow rate.

### 2.4. *Air pressure loss estimation*

When calculating forced air cooling, it is important to know the amount of pressure loss of the airflow as it passes through the battery pack in order to further select the fan operating characteristics.

The Darcy-Weisbach equation was used to calculate pressure losses:

$$\Delta p= 0.5\cdot\xi\cdot w^{2}\cdot\rho, \quad (2)$$

where $\xi$ is the coefficient of local resistance.

For a corridor beam of smooth pipes, the local resistance coefficient for a row of a beam is determined by the following dependence [17,18]:

$$\xi_0 = 0.38\cdot(\sigma_1-1)^{-0.5}\cdot(\Delta-0.94)^{-0.59}\cdot \mathrm{Re}^{-0.2\cdot\Omega\cdot\Omega}, \quad (3)$$

where $\Delta=(\sigma_1-1)/(\sigma_2-1)$, $\Omega=(\sigma_2-1)/(\sigma_1-1)$, $\sigma_1=s_1/d$, $\sigma_2=s_2/d$.

## 3. Results and Discussion

### 3.1. *Thermal calculation*

The temperature distribution diagram with assumptions considered is shown in Figure 4.

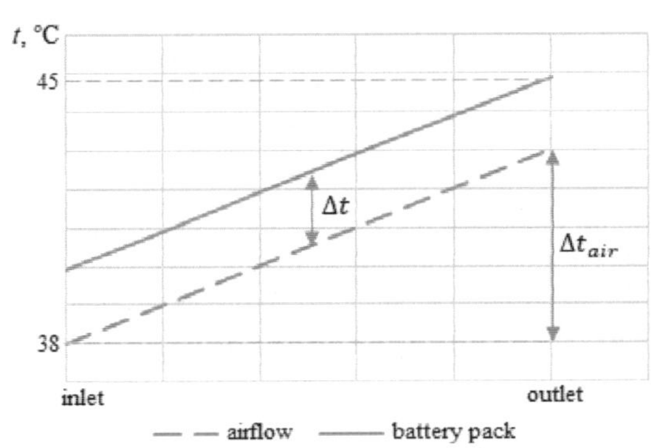


**Figure 4.** The temperature distribution over surface.

The highest temperature of the battery in steady state consists of the ambient temperature, the inlet and outlet airflow temperature difference $\Delta t_{air}$ and the battery and airflow temperature difference $\Delta t$.

$\Delta t_{\mathrm{air}}$ value is set in the first approach to calculate airflow rate which is enough to heat the air to the temperature $t_{\mathrm{air}}^{\mathrm{outlet}}$, that equals to:

$$t_{\mathrm{air}}^{\mathrm{outlet}} < t_{\mathrm{bat}}^{\mathrm{max}} - \Delta t,$$

where $t_{\mathrm{bat}}^{\mathrm{max}}$ is maximum temperature of the battery pack.

Assuming that $\Delta t$ will not exceed 2°C, the air temperature at the outlet is set as 43°C and $\Delta t_{\mathrm{air}}$ is set as 5°C, so applying the (1) dependence the airflow rate is calculated as:

$$G = \frac{Q_{\mathrm{v}}}{c_p \cdot \Delta t_{\mathrm{air}}} = 0.021 \text{ kg/s}$$

$\Delta t$ is determined using Newton's law of cooling

$$Q_\alpha = \alpha \cdot F \cdot \Delta t,$$

where $\alpha$ is heat transfer coefficient.

To find $\alpha$, the flow mode is determined and the convective heat transfer criterion equation is resolved.

The narrowest cross section area of the battery assembly:

$$f = (s - d) \cdot l \cdot (n - 1) = 1.09 \cdot 10^{-2} \text{ m}^2.$$

Airflow velocity:

$$w = G / (\rho f) = 1.75 \text{ m/s},$$

Reynolds number:

$$\mathrm{Re} = w \cdot d / \nu = 1.82 \cdot 10^3,$$

where $d$ is the cell diameter, that is used to calculate Reynolds criterion for smooth tubes, $\nu$ is air kinematic viscosity coefficient.

At advanced turbulent regime ($\mathrm{Re} = 10^3 \ldots 10^5$) for square pitches of tubes, the heat transfer criterion equation is [19,20]:

$$\mathrm{Nu} = 0.27 \cdot \mathrm{Re}^{0.63} \cdot \mathrm{Pr}^{0.33} = 27.1,$$

where Nu is the non-dimensional heat transfer coefficient, Pr – Prandtl criterion.

### *3.2. The discussion of the results*

#### *3.2.1. Thermal calculation*

The following conclusions can be drawn based on the calculation results analysis.

1. There is a lack of heat transfer data of narrow square pitch bundles.
2. Numerous researchers recommend using as a characteristic velocity:

- the most constrained cross section average velocity;
- the narrowest section velocity;
- velocity in a narrow cross section of the studied beam row.

As a characteristic speed to clearly compare the results, the velocity in the narrowest beam cross section is adopted.

Heat transfer coefficient equals to:

$$\alpha = \mathrm{Nu} \cdot \lambda / d = 42.2 \text{ W/(m}^2 \cdot \text{K)}.$$

The airflow and battery pack temperature difference is

$$\Delta t = \frac{Q}{\alpha \cdot F} = 1.9 \text{°C}.$$

Calculated maximum battery temperature is:

$$t_{bat}^{max} = t_{amb} + \Delta t_{air} + \Delta t = 44.9°C.$$

As long as the maximum battery temperature must not exceed 45°C under the conditions of safe battery operation, fan with airflow rate of $G = 0.021$ kg/s ensures adequate battery pack cooling.

*3.2.2. Hydraulic calculation*
With equations (2) and (3), the pressure loss for 28 rows is 8.02 Pa.

The dependence of the pressure loss on airflow rate at various relative transverse pitches is shown in Figure 5. The points are indicated at the figure at which the airflow is minimal to ensure the required battery temperature.

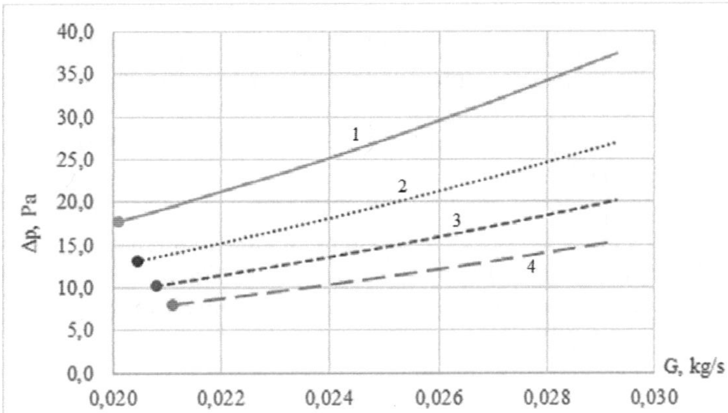


**Figure 5.** The dependence of the pressure loss on airflow rate at different relative transverse steps. 1. $\sigma_1 = 1.5$ 2. $\sigma_1 = 1.56$ 3. $\sigma_1 = 1.61$ 4. $\sigma_1 = 1.67$

It can be seen that with an increase in the relative transverse pitch $\sigma_1$ the required minimum airflow rate increases slightly, while the pressure loss decreases significantly. At the same time, as the pitch increases, the airflow rate decreases, as well as the Reynolds number. Therefore, it is advisable to increase the step $\sigma_1$ to such a value that the Reynolds criterion remains within the limits of the turbulent flow of air. When $\sigma_1 = 1.67$, the Reynolds number is $Re = 1.8 \cdot 10^3$.

## 4. Conclusions
Following conclusions can be made:

1. The heat transfer coefficient is determined for battery pack of stated parameters forced air cooling.
2. According to the results of the calculation, a fan with an airflow rate of 0.021 kg/s is selected. With steady state heat transfer and under the most powerful battery discharging mode it can keep the battery temperature that is suitable for maintaining reliability of lithium-ion battery.
3. Comparison of the dependences of the pressure loss on airflow at different relative transverse steps was made. It is advisable to use possibly large transverse pitch values; however, it is necessary to provide a turbulent flow regime.

## References
[1] Butuzov V 2018 *Thermal Engineering* **10** 78–88
[2] Zaripov R 2017 *Science and Technology of Kazakhstan* **3–4** 36–45
[3] Vandishev D, Goldina I 2016 *Youth and science* **12**
[4] Betancur E, Osorio-Gómez G, Rivera J 2017 *Sustainability* **9(10)** 1576
[5] Tamura S 2016 *Reinforced Plastics* **60(3)** 160–163
[6] Giannouli M, Yianoulis P 2012 *Sol. Energy* **86(1)** 441–451
[7] Minak G, Fragassa C, Vannucchi de Camargo F 2017 *Int. Conf. on Sustainable Design and*

*Manufacturing* 847–856
[8] Vinnichenko N, Uvarov A, Znamenskaya I, Ay H, Wang T 2014 *Sol. Energy* **103** 183–190
[9] Paterson S, Perera C, Vijayaratnam P, Doig G 2016 *Proc. of the Institution of Mechanical Engng., part d: J. of Automobile Engineering* **230(14)** 1972–1986
[10] Doig G, Beves C 2014 *Int. J. Veh. Engng.* **66(2)** 143–167
[11] Zhoujian A, Li J, Yong D, Chao D, Xuejia L 2017 *J. of Thermal Science* **26(5)** 391–412 (2017)
[12] Waldmann T, Wilka M, Kasper M, Fleischhammer M, Wohlfahrt-Mehrens M 2014 *J. of Power Sources* **262** 129–135
[13] Barré A, Deguilhem B, Grolleau S, Gérard M, Riu D 2013 *J. of Power Sources* **241** 680–689
[14] Borisevich A 2014 Modelling of Lithium-Ion Batteries for Battery Management Systems: a Review of the Current State. *Modern technics and technologies* **5** 3542
[15] Kasatkin I, Egorov M, Kotov E, Zakhlebaev E 2018 *MATEC Web of Conferences* **245** 15003
[16] Bhatia P 2013 Thermal analysis of lithium-ion battery packs and thermal management solutions *Doctoral dissertation, The Ohio State University.*
[17] Vilemas J, Voronin G, Dzyubenko B 1988 *Heat transfer intensification. Heat transfer successes* vol 2 (Vilnius: Mokslas) p 188
[18] Gavra T, Mihaylov P, Ris V 1982 *Thermal and hydraulic calculation of heat exchangers of compressor units* (Leningrad: Leningrad Polytecnical University) p 72
[19] Wang Q, Jiang B, Li B, Yan Y 2016 *Renewable and Rustainable Energy Reviews* **64** 106–128
[20] Kirillov P, Bobkov V, Zhukov A, Yuryev Y 2010 *Handbook on thermohydraulic calculations in nuclear power* vol 1 (Moscow: IzdAT) p 270

5th International Workshop on Heat/Mass Transfer Advances
for Energy Conservation and Pollution Control
August 13-16, 2019, Novosibirsk, Russia

# Experimental investigation of freon mixture spreading over the structured Sulzer 500X packing at single drip point irrigation

**V E Zhukov[1], A N Pavlenko[1], E Yu Slesareva[1], N I Pecherkin[1], G Meski[2], and P. Houghton[2]**

[1]Kutateladze Institute of Thermophysics SB RAS, Novosibirsk, Russia

[2] Air Products, Inc., 7201 Hamilton Boulevard, Allentown, PA 18195 USA

zhukov@itp.nsc.ru

Abstract. The paper presents the results of experimental studies on liquid spreading over the Sulzer 500X packing at single drip point irrigation. The experiments were carried out using a 10% mixture of R114/R21 freons on one packing layer with a diameter of 0.6 m. The flow rate of liquid flowing from the lower edge of the packing was measured automatically by a flow meter moving along the packing sheets with a given step. The flow meter used volumetric measurement method. The receiving collector of flow meter is 50x50 mm in size, operating range of flow measurement is 0.05–50 ml/s. Distributions of liquid over the packing were measured at single-jet irrigation in the range of jet flow rates $2.5 < q_{irrig} < 60$ ml/s. Spreading zone in the case of irrigation by a single jet is limited by the corrugation channels coming from the irrigation point. The pattern of spreading at low-flow jet irrigation has the shape of a dome with a low flow rate at the outlet of corrugation channels, which limit the spreading zone. With an increase in the flow rate, the share of liquid in these channels increases and dominates over other parts of the spreading zone.

## 1. Introduction

The conditions of uniform liquid spreading over a structured packing and uniform distribution of vapor in the packing are the main requirements for highly efficient operation of distillation columns. Achieving a good quality of liquid distribution is crucial for performance of the structured packing with corrugated sheet. According to practical experiences and research work, the sources and nature of large scale liquid maldistribution have been revealed [1]. Corrugated sheets of the structured packing have complex geometry, characterized by a large number of parameters, which greatly complicates modeling of the liquid spreading process [2-3]. In [4], liquid nitrogen spreading over a single sheet of corrugated Sulzer 500Y packing was studied for the case of single jet irrigation. Therefore, a reliable and extensive experimental database on various liquids spreading in various types of structured packings is a necessary condition for the development and verification of models. The purpose of this work is to obtain experimental data on spreading a high wetting liquid (freon mixture) over the sheets of Sulzer 500X packing at single drip point irrigation.

## 2. Setup and description method

The experiments were carried out on the large-scale research setup "Large Freon Column", whose detailed description is presented in [5]. A 10-% mixture of R114/R21 freons was used as a working fluid. Studies were carried out on a Sulzer 500X packing with a diameter of 600 mm. The setup was equipped with a controlled liquid distributor [6] and automated system for measuring the flow rate of liquid flowing from the structured packing sheets. The controlled liquid distributor allows activation of specified irrigation points with the required diameter in the range of 1.5 - 5.2 mm. The automated system for measuring the liquid flow consisted of a coordinate system for moving the gauge along the packing sheets and flow meter with a square receiving collector of 50x50 mm. The scheme of collector and its position relative to the packing are shown in figure 1.

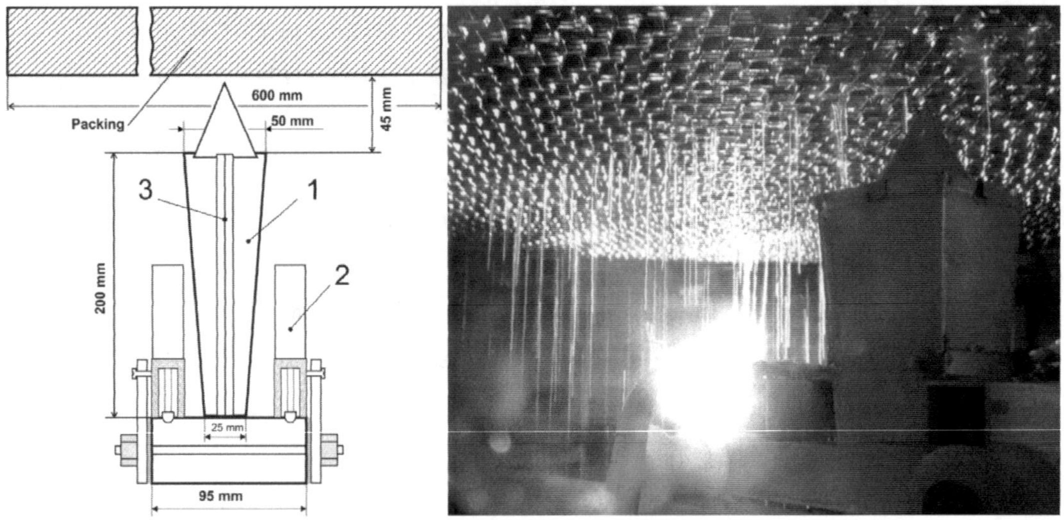


(a) (b)

Figure 1. Layout of collector relative to the packing (a) and picture of measurement process (b).

The flow meter consists of collector (1) and drainage solenoid valves (2). The collector is emptied with the help of valves before a new cycle of liquid flow measurements. The collector has a square cross-section; in the upper part, the face size is 50 mm, in the lower part, it is 25 mm. The height of collector is 200 mm. To measure the collector filling with liquid, two parallel lines (3), equipped with four thermometers, are installed. The scale bar is placed coaxially in a glass tube with an inner diameter of 13 mm. This tube is required to suppress oscillations of the free surface of liquid when measuring the flows and velocity. The bar construction provides overheating of thermometers by 5–10 degrees with respect to the ambient temperature. In the process of collector filling with liquid, the temperature of sensor, reached by the level of liquid free surface, decreases to the temperature of liquid, which is in equilibrium with vapor. The filling of this volume is registered by a sharp change in the temperature of sensors. The range of measured flow rates is 0.05 - 50 ml/s. The flow meter is moved with a coordinate device along the packing sheets, according to the predetermined program. The step of transportation of the flow meter collector can be 10, 25 and 50 mm. The measurement step is preferably 10 mm.

## 3. Results and discussions

Measurements were carried out at liquid flows through a single drip point in the range $2.5 < q < 60$ ml/s. To ensure such a range of changes in the liquid flow, five nozzles with different hole diameters from 1.5 to 5.2 mm were used. In addition, the flow through the hole was changed by varying the liquid level in the distributor. The experiments were carried out at liquid levels in the distributor of 460 and 120 mm. Figures 2 show the photographs of liquid flow from the lower edge of the packing at irrigation by a single drip point with different flow rates. The photographs show that with an increase in the flow rate through a single drip point, an increase in the flow at the edges of the spreading area is observed. At the maximal flow rate through the drip point, there are the strong jets at the edges of the spreading area.

(a)

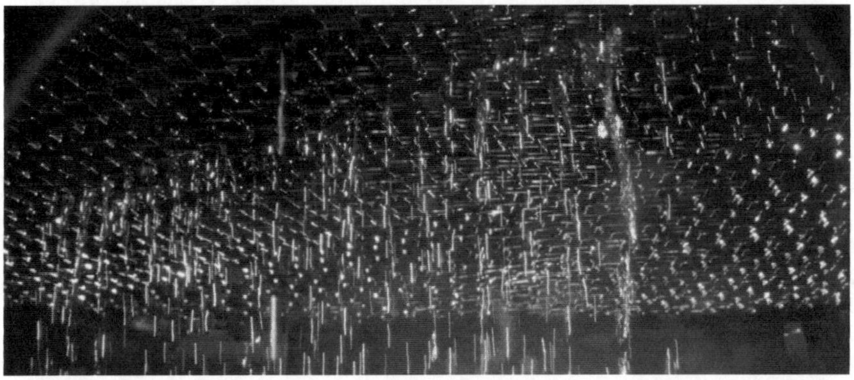

(b)

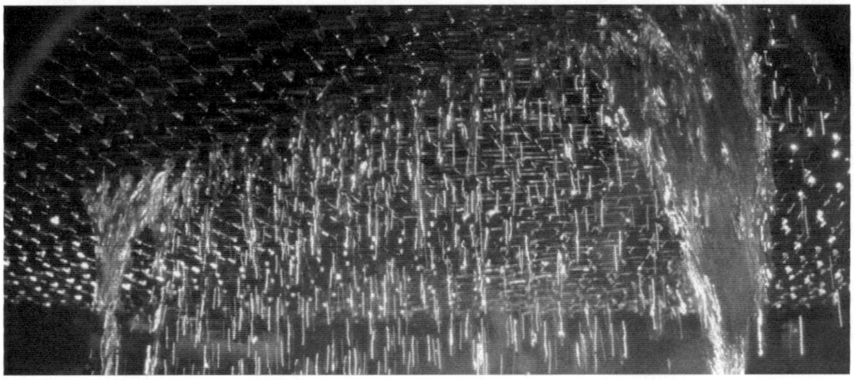

(c)

Figure 2. Liquid flow from the lower packing edge. Flow rate through a single drip point:

(a) - 4.9; (b) - 21; (c) - $q_{irrig}$ = 59 ml/s.

Journal of Physics: Conference Series **1369** (2019) 012049 doi:10.1088/1742-6596/1369/1/012049

The scheme of corrugated sheets in the packing layer is shown in figure 3. Solid lines are vertices of corrugation, dotted lines are depressions of corrugation, red and black lines belong to the adjacent packing sheets, and green cone is the area bounded by the irrigated canals. The zero coordinate at the lower edge of the packing sheets corresponds to drip point projection.

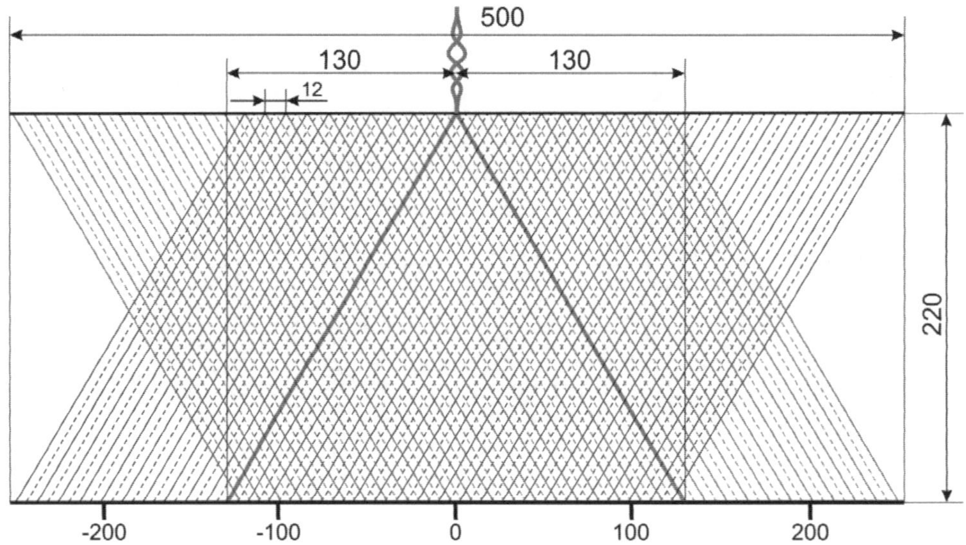


Figure 3. Scheme of corrugated sheets in the packing layer.

Figure 4 shows the results of measuring the distribution of liquid flow density over the 500X packing for different flow rates through a single irrigation point. Coordinates of the drip point (xn), hole diameter (d), level of liquid in the distributor (h) and flow rate in the jet (q) are shown in notation to the diagram. Data in the diagram are presented as dependence of the density of relative fraction of the jet flow rate (mean-integral on the collector scale) on the collector center coordinate.

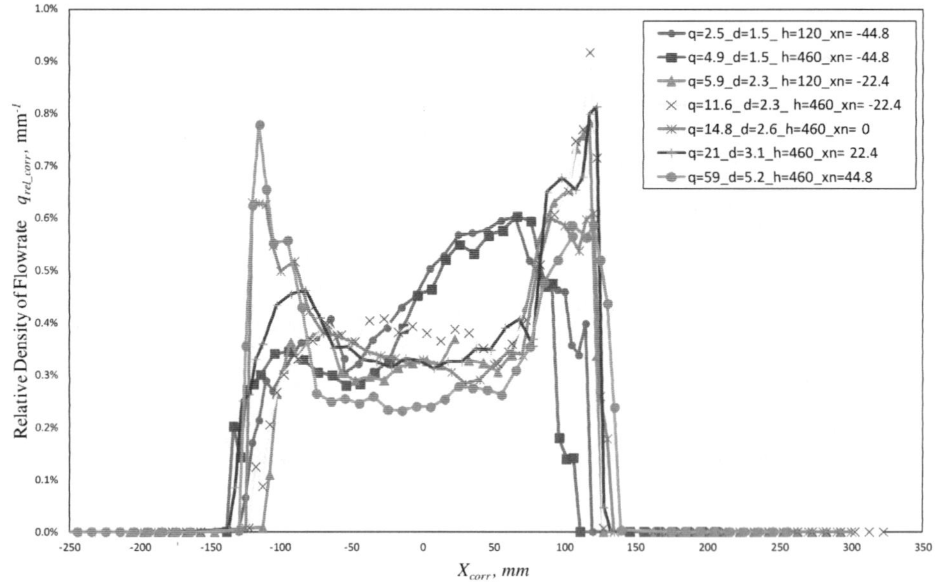


Figure 4. Distribution of relative density of liquid flow from the lower edge of the packing.

To show correctly the nature of spreading at the edges of the zone, where extreme jets fall only in a partial collector area and the coordinate of collector center can be beyond the spreading area, the measurement results were processed according to the following algorithm. The minimal flow rate measured at the spreading zone edge (usually, this value is in the range of 0.01-0.05 ml/s) was taken as the boundary of the spreading zone. The boundary coordinate corresponded to the coordinate of the collector edge nearest to the spreading zone. Displacement by one step (10 mm) towards the spreading area was interpreted as a flow rate measurement by a virtual collector with a size of 10 mm along the axis of displacement. The mean-integral density of the flow rate was determined at this size and result was reduced to the coordinate corresponding to the middle of this virtual collector. Since the nozzles with different hole diameters have different coordinates, the measurement data are presented in a modified coordinate along the abscissa axis for more visual comparison of the spreading patterns. Zero value of the coordinate corresponds to the position of a nozzle irrigating the packing.

As it can be seen in the diagram, a single jet spreads along the Sulzer 500X packing to a width of 250 - 270 mm for all flow rates through a single drip point. Thus, liquid spreads within a cone (green line) formed by corrugation channels of the adjacent sheets following from the point irrigated by a single jet. The nature of liquid spreading over the packing is different for different flows through the drip point.

At low flow rates (2.5, 4.9 ml/s), asymmetrical liquid spreading is observed. Liquid flows predominantly into the right (in the figure) part of the packing. In this part of the packing, we see a fairly large proportion of the flow from the projection of irrigation jet almost to the lower edge of the corrugation channel, which comes from the point of irrigation. However, along the extreme channel and the channel adjacent to it (the distance between the channels is 12 mm), the flow density is substantially less than that in the area closer to the middle of the spreading zone. To the left of projection of the irrigation hole, the proportion of flow is twice as less as in the right side. The proportion of flow in two leftmost channels is almost the same as in two rightmost channels. At a distance of 100 mm to the left of the drip point projection, a local maximum of the flow rate is observed. The asymmetrical nature of distribution and small proportion of the flow in the extreme channels can be connected with the fact that these flow rates through the drip point (2.5 and 4.9 ml/s) are achieved through a nozzle with a hole diameter of 1.5 mm. A fairly thin jet, falling into the packing cell, irrigates mainly a channel of one sheet, directed to the right. Thus, spreading is largely implemented in this direction. Under the conditions of small flow through the irrigation hole, the proportion of liquid transported through the contact points and holes is commensurate with the total flow rate and, as a result, an increased flow density is observed in the middle part of the zone of liquid spreading.

In the area of high flow rates (11.6 - 59 ml/s), the pattern of spreading is significantly different from spreading at low flow rates. In the middle part of the spreading zone, a fairly uniform flow density is observed. The proportion of flow in the middle part of the spreading zone is within 0.25 - 0.4% per mm. Moreover, a smaller proportion of flow density in the middle part of the spreading area corresponds to a larger flow rate through the irrigation hole. At the edges of the spreading zone there is a significant increase in the density of the liquid flow. This pattern of spreading can be explained by the fact that a limited amount of liquid transferred through the contact points and holes is non-linearly related to an increase in the flow through the outermost channels limiting the spreading zone. The greater the flow through the irrigation hole, the smaller the proportion of liquid that is redistributed to the middle region of spreading area and the larger the proportion of liquid that flows through the utmost corrugation channels, limiting the spreading area.

## 4. Conclusions

Liquid distributions over the structured 500X packing were measured at single-jet irrigation in the range of jet flow rates $2.5 < q_{irrig} < 60$ ml/s. The spreading zone in case of irrigation with a single jet is limited by corrugation channels coming from the point of irrigation. The pattern of spreading at a low-flow jet irrigation represents the shape of a dome with a small flow rate at the outlet of corrugation channels, which limit the spreading zone. With an increase in the flow rate, the portion of liquid in these channels increases and dominates over other parts of the spreading zone. The measurements showed that the character of liquid distribution under the studied unit element of the packing is significantly different at low (2.5, 5 ml/s) and high (15 - 59 ml/s) flow rates at the point of irrigation. At low flow rates, an increased flow density is observed in the middle part of the zone of liquid spreading; at high flow rates, an increased density is observed at the edges of the spreading zone.

**Acknowledgment**
The liquid distributions over the structured packing at single-jet irrigation for the different flow rates was studied within the joint project of IT SB RAS and Air Products Inc., USA and the specified irrigation systems for the structured packing, measurement methods of local flow rate distribution were developed in the framework of the state contract with IT SB RAS (AAAA-A17-117030310025-3, project No. III.18.2.3).

**References**
[1] Stoter C F 1993 Thesis, Delft University of Technology, Delft
[2] Spiegel L, Duss M 2014 *Distillation, Equipment and Processes* 145-181
[3] Olujić Ž 2014 *Distillation, Equipment and Processes* 1-34
[4] Pavlenko A N, Volodin O A, Surtaev A S 2017 A *Applied Thermal Engineering* **114(5)** 1265–1274
[5] Pavlenko A N, Zhukov V E, Pecherkin N I, Chekhovich V Yu, Volodin O A, Shilkin A, Grossmann C 2014 *AIChE* **60** 690–705
[6] Pavlenko A N, Zhukov V E, Pecherkin N I, Nazarov A D, Serov A F, Mis'kiv N B, Li X, Jiang B, Sui H, Li H, Gao X 2017 B *Optoelectronics, Instrumentation and Data Processing* **53(1)** 15–20

5th International Workshop on Heat/Mass Transfer Advances
for Energy Conservation and Pollution Control
August 13-16, 2019, Novosibirsk, Russia

# Dynamics of propagation of a self-sustained evaporation front on the flat and cylindrical heaters

**V E Zhukov, M I Moiseev, and E Yu Slesareva**

Kutateladze Institute of Thermophysics SB RAS, Novosibirsk, Russia

zhukov@itp.nsc.ru

Abstract. The paper presents the results of experiments on freons R21 and R114 and their mixtures. The variation range of volatile component concentration is 0 <C <1. The propagation velocity of a self-sustaining evaporation front was studied. The experiments were carried out under the conditions of free convection at a temperature of saturation. The stainless steel tubes with the diameters of 3 and 8 mm and a flat section of 25 x 50 mm were used as the working sections. The experimental results showed that in the field of development of hydrodynamic instability of a vapor front, a higher velocity of front propagation corresponds to a higher concentration of volatile component. Experiments with different heater geometry showed that dependence of the evaporation front velocity on the temperature head for a cylindrical heater with a diameter of 3 mm almost coincides with the dependence for a flat heater. In the region of temperature heads corresponding to the loss of hydrodynamic stability of the evaporation front interface, an earlier development of instability is observed on a heater with a diameter of 8 mm.

## 1. Introduction

Unsteady heat generation in a local section of the heat exchanger can cause significant overheating of liquid heat carrier contacting with the heated wall and lead to formation and propagation of a vapor film. The vapor film spreads due to heat accumulated in the metastable liquid layer. Many authors suggest various models for propagation of a self-sustaining evaporation front. The models, as a rule, consider propagation of the undisturbed evaporation front [1, 2]. Only in some cases, the authors considered the influence of small-scale perturbations of the interface on propagation dynamics of the evaporation front [3]. In [4, 5], the authors consider the shape of interface of a self-sustaining evaporation front without taking into account small-scale disturbances. In recent years, there is a tendency of using mixtures as the working fluids in refrigeration machines, heat pumps and thermal transformers instead of single-component liquids [6]. The ozone-safe non-azeotropic mixtures of refrigerants R32/R134a and R32/R152a are considered in [7] as well as heat transfer at boiling in the horizontal tubes. The purpose of this work is an experimental study of dynamics of a self-sustaining evaporation front and development of hydrodynamic instability of disturbances on the interface in R21, R114 freons and mixtures of R21 - R114 freons on the heaters of different geometries.

## 2. Setup and description method

The experimental setup is a cylindrical vessel with a diameter and height of 250 mm with built-in windows for visualization of processes. Freon R21 on the saturation line under the pressure of 0.27 - 0.28 MPa was used as a working fluid. In the experiments, horizontally oriented cylindrical working sections made of tubes (diameter of 8 mm, wall of 0.2 mm, length of 50 mm, and diameter of 3 mm, wall of 0.5 mm, and length of 50 mm) and vertically oriented flat rectangular working section with a width of 25 mm, length of 50 mm, and thickness of 0.2 mm were used. The material was stainless steel. The roughness of the section with a diameter of 3 mm was represented by multidirectional risks with a width of not more than 10 μm and individual smoothed cavities with a diameter of 30-50 μm. The roughness of the flat and 8-mm diameter working sections is represented by randomly oriented grains of 10-100 μm in size with an arbitrary shape. A high-speed video camera was used to observe the dynamics of evaporation front. The shooting speed was 25,000 frames per second with an exposure of 25 μs. The heat-generating surface was heated by a rectangular current pulse of a given duration and amplitude. To illuminate the object in transmitted and reflected light, the LED assemblies with a luminous flux of 9000 lm were used; the moment of their switching-on was synchronized with the moment of heat generation beginning. A detailed description of setup and experimental technique is given in [8].

## 3. Results and discussions

The dependence of propagation velocity of a self-sustaining evaporation front $V_{fr}$ on the temperature of wall overheating with respect to the temperature of liquid saturation $\Delta T = T - T_s$ is shown in Figure 1. Data are given for a cylindrical heater with a diameter of 3 mm. The experiments were carried out using the mixture of R114/R21freons. Molar concentration (C) of R114 freon in the mixture was varied from 0 (C = 0, pure R21) to 1 (C = 1, pure R114). Concentration C = 0.62 corresponds to the azeotropic point. For the given concentration, volatilities of both components of the mixture are equal.

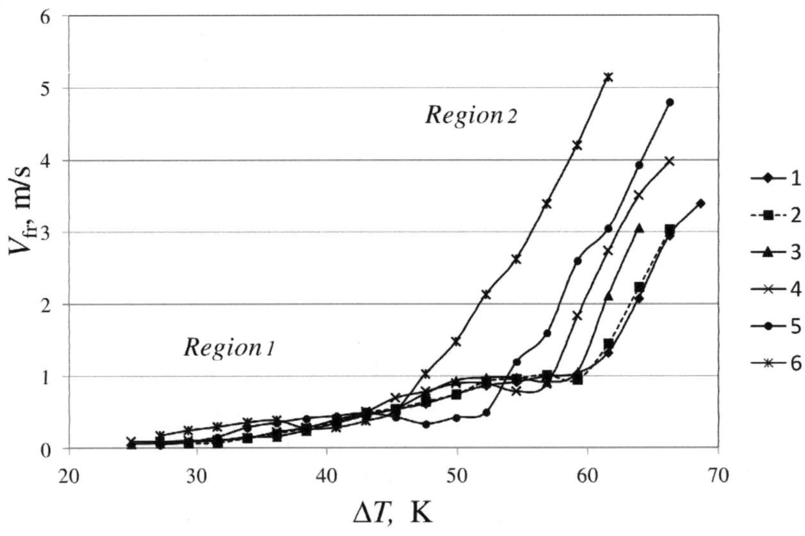


Figure 1. Evaporation front velocity vs. temperature difference. $P = 0.27$ MPa: a – cylindrical heater of 3 mm, R114/R21 mixture, concentrations C = 0 (1); 0.1 (2); 0.2 (3); 0.35 (4); 0.62 (5); 1.0 (6)

As it can be seen from the diagram, dependence $V_{fr}$ ($\Delta T$) has two characteristic regions: the region of weaker dependence (*region 1*) and the region of stronger dependence (*region 2*). In [8,9], the presence of *region 2* is associated with the development of hydrodynamic instability of Landau. The loss of stability by interface perturbations on the scale of the thermal layer thickness leads to heat transfer intensification through the interface and, as a consequence, to an increase in the front velocity.

In the region of weaker dependence, there is no significant difference in the velocity of front propagation. In *region 2*, the experimental data are substantially stratified, depending on concentration of the R114 component in the mixture. At the same values of temperature head $\Delta T$, the velocity of evaporation front propagation is greater in the mixture with a higher concentration of the R114 component. From the standpoint of development of instability of small-scale disturbances, the effect of a higher front velocity in the mixture can be explained by the fact that the vapor counter-pressure is a destabilizing factor, and capillary forces are a stabilizing factor. For freon R21 under the conditions of experiments, the vapor density is 11.46 kg/m$^3$, and the surface tension coefficient is 0.0163 N/m. For the 62% mixture of R114 - R21, it is 16.84 kg/m$^3$ and 0.012 N/m, respectively. That is, in the mixture, instability develops at lower vapor velocities, and, consequently, lower temperature heads $\Delta T$.

Dependence $V_{fr}(\Delta T)$ for the cylindrical heaters with the diameters of 3 and 8 mm is shown in Figure 2. As it can be seen in the diagram, the development of hydrodynamic instability is observed on the heater with the diameter of 8 mm at smaller temperature head $\Delta T$ than on the heater of the 3-mm diameter. This effect is observed for freon R21 and freon mixtures at the azeotrope point. Earlier development of instability on the heater of the 8-mm diameter may be related to the curvature of the working section and requires instability analysis not in the Landau formulation for an infinite plane, but in a formulation that takes into account interface curvilinearity of the interfacial surface. To test this hypothesis, the experiments were carried out on a flat working section.

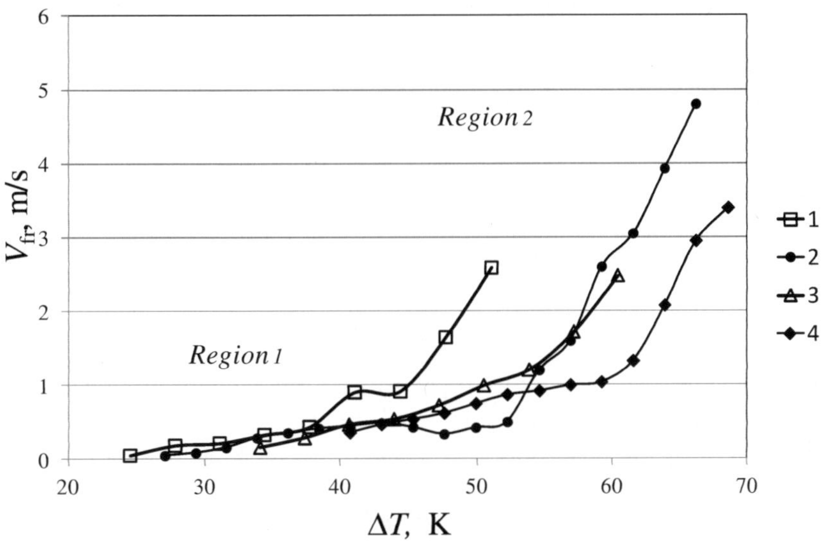


Figure 2. Evaporation front velocity vs. temperature difference. $P = 0.27$ MPa. Cylindrical heater, R114/R21 mixture, concentrations $C$: 1 – $D = 8$ mm, $C = 0.62$; 2 – $D = 3$, $C = 0.62$; 3 – $D = 8$, $C = 0$; 4 – $D = 3$, $C = 0$

Dependence of the velocity of evaporation front propagation for the flat and cylindrical heaters is presented in Figure 3. The experiments were carried out on Freon R21 under equilibrium pressure $P = 0.167$ MPa. As it can be seen from the diagram, dependence $V_{fr}(\Delta T)$ for a flat heater and a heater with a diameter of 3 mm almost coincide. For all d series of experiments in *region 1*, there is no significant difference in the velocity of front propagation. At the front velocity of more than 2.5 m/s, the data for the flat and cylindrical heaters coincide also. The difference is observed only in the zone of hydrodynamic instability development. On the 8-mm heater, the development of instability starts earlier, when the temperature head reaches value $\Delta T = 57$ K. For the flat and cylindrical heaters with a diameter of 3 mm, transition to *region 2* is observed at higher temperature heads ($\Delta T = 62$ K) than for the cylindrical heater with a diameter of 8 mm.

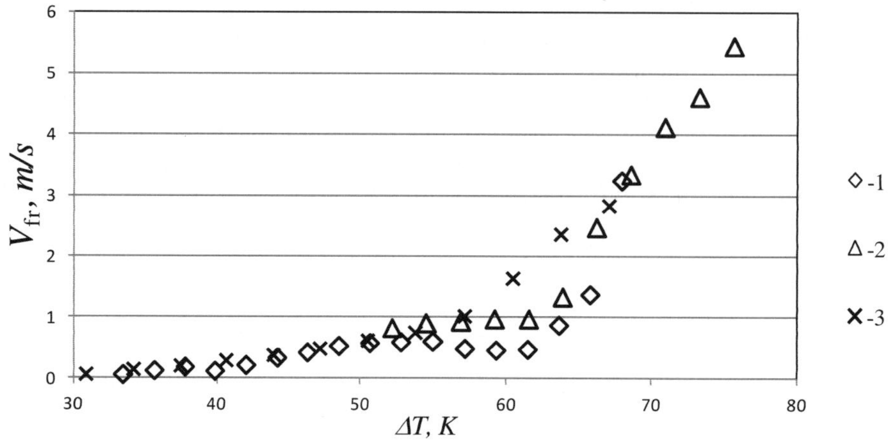


Figure 3. Evaporation front velocity $V_{fr}$ vs. temperature head $\Delta T$ for the flat and cylindrical heaters. Freon R21, $P$ = 0.167 MPa: 1 – flat heater; 2 – cylindrical heater, $D$ = 3 mm, 3 – cylindrical heater, $D$ = 8 mm.

In *region 2*, at $\Delta T$ = 67 K, the front velocity on all three heaters depends equally on the temperature head. Experimental data for different working sections are significantly stratified in the range of temperature heads 56 < $\Delta T$ < 62 K. On the flat and cylindrical heaters with a diameter of 3 mm, stabilization of propagation velocity of the evaporation front is observed in this temperature range, and only when $\Delta T$ = 62 K, there is a loss of interface stability and more intense increase in velocity with increasing temperature head. On a cylindrical heater of the 8-mm diameter, loss of stability is observed at $\Delta T$ = 56 K.

## 4. Conclusions

Experiments on propagation dynamics of a self-sustaining evaporation front in a mixture of R114/R21 freons showed that for the same temperature head $\Delta T$, the velocity of evaporation front propagation is greater in a mixture with higher concentration of the R114 component. From the standpoint of the development of instability of small-scale disturbances, the effect of a higher front velocity in the mixture can be explained by the fact that hydrodynamic instability in a mixture with higher concentration of R114 develops at lower values of temperature head $\Delta T$. The reason for the earlier development of instability is the difference in density and surface tension coefficient for different mixtures. For hydrodynamic stability of the interface, the vapor counter-pressure is a destabilizing factor, and capillary forces are a stabilizing factor. Experiments with different heater geometries showed that dependence of the evaporation front velocity on the temperature head for a cylindrical heater with a diameter of 3 mm almost coincides with dependence for a flat heater. In the region of temperature heads corresponding to the loss of hydrodynamic stability of evaporation front interface, the earlier development of instability is observed on a heater with a diameter of 8 mm.

## Acknowledgment

This work was carried out by the BSI SAS Program for 2017-2020 (project No. III.18.2.3), registration No AAAA-A17-117030310025-3.

## References

[1] Avksentyuk B P 1995 *Russian Journal of Engineering Thermophysics* **5** 1-8
[2] Pavlov P A, Vinogradov V E 2010 *High Temperature* **48(5)** 683-690
[3] Pavlenko A N, Lel V V 1999 *Thermophysics and Aeromechanics* **6(1)** 105-117

[4] Aktershev S P, Ovchinnikov V V 2011 *Journal of Engineering Thermophysics* **20(1)** 77-88
[5] Sharypov O V 2018 *Journal of Engineering Thermophysics* **27(2)** 137-144
[6] Mezentseva N N, Mezentsev I V 2018 *Journal of Engineering Thermophysics* **27(4)** 554-559
[7] Mezentseva N N 2011 *Thermophysics and Aeromechanics* **18(2)** 323-330
[8] Pavlenko A N, Tairov E A, Zhukov V E, Levin A A, Tsoi A N 2011 *Journal of Engineering Thermophysics* **20(4)** 380-406
[9] Pavlenko A N, Tairov E A, Zhukov V E, Levin A A, Moiseev M I 2014 *Journal of Engineering Thermophysics*. **23(3)** 173-193

5th International Workshop on Heat/Mass Transfer Advances
for Energy Conservation and Pollution Control
August 13-16, 2019, Novosibirsk, Russia

# Pore scale numerical simulation of heat transfer in propagating thermal wave during filtration combustion of rich and lean methane-air mixtures

**I A Yakovlev and S D Zambalov**

Tomsk State University, 36, Lenina Ave, Tomsk, 634050, Russian Federation

E-mail: yakovlev-i-a@yandex.ru

**Abstract**. The porous media combustion phenomenon was numerically studied with focus on the heat transfer effect in propagating combustion wave for lean and ultra-rich mixture compositions using the approach when the three-dimensional porous structure and interstitial flow are simulated directly at pore scale with explicit consideration of the thermal interaction between fluid and solid phases including the detailed chemical kinetics model and solid-to-solid radiation. The results demonstrate that irregularity of the porous structure leads to a large spatial variation of the process parameters and the flow inhomogeneity. The most intensive heat sources placed within the cavities where the flow is developed. During the process, the interface heat transfer and radiation contribute to heat recuperation mechanism that leads to thermal non-equilibrium in the combustion wave. The heat is transferred through the bed via radiation layer-by-layer due to restricted visibility of the particles. The numerical data about a local variation of the process parameters were presented. The data of this type can be used for modification of the volume-averaged models within the context of spatial variation consideration. It was shown that there is a correlation between the heat release rate, the interface heat flux, the radiative heat flux, and its root mean square values.

## 1. Introduction

The porous media combustion (PMC) phenomenon is studied intensively during the last decades because extended flammability limits, low pollutants emission and high stability of combustion that can be achieved during the process [1]. Filling a combustion region with inert porous media leads to energy feedback from the hot products to the incoming mixture via conductivity of the solid matrix and radiation. Such heat recuperation is accompanied by combustion wave formation that stabilizes within porous media or propagates through it. In accordance with the classification proposed by Babkin [2], the wave can propagate through the porous matrix with significantly different velocities. The regime when the wave velocity is the order of $10^{-4}$ m/s was called as the low-velocity regime (LVR). In contrast to the high-velocity regime, the LVR is characterized by strong heat coupling between the flow and solid matrix whereas hydrodynamic interaction is less important. This regime was intensively studied using experimental [3] and numerical [4] techniques.

During the experiments the limited list of measurement methods is available due to non-transparence of the porous media. The most widespread way to analyze the thermal structure near the flame front is based on thermocouples measures. Hence, this type of data does not allow observing local effects such as a difference between fluid and solid temperatures because the registered temperature of covered junction or laid on solid matrix surface junction correlates with only solid temperature. Zheng et al. [3] developed the method to register fluid and solid temperatures separately.

They showed that the difference between $T_f$ and $T_s$ is sufficient. Analysis of the mixture composition in the combustion wave without impact on the process by probe pipes is also unavailable. Alternative non-intrusive methods were used for analysis by Kiefer [5] and Stelzner [6]. Hence, these methods have specific limits and cannot provide the three-dimensional data with high resolution.

Another way to estimate the spatial variation of the heat transfer parameters at the pore scale caused by the structural irregularity of the porous media relates to numerical simulation methods [7]. Most of the numerical investigations are based on a two-temperature volume-averaged model describing the system behavior with the assumption that all parameters are averaged over a macroscopic scale and temporal scale of several hydrodynamic pulsations. This approach provided many important data about the PMC in a wide range of operational parameters [8]. Hence, this approach has a principal limitation that cannot be used for estimation of the local variation of the process parameters such as the velocity or temperature.

To overcome the limitation the approach of representation of the realistic porous media structure at pore scale was developed. Sahraoui and Kaviany [9] showed that the volume-averaged model could not predict local high-temperature spots and flame speed with respect to the flame location. Hackert et al., Jouybari et al., Shi et al. and Sirotkin et al. [10–13] advanced the approach of the pore-scale simulation taking into account solid-to-solid radiation, turbulence effects and randomness of particles packing using two-dimensional models. Authors showed that local variation of process parameters at the pore scale cannot be predicted using the volume-averaged model. Bedoya et al. [14] experimentally and numerically studied lean methane-air flame stabilization in porous media of different geometry. Authors compared the numerical results of volume-averaged and three-dimensional pore-scale simulation approach. It was shown that the volume-averaged approach provides significantly lower thermal flame thickness than in experiments and this approach cannot quantitatively predict temperature profile and burning velocity.

This study is devoted to numerical simulation of the PMC in the low-velocity regime with explicit three-dimensional representation of realistic porous media in the form of a packed bed of spherical particles at the pore scale. The overall objective is to visualize the distribution of the interfacial and radiative heat fluxes at pore scale and estimate heat transfer intensity in the propagating combustion wave depending on the input mixture composition.

## 2. Methods

The first stage of the study was devoted to the problem of realistic porous structure representation. There are two main approaches when a real sample of the media is scanning using three-dimensional computer tomography technique (sponge-like structures mainly) or generating synthetically using discrete elements method (packed beds mainly). Since in the paper the packed bed of spheres was studied, the second method was used. During the calculation based on the discrete element model to describe particle-to-particle interaction, 5-mm spherical particles were falling into the box under gravity. For verification purposes, the radial porosity profile of the calculated structure was compared with experimental data and empirical equation [15]. The maximum difference was less than 5%.

The next stage related to the computational mesh generation. For packed beds, there is a specific problem with particle-to-particle contact points. Near this region, the mesh cells have a tendency to be distorted due to small angles. Widespread way to overcome the problem is to create an additional geometrical objects near point contacts that have bridge-like shape. Partopour and Dixon [16] developed integrated workflow for pore-scale models based on wrap method when an initial set of surfaces are replacing with a continuous wrapper surface. After the procedure, the conformal fluid-solid volumetric mesh was generated. The total number of cells in the model was about 6 millions.

The mathematical model of the methane-air combustion process within porous media at pore scale should simultaneously describe multicomponent gas flow with chemical reactions, heat transfer between gas and solid porous matrix in which heat transfer by conduction and radiation occurs. In the fluid region, the conservation equations for mass, momentum, energy and species transport equation with unsteady terms were solved. In the solid region, the energy transport equation is solved.

Solid-to-solid radiation significantly contributes to total heat transfer in porous media. Upstream energy transport by radiation influences to heat recuperation from high-temperature products to income reactants [17]. Heat exchange between solid surfaces was simulated with the assumptions that all surfaces of the alumina spheres are diffuse and gray without the participation of the gas phase in radiation heat transfer. To simulate radiative heat transfer under the assumptions the model based on view factor calculation was used [18]. Detailed chemical mechanism is a common way to increase the accuracy of major combustion products prediction. The San Diego Mechanism was chosen as a detailed chemical kinetic mechanism of methane-air combustion [19]. The mechanism consists of 57 species and 268 reactions.

Ansys Fluent software was used to solve the governing equations using a finite volume method. The pressure-based solver was used due to low fluid velocity. The coupled algorithm was chosen to solve the momentum and continuity equations together. The second-order upwind interpolation method was chosen for convection terms in all the equations. All diffusion terms were interpolated using a second-order scheme. The second-order accurate implicit temporal discretization scheme was used. Fully implicit time stepping with dual timing formulation was used [20]. This formulation is based on the method when the fluid and solid domains are solving simultaneously with different time steps. Time step in fluid region was set to be $8 \cdot 10^{-5}$ s, in solid region 0.05 s.

## 3. Results and discussion

In the study, the porous media combustion of two methane-air mixture composition corresponding to equivalence ratio $\varphi$ = 0.8 (I case) and 2.5 (II case) was simulated with input velocity of 0.14 m/s (along with $y$ axis). In figure 1, the temperature distribution at the interface (a), the heat release rate ($r$) in the fluid region (b) and the interface (c) and radiative (d) heat fluxes at the interface are shown.

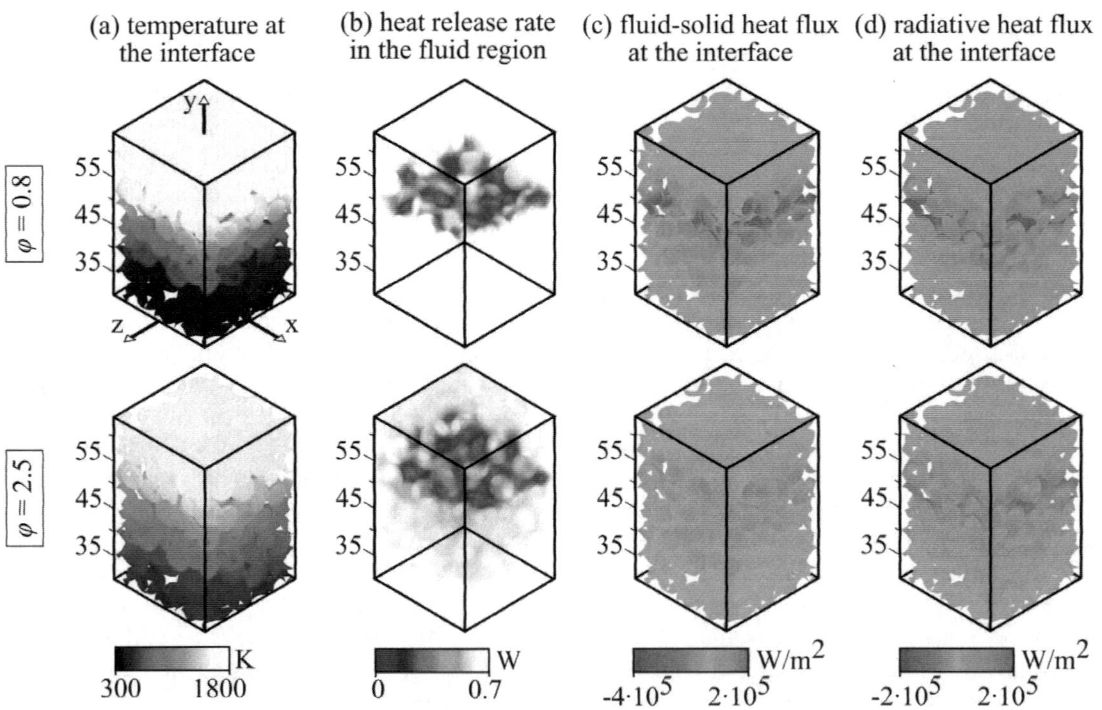


**Figure 1.** Distribution fields of temperature (a), velocity (b), heat of reaction (c) and $CH_4$ mole fraction (d) for the case of $\varphi$ = 0.8.

It can be seen that the rich mixture composition is characterized by a more sloping temperature profile near the combustion region. This difference relates to the heat release rate of the chemical reaction. The ultra-rich mixture produces less energy due to incomplete combustion with oxygen deficit. The heat release rate for $\varphi$ = 2.5 is about 43.8% of the $r$ that was predicted for lean mixture composition with $\varphi$ = 0.8. The combustion wave in the first case is upstream with the propagation velocity of –0.031 mm/s whereas in the second case is downstream with the velocity of 0.012 mm/s. Mixture composition has a great influence on the structure of the heat fluxes near the reaction region. Figures 1c and 1d show a distribution of the interface heat flux and the radiative heat flux at the surface of the porous matrix. It can be seen that the parameters are distributed irregularly with strong matching with the porous structure. The region position where the interface heat flux $q_{fs}$ has negative values correlates with the position of the heat release region whereas in lower region heat is transferring from the solid to the mixture. This distribution indicates the mechanism of income mixture preheating in the region before ignition point. Distribution of the radiative heat flux $q_r$ has a more complex structure that explains by used radiation model based on view factor calculation. The model allows considering the visibility of the surfaces of the particles between each other explicitly. Due to that, the radiative heat flux distribution has the wave-shaped form. In the region of the temperature gradient, several nominal layers of the particles are placed. The radiative heat transfer from the most heated region to the preheating region occurs layer by layer with damping. Such structure of the heat fluxes distribution takes place for both cases. But the intensity of the heat transfer is smaller for the case of an ultra-rich mixture.

In figure 2a the fluid and solid temperature profiles are shown for both cases ($T_f$ and $T_s$ profiles are shifted to the right by 20 mm for better visibility). These profiles are the temperature values averaged over cross sections. It can be seen that there is sufficient thermal non-equilibrium between the phases. This difference is a specific feature of the PMC and is the result of heat recuperation mechanism. In figure 2a the preheating region is denoted by I. Here, the heat stored by solid matrix transfers to the fluid. In the region II the reaction is occurring. The heat energy released from the chemical reaction transfers to the solid matrix and then to the preheating region via solid conductivity and radiation. In the region III a thermal relaxation occurs. The same mechanism can be observed for ultra-rich conditions. Hence, the length of the thermal wave in this case is larger. The thermal flame thickness based on averaged fluid temperature $T_f$ is 6.75 mm for lean mixture and 13.5 mm for the ultra-rich mixture. So, the thermal flame thickness correlates with averaged heat release rate: $\delta \propto r$ with the accuracy of 12.5%. The RMS (root mean square) values of $r$ describe it spatial variation at the cross section. The RMS profiles for both cases are shown in figure 2b by dotted curves and denoted as $r_{rms}$. The RMS values are proportional to their magnitude with coefficient about unity.

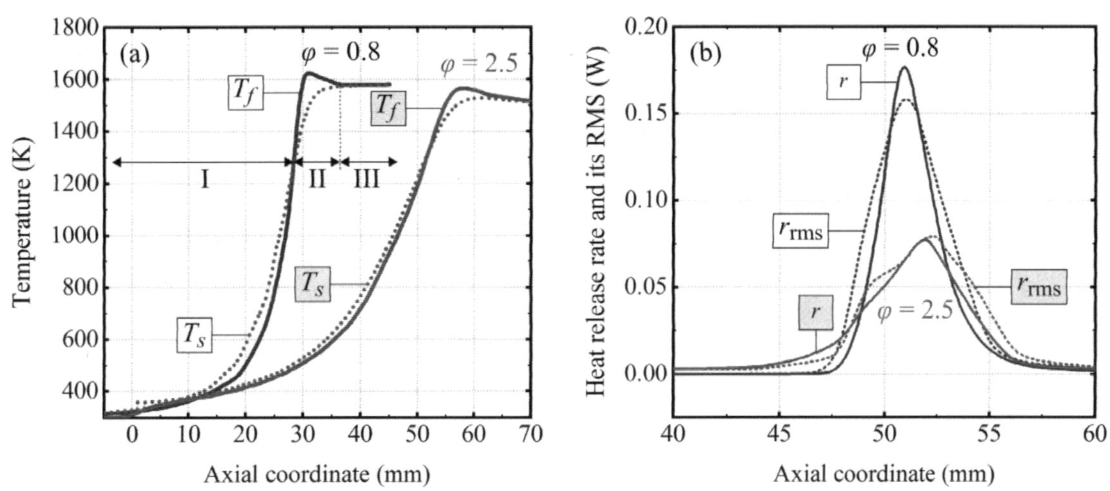


**Figure 2.** Temperature profiles (a) and heat release rates (b) for the cases of $\varphi$ = 0.8 and 2.5.

Figure 3a shows the averaged interface heat flux profiles $q_{fs}$. The profiles have wave-shaped form in the combustion region with the direction changing near the ignition point. Such form is the result of heat recuperation mechanism: before the ignition, the heat flux is directed from the solid matrix to the fluid and vice versa. The magnitude of the interface heat fluxes before and after the flame front is not equal due to heat release region position. The maximum flux is located near the maximum heat release rate. In the preheating region, the heat energy does not release but only transfer from the reaction zone. The maximum RMS value $q_{fs,rms}$ is equal maximum absolute value with 5% accuracy for both cases as it was observed analogously to the heat release rate $r$. The averaged radiative heat flux profiles have a similar but reversed form. Broken shape of the curves is the result of the local porous structure. The RMS values of $q_r$ are larger than their absolute values. For the lean mixture $q_{r,rms} \approx 2q_r$ and $q_{r,rms} \approx 2.8q_r$ for the ultra-rich mixture composition near reaction zone. The radiative heat flux demonstrates the largest spatial variation in comparison with other estimated parameters.

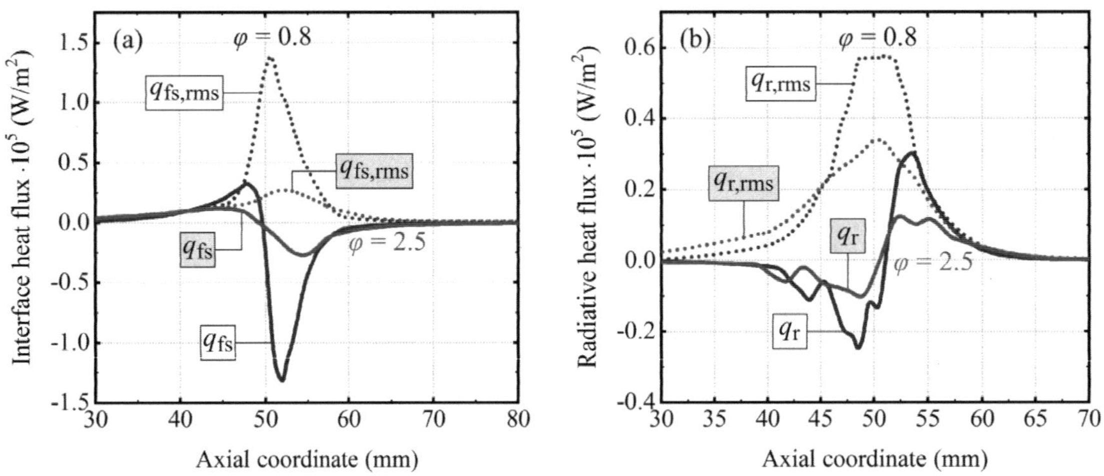


**Figure 3.** Interface (a) and radiative (b) heat fluxes for the cases of $\varphi = 0.8$ and 2.5.

## 4. Conclusion

At the local scale, the parameters distribution has large spatial variation due to an irregularity of the porous structure. The flame front has the shape of combustion cells that form the structure different from planar. The difference in the local wave shape does not lead to dissociation of the front under considered regimes due to compensation of the thermal inhomogeneity by the radiative and convective heat fluxes. Spread of the local heat sources along flow direction as well as the porous structure irregularity leads to the local variation of the interface heat flux that has a strong coupling with a shape of the flame front. The largest spatial variation is observed for the radiative heat flux due to the complex shape of the porous matrix surface. The averaged radiative heat flux profile reflects the local structure of the particles. The interface heat transfers, the conductivity of the porous matrix and the radiation have an important role in heat recuperation mechanism. In the combustion wave, the thermal non-equilibrium between fluid and solid phases is obvious. Averaged thermal flame thickness is proportional to the averaged heat release rate.

Input mixture composition has a great influence on the combustion process. The lean mixture is characterized by total methane consumption and the high heat release rate whereas during ultra-rich mixture combustion major species in the exhaust gas are $H_2$ and CO diluted by nitrogen. Oxygen deficit leads to the significantly lower heat release rate. Such a difference leads to different behavior of the combustion wave. In the case of the lean mixture the wave is upstream whereas in the case of the ultra-rich mixture the wave is downstream. The simplest conception about the balance between heat fluxes in the combustion wave can explain this dependence. Input fluid flow takes the heat energy in the preheating region via convective heat transfer with intensity $q_1$. Conductive heat flux in the porous matrix and radiation deliver the energy from the high-temperature region in the preheating region with intensity $q_2$. If $q_1 > q_2$ the wave is downstream, $q_1 < q_2$ the wave is upstream [8]. If the fluxes are equal the wave stabilizes within porous media.

The pore-scale approach provides numerical data about the local variation of the process parameters. The data of this type can be used for modification of the volume averaged models within the context of spatial variation consideration. It was shown that there is a correlation between the heat release rate, the interface heat flux, the radiative heat flux and their RMS values. As it was proposed by Bedoya et al. [14] spatial fluctuations of the process parameters can be taken into account within the framework of the volume-averaged approach using a presumed shape probability density function which can be constructed based on the pore-scale simulation data.

**Acknowledgment**
The reported study was funded by Russian Foundation for Basic Research according to the research project № 18-31-00071.

**References**

[1] Mujeebu M A, Abdullah M Z, Mohamad A A and Bakar M Z A 2010 *Prog. Energy. Combust. Sci.* **36** 627–50
[2] Babkin V S 1993 *Pure Appl. Chem.* **65** 335–44
[3] Zheng C, Cheng L, Saveliev A, Luo Z and Cen K 2011 *Proc. Combust. Inst.* **33** 3301–8
[4] Dobrego K V, Gnezdilov N N, Lee S H and Choi H K 2008 *Int. J. Hydrogen Energy* **33** 5501–9
[5] Kiefer J, Weikl M C, Seeger T, von Issendorff F, Beyrau F and Leipertz A 2009 *Proc. Combust. Inst.* **32** 3123–9
[6] Stelzner B, Keramiotis C, Voss S, Founti M A and Trimis D 2015 *Proc. Combust. Inst.* **35** 3381–8
[7] Loukou A, Mendes M A A, Frenzel I, Pereira J M C, Ray S, Pereira J C F and Trimis D 2017 *Int. J. Hydrogen Energy* **42** 652–63
[8] Dobrego K V, Koznacheev I A and Shmelev E S 2008 *J. Eng. Phys. Thermophy.* **81** 456–64
[9] Sahraoui M and Kaviany M 1994 *Int. J. Heat Mass Transfer* **37** 2817–34
[10] Sirotkin F, Fursenko R, Kumar S and Minaev S 2017 *Proc. Combust. Inst.* **36** 4383–9
[11] Shi J, Xiao H, Li J, Li N, Xia Y and Xu Y 2017 *Combust. Sci. Technol.* **189** 1260–76
[12] Hackert C L, Ellzey J L and Ezekoye O A 1999 *Combust. Flame* **116** 177–91
[13] Jouybari N F, Maerefat M and Nimvari M E 2016 *Heat Mass Transfer* **52** 269–80
[14] Bedoya C, Dinkov I, Habisreuther P, Zarzalis N, Bockhorn H and Parthasarathy P 2015 *Combust. Flame* **162** 3740–54
[15] Benenati R F and Brosilow C B 1962 *AIChE J.* **8** 359–61
[16] Partopour B and Dixon A G 2017 *Powder Technol.* **322** 258–72
[17] Barra A J and Ellzey J L 2004 *Combust. Flame* **137** 230–41
[18] Modest M F 2013 *Radiative heat transfer. 3rd ed.* (New York: Academic Press)
[19] Kennedy L A, Bingue J P, Saveliev A V, Fridman A A and Foutko S I 2000 *Proc. Combust. Inst.* **28** 1431–8
[20] He L and Fadl M 2017 *Int. J. Numer. Methods Fluids* **83** 887–904

5th International Workshop on Heat/Mass Transfer Advances
for Energy Conservation and Pollution Control
August 13-16, 2019, Novosibirsk, Russia

# Heat transfer and dynamics of transient processes at liquid film flowing on smooth and modified surfaces

**A N Pavlenko[1], D V Kuznetsov[1], O A Volodin[1] and N N Zubkov[2]**

[1] Kutateladze Institute of Thermophysics, Acad. Lavrentiev ave., 1, Novosibirsk, Russia

[2] Bauman Moscow State Technical University, Baumanskaya 2-ya, 5, Moscow, Russia

E-mail: kuznetsov_9308@mail.ru

**Abstract.** The paper presents the results of an experimental study on propagation dynamics and structure of the wetting front under adiabatic conditions and on a preheated surfaces with a needle-like microstructure, as well as heat transfer and critical heat fluxes at a film flow of liquid nitrogen. The surface was structured by the method of deforming cutting (MDC). A comparison with the data obtained on a smooth plate was performed. Data of high-speed video of the processes are presented.

## 1. Introduction

As it is known, the film flows, being an effective method of heat and mass transfer, are widely used in various technological processes such as absorption, rectification, evaporation, cooling, etc. To date, there are quite a large number of works devoted to the theoretical and experimental study of wave flow hydrodynamics, heat transfer of evaporating and boiling liquid films under the conditions of stationary heat release. Recently, the increased interest of researchers is focused on the ways to intensify heat transfer and increase critical heat fluxes. The most effective methods for these tasks are the methods for creating modified heat transfer surfaces: various micro and nano-coatings, porous and structured surfaces, etc. [1].

The authors of this paper have carried out a series of research to study the effect of various coatings on rewetting dynamics of strongly overheated plates, as well as heat transfer under stationary heat release during film cooling by liquid nitrogen. Thus, it was shown in [2] that the presence of a low-heat-conductive coating with a thickness of 250 μm has a significant effect on the character of temperature curves and reduces the total time of plate cooling from the maximum temperature to the saturation temperature of liquid nitrogen by more than four times as compared to an uncoated heater. Based on a comparison of experimental data, it has been found that the presence of a low-heat-conductive coating leads to the formation of a cellular front structure with a regular boiling jet, typical of rewetting of superheated thin-walled plates. The results of an experimental study of the effect of a structured capillary-porous coating obtained by the method of directional plasma spraying on heat transfer at a film flow of liquid nitrogen are presented in [3, 4]. It was shown that the presence of such coating also leads to a decrease in time of complete cooling of the overheated working section by more than three times; however, it slightly affects heat transfer under the conditions of stationary heat release in the studied range of irrigation. Based on the analysis of synchronized measurements of the plate temperature and high-speed video recording of transient processes, the authors of [3, 4] showed that a higher cooling rate in the presence of coatings is associated with the development of intense boiling in the rewetting front at a significantly higher plate temperature.

Another method for making the modified surfaces that has proven itself in relation to the tasks of heat transfer enhancement and increasing critical heat fluxes under the conditions of film flow of liquids is the method of deforming cutting (MDC). For example, experimental results on heat transfer in a film flow of a binary mixture of R114/R21 freons on vertical cylinders with microstructures, obtained using the above method, are presented in [5, 6]. The authors have shown that the heat transfer coefficients at nucleate boiling on such surfaces depend on the characteristic parameters of the structure and can exceed the corresponding values for a smooth heater up to three times. In addition, a significant increase (up to 2 times) in the critical heat fluxes under the conditions of stationary heat release was revealed for the modified samples.

Based on the analysis of these works, it is of direct interest to study the effect of structured surfaces, obtained by the MDC, on wetting transients, as well as the effect of characteristic parameters of these coatings on heat transfer in flowing liquid films. Thus, the goal of this work is an experimental study of propagation dynamics and structure of the wetting front under the adiabatic conditions and a preheated surfaces with a needle MDC-microstructure, as well as heat transfer and critical heat fluxes at film flow of liquid nitrogen.

## 2. Experimental setup and techniques

To perform the experiments with film flow of liquid nitrogen on the saturation line under the atmospheric pressure, an experimental setup was used: a cryogenic circuit, whose detailed description is presented in [3, 7]. The film Reynolds number $Re = 4\Gamma/\nu$ in experiments varied in the range from 200 to 1500 (here: $\Gamma$ is irrigation density, $\nu$ is kinematic viscosity of liquid). The processes were recorded through the optical windows of cryostat using a high-speed digital video camera Phantom v.7.0.

A smooth copper plate and a plate with micro-needle fins, obtained using MDC, were used as the working sections in experiments. This method of mechanical processing is based on cutting the surface layer of the workpiece material and subsequent deformation of the cut layer with the formation of macro- and micro-relief in the form of edges, spikes, and cells. The description of this technology of micro-surface treatment is presented in [8–10]. The thickness of the heated copper plate was 2.5 mm, the height and width were 50 and 75 mm, respectively. The characteristic parameters of the investigated coating are as follows: the needle microstructure height is 500 μm; the step of the micro-pins alternation is 70 μm; the characteristic transverse size of micro-pins is 70 μm.

The working sections were heated by passing an electric current through a constantan foil pressed to the backside of the copper plate without structuring, insulated by a thin dielectric gasket using a controlled power source. The plate temperature was measured by four copper-constantan thermocouples. The control of current sources, measurement of signals of thermocouples and liquid level gauges that determine the flow rate were carried out using two ADC/DAC boards and the LGraph program of experiment control. For experiments under the conditions of stationary heat release, the heat flux density was determined by the values of current passed through a constantan foil and potential difference between the terminals located at its ends.

To perform the experiments on the study of rewetting dynamics after the supply of liquid nitrogen with a given flow rate, the flow lock was turned, and the flow of nitrogen along the working section was stopped. While, the flow rate of liquid nitrogen, output by the lock, was maintained. Then, the plate was heated by a constantan foil to a predetermined maximum (initial) temperature $T_{max}$. For each Re number, the experiments were carried out at different maximum temperatures (105, 150, 215, and 300 K). At the next stage, the heater was turned off and simultaneously the lock went to its original position. After that, the uniform film flow of liquid nitrogen was renewed and the process of working section rewetting was carried out. The experiments under adiabatic conditions (without plate heating) also were carried out synchronizing video recording and restoration patterns of the liquid flow.

## 3. Experimental results and discussion

The frames of high-speed video recording of the wetting of the microstructured working section with a liquid nitrogen film under adiabatic conditions at Re = 1250 are presented in figure 1. Dashed lines indicate the boundaries of the wetted surface. As it can be seen, the wetting front is not flat, but it is formed by the separate liquid jets and inter-jet zones. Such a dynamics of wetting front formation under the adiabatic conditions is associated with the development of instability when the large waves in a liquid nitrogen film surge towards the surface of the studied plate. According to the analysis of experimental data, the number of individual rivulets (jets) increases with increasing degree of irrigation. A similar picture of the process development was observed for a smooth working section. At the same time, the structure, formed on the heater surface, leads to additional distribution of liquid due to the capillary forces in the transverse direction relative to the nitrogen flow and, as a result, to transverse expansion of the individual rivulets. This effect is most pronounced for high values of the film Reynolds number.

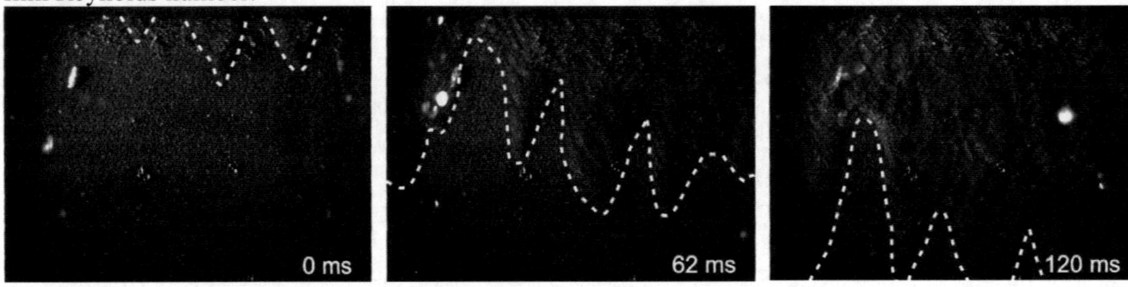


**Figure 1.** The rewetting front under the adiabatic conditions. Re = 1250.

Based on the processing of experimental data of high-speed video recording, the propagation velocities of individual rivulets, inter-rivulet zones, as well as the propagation velocity of large waves in a liquid nitrogen film after full wetting of the working section at various Reynolds numbers were obtained. As it follows from the analysis (figure 2), the average propagation velocities of liquid rivulets and inter-rivulet zones over the unwetted surface with a fixed Re number almost coincide and they are significantly below the values of velocity of large wave propagation, which is caused by the influence of a dynamic wetting angle between the oncoming film of liquid nitrogen and plate surface. It is also worth noting that for a modified working section the velocity of jets propagation is less than the similar values for a smooth heater at high Re numbers. The above-described expansion of the transverse size of rivulets on the surface of a structured heater leads to a decrease in local liquid flow and, as a consequence, to a decrease in the velocity of the wetting front in the zone of rivulet propagation.

The time dependences of the smooth plate temperature and temperature of a plate with microstructure in the process of rewetting by a flowing nitrogen film at different Reynolds numbers and fixed maximal initial temperature $T_{max}$ are compared in figure 3. As one can see, there are two main characteristic stages of cooling the highly overheated plate: the first stage with slow cooling of the working section and the second stage with a fast rate of cooling, the beginning of which is accompanied by a bend in the temperature curve, reflecting a change in the heat transfer regime during rewetting front propagation. The characteristic temperatures at the bend points correspond to the plate threshold temperatures when the transition from low-intensity heat transfer on the dried surface to its wetting and development of heat transfer in a boiling regime in the process of front propagation with liquid film fall begins. While turning to the second fast stage, the heater temperature, when the front begins to spread rapidly across the plate, depends weakly on both the degree of irrigation throughout the entire range of Reynolds numbers and initial temperature. In addition, there is a decrease in time of complete heater cooling with increasing degree of film irrigation for all studied plate overheating. This structure on the surface of the working section has a little effect (as compared with the results obtained in [3, 4] when using capillary-porous coatings) on the time of complete plate cooling. However, for a fixed degree of irrigation, the curve zone corresponding to the second stage of cooling for a smooth heater has a steeper slope than for a modified sample, indicating a decrease in the velocity of rewetting front propagation in the presence of a microstructure. This effect can be explained by partial steaming of the lower layers of the micro-needles structure (micro-needles zones adjacent to the plate) in the process of rewetting and by an increase in thermal resistance.

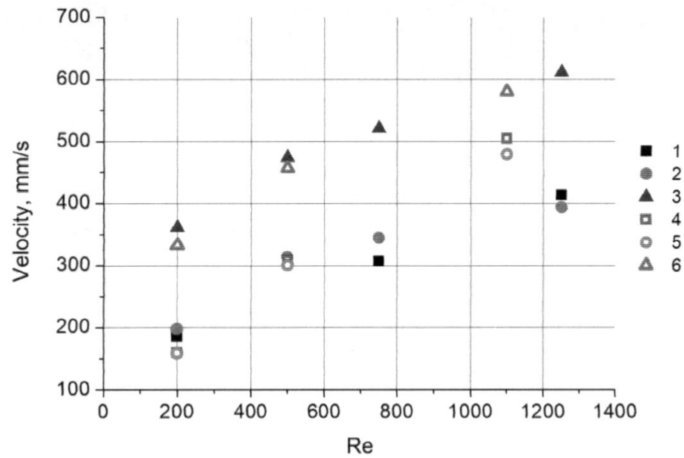


**Figure 2.** The average propagation velocities of characteristic zones of wetting front. Structured heater: 1 – rivulets, 2 – inter-rivulet zones, 3 – waves on a wetted surface. Smooth heater: 4 – rivulets, 5 – inter-rivulet zones, 6 – waves on the wetted surface.

According to the analysis of high-speed video data (figure 4), the rewetting front for both a smooth and structured heater is quasi-flat and differs little in the form and structure over the entire range of the studied Reynolds numbers. The velocity of the rewetting front propagation, average over the plate height, at the second stage of the process increases with an increase in the degree of irrigation and does not depend on the initial maximal temperature (see table). It can be seen that for a smooth heater the velocity is almost twice as high as the similar values for the modified working section at a fixed Re number.

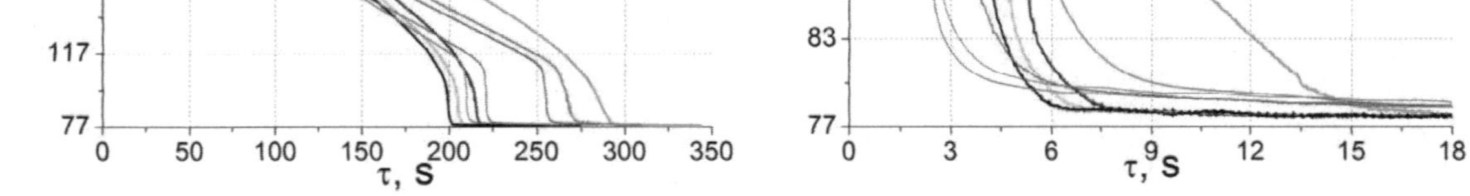


**Figure 3.** Dependence of plate temperature on time for different Reynolds numbers at initial temperature: (a) – 300, (b) – $T_{max}$ = 105 K. Structured heater: 1 – 200, 2 – 500, 3 – 750, 4 – Re = 1250; smooth heater: 5 – 200, 6 – 500, 7 – 750, 8 – Re = 1100.

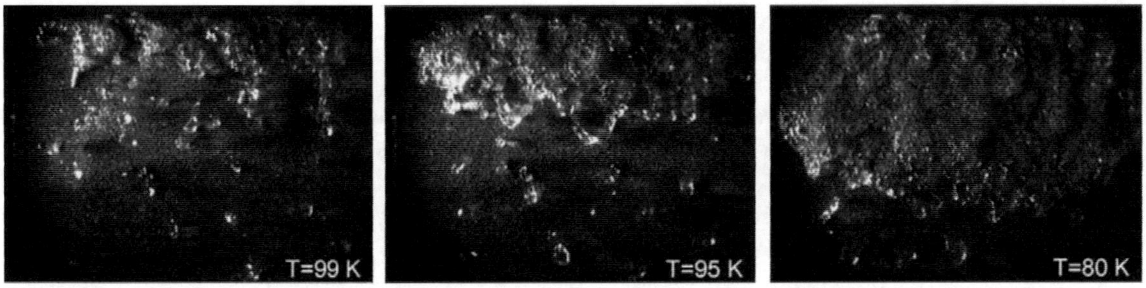


**Figure 4.** The process of rewetting of the structured plate at Re = 1250 and initial temperature $T_{max}$ = 300K.

**Table**. Average velocities of the rewetting front (mm/s).

| **Re** | **200** | **500** | **750** | **1250(1100)** | **1500** |
|---|---|---|---|---|---|
| Structured heater | 5 | 8 | 9 | 11 | - |
| Smooth heater | 8 | 14 | 16 | 25 | 35 |

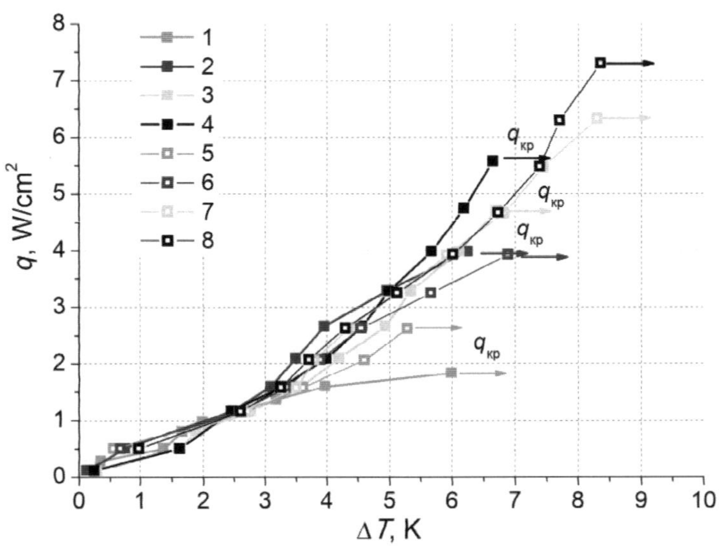


**Figure 5.** Heat transfer curves at stationary heat release. Structured heater: 1 – 200, 2 – 500, 3 – 750, 4 – Re = 1250; smooth heater: 5 – 200, 6 – 500, 7 – 750, 8 – Re = 1100.

The results of an experimental study of heat transfer intensity and critical heat fluxes for a smooth plate heater and heater with micro-needle finning are presented in figure 5 depending on the degree of irrigation with a liquid nitrogen film under the conditions of stationary heat release. This structure on the copper plate surface does not lead to a significant change in the heat transfer coefficients as compared to a smooth plate in the entire range of irrigation levels. An exception is Re = 200 when the heat transfer coefficients for the modified sample are significantly reduced in comparison with a smooth plate in the pre-crisis boiling regimes. Also, a decrease in the critical heat flux is observed at Re = 750 and 1250 with respect to a smooth heater.

## 4. Conclusions

The experimental results on propagation dynamics and structure of the wetting front under adiabatic conditions and on a preheated surface with a needle MDC-microstructure, as well as heat transfer and critical heat fluxes in the film flow of liquid nitrogen are presented. It is shown that the presence of a MDC structure on the heater surface leads to a change in the character of the wetting front propagation of an unheated plate for high Reynolds numbers, reducing the velocity of liquid jets due to a decrease in the local flow of the film. In the rewetting regimes, the average propagation velocity of the front along the heater height is halved for a modified sample as compared to a smooth heater. However, the time for complete cooling of a coated plate does not change significantly over the entire range of studied Re numbers. Experiments under the conditions of stationary heat release showed the absence of a significant effect of micro-needle finning on heat transfer in a film flow of liquid under the boiling regime. At the same time, a certain decrease in the critical heat flux at boiling on a heater with micro-needle finning is shown.

**Acknowledgements**
The study was performed at Kutateladze Institute of Thermophysics (IT SB RAS) with the support of the BSI SAS Program for 2017-2020 (project III.18.2.3, reg. no. AAAA-17-117030310025-3) and Russian Foundation for Basic Research (RFBR) (project no. 18-08-00402-a).

**References**
[1] Surtaev A S, Serdyukov V S and Pavlenko A N 2016 *Nanotech. In Russia* **11** (11-12) 696
[2] Pavlenko A N, Tsoi A N, Surtaev A S, Kuznetsov D V and Serdyukov V S 2016 *High Temp.* **54** (3) 370
[3] Pavlenko A N, Tsoi A N, Surtaev A S, Kuznetsov D V, Kalita V I, Komlev D I, Ivannikov A Yu, Radyak A A 2018 *High Temp.* **56** (3) 404
[4] Pavlenko A N and Kuznetsov D V 2018 *J. Phys.: Conf. Ser.* **1105** (1) 012053
[5] Volodin O A, Pecherkin N I, Pavlenko A N and Zubkov N N 2017 *Interf. Phenomena and Heat Transf.* **5** 215
[6] Volodin O A, Pecherkin N I, Pavlenko A N, Zubkov N I and Bityutskaya Yu L 2017 *J. Phys.: Conf. Ser.* **891** 012035
[7] Pavlenko A N, Surtaev A S, Tsoi A N, Starodubtseva I P and Serdyukov V S 2014 *High Temp.* **56** (6) 861
[8] Zubkov N N, Ovchinnikov A I 1996 European Patent EP 0727269
[9] Thors P and Zoubkov N 2007 U.S. Patent 7311137
[10] Thors P and Zoubkov N 2013 U.S. Patent 8573022

Journal of Physics: Conference Series **1369** (2019) 012053 doi:10.1088/1742-6596/1369/1/012053

5th International Workshop on Heat/Mass Transfer Advances
for Energy Conservation and Pollution Control
August 13-16, 2019, Novosibirsk, Russia

# Experimental study of heat transfer in a heat pipe

**V V Cheverda, F V Ronshin**

Institute of Thermophysics SB RAS, Novosibirsk, Russia
Novosibirsk State University, Novosibirsk, Russia

slava.cheverda@gmail.com

Heat pipes are a promising cooling system due to the fact of high efficiency and non-availability of moving parts such as pumps. The experimental investigation of heat transfer in a flat heat pipe has been carried out. The temperature field on the heat pipe has been evaluated using an infrared camera and K-type thermocouples. The IR-camera data and K-type thermocouple data have been compared. The linear dependence of the power supplied to the evaporator on the temperature difference between the evaporator and the condenser is shown.

## 1. Introduction

One of the most important research topics in our days, which will remain urgent for a long time, is the problem of efficient heat transfer. The use of heat pipes is a promising direction in the fields of solar energy, geothermal energy, aerospace industry, as well as cooling electronic components due to the high thermal conductivity.

Heat pipes, as an invention, were created relatively recently, several decades ago, but despite this, they gained quite a lot of popularity. So, for example, when creating an electronic device, be it a personal computer, laptop, or mobile phone, one of the most important engineering tasks is the development of the cooling system for this device. Currently, to solve the problem of cooling microelectronics, heat pipes are widely used. Simplicity, long service life and the absence of moving parts, allows this type of device to serve humanity for decades.

Heat pipes are used in the space industry for cooling aircraft components, in atomic energy for thermal stabilization of nuclear reactor elements, and in everyday life for cooling microprocessors [1]. These applications of heat pipes can be expanded and complimented, but it can be argued that the practical application of heat pipes in industry and in everyday life will only increase [2].

The overview of heat pipe studies is presented in [3]. For such a short period of time, extensive tests and studies were carried out, and theoretical models were constructed describing the operation of heat pipes. The efficiency of the heat pipe was investigated depending on the wall and wick materials used, as well as the working fluid. The analysis of the compatibility of materials and working fluids is made, the design is improved. Optimal parameters and materials have been developed for specific temperature ranges and long service life. Due to the urgency of the problem, an increase in the number of publications on this topic is noticeable in the literature. Some authors attempt to summarize the available data and build a theoretical model that fully describes the physical processes inside a heat pipe.

The term "heat pipe" was proposed in the patent by Grover G.M. on behalf of the United States Atomic Energy Commission in 1963. This patent includes a theoretical analysis of the ongoing thermal processes. Under the leadership of Grover, a series of research and development was carried out to create heat pipes. The first article on these studies was published in [4]. Grover's patent was not the first on this kind of device. Earlier, a similar development was proposed by an employee of General Motors Corporation, Gogler in 1942. The documents stated the following purpose of the invention: "... ensuring the absorption of heat, or in other words, the evaporation of a liquid at a point above the condensation area or the heat removal zone, without additional costs for raising the liquid from the level of the condenser". Later, the properties of such devices were evaluated, and in the future, scientists continued to study heat pipes. In the laboratory of Atomic Energy in Harwell (UK), researchers have been working on sodium heat pipes. This work was aimed at the use of heat pipes in nuclear power. Such studies were conducted under the direction of Nei and Busse at the Joint Nuclear Research Center in Ispra (Italy), which started in 1965. This laboratory was one of the most active in the study of heat pipes. In 1969 the first article devoted to the review of heat pipes in the USSR was published [5].

The heat pipe was first tested on an artificial Earth satellite in 1967 (case - stainless steel, working fluid - water, electrical heating), and in 1968 - used for the first time for thermoregulation of the Geos - B satellite (two heat pipes, case - aluminum alloy, wick - aluminum mesh, working fluid - freon- 11; used for temperature control of the equipment on board the satellite). Later, the practical application of heat pipes in the industry, from aviation, radio electronics to household appliances and medicine, was significantly expanded. This is due to the widespread use of fuel or heat consuming devices with multi-stage energy conversion processes. Commercial production of heat pipes began with RCA, from 1964. The company used the following materials to produce heat pipes: glass, copper, nickel, stainless steel, molybdenum. The following fluids were used as working fluids: water, cesium, sodium, lithium, and bismuth.

## 2. Experimental setup

In the present work, an experimental study of heat transfer in a flat heat pipe has been carried out. Heat pipe length 220 mm long with a porous filling was selected for the study. The cross section of the pipe is shown in Fig. 1. The thickness of the copper wall of the heat pipe is 0.33 mm. Next is a layer of porous filler (copper powder). The thickness of this layer is 1.4 mm and 0.9 mm along the long and short sides of the heat pipe, respectively. In the center of the heat pipe, there is a gas chamber with characteristic dimensions of 4.8 mm and 0.84 mm along the long and short sides of the heat pipe, respectively.

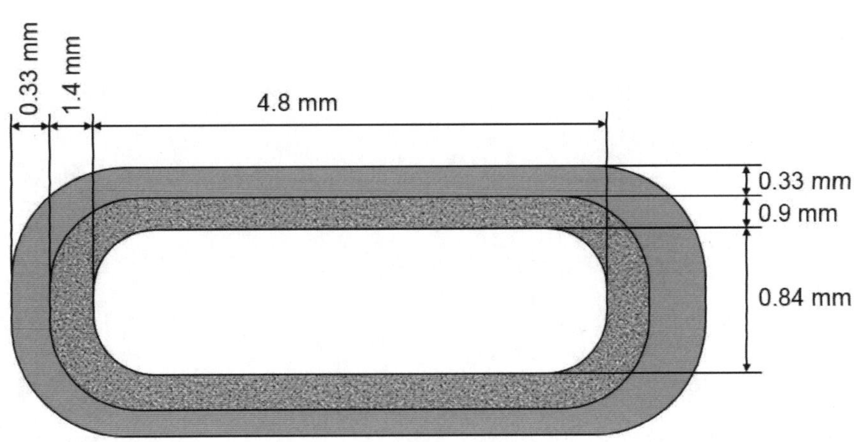


Fig. 1. Heat pipe cross-section.

The experimental setup is shown in Fig. 2. A heater is wound on the evaporator and is connected to a TTi QPX 1200L power supply. The condenser is cooled by a liquid heat exchanger into which water enters from the LOIP FT-211-25 thermostat. The condenser temperature was kept constant at 20°C.

Temperature is measured by thermocouples in the area of the evaporator, condenser and isothermal part. For accurate calculation of temperatures, 6 K-type thermocouples were used, one for each in the part of the evaporator and one for the condenser, and four for the isothermal part. The temperature field on the heat pipe was investigated using an infrared camera Titanium HD 570M.

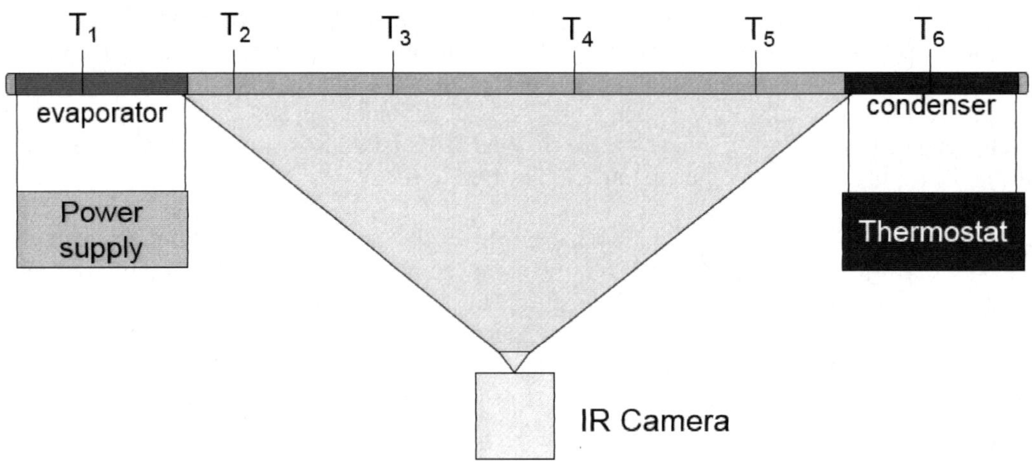


Fig. 2. Schematic diagram of the experimental setup.

## 3. Experimental results

The heat exchange of a heat pipe under various heat flows has been experimentally investigated. Figure 3 shows a comparison of data obtained from thermocouples with an infrared camera. It is seen that in the area of the evaporator and condenser, the temperatures are different.

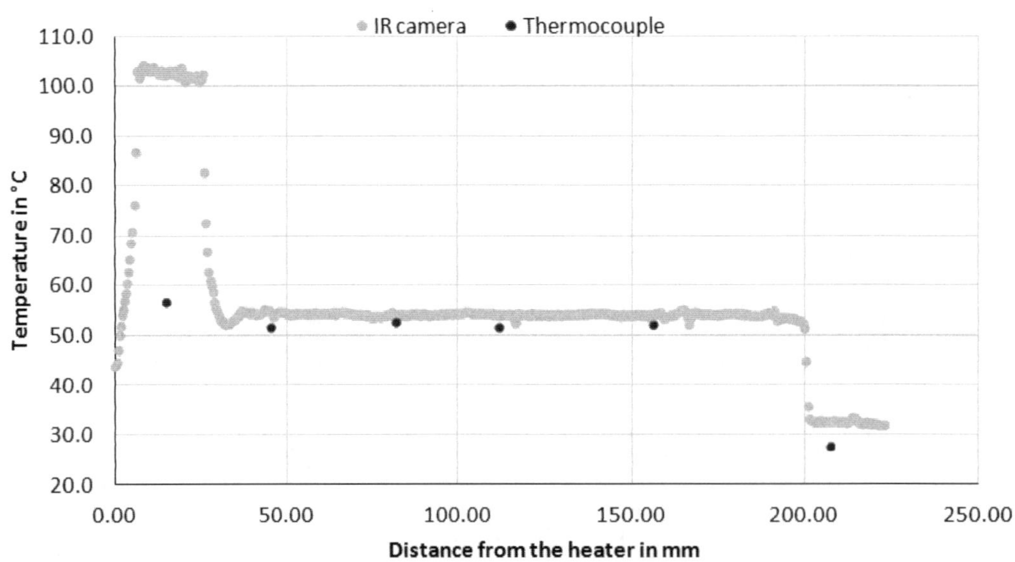


Fig. 3. Temperature field along the heat pipe at P = 20 W, measured using an IR camera and K-type thermocouples.

This is explained by the fact that in the area of the evaporator the infrared camera measures the temperature of the heater, and in the area of the condenser, it measures the temperature of the heat exchanger installed on the pipe. Thermocouples always show the temperature of the heat pipe itself. It is also seen that the temperature along the heat pipe is constant and slightly lower than the temperature in the area of the evaporator. Figure 4 shows the dependence of the power supplied to the evaporator on the temperature difference between the evaporator and the condenser. It can be seen that the power increases monotonically with increasing temperature differences.

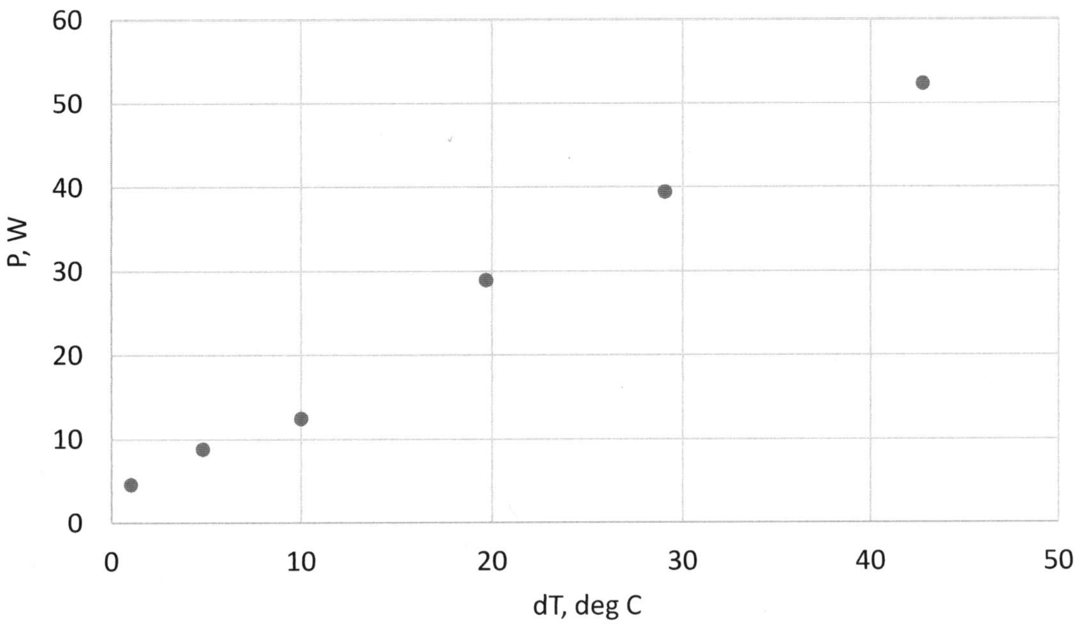


Fig. 4. The dependence of the power supplied to the evaporator on the temperature difference between the evaporator and the condenser.

## 4. Conclusion

In this work, an experimental study of heat transfer in a heat pipe with a porous filler was carried out using two independent methods, namely, by means of an IR scanner and K-type thermocouples. An almost linear dependence of the heat transfer capacity for a heat pipe with a porous filler has been established.

This work was performed in the Kutateladze Institute of Thermophysics SB RAS by a state contract AAAA-A19-119083090009-6.

## References

[1] Sundén B, Fu J, 2017 Heat Pipes for Aerospace Application. Heat Transfer in Aerospace Applications, 117.
[2] Faghri A, 2016 Heat pipe science and technology, 2nd edn. Global Digital Press, Columbia, Missourie
[3] Lips S, Sartre V, Lefevre F, Khandekar S, Bonjour J, 2016 *Interfacial Phenomena and Heat Transfer* **4(1)** 33
[4] Grover G M, Cotter T P, Frikson G F, 1964 *J. Appl. Phys.* **35 (6)** 1990
[5] Moșkvin Yu V, Filippov Yu A, 1969 *Teplofizika Vysokikh Temperatur* **7(4)** 766

5th International Workshop on Heat/Mass Transfer Advances
for Energy Conservation and Pollution Control
August 13-16, 2019, Novosibirsk, Russia

# Heat pipe: a simple one-dimensional model and an experimental study

**V V Cheverda, I S Vozhakov, F V Ronshin**

Institute of Thermophysics SB RAS, Novosibirsk, Russia
Novosibirsk State University, Novosibirsk, Russia

slava.cheverda@gmail.com

In this work the authors proposed to use 1-D model for heat and masstransfer calculations inside a porous heat pipe. Several limits are taken in to account: sound, boiling, capillary and dry out of evaporator. To calculate the heat flow, we analyzed processes in four zones: the evaporator, the condenser, the porous body in which fluid flows, and the cavity in which steam flows. We assume that in the evaporator all the incoming liquid evaporates, and in the condenser, all the vapor condenses. The liquid and the vapor are in a state of saturation and, therefore, the pressure in the condenser and the evaporator is equal to the saturation pressure. The good agreement between theoretical and experimental results is observed.

## 1. Introduction

A heat pipe is a passive device that can dissipate a large amount of heat due to evaporation and condensation of the working fluid. The working fluid is evaporating in the heater part. Due to the pressure difference created, the steam is supplied to the condenser, where it releases heat and changes its phase to liquid. The liquid is returned to the evaporator zone using the wick capillary force. Since the latent heat of evaporation is usually large, heat pipes can efficiently transfer a significant amount of heat over a long distance with a relatively small temperature drop. The choice of working fluid depends on the working temperature of the heat pipes. Because of the wide range of operating temperatures, heat pipes can be used in many areas, such as the accumulation of thermal energy, the cooling of electronic and electrical equipment [1, 2], heat exchangers and heat pumps [3, 4].

The phenomenon of liquid penetration into a porous medium under the action of capillary forces is widely used, including microfluidic devices [5, 6], fuel cells [7, 8], inkjet printing [9] and much more. Such fluid movement in a porous medium is due to negative capillary pressure. To create a model, a porous medium can be simplified to a hollow tube with an effective capillary radius. The first description of the dynamics of capillary movement in the pipe was made by Lukas [10] and Washburn [11] at the beginning of the last century. They suggested that the distance of penetration of a liquid into a porous medium $L$ and the time of penetration $t$ satisfy the diffusion relation $L=\sqrt{Dt}$ , where $D$ is the diffusion coefficient, depending on the size of the pipe and the properties of the liquid. It was also established that the diffusion correlation between the distance and the time of fluid penetration in porous media corresponds to the Lukas-Washburn law for both unidirectional [12, 13] and radial penetration [14, 15, 16, 17].

In the present work, one-dimensional analysis of the steady state of a heat pipe was performed. To calculate the maximum heat flux, it is necessary to consider the main constraints that impede its growth. First of all, these are capillary restrictions, sound effects, liquid entrainment from the surface of the wick and boiling up of the coolant. These restrictions, in turn, depend on the various properties of the coolant, the structure of the wick and the geometry of the heat pipe.

## 2. Theory description

A detailed theoretical and experimental interpretation of the sound limit is presented in the works of Levy [18], Kemme [19] and Deverall [20]. In a heat pipe with a constant diameter of the vapor channel, the flow is accelerated and slowed down due to the supply of steam in the evaporator and drainage in the condenser. The sound limit is reached when the steam velocity at the evaporator outlet becomes sound. The description of the sound limit can be obtained from the one-dimensional theory of steam flow, in which it is assumed that the properties of steam obey the ideal gas law, inertial effects are predominant, friction effects can be neglected. Then the sound limit of the heat pipe can be expressed as follows [18]:

$$Q_{s,\max} = A_v \rho_v \lambda \left[ \frac{\gamma_v R_v T_v}{2(\gamma_v + 1)} \right]^{1/2} \tag{1}$$

Here $A_v$ is the cross-sectional area of the steam channel, $\rho_v$ is the vapor density, $\lambda$ is the latent heat of vaporization, $R_v$ is the vapor constant, $T_v$ is the vapor temperature.

Since in the heat pipe the liquid and vapor move in opposite directions, a shearing force arises at the interface. If the vapor velocity is high enough, then the limit can be reached when the liquid will break away from the surface of the wick and be carried away by the flow of steam. The occurrence of this phenomenon leads to instant drying of the wick in the evaporation zone. The phenomenon of liquid entrainment was detected by the sound emitted by liquid droplets hitting the end of the condenser, and the evaporator overheating sharply [21]. The maximum heat flux, taking into account the restriction on the entrainment of liquid is expressed as follows [22]:

$$Q_{e,\max} = A_v \lambda \left( \frac{\sigma \rho_v}{2 r_h} \right) \tag{2}$$

Here $\sigma$ is the coefficient of surface tension, $r_h$ is the hydraulic radius of the pore surface of the wick.

The main mechanism of heat transfer in the evaporator and the heat pipe condenser is thermal conductivity with evaporation and condensation. The passage of heat through a wick saturated with liquid is accompanied by the appearance of a radial temperature gradient in the liquid. With an increase in heat flow in the wick, vapour bubbles may begin to form. Their appearance is undesirable because they lead to the occurrence of overheated areas and obstructing the circulation of fluid. The heat flow restriction associated with vaporization in a heat pipe is called boiling restriction. There is a fundamental difference between this restriction and the rest, since here the restriction is imposed on the radial heat flux, and in the rest on the axial one. The expression for limiting boiling power is:

$$Q_{b,\max} = \frac{2\pi L_e k_e T_v}{\lambda \rho_v \ln(r_i / r_v)} \frac{2\sigma}{r_n} \tag{3}$$

Here $L_e$ is the length of the evaporation zone, $k_e$ is the effective thermal conductivity of the wick, $r_i$ is the inner radius of the tube envelope, $r_v$ is the radius of the vapor channel, $r_n$ is the critical radius of the nucleus.

In the course of steady-state operation of the heat pipe, the coolant flows as a vapor phase from the evaporation zone to the condensation zone and returns to the evaporator in the form of a liquid phase. Since steam moves from the evaporator to the condenser, there is a pressure gradient along the vapor channel in the steam flow. There is also a pressure gradient in the fluid, which drives it back from the condenser to the evaporator. For the required pressure balance, it is necessary that the pressure from the liquid at the liquid-vapor interface over the entire length of the pipe is different from the pressure from the vapor, except for the point where this difference is minimal and equal to zero. This pressure difference between the phases is called capillary pressure and is created by the surface tension and the porous structure of the wick. There is a maximum capillary pressure that can be achieved with the selected coolant-wick pair. If the heat pipe is operating continuously without drying the wick, the working capillary pressure should not exceed its maximum value. This restriction, which is superimposed on the transferred thermal power, is known as a capillary restriction:

$$Q_{c,\max} = \frac{(QL)_{c,\max}}{0.5L_c + L_a + 0.5L_e} \tag{4}$$

Here $(QL)_{c,\max} = \frac{2\sigma / r_c}{F_l + F_v}$ is the capillary restriction of the transmitted power factor, $F_l = \frac{\mu_l}{KA_w \rho_l \lambda}$ is the liquid friction coefficient, $F_v = \frac{(f_v \mathrm{Re}_v)\mu_v}{2A_v r_h^2 \rho_v \lambda}$ is the vapor friction coefficient, $r_c$ is the capillary radius, $L_c$ is the condenser length, $L_a$ is the transport zone length, $(f_v \mathrm{Re}_v)$ is the hydraulic resistance coefficient, $\mu_v$ is the vapor viscosity, $\mu_l$ is the liquid viscosity, $K$ is the wick permeability, $A_w$ is the area cross section wick.

### *2.1. Simple 1d model*

To calculate the heat flow, we will analyze the processes in each zone of the heat pipe. We distinguish four such zones: the evaporator, the condenser, the porous body in which fluid flows, and the cavity in which steam flows. To estimate the speed of steam, we use the well-known dependency:

$$\Delta p = \lambda \frac{L}{D_h} \frac{\rho_v u_v^2}{2} \tag{5}$$

In the case of laminar motion, the drag coefficient:

$$\lambda = \frac{64}{\mathrm{Re}_v} = \frac{64\mu_v}{\rho_v u_v D_h} \tag{6}$$

For the turbulent case:

$$\lambda = \frac{0,316}{\mathrm{Re}_v^{0,25}} \tag{7}$$

Substituting (6) or (7) into (5), we find the stationary velocity of the vapor.

To estimate the fluid velocity in a porous body, we turn to the theory of Lukas-Washburn:

$$\frac{dl}{dt} = \frac{2\sigma\cos\theta / r - (\mathrm{P}_1 - \mathrm{P}_2)}{8l\mu}\left(r^2 + 4\varepsilon r\right) \tag{8}$$

$$\frac{l^2}{t} = \frac{2\sigma\cos\theta / r - (\mathrm{P}_1 - \mathrm{P}_2)}{8\mu}\left(r^2 + 4\varepsilon r\right) = C \tag{9}$$

$$l = \sqrt{C \cdot t} \tag{10}$$

$$\frac{dl}{dt} = u_l = \frac{\sqrt{C}}{2\sqrt{t}} \tag{11}$$

$$t_f = \frac{L^2}{C} \tag{12}$$

$$u_l = \frac{C}{2L} \tag{13}$$

We assume that in the evaporator all the incoming liquid evaporates, and in the condenser, all the vapor condenses. In this case, the liquid and vapor are in a state of saturation and, therefore, the pressure in the condenser and the evaporator is equal to the saturation pressure.

When calculating the heat flow, it is necessary to keep track of it so that it is not larger than the maximum possible. Thus, a mathematical model is presented for calculating the maximum heat flux in a heat pipe depending on the properties of the coolant and the characteristics of the porous medium.

## 3. Experimental results

A heat pipe 22 cm long was used to study heat transfer, where the measurements of the condenser, evaporator, and transport zones are 2.5 cm; 2.2 cm and 17.3 cm, respectively. K-type thermocouples are installed in the evaporator and condenser zones, which fix the temperature. Sintered copper particles with a characteristic size of 50 μm were used as a porous medium. Figure 1 shows a schematic representation of a heat pipe and an experimental setup. The evaporator power was set by the power source to which the heater is connected. The condenser zone was cooled with a thermostat.

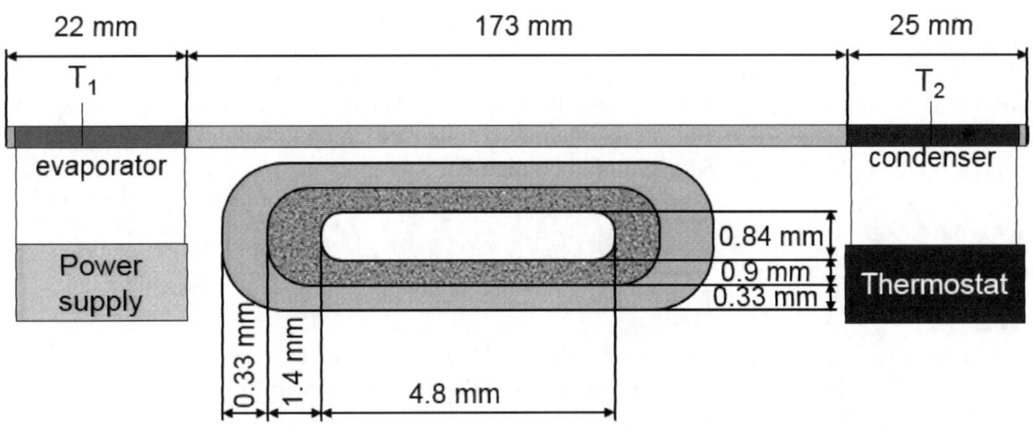


Fig. 1. Schematic representation of the heat pipe and experimental setup.

Figure 2 shows a comparison of the calculated heat flux and experimental data at the same values of the evaporator and condenser temperatures. It can be seen from the figure that as the heat flux increases, the temperature difference between the evaporator and the condenser also increases. In this case, it is possible to distinguish two zones with different types of dependence. In the first zone, with a small heat flux (up to 15 W), the derivative of the graph is larger, and the heat pipe operates without any restrictions. In this area, one can see a fairly good agreement between experimental and calculated data.

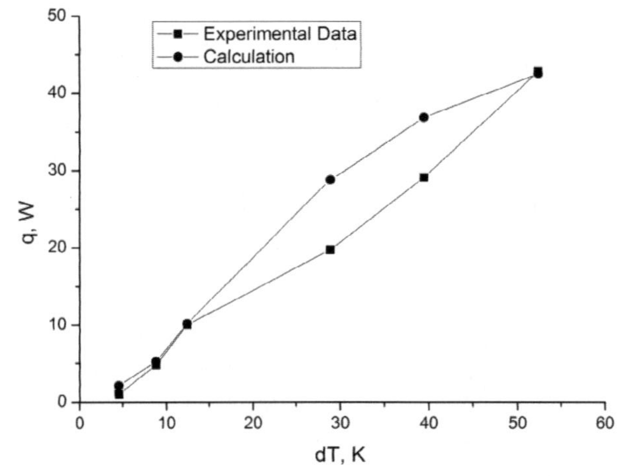


Fig. 2. Experimental and calculated dependences of the heat flux on the temperature difference between the condenser and the evaporator.

In the second part of the graph, with an increased heat flux (more than 15 W), the derivative of the graph decreases, which indicates the operation in the limiting mode. Based on the calculation data, it can be said that in this area the velocity of the steam, which reaches the sound velocity, significantly increases. However, the results of the calculation give overestimated values of the heat flux under the same conditions, which may indicate that some other restrictions that are not taken into account in the numerical model are starting to impact.

## 4. Conclusion

In this paper the 1-D model is developed for mass- and heat transfer inside the heat pipe with porous media. A comparison of theoretical and experimental results is carried out. The satisfactory agreement between theoretical and experimental results are demonstrated for heat flux versus temperature difference between evaporator and condenser.

This work was performed in the Kutateladze Institute of Thermophysics SB RAS by a state contract AAAA-A19-119083090009-6.

## References

[1] Chernysheva M A, Yushakova S I, Maydanik Y F 2014 *Energy* **69** 534
[2] Dobre T, Parvulescu O C, Stoica A, Iavorschi G 2010 *Appl. Therm. Eng.* **30** 2435
[3] Vasiliev L L 2005 *Appl. Therm. Eng.* **25** 1
[4] Jung E G, Boo J H 2014 *Appl. Energy* **135** 586
[5] Li X, Tian J F, Shen W 2010 *ACS Appl. Mater. Interfaces* **2** 1
[6] Owens T L, Leisen J, Beckham H W, Breedveld V 2011 *ACS Appl. Mater. Interfaces* **3** 3796
[7] Ge J, Schirhagl R, Zare R N 2011 *J. Chem. Educ.* **88** 1283
[8] Narvaez Villarrubia C W, Lau C, Ciniciato G P M K, Garcia S O, Sibbett S S, Petsez D N, Babanova S, Gupta G, Atanassov P, 2014 *Electrochem. Commun.* **45** 44
[9] Clarke A, Blake T D, Carruthers K, Woodward A, 2002 *Langmuir* **18** 2980
[10] Lucas R 1918 *Kolloid-Z* **23** 15
[11] Washburn E W 1921 *Phys. Rev.* **17** 273
[12] Delker T, Pengra D B, Wong P Z 1996 *Phys. Rev. Lett.* **76** 2902
[13] Lago M, Araujo M, 2001 *J. Colloid Interface Sci.* **234** 35

[14] Marmur A 1988 *J. Colloid Interface Sci.* **124** 301
[15] Borhan A, Rungta K K 1992 *J. Colloid Interface Sci.* **154** 295
[16] Danino D, Marmur A 1994 *J. Colloid Interface Sci.* **166** 245
[17] Conrath M, Fries N, Zhang M, Dreyer M E 2010 *Transp. Porous Media* **84** 109
[18] Levy E K 1968 *Journal of Engineering for Industry* **90(4)** 547
[19] Kemme J E 1969 *IEEE Transactions on Electron Devices* **16(8)** 717
[20] Deverall J E, Kemme J E, Florschuetz L W Sonic limitations and startup problems of heat pipes 1972
[21] Kemme J E 1968 HIGH-PERFORMANCE HEAT PIPES. – Los Alamos Scientific Lab., N. Mex. – №. LA-DC-9027; CONF-671045-4.
[22] Chi S W Heat pipe theory and practice. Washington, DC, Hemisphere Publishing Corp.; New York, McGraw-Hill Book Co., 1976. 256 p. 1976.

5th International Workshop on Heat/Mass Transfer Advances
for Energy Conservation and Pollution Control
August 13-16, 2019, Novosibirsk, Russia

# The heat transfer crisis in shear driven water film

**V V Cheverda, T G Ponomarenko**

Institute of Thermophysics SB RAS, Novosibirsk, Russia
Novosibirsk State University, Novosibirsk, Russia

slava.cheverda@gmail.com

In this work the authors demonstrate efficiency of thin water film, shear driven by gas flow in the mini-channel with local heater. The effect of microgrooves on heating surface is demonstrated. It is shown that microgrooves on the heating surface can provide an increase in the critical heat flux due to the capillary effect of microgrooves.

## 1. Introduction

In semiconductor technology, the size of transistors is decreasing, so density of transistors for each processor is increasing. Unfortunately, this causes higher heat fluxes and power dissipation from a microprocessor. For a microprocessor, the local heat flux could be around 1 kW/cm$^2$ [1]. A microprocessor can be cooled by different methods: by single-phase flow, as well as by the flow with phase transitions during boiling. In [2] it is demonstrated that in a gasshear-driven liquid film the critical heat flux is five times larger than that in the falling liquid film for the sameliquid flow rate. So, authors, in the papers [2, 3] proposed to use thin shear driven liquid film flow in a minichannel. In [4, 5] a stable shear-driven liquid film flow can be realized in micro- and minichannels, at different inclination angles of the setup to the vertical axis. The critical heat flux up to 1200 W/cm$^2$ from the heating zone of 10x10 mm$^2$ can be reached by means of stratified flow [6].

Aheat transfer coefficient could be increased byvarious kinds of structures on the heating surface [7]. Stabilization of the falling liquid film flow on a horizontal heated tube by using microstructures along the flow was studied in [8]. Acritical heat flux forshear-driven liquid film of FC-72 flowing inside a minichannel along the surface with micro-structures was studied in [9]. It is shown that the critical heat flux on the surfacewith microgrooves is higher incomparison with thesmooth surface. This effect is due to additional liquid flow in microstructures of the heated surface due to the capillary effect.

## 2. Experimental setup and techniques

The experiments to study critical heat flux in shear-driven water film with structured heater were carried out using the experimental setup (Fig. 1). For experiment, water and gas were supplied from the water and gas tanks. By means of gas flow controller (Bronkhorst) gas was supplied to the test cell. Using the syringe pump, degassed water was pumped to the test cell. A two-phase mixture was evacuated from the test cell through the two-phase outlet. Constant temperatures of the liquid and gas were controlled by temperature stabilization system. The program developed in the LabView system was used to measure temperature inside the test cell using K-type thermocouples.

The experimental test cell is demonstrated in Fig. 2. It consists of stainless steel and textolite plates. The minichannel is mounted by fixing textolite frame to the textolite plate, and it is covered from the top side by the transparent optical window, so a minichannel with a height of 1.5 mm and a width of 30 mm was created. Water was supplied to the minichannel through liquid nozzle (with a height of 150 microns, and width of 30 mm). The gas flow was supplied to the minichannel and forced the water film to move along the minichannel. A rod made of copper (sized 10 x 10 $mm^2$) was pressed to the stainless steel plate. A ceramic heater was attached to the copper rod. The surface of the copper rod in the minichannel could be smooth or with microgrooves (period of the structures was equal to 0.5 mm).

Before experiments with liquid film flow inside a minichannel a wetting contact angle was measured by Kruss DSA-100 system (Fig.3). The contact angle was about 90±3°.

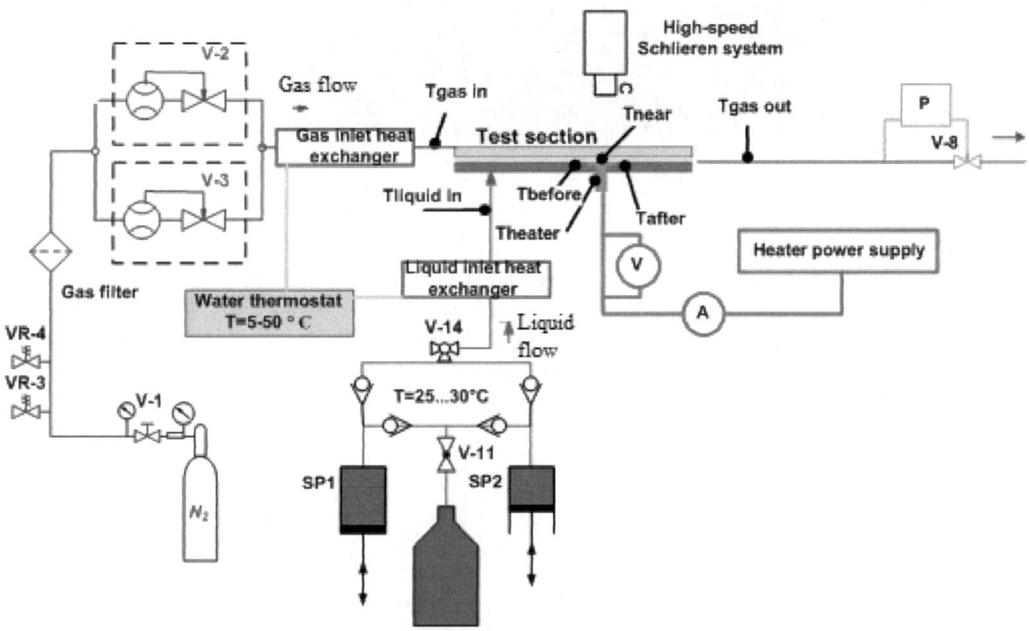


Fig. 1. Experimental setup.

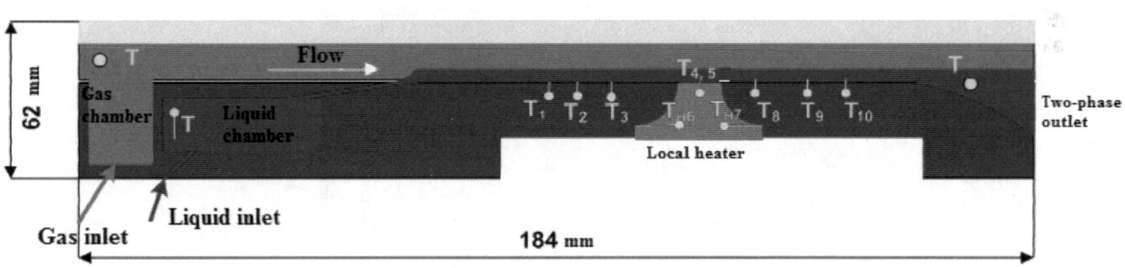


Fig. 2. Cross view of the experimental test cell with the position of thermocouples.

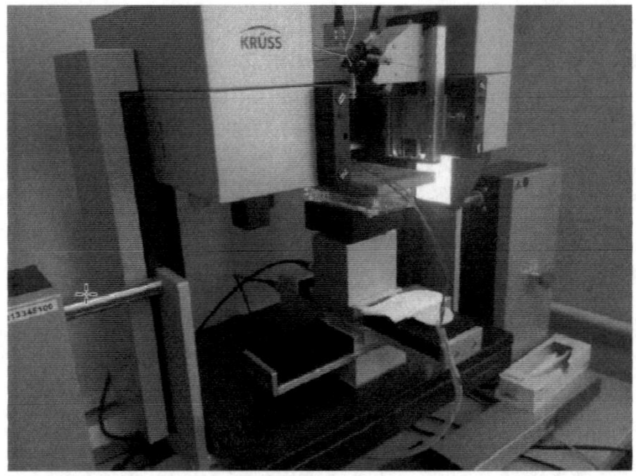


Fig. 3. Experimental setup Kruss DSA-100 to measure wetting contact angle.

## 3. Experimental results

The critical heat flux during the flow of a water liquid film in a minichannel with smooth heater or that with microgrooves (the width and height of grooves were 0.5 mm), shear-driven by the nitrogen flow was studied experimentally. The critical heat flux for the heaters with a smooth and microgrooves surface was measured for various gas and liquid Reynolds numbers. The experiments have shown that the critical heat flux on a heater with microgrooves was larger than that for the smooth surface (Fig. 4).

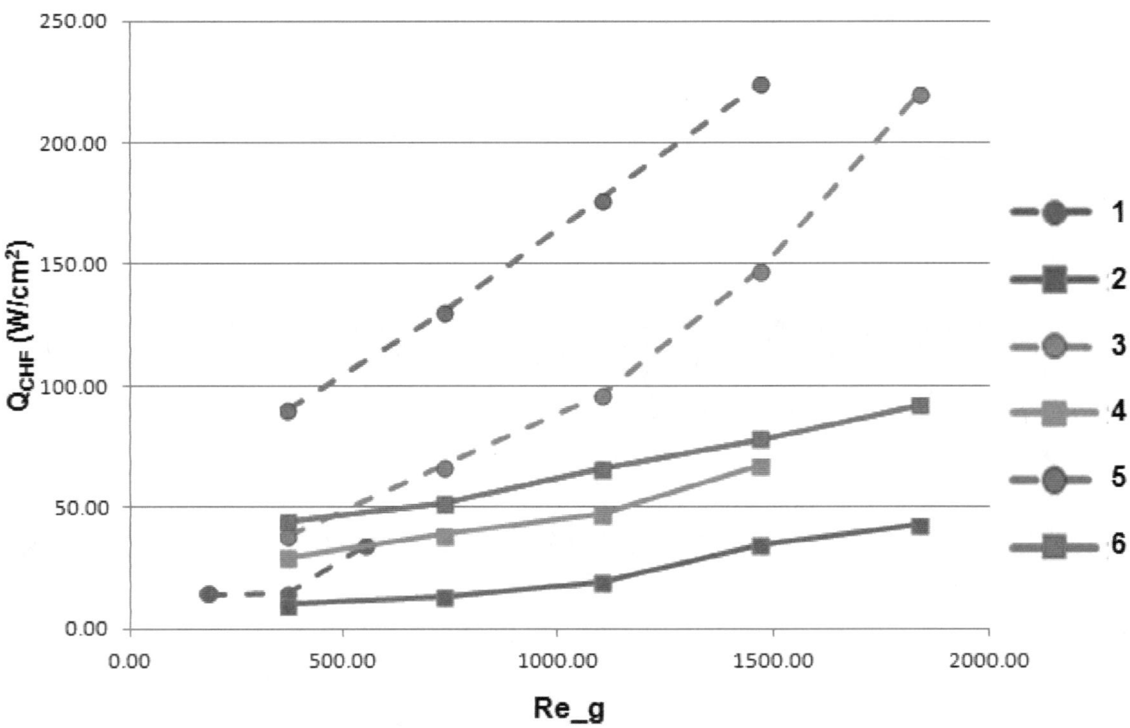


Fig. 4. The critical heat flux for the gas-driven water film flowing in a horizontal minichannel with smooth surface and microgrooves (0.5 mm).

1 — microgrooves, Rel = 14.5; 2 — smooth heater, Rel = 11.7; 3 — microgrooves, Rel = 20.5; 4 — smooth heater, Rel = 18.0; 5 — microgrooves, Rel = 29.0; 6 — smooth heater, Rel = 29.0.

For some flow rates, critical heat flux for surface with microgrooves can be up to twotimes higher than that for a smooth surface at the same flow rates. It can be seen that the criticalheat flux increases with increase in gas and liquid flow rates.

## 4. Conclusion

The critical heat fluxes for shear-driven water liquid film in a minichannel with smooth heating surface and the surface with microgrooves have been measured. Due to capillary effect a critical heat flux for the surface with microgrooves is higher than that for smooth surface.

The contact angle of the substrate was measured in the framework of the state task contract to IT SB RAS , the study of the heat transfer crisis on the micro-finned heater was supported by the Russian Science Foundation RSF grant (agreement No. 18-79-10258)

## References

[1] Bar-Cohen A., Wang P. 2009 Microgravity Sci. and Technology **21** 351–359.
[2] Kabov O A, Lyulin Yu V, Marchuk I V and Zaitsev D V 2007 International Journal of Heat and Fluid Flow **28** 103–12
[3] Kabov O A, Zaitsev D V, Cheverda V V and Bar-Cohen A 2011 Experimental Thermal and Fluid Science **35** 825–31
[4] Zaitsev D and Kabov O 2016 MATEC Web of Conferences **84** 00043
[5] Tkachenko E M and Zaitsev D V 2016 MATEC Web of Conferences **72** 01114
[6] Zaitsev D., Tkachenko E. and Kabov O. 2017 EPJ Web Conf. **159**, 00054.
[7] Kim J.K., Park C.W., Kang Y.T. 2003 Int. J. Refrigeration **26** 575–585.
[8] Gogonin I. I. 2014 Theoretical Foundations of Chemical Engineering **48** 96–103
[9] Cheverda V. V. And Kabov 2019 O. A. Thermophysics and Aeromechanics **26** 461–464

5th International Workshop on Heat/Mass Transfer Advances
for Energy Conservation and Pollution Control
August 13-16, 2019, Novosibirsk, Russia

# Heat and mass transfer in the boundary layer during evaporation of aqueous solutions of ethanol and acetone into air and superheated vapor components of solutions

**M S Makarov and S N Makarova**

Kutateladze Institute of Thermophysics, Siberian Branch
of the Russian Academy of Sciences, Ac. Lavrentyev ave. 1, Novosibirsk, Russia

E-mail: msmakarov@itp.nsc.ru

**Abstract.** A stationary flow of three-component gas in a laminar boundary layer on a flat plate is considered at adiabatic evaporation of a two-component liquid film. Numerical modelling were carried out at atmospheric pressure, air flow velocity of 2 m/s and temperature from 20 to 500 °C in the whole range of concentrations of a low-boiling component of binary solutions: ethanol/water, acetone/water, and acetone/ethanol. Data on the local specific evaporation rate of ethanol/water and acetone/water solutions into dry air and superheated mixture of ethanol, acetone and water vapors were obtained depending on the temperature of the inlet flow. Composition of superheated vapor mixture was taken to be equilibrium relative to the liquid solution. With an increase in the molar fraction of ethanol or acetone in the water solution, the evaporation rate increases, and when evaporated into superheated vapor, the growth of evaporation rate with increasing vapor temperature is significantly higher than at evaporation into air. With an increase in ethanol or acetone concentration in the liquid solution, the inversion temperature decreases from down to 155...165°C.

## 1. Introduction

It is necessary to take into account the heat and mass transfer processes with phase and chemical transformations to designing the power equipment [1, 2]. The describing of the process of liquid mixture evaporation as a part of heat and mass transfer problem is of interest to a number of modern technologies. To obtain a new energy carrier (alternative to oil and gas) in the fuel industry the mixtures of liquid are used [3]. For example, ethanol and methanol are added to gasoline. The fuel companies develop the biofuel, whose advantage is renewability of the raw materials base. The evaporative cooling is important for the chemical industry, when preparing the supersaturated water solutions in a new method of nanoparticles production [4, 5]. Advance of distillation devices of chemical technologies depends on optimization of evaporation process, which in turn depends on the geometry, thermodynamic parameters, composition of liquid and gas phases, etc.

The heat and mass transfer dependencies during evaporation of pure liquids are widely known. The current state of the problem is set out in a number of reviews and monographs [6-7]. At the same time, the effects of heat and mass transfer at evaporation of liquid mixtures poorly understood despite their widespread in industrial processes. The task becomes more complicated at evaporation of liquid mixtures because it is necessary to account for the non-ideality, azeotropy, and mutual miscibility of liquid components. All of these factors are dependent on temperature, pressure and composition. In the transient regime, a more intensive decrease in concentration of volatile substances in solution occurs due to the difference of component fugacity. The droplets of ethanol/water, methanol/water, and acetone/water evaporate with temperature maldistribution over the droplet surface [8]. The additional actions are taken to ensure the constancy of liquid mixture composition in time. In [9], steady evaporation of ethanol/water, methanol/water and acetone/water from the porous sphere are studied experimentally at continuous supply of the liquid phase. New correlation for temperature of adiabatic evaporation of binary liquid is proposed. The two-component liquid film flowing on the inner wall of a vertical cylinder is investigated numerically in [10]. It is shown that the addition of methanol to benzene leas to an increase in evaporation. The influence of geometry of evaporation surface on the heat transfer coefficients in the flow of the film of a binary Freon mixture is presented in experimental works [11, 12]. It is found out that at evaporation in the region of nucleate boiling development, the heat transfer coefficient for the surface with a mesh coating is higher than that for the smooth surfaces.

Heat and mass transfer at evaporation of mixtures is often simulated using the Raoult law. For example, in [13-15] the evaporation of a thin film of an idealized water/ethylene glycol solution to the laminar boundary layer on the vertical or horizontal surfaces are simulated numerically. It is shown that with increasing concentration of ethylene glycol, the temperature of evaporation surface increases due to the less fugacity of its vapors. The existence of inversion temperature at evaporation of binary liquid, observed previously only at evaporation of pure liquids [16, 17], is confirmed. In [18] the heat and mass transfer in a laminar boundary layer on a flat plate at stationary adiabatic evaporation of binary liquid mixtures was simulated numerically. Evaporation of mixtures with the properties of perfect solution and real zeotropic solution with positive deviation from the Raoult law was considered. For the real solutions, equilibrium between liquid and vapor is described using the Carlson-Colburn correlation. It is shown that at adiabatic evaporation of binary liquids, the wall temperature can be determined based on the similarity of heat and mass transfer for each component of the mixture. The calculation method was proposed for the engineering analysis of the processes of adiabatic evaporation of non-ideal binary liquids.

In this work, heat and mass transfer in a laminar boundary layer on a flat plate at stationary adiabatic evaporation of binary liquids of water, ethanol and acetone was simulated numerically at atmospheric pressure and temperature up to 500 °C. The data on the evaporation rate of ethanol/water, acetone/water, acetone/ethanol solutions into dry air and superheated mixture of ethanol, acetone and water vapors were obtained depending on the temperature of the inlet flow to determine the inversion temperature.

## 2. Problem statement and solution method

The scheme of the flow and the main parameters of the problem are presented in figure 1. Calculations were carried out at atmospheric pressure, air flow velocity of 2 m/s and temperature from 20 to 500 °C in the whole range of concentrations of a low-boiling component of binary solutions: ethanol/water, acetone/water, and acetone/ethanol.

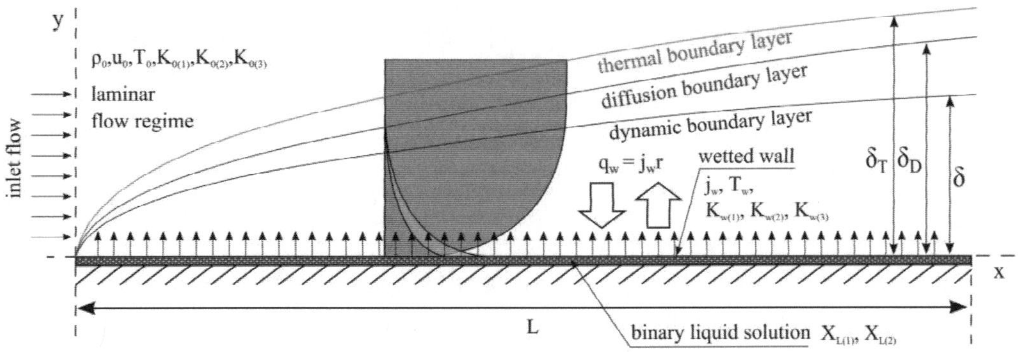


**Figure 1.** The scheme of laminar boundary layer during an evaporation of aqueous solutions on a wetted wall.

Journal of Physics: Conference Series **1369** (2019) 012056 doi:10.1088/1742-6596/1369/1/012056

Composition of the vapor-gas mixture at the interface was determined according to the Dalton law and equation of state of an ideal gas. The pressure of saturated vapor of components was determined from relationships based on the Margules equations: formulas of S E Harin [19] and I A Semyonov [20]. The results of calculations were verified by the known experimental data (figure 2).

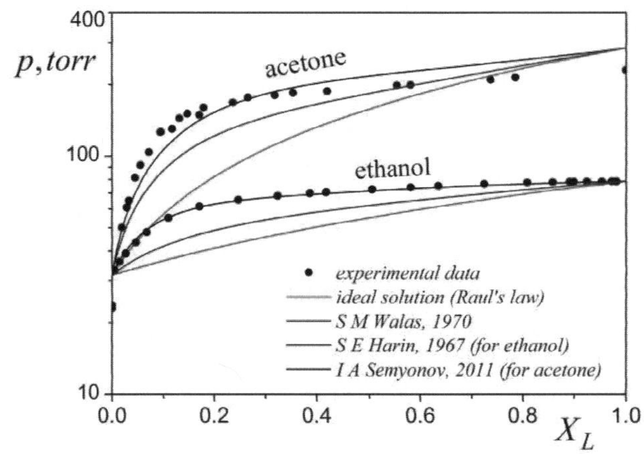


**Figure 2.** Saturated vapor pressure of the ethanol/water and acetone/water solution mixture at 30°C calculated by various models.

The problem was solved by the numerical methods based on the system of differential equations of the two-dimensional boundary layer in physical coordinates. The secondary effects, such as thermal diffusion and diffusion thermal effect, are not taken into account. To close the system of differential equations, the boundary conditions are set at the interface and in the inlet flow. Based on the condition of viscous gas adhesion, the longitudinal velocity component on the evaporation surface is equal to zero. The transverse velocity component and heat flux at the gas-liquid interface is expressed through the balance of matter and energy at adiabatic conditions. More detail of the equations, boundary conditions, and solution method is presented in [18].

## 3. Results and discussion

Data on the local specific evaporation rate of ethanol/water, acetone/water, and acetone/ethanol solution into dry air and superheated mixture of its vapors were obtained depending on the temperature of the inlet flow. Composition of superheated vapors mixtures was taken to be equilibrium relative to the solutions.

As it show in figure 3, with an increase in the molar fraction of ethanol in the solution, the evaporation rate increases, and when evaporated into superheated vapor, the growth of evaporation rate with increasing vapor temperature is significantly higher than at evaporation into air. Equality of the rate of evaporation into superheated vapor and dry air occurs at a certain temperature: the temperature of inversion, which depends on the composition of solution, and for an aqueous solution of ethanol, it does not depend on the evaporation rate or local coordinate. With an increase in ethanol fraction in the solution, the inversion temperature decreases from 460 to 166°C.

For an aqueous solution of acetone, it is shown that the inversion temperature decreases slightly along the plate length, which may be related to a larger temperature gradient across the boundary layer thickness, which increases the effect of acetone vapors on the density and viscosity of mixture near the evaporation surface (figure 4). With an increase in the fraction of acetone, the average-length inversion temperature decreases from 460 to 154°C.

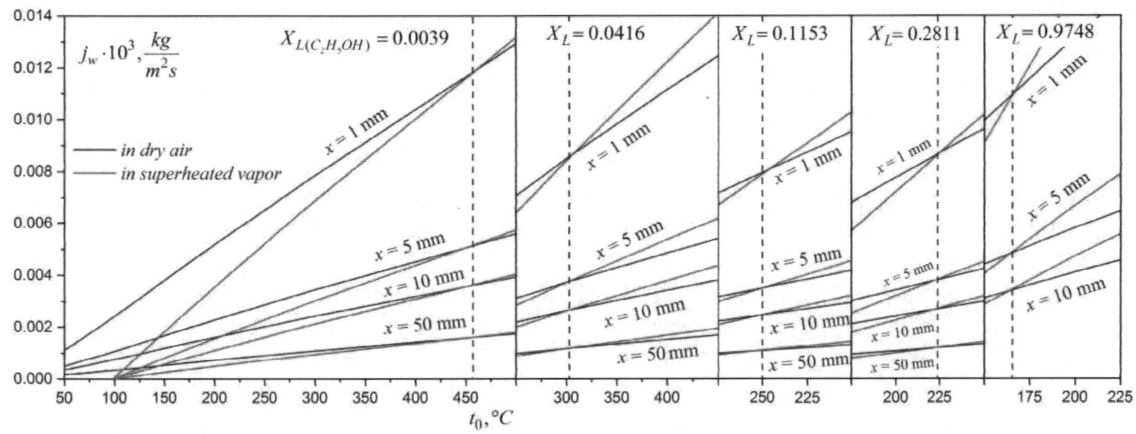


**Figure 3.** Evaporation rate depending on the inlet temperature of dry air and superheated vapor with an equilibrium concentration corresponding to the molar fraction of ethanol in liquid solution ethanol/water (the dash line shows the inversion temperature).

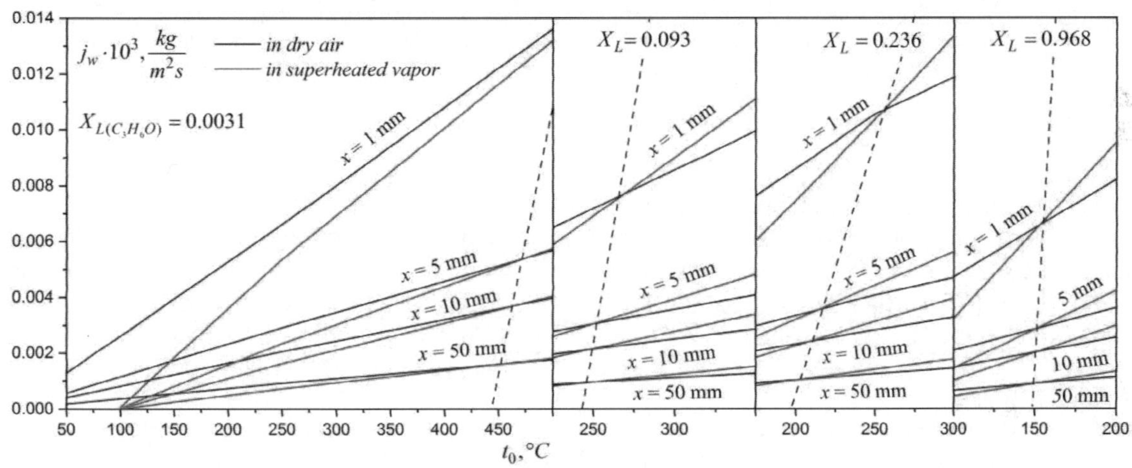


**Figure 4.** Evaporation rate depending on the inlet temperature for acetone/water solution.

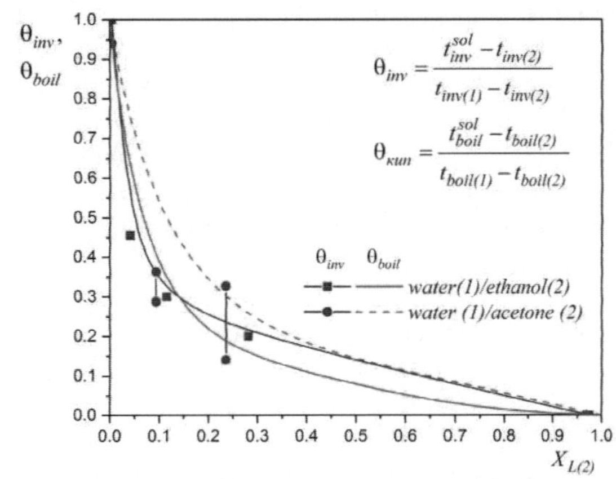


**Figure 5.** The relative inversion and boiling temperatures depending on the molar fraction of the light-boiling component of liquid for acetone/water and ethanol/water solutions.

A decrease in the boiling point of solution correlates with a decrease in the inversion temperature as it shown in figure 5 in the dimensionless form. It should be mentioned that the correlation of the inversion temperature with the boiling temperature was noted earlier in [17], where the influence of the pressure of the drying medium on the intensity of water evaporation into the steam-air mixture was studied.

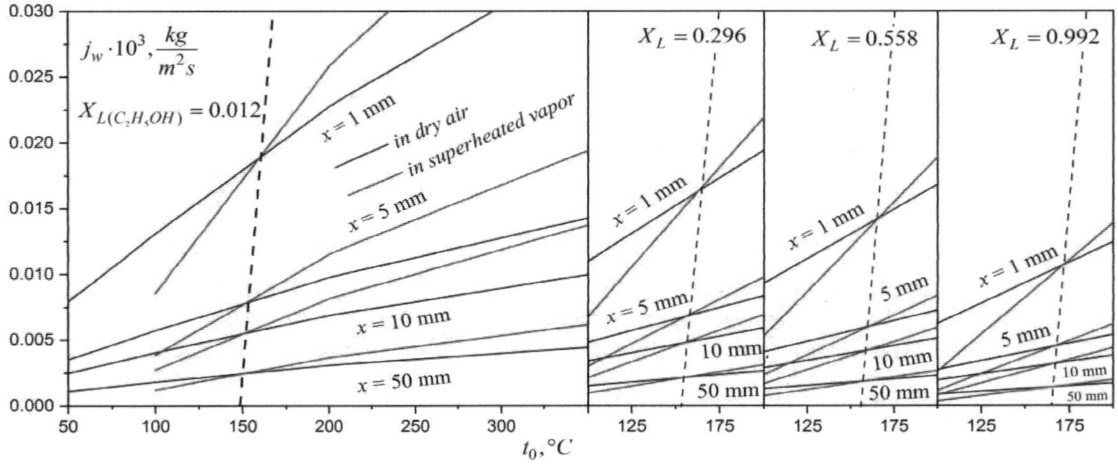


**Figure 6.** Evaporation rate depending on the inlet temperature for acetone/ethanol solution.

The obtained data on the rate of evaporation of the acetone/ethanol mixture show in figure 6. The boiling points of components are close (78.3 for ethanol and 56.3 for acetone); the heat capacities of liquids are close to. With an increase in ethanol concentration, the inversion temperature increases slightly from 155 to 165°C, which, as in previous cases, is consistent with an increase in the boiling point of the solution.

## 4. Conclusions

The heat and mass transfer on the wetted plate with adiabatic evaporation of ethanol/water, acetone/water, and acetone/ethanol binary solutions into the laminar boundary layer of the vapor-air mixture was studied. The values of the inversion temperature for the binary solutions under study were determined. It is shown that the dependence of the inversion temperature for water on the boiling point obtained earlier is maintained for solutions in which the boiling point changes due to changes in composition. An increase in the boiling point leads to an increase in the inversion temperature. The minimum value of the inversion temperature at atmospheric pressure for the solutions under studies is about 150 °C, which is significantly less than that of water. Lowering the pressure should also contribute to lowering the temperature of the inversion. The obtained results can be used in the development of effective technologies for separation of mixtures containing these components, in the absorption systems of refrigeration equipment, or in evaporative cooling systems.

**Acknowledgments**
This work was carried out under state contract with IT SB RAS.

**References**
[1] 2014 *Handbook of Industrial Drying*, Ed. by Mujumdar A S (CRC Press) p 1348
[2] Basharov M M and Laptev A G Energy saving at rectification installations during phenol and acetone production 2014 *Khimicheskaya Promyshlennost Segodnya* **4** 15 (In Russian)
[3] Markov V A and Patrakhaltsev N N Alcohol fuels for diesel engines 2010 *Transport na*

*Alternativnom Toplive* **1** 22 (In Russian)
[4] Efimenko Yu A and Fisenko S P 2008 *J. Eng. Phys. Thermoph.* **81**(2) 256
[5] Fisenko S P and Khodyko J A 2009 *Int. J. Heat Mass Transf.* **52** 3842
[6] Erbi H Y 2012 *Adv. Colloid Interface Sci.* **170** 67
[7] Eames I W, Marr N J and Sabir H 1997 *Int. J. Heat Mass Transf.* **40** 2963
[8] Terekhov V I and Shishkin N E Surface temperature of evaporating drops of binary solutions 2010 *Polzunovskiy Vestnik* **1** 55 (In Russian)
[9] Terekhov V I and Shishkin N E 2009 *Thermophysics and Aeromechanics* **16**(2) 239
[10] Baumann W W and Thiele F 1990 *Int. J. Heat Mass Transf.* **33** 267
[11] Pecherkin N I, Pavlenko A N and Volodin O A 2011 *Thermophysics and Aeromechanics* **18**(4) 579
[12] Volodin O A, Pavlenko A N and Pecherkin N I 2013 *High Temp.* **51**(6) 785
[13] Nasr A, Belhadj A M, Orfi J, Debissi C and Ben Nasrallah S Evaporation of a thin binary liquid film covering one plate of a vertical channel 2008 *Revue des Energies Renouvelables* **11**(4) 611
[14] Nasr A, Debissi C and Ben Nasrallah S 2010 *J. Thermal Sci.* **19**(4) 346
[15] Nasr A, Debissi C and Ben Nasrallah S 2011 *J. Thermal Sci.* **15**(3) 773
[16] Volchkov E P, Leontiev A I and Makarova S N 2007 *Int. J. Heat Mass Transf.* **50** 2101
[17] Volchkov E P, Leontiev A I and Makarova S N 2007 *Thermophysics and Aeromechanics* **14**(4) 493
[18] Makarov M S and Makarova S N 2016 *Thermophysics and Aeromechanics* **23**(1) 23
[19] Harin S E, Perelygin V M and Remizov G P On the equilibrium of liquid-vapor of water-alcohol solutions 1967 *Proceedings of USSR Universities. Food technology.* **3** 132 (In Russian)
[20] Semyonov I A, Ulyanov B A and Sitnikov D N 2011 *Proceedings of Universities. Applied Chemistry and Biotechnology* **1** 130

5th International Workshop on Heat/Mass Transfer Advances
for Energy Conservation and Pollution Control
August 13-16, 2019, Novosibirsk, Russia

# The effect of viscosity on rupture dynamics in the non-uniformly heated horizontal liquid film

**D Y Kochkin[1, 2], D V Zaitsev[1] and O A Kabov[1]**

[1] Kutateladze Institute of Thermophysics, Russian Academy of Sciences, Prosp. Lavrentyeva 1, Novosibirsk, 630090, Russia

[2] Novosibirsk State Technical University, Prosp. K. Marksa 20, Novosibirsk, 630073, Russia

kochkin1995@mail.ru

Abstract. In the present work, we study the rupture dynamics of a horizontal liquid film non-uniformly heated from below in a wide range of liquid viscosity. To visualize the liquid surface deformations and disruption of the film, we use a Nikon digital camera (shot at 60 fps) coupled with an optical schlieren system. In order to measure the instantaneous film thickness, we use the confocal Micro-Epsilon chromatic sensor. It was found that the process of film rupture can be divided into three stages: 1) film thinning down to a residual film on the heater; 2) existence of a stable residual film for some time; 3) rupture and dryout of the residual film. The thickness of the residual film was found to strongly depend on the liquid viscosity: for water, it is about 10 μm, whereas for PMS-200 it is about 275 μm.

## 1. Introduction

Thin liquid films are widely used in different branches of industry and may provide very high heat transfer intensities at comparatively low flow rates. Also, liquid films can be used for cooling of high-power computer chips of electronics devices whose power dissipation has been increasingly growing in recent years.

A subcooled liquid film heated from the substrate is susceptible to thermocapillary instabilities arising from the temperature dependence of the surface tension (the so-called Marangoni effect), which may lead to film rupture. To avoid the loss of systems performance by film rupture, it is of paramount importance to understand when, why and how the film rupture occurs. Liquid films can be of different configurations: films flowing under the action of gravity, films flowing under the action of hydrostatic pressure, shear-driven films, but the simplest case is a still horizontal layer.

Papers [1, 2] were pioneering in studying of thermocapillary rupture of a horizontal liquid film non-uniformly heated from below. In 0.85–1.3 mm thick layers of ethanol [1] the appearance of a dry patch is preceded by the formation of a thinned region in the film with convective cells pattern. At the threshold heat flux, a dry patch quickly originates as a pinhole within the thinned region. In 0.13–1.68 mm thick silicon-oil layers [2] the heating causes a steady state thermocapillary dimple above the heater, gradually deepening with the heat flux, until finally the edge of the dimple touches the substrate, with formation of a nearly dry patch, remaining covered with a thin layer of oil less than 1 μm thick. In [3, 4] it was found that thermocapillary film rupture occurs through the formation of a thin residual film and its subsequence disruption. In [5] the authors used a confocal method trying to measure the thickness of the residual film, but they failed due to the reflection of the probing light from the substrate.

In experiments with liquid films, it is important to measure the instantaneous liquid film thickness. Film thickness measurement methods are divided into intrusive and non-intrusive. In turn, non-intrusive methods are divided into the field (schlieren method [6], fluorescence method [7, 8]) and local (electric conduction method, capacitance method [8], fiber-optical method [4, 9-11], and confocal method [12-13]). The schlieren method is suitable for measuring film surface deformations with a tilt angle of no more than 3-5° and requires a reference point for measuring the absolute film thickness. The fluorescent method implies the use of a dye and is not suitable for measuring the thickness of evaporating liquid films (the concentration of the dye changes). Capacitance method and electric conduction method have low spatial resolution and require installation in the substrate. The fiber-optical method has bad angular characteristic and is appropriate for measuring the thickness of nearly flat films (with an inclination angle less than 1-2°). Against the background of the shortcomings of the methods described above, the confocal method seems to be most suitable for measuring thin liquid films since it has a high spatial and temporal resolution, high accuracy and good angular characteristic. In the present work, we use the confocal method for measuring rupture of heated liquid films in a wide range of liquid viscosity. In order to eliminate the reflection from the substrate, we applied a special black coating. The use of the thus improved confocal method has allowed us for the first time to measure the thickness of the residual film for water.

## 2. Experimental equipment and methods

The test section is a textolite plate with an embedded copper rod having a diameter of 12 mm (Fig. 1). The substrate has a cooling circuit over the perimeter with the temperature of the cooler (water) kept at 5°C. A ceramic electrical heater is attached to the bottom edge of the copper rod. In the experiments, the electrical power on the ceramic heater was 75 W. At that, the temperature of the surface of the copper rod, measured with the thermocouple, did not exceed 85°C. A plexiglass cylinder is mounted on the test section. To reduce the reflection coefficient, the substrate is coated with a special black paste. The experiments are carried out at an ambient temperature of 23–25°C. The test section is open to the atmosphere.

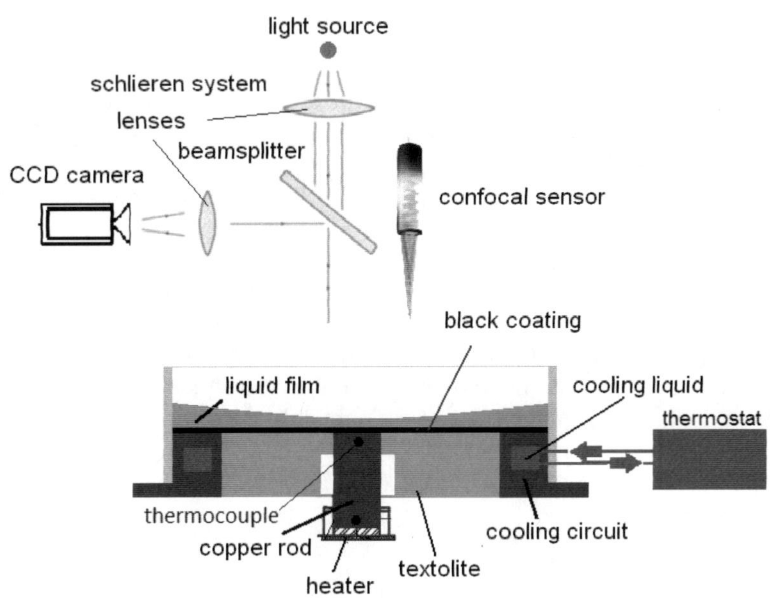


Fig. 1. Schematic diagram of the experimental setup.

Four different working liquids with the viscosity changing 200 times were used: ultrapure water (Milli-Q), silicon oils PMS-5, PMS-100, and PMS-200. PMS silicone oils are chosen as working liquids since they have practically the same basic thermophysical properties with a large variation of viscosity. A predetermined volume of the working liquid with the initial temperature of 25°C is fed to the substrate via a syringe pump to form a liquid film with the initial thickness varied from 400 to 700 μm.

To visualize the liquid surface deformations and disruption of the film, we used a Nikon camera (shot at 60 fps) coupled with an optical schlieren system. The main purpose of the schlieren system is to determine the curvature of the liquid surface. The schematic diagram of the system is shown in Fig. 1. The Lens 1 is placed in front of incoherent Light source at the focus distance. Collimated light generated by a collimating Lens 1 passes through 50% Beam splitter towards the liquid layer. Then part of the light reflects from the surface of the liquid. The reflected light goes again through the beam splitter and then focuses by the Lens 2 in its focal plane. Finally, the intensity of the images obtained is related to the inclination angle of the film surface. In our work, the schlieren method was used only for visualization without measuring surface deformation. The field of view of the camera is 32×18 mm. Resolution of the image is 1920×1080 pixels.

To measure the instantaneous local film thickness, Micro-Epsilon controller IFC2451 with confocal chromatic sensors IFS2405-0.3 and IFS2405-3 were used. The sensor is positioned on the free surface side of the film and oriented perpendicularly to the substrate surface. Film thickness measurements are performed at a fixed point located above the heater.

To determine the instantaneous local film thickness during the rupture, we measure the distance to the film surface during the film rupture and the distance to the substrate surface after the rupture. To minimize reflection of the incident light at the substrate, the heater is covered with a thin layer of special black paste and is carefully polished. The reflection from the heater surface without black coating is an order of magnitude larger than the useful reflection from the surface of the film. Using black coating it is possible to reduce the reflection from the substrate by almost two orders of magnitude. An estimated thickness of the coating is 50 μm. The method of treatment of the coating provides the surface roughness of the order of 1 μm. The wettability of the working surface was measured at different points using a drop shape analysis system (DSA100 by Krüss GmbH). At room temperature, the static advancing contact angle was 85±5°, whereas the static receding contact angle was 13±4°.

## 3. Experimental results and discussion

Powering on the heater causes thinning of the liquid film and formation of a dry patch in the film within a period of several tens of second (depending on liquid). Figure 2 shows the typical change of the film thickness with time for water.

It was found that the process of film rupture can be divided into three stages: 1) film thinning down to a residual film on the heater (period of 0-12 s in Fig. 2); 2) existence of a stable residual film (period of 12-16 s in Fig. 2); 3) rupture and dryout of the residual film (period of 16-19 s in Fig.2). Rupture of the residual film usually begins at the heater edge and is accompanied by the formation of a drop in the heater center. Measuring the water film thickness (Fig. 2) was performed using an IFS2405-0.3 sensor at a point where no drops were formed. 10 runs were carried out and the measured residual film thickness for water was found to be 9.5 ± 3 μm. Schlieren images of the residual film allow us to state that it is flat and its thickness is approximately the same all over its area.

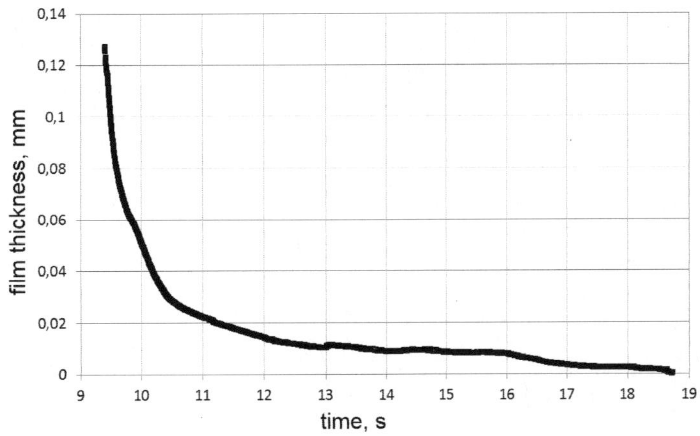


Fig. 2. Change of water film thickness in time, measuring rate is 0.3 kHz, the initial film thickness is 400 mm, heating power is 75 W.

Figure 3 shows the rupture process of a horizontal layer of silicone oil PMS-5, which also includes the same three stages as in the case of water. However, the residual film of silicone oil exists for a much longer time.

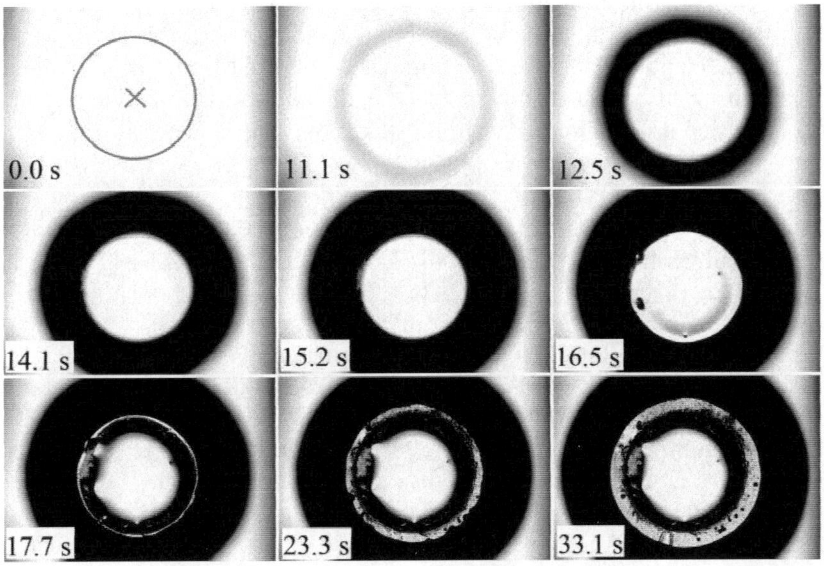


Fig. 3. The sequence of schlieren images showing the dynamics of silicone oil (PMS-5) film rupture. Initial film thickness is 600 μm, the heating power is 75 W. A circle indicates the position of the heated rod (of diameter 12 mm).

It is interesting to note that the residual film thickness is virtually not dependent on the initial film thickness in the range studied (from 400 to 750 μm). The rupture of PMS-100 and PMS-200 silicone oil films proceeds in accordance with the same scenario as the case of PMS-5 film rupture. The thickness of the residual film was found to strongly depend on the liquid viscosity: for water, it is about 10 μm, whereas for PMS-200 it is about 275 μm (Fig. 4).

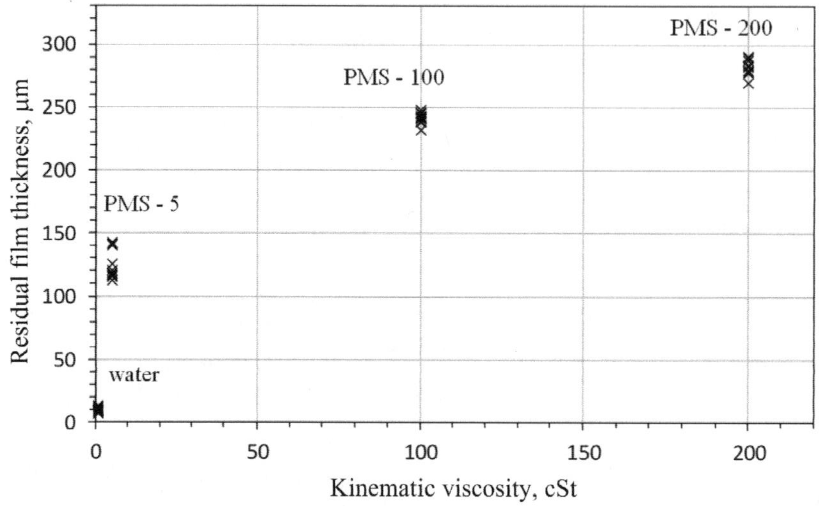


Fig. 4. Residual film thickness vs. kinematic viscosity for different working liquids. Initial film thickness varies from 400 to 750 μm, the heating power is 75 W.

## 4. Conclusion

It was found that the process of rupture of a horizontal liquid layer, non-uniformly heated from the substrate, is accompanied by the formation of a residual film on the heater. Schlieren images of the residual film allow us to state that the residual film is rather flat. Residual film thickness was measured using a confocal sensor. To minimize reflection of the incident light from the substrate, the heater was covered with a thin layer of special black paste. The thickness of the residual film was found to strongly depend on the liquid viscosity: for water, it was about 10 μm, whereas for PMS-200 oil it was about 275 μm.

This work was supported by the Russian Science Foundation (Project No. 19-19-00695). The test section was designed and built under the support of IT SB RAS (AAAA-A17-117022850022-0).

## References

[1] Orell A, Bankoff S G 1971 Int. J. Heat and Mass Transfer **14** (11) 1835–1842
[2] Burelbach J P, Bankoff S G, Davis S H 1990 Physics of Fluids A: Fluid Dynamics **2** (3) 321
[3] Zaitsev D V, Kabov O A 2007 Microgravity Science and Technology **19** (3–4) 174–177
[4] Zaitsev D V, Rodionov D A, Kabov O A 2007 Microgravity science and technology **19** (3-4) 100-103
[5] Lyulin Y V, Spesivtsev S E, Marchuk I V, Kabov O A 2015 Tech. Phys. Lett. **41** (11) 1034-37
[6] G.S. Settles, Schlieren and Shadowgraph Techniques: Visualizing Phenomena in Transparent Media, Springer-Verlag, Berlin, 2001.
[7] Chinnov E A, Kharlamov S M, Nazarov A D, Sokolov E E, Markovich D M, Serov A F, Kabov O A 2008 High Temperature **46** (5) 647–653
[8] Hewitt G F, Lovegrove P C and Nicholls B 1964 AERE-R 4478
[9] Zaitsev D V, Kabov O A and Evseev A R 2003 Experiments in Fluids **34** (6) 748
[10] Zaitsev D V and Kabov O A 2005 Experiments in Fluids **39** (4) 712
[11] Zaitsev D V, Chinnov E A, Kabov O A, Marchuk I V 2004 Technical Physics Letters **30** (3) 231-233
[12] Zhou D W, Gambaryan-Roisman T, Stephan P 2009 Experimental Thermal and Fluid Science **33** (2) 273-283
[13] Gong S, Ma W, Wang C, Mei Y, Gu H 2015 Int. J. Heat Mass Transfer **90** 636–644

5th International Workshop on Heat/Mass Transfer Advances
for Energy Conservation and Pollution Control
August 13-16, 2019, Novosibirsk, Russia

# Study of total heat losses in experiments with shear-driving liquid films in a mini-channel with intense heating

**D V Zaitsev[1,2], E M Tkachenko[1,2]**

[1]Kutateladze Institute of Thermophysics SB RAS, 1, Lavrentiev Ave, Novosibirsk, 630090, Russia
[2]Novosibirsk State University, 2, Pirogova str., Novosibirsk, 630090, Russia

E-mail: zaitsev@itp.nsc.ru

**Abstract.** Recent works by the authors have shown the high efficiency of using thin liquid films, moving under the action of gas in a flat mini-channel, for removing high heat fluxes. In the present work, we carefully study the total heat losses in such a system. Total heat losses are divided into two portions: heat losses from the heating coil to the atmosphere and heat spreadings from the heater to the substrate. The relative values of the total heat losses prior to the crisis are decreasing with increasing critical heat flux and do not exceed approximately 20% for the critical heat flux between 500 and 900 W/cm$^2$.

## 1. Introduction

The problem of microelectronics cooling is nowadays one of the most challenging problems of thermophysics. At this point in the production of processors, a technical process with a resolution of 22 nm is used. The number of transistors on a chip reaches an amount of about 2x10$^9$. The heat flux in electronic converters and inverters in hybrid cars now reaches up to 100-200 W/cm$^2$. The average heat flux on mass-produced computer chips and other electronic devices is up to 200-300 W/cm$^2$. While localized in the area of ~100 micron$^2$ heat flux reaches 1 kW/cm$^2$ and higher [1]. A number of military and technical applications already require heat flux removal from localized zones of up to 3-5 kW/cm$^2$ and more (i.e. GaN transistors). It should be noted that the heat flux on the sun surface is about 6 kW/cm$^2$. The situation with heat removal in microelectronics is even more complicated due to the transition of the electronics industry to the production of three-dimensional chips where multiple substrates with electronic components are installed in parallel with a distance of 50-100 microns. Therefore, development of advanced cooling techniques, capable to cope with such thermal loads, is needed.

There are three well-known ways of cooling intense localized heat sources: 1) microchannel boiling [2], 2) spray cooling [3], and 3) micro-jet cooling [4]. In [5, 6], a new way of efficient cooling was proposed. In this way of cooling the intense evaporation of a thin liquid film, moving under the action of gas flow friction in a flat micro-/mini-channel is used for heat removal. In [7], using this technique the possibility of removing heat fluxes with a density of up to 1 kW/cm$^2$ from a heating area of 1x1 cm$^2$ was shown, which is an order of magnitude higher than critical heat flux in falling liquid films [8]. In [9-11] stable operation of such system at different heights of the channel (from 0.17 to 2.0 mm) and at different angles of inclination to the horizon (from 0 to 360°) was demonstrated. In all works [5-7, 9-11], it was assumed that the heat losses were negligible, that is, all the electrical heat from the heating coil goes through the heater surface. However, for small flow rates of liquid and gas (i.e. when the heat transfer rates from the heater to the liquid film are comparatively low), the heat losses can be significant. In [5-7, 9-11] the critical heat fluxes are given without taking into account heat losses since the test sections in [5-7, 9-11] did not allow determining the local heat flux in the heater. In the current work we use a new test section with a similar design as in [5-7, 9-11] with the exception that, using two thermocouples with junctions placed one above the other in the heater, we can determine the local heat flux. The main task of the current work is to investigate the total heat losses in such experiments.

## 2. Experimental setup

The experimental setup consists of a test section, closed liquid line, open gas line, measuring and control systems. As a working gas, atmospheric air is used. The relative humidity of air is 20–40%, whereas the initial temperature is 25 °C. Ultrapure distilled water Milli-Q with the initial temperature of 25 °C is used as the working liquid. Figure 1 shows a schematic of the test section. The substrate is a plate of stainless steel. The copper cylinder, serving as a heater, is flush-mounted in the steel plate. Cylinder heating is carried out by passing an electric power through the nichrome coil, attached to its bottom part. For the thermal insulating of the heater, a layer of mineral wool is used. The copper heater surface is designed to simulate the surface of a typical electronic chip and has a square shape of 10x10 mm$^2$. Such construction of the heater provides a condition of the heater surface temperature to be nearly constant. The working surface (stainless steel plate together with the heater surface) was roughly polished so that the root mean square (RMS) roughness is 0.5 μm. The test section is covered by a transparent optical glass plate. The height of the channel is 1 mm.

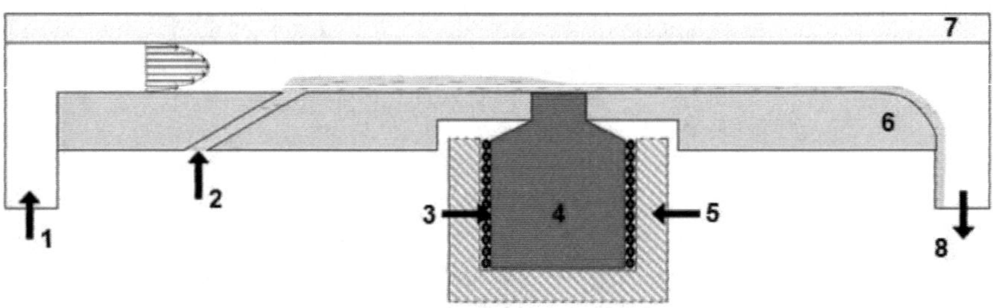


**Figure 1.** Schematic of the test section. 1 – gas inlet; 2 – liquid inlet; 3 – nichrome coil; 4 – copper cylinder; 5 – thermal insulation; 6 – steel plate; 7 – glass plate; 8 – outlet of liquid and gas.

To measure the temperature, a system of chromel-copel thermocouples was built into the test section as shown in Fig. 2 (left). Thermocouples 1 and 3 are used to determine the local heat flux, $q_{loc}$, whereas thermocouple 2 is used to determine the temperature of the surface of the heater, $T_w$, (taking into account the depth of the thermocouple junction location). The thermocouples 4 and 5 are used to determine the temperature of the steel plate, as well as the heat spreadings from the heater to the substrate. The diameter of the thermocouple junction is 0.5 mm.

The test section was specially designed to achieve high heat fluxes, so that embedding of thermocouples in the heater could not be made along the isotherms and was made through the holes along the copper cylinder from the bottom up, as depicted in Fig. 2 (left). In order to reduce inaccuracies in the temperature measurement (due to heat transport from thermocouple wires to the junction), the thermocouples are wound with a layer of insulating material along the entire length, with the exception of the thermocouple junctions which are in contact with the metal (through a layer of thermal paste), Fig. 2 (left). To verify the correctness of the temperature measurements by thermocouples, we compare temperatures of the heater surface obtained in the current work with those obtained in experiments [11] where thermocouples were embedded along the isotherms, as shown in Fig. 2 (right). In Fig. 3 the comparison of the heater surface temperatures, $T_w$, measured just prior to the heat transfer crisis, vs. the critical heat flux $q_{total,cr}$ (calculated without taking into account the heat losses) is presented (for different Reynolds numbers of liquid $Re_l$ and different superficial gas velocities $U_{sg}$). A good agreement (in view of data scattering) is seen for $q_{total,cr}$>150 W/cm$^2$ (where the heater surface temperature is weakly dependent on $q_{total,cr}$ for both experiments).

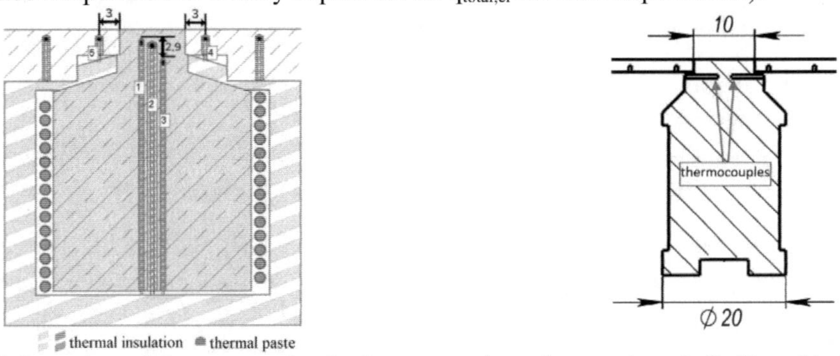


**Figure 2.** Location of thermocouples in the test section of current work (left) and in that of [11] (right).

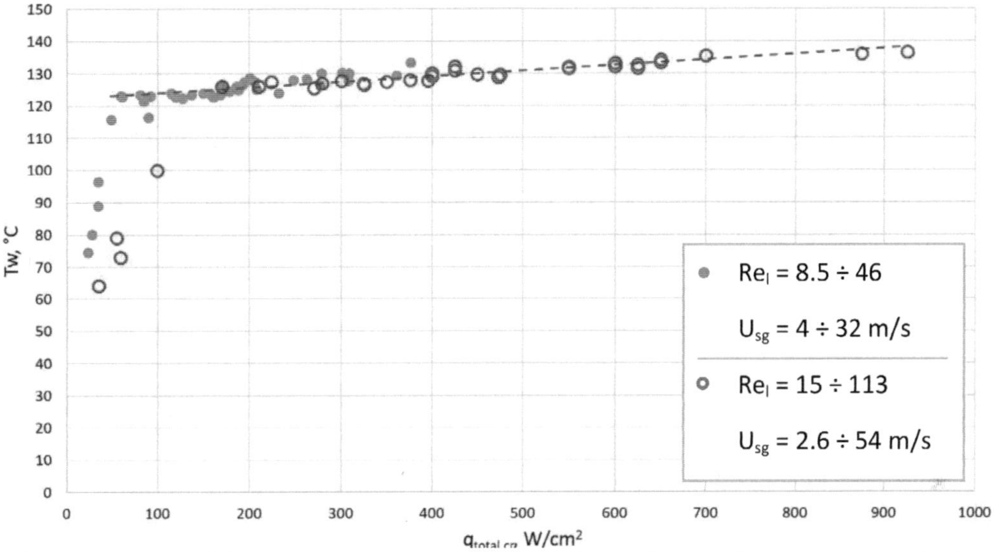


**Figure 3.** Comparison of the heater surface temperatures, just prior to crisis occurrence, obtained in the current work (red empty circles) with those obtained in [11] (blue filled circles).

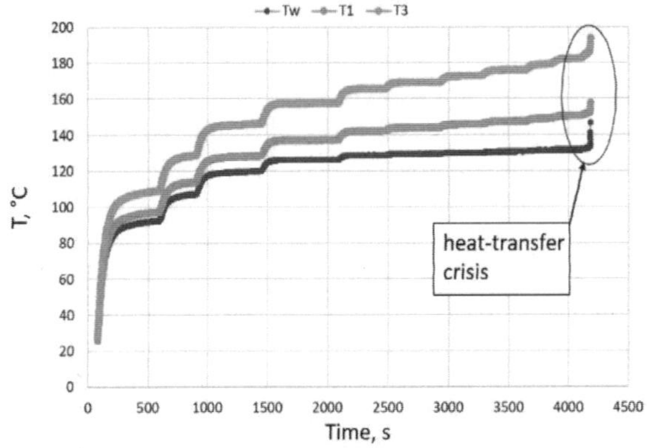


**Figure 4.** Time variation of temperatures at different depths in the heater for $Re_l$ = 70, $U_{sg}$ = 26.9 m/s.

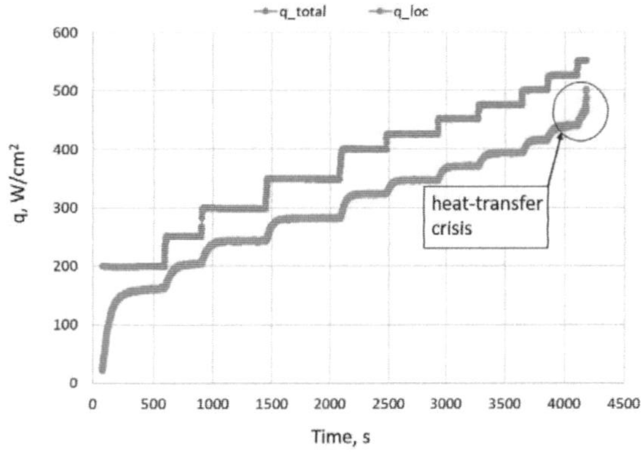


**Figure 5.** The dependence of the heat flux on time for $Re_l$ = 70, $U_{sg}$ = 26.9 m/s.

It should be noted that the calculation of the local heat flux, $q_{loc}$, is based on the difference between readings of thermocouples 1 and 3, Fig. 2 (left), therefore the uncertainty in $q_{loc}$ is much lower than uncertainty in temperature measurements, since systematic uncertainties in temperatures measured by thermocouple 1 and 3 are subtracted from each other. Thus, we can conclude that correctness of temperature measurements and especially correctness of local heat flux determination is enough justified in the current experiment.

Figure 4 shows a typical change of temperatures in the course of increasing heat flux up to the crisis; shown are the heater surface temperature, $T_w$, and temperatures T1 and T3 (measured by thermocouples 1 and 3, respectively). Figure 5 shows typical time dependencies of the total heat flux, $q_{total}$, (calculated based on Joule heat generated on the nichrome coil) and the local heat flux, $q_{loc}$, (determined based on measurements by thermocouples 1 and 3) in the process of increasing heat flux.

## 3. Determination of total heat losses

### *3.1. Heat losses to the atmosphere*

The heat losses to the atmosphere are the part of the generated heat that is going to the atmosphere from the nichrome coil, not reaching the heater surface. We determine the relative heat losses into the atmosphere as $\left(1-\frac{q_{loc}}{q_{total}}\right)\cdot 100\%$, where the local heat flux, $q_{loc}$, is determined based on the measurement of two thermocouples, embedded in the heater (thermocouples 1 and 3 in Fig. 2 (left), the distance between the thermocouples being 2.9 mm); while the total heat flux, $q_{total}$, is determined based on total Joule heat generation (i.e. without taking into account heat losses). Figure 6 shows the relative heat losses prior to the crisis, vs. the critical heat flux, $q_{total,cr}$. For data in Fig. 6, both $q_{loc}$ and $q_{total}$ are taken at the next to the last step of the increasing heat flux (i.e. just prior to the crisis occurrence), while $q_{total,cr}$ is taken at the very last step of the increasing heat flux (i.e. at the crisis occurrence). In the experiment, the difference between the two last steps is usually less than 5% (see Fig. 5). Different points in Fig. 6 correspond to different Reynolds numbers of liquid and different superficial gas velocities. From Fig. 6 it is seen that the relative values of heat losses into atmosphere generally decrease with increasing critical heat flux and for 250 W/cm$^2$ < $q_{total,cr} \le$ 935 W/cm$^2$; the relative heat losses are within 16 ± 4%.

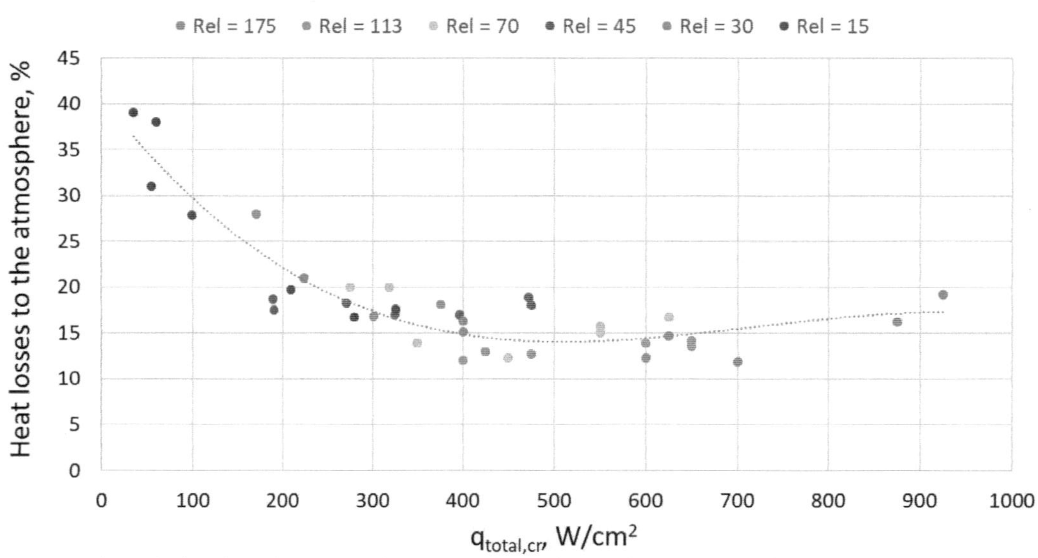


**Figure 6.** The relative heat losses to the atmosphere (just prior to the crisis) versus the critical heat flux (determined based on the electrical power generated on the nichrome coil) for different Reynolds number of liquid (shown in legend) and different gas velocities ($U_{sg}$ = 2.6 ÷ 54 m/s). Blue dotted line shows a generalization of the data.

Journal of Physics: Conference Series **1369** (2019) 012058 doi:10.1088/1742-6596/1369/1/012058

### *3.2. The heat spreading to stainless steel plate*

Thermocouples 4 and 5 embedded in the substrate, Fig. 2 (left), were used to determine the part of heat, "spreading" from the heater into the stainless steel plate. The thermal conductivity of the stainless steel is 15 W/mK which is about 30 times lower than that of copper (400 W/mK). We determine relative heat spreadings as a ratio between the heat that goes to the steel plate, $q_{steel}S_{cont}$, and the heat that goes to the heater surface, $q_{loc}S_{heater}$. Where $q_{steel}$ is average heat flux from the heater to the steel plate, determined using thermocouples embedded in the steel plate and in the heater, $S_{cont}$ – area of contact between the copper heater and the stainless steel plate, $S_{heater}$ – area of the heater surface.

Fig. 7 shows the relative heat spreadings to the substrate, $\frac{q_{steel}}{q_{loc}} \frac{S_{cont}}{S_{heater}} \cdot 100\%$, prior to the crisis, depending on the critical heat flux, $q_{total,cr}$, for different Reynolds numbers of liquid and different superficial velocities of gas. For data in Fig. 7, both $q_{steel}$ and $q_{loc}$ are taken at the next to the last step of the increasing heat flux (i.e. just prior to the crisis occurrence, see Fig. 5). From Fig. 10 it is seen that the relative values of heat spreadings decrease with increasing critical heat flux as $\sim 1/q_{total,cr}$. This is because the absolute value of the heat spreadings to the substrate, $q_{steel}S_{cont}$, prior to the crisis is almost non-dependent on the critical heat flux and does not go beyond 35 W/cm$^2$ in our experiments. According to thermocouple measurements, the temperature of the heater surface prior to the crisis is also weakly dependent on the critical heat flux and does not exceed 136 °C, Fig. 3. From Fig. 10 one can see that for $q_{total,cr}$>400 W/cm$^2$, the relative spreadings to the substrate do not exceed about 10%, while for $q_{total,cr}$>700 W/cm$^2$, the spreadings are less than about 5%.

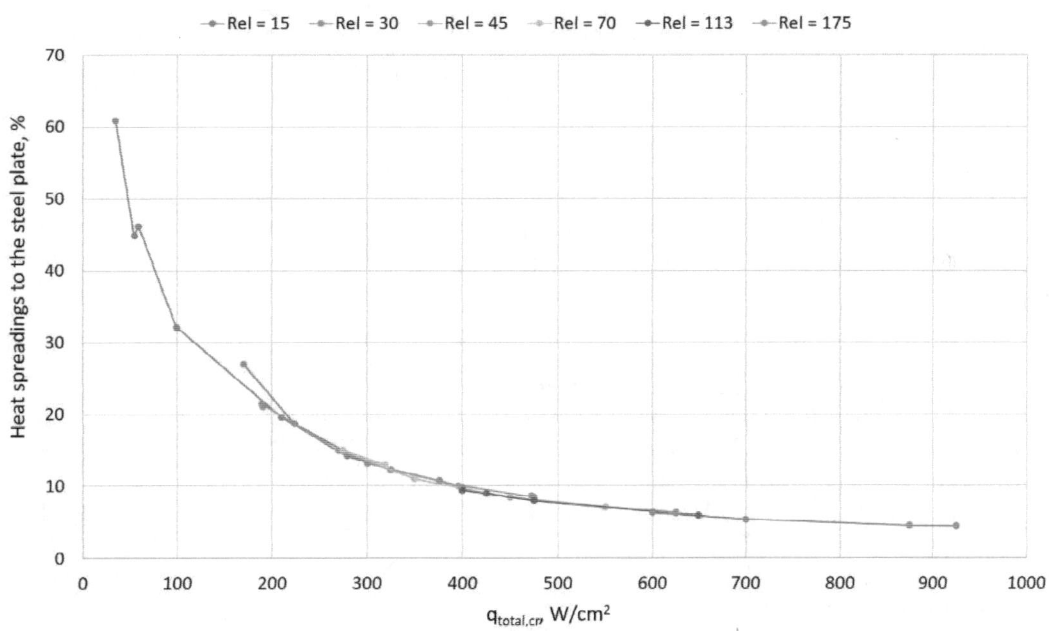


**Figure 7.** The relative heat spreadings to the substrate (just prior to the crisis) versus the critical heat flux (determined based on the electrical power generated on the nichrome coil) for different Reynolds numbers of liquid (shown in legend) and different gas velocities ($U_{sg}$ = 2.6 ÷ 54 m/s).

## 4. Critical heat flux with account for total heat losses

The equation for "true" heat flux $q_{whl}$ (*with* account for total *heat losses*) can be represented as:

$$q_{whl} = q_{loc} - \frac{q_{steel} S_{cont}}{S_{heater}}.$$

However, this equation cannot be directly used for calculation of "true" value of the *critical* heat flux, as at the moment of the crisis the heater is dried out so that the temperature of the heater starts to raise very sharply (see the last step in Fig. 4) leading to incorrect determination of $q_{loc}$ (see the last step in Fig. 5). To obtain "true" value of the critical heat flux, $q_{whl,cr}$, we are based on the critical heat flux, $q_{total,cr}$ (without taking into account heat losses), determined at the moment of the crisis, and adjust it for the total heat losses, using $q_{whl}$ determined at the previous step (just prior to crisis), where we can correctly determine $q_{loc}$ (see the next to the last step in Fig. 5). Thus, we get the following equation for "true" value of the critical heat flux:

$$q_{whl,cr} = q_{total,cr} \frac{q_{whl}}{q_{total}} = q_{total,cr} \frac{q_{loc} S_{heater} - q_{steel} S_{cont}}{q_{total} S_{heater}},$$

where $q_{total,cr}$ is taken at the moment of the heat transfer crisis occurrence, whereas $q_{whl}$, $q_{total}$, $q_{loc}$, $q_{steel}$ are taken just prior to the crisis occurrence. It should be noted, that during our experiments, when approaching to the critical heat flux, the increment in the heat flux is usually equal or less than 5%, so the above-mentioned equation can be considered as a good estimation of "true" value of the critical heat flux.

## 5. Conclusion

The total heat losses in our experiments with locally heated liquid films moving under the action of the gas flow in a mini-channel have been determined. Total heat losses are divided into heat losses into the atmosphere and heat spreadings into the substrate. For the critical heat flux between 250 and 935 W/cm$^2$ the heat losses into the atmosphere are within 16 ± 4%. Relative values of the heat spreadings into the substrate decrease in inverse proportion to the critical heat flux. For the critical heat flux above 400 W/cm$^2$, the heat spreadings do not exceed 10%. The relative values of the total heat losses decrease with increasing critical heat flux and do not exceed about 20% for the heat flux between 500 and 900 W/cm$^2$.

This work was supported by the Russian Science Foundation (Project No. 19-19-00695). The test section was designed and built under the state contract with IT SB RAS (AAAA-A17-117022850022-0).

## References

[1] A Bar-Cohen and P Wang 2012 J. Heat Transfer 134 051017
[2] Thome J R 2004 International Journal of Heat and Fluid Flow 25 128–39
[3] Kim J 2007 International Journal of Heat and Fluid Flow 28 753–67
[4] Robinson A J and Schnitzler E 2007 Exp. Therm Fluid Sci. 32 1–13
[5] Kabov O A, Lyulin Yu V, Marchuk I V and Zaitsev D V 2007 International Journal of Heat and Fluid Flow 28 103–12
[6] Kabov O A, Zaitsev D V, Cheverda V V and Bar-Cohen A 2011 Experimental Thermal and Fluid Science 35 825–31
[7] Zaitsev D, Tkachenko E and Kabov O 2017 EPJ Web of Conferences 159 00054
[8] Zaitsev D V, Semenov A A, Kabov O A 2016 Thermophysics and Aeromechanics 23 625-628
[9] Zaitsev D and Kabov O 2016 MATEC Web of Conferences 84 00043
[10] Tkachenko E M and Zaitsev D V 2016 MATEC Web of Conferences 72 01114
[11] Kabov O A, Zaitsev D V, Kabova Y O, Cheverda V V 2014 Proceedings of the 15th International Heat Transfer Conference IHTC-2014 119334

5th International Workshop on Heat/Mass Transfer Advances
for Energy Conservation and Pollution Control
August 13-16, 2019, Novosibirsk, Russia

# Vapour bubble dynamics in a flat mimi-channel under intense heating

**V V Belosludtsev**[1,2]

[1]Kutateladze Institute of Thermophysics SB RAS, 1, Lavrentiev Ave, Novosibirsk, 630090, Russia
[2]Novosibirsk State University, 2, Pirogova str., Novosibirsk, 630090, Russia

E-mail: v.belosludtsev@g.nsu.ru

Abstract. Investigation of subcooled nucleate boiling in minichannels was conducted with high speed recording optical system also equiped with high spatial resolution optics. Obtained data on the bubble dynamics is in good agreement with well-known correlations from the literature. The final stage of bubble dynamics is in good agreement with the Labuntsov correlation for spherical vapor bubble growth on a heated wall.

## 1. Introduction

The rapid development and miniaturization of computational tools, in particular, the use of three-dimensional chip packages, as well as extreme heterogeneity of heat flux on the electronic component, for example, a computer chip, poses new scientific tasks for developers and researchers [1, 2]. The local heat flux in certain areas of the chip of about 0.1-2 mm in size (the so called "Hot spots") can exceed the average heat flux density on the chip by more than an order of magnitude and can equal 1 $kW/cm^2$ and more, which puts the task of studying heat removal of high and simultaneously uneven heat fluxes.

There are four well known methods of removing ultra high heat fluxes from localized heat sources : 1) boiling in mini and micro-channels [3], 2) spray cooling [4], 3) micro-jet cooling [5], and 4) evaporation of shear-driven thin liquid film in a channel [2,6,7]. Boiling in mini- and micro-channels is considered very promising for cooling in high heat generation systems, especially in future's 3D stacked microelectronics. Those type of systems is capable to cool high heat fluxes with minimum required flow rate of the working fluid and maintaining low temperature of the channel walls [8]. Likely, microchannels is the most effective solution for thermal management of future microelectronics with 3D stacked architectures.

In the majority of studies on flow boiling in mini- and microchannels known from literature the vapour bubbles was studied at relatively low heat fluxes and the majority of authors studied dynamics of rather large bubbles - millimetre in size and larger (see for example [9]). In the recent work of [10] there have been observed two previously unknown stages of vapour bubble growth and the first stage was characterized by growth with rate exceeded 12.5 m/s and life time less than 10 μs, which means those type of bubbles can lead to blockage of microchannel. Data about bubble dynamics in microscale channels is quite limited and will be crucial for design of microchannel cooling systems.

The main intention of this paper is to study the subcooled flow boiling in a flat mini-channel under localized heating of the wall. Investigation of microbubble dynamics during subcooled nucleate boiling was conducted with high-speed recording optical system also equiped with high spatial resolution optics.

## 2. Experimental setup

Research were conducted on the experimental facility shown in Fig. 1. The facility consisted of the following: test section, liquid loop, thermal stabilization circuit, optical system with high speed camera equipped with high spatial resolution optics. Facility also equipped with automatic data acquisition system.

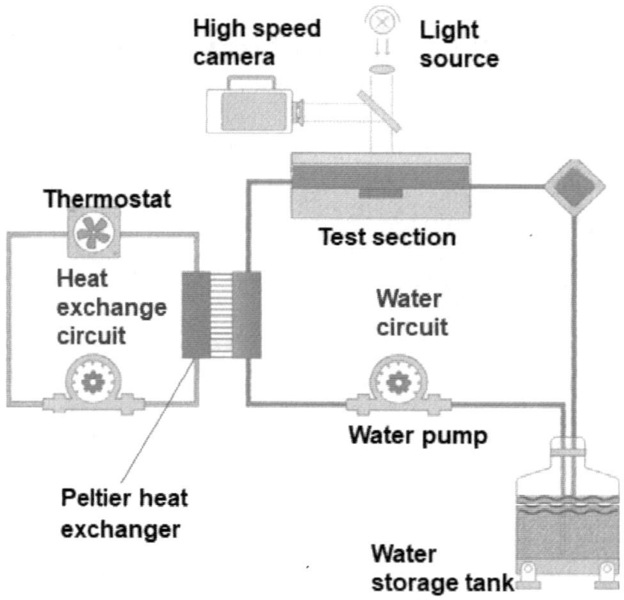


Fig. 1. Schematic of the experimental facility.

Milli-Q degassed ultra-pure water which was used as working liquid was supplied to the test section by the Ismatec Reglo Z-183 gear pump. The liquid flow rate was controlled manually by means of the pump in the range from 4.2 to 420 ml/min. The working liquid was thermally stabilized in a plate heat exchanger with the aid of a thermostat to the temperature of 25 °C and after that it was supplied to the test section. The temperature of the water at the inlet to the test section was controlled by a thermocouple integrated into the test section. The deviation of the temperature of the working liquid at the inlet was not exceeding ± 0.4 °C during the experiments. Pressure at the inlet and outlet of the channel was measured by pressure sensors, experiments were carried out under normal pressure The pressure drop along the channel of test section did not exceed 0.2 Bar.

The test section consisted of a thin flat plate made of stainless steel with a copper heater rod having a square head of 1 × 1 cm serving as a heater surface (Fig. 1), heater and plate was grinded to form uniformily flat plate. Joule heat was generated by electricity supplied to a nichrome wire which was wrapped around heater rod. The design of the heater provides constant temperature on the surface of the rod, $T_w$=const (as confirmed by thermocouple measurements). The top wall of test section is made from enlightened glass. The height of a channel is adjusted by placing changeable inserts between the working substrate and the glass top. The height of the channel H varies from 1.0 to 2.0 mm. The width of the channel is 30 mm, i.e. 3 times the width of the heater. The working surface (stainless steel plate with copper rod) was roughly polished and the morphology of the working surface was analyzed by an atomic force microscope. The root mean square (RMS) surface roughness was found to be 0.79 μm. Dynamics of the process of boiling on the heater was visualized by a high-speed camera FASTCAM SA1.1 made by Photron. The frame rate was varying from 5400 frames per second at the resolution of 1024x1024 pixels to 100000 frames per second at the resolution of 300x128 pixels. Optical system with high spatial resolution (up to 2.5 μm per 1 pixel of the camera sensor) was equipped on the camera. The distance from the upper edge of the heater to the center of the field of view of the camera view is about 1.5 mm..

## 3. Results

Data on the nucleate flow boiling in the channel was obtained by high-speed visualization. Images obtained from top view was processed with help of ImageJ software. Image of typical bubble evolution obtained with frame rate of 54 000 fps (one frame every 18.5 μs) are presented in Fig. 2. The zero time of bubble growth is taken at the very first frame when bubble is observed. Because of existence of dissolved gas in the liquid an additional nucleus can be observed near pure vapour bubble shown in Fig. 2.

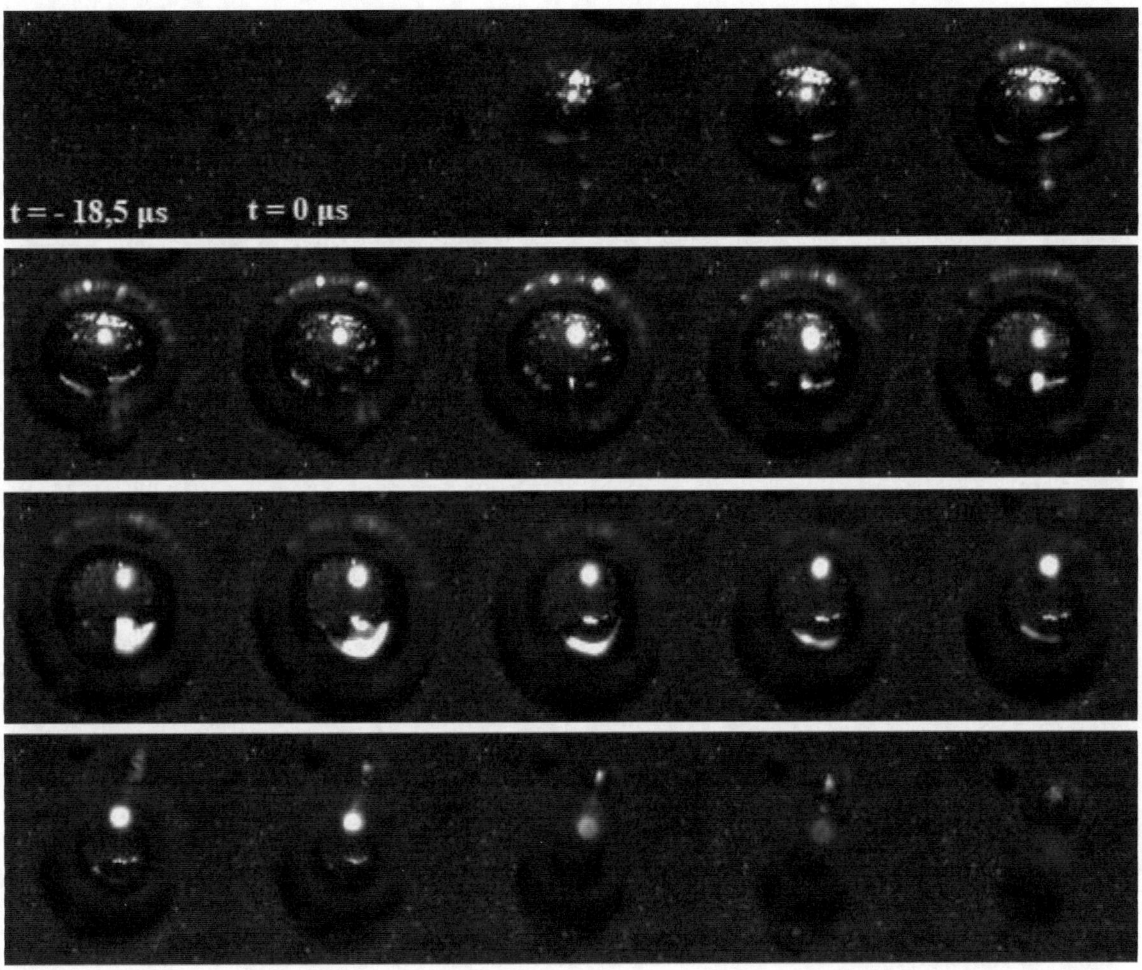


Fig. 2. Evolution of typical vapour bubble. Interval between frames is 18.5 μs.

Our data on boiling in flat minichannel with height of 1 mm is presented in color in Fig. 3. Data has been compared with data from literature obtained for water at atmospheric pressure, see Table 1 for exact conditions. Data has been compared with Labuntsov [11] equation for spherical bubble growth on the heated wall and Mikic-Rohsenow [12] equation for spherical and hemispherical bubble on a heated wall. At the final stage of growth, dynamics of our vapor bubbles is in good agreement with the Labuntsov [11] correlation dependence for vapor bubble growth on a heated wall, while the initial growth stage is satisfactorily described by the Mikic-Rohsenow correlation for hemispherical bubble on a heated wall [12].

| Author | Experimental setup | Conditions |
|---|---|---|
| [9] | Stainless steel tube d=12,7 mm; L=48 mm | p= 1,05 atm, G=78-800 kg/(m$^2$·s), $\Delta T_{sub}$ =20 – 30 K, q = 30 – 120 W/cm$^2$ |
| [10] | Silicon substrate with gold plated spheres d= 45 nm | p= 1 atm, pool boiling, $\Delta T_{sub}$ = 75 K, q = 80 – 100 W/cm$^2$ |
| [13] | Stainless steel tube d = 10 mm inside of glass tube with d = 30 mm, L = 100 mm | p= 1 atm, G=100 kg/(m$^2$·s), $\Delta T_{sub}$ = 20, q = 70 W/cm$^2$ |
| [14] | - | p= 1 atm, pool boiling, ΔTsup = 38 K |

Table 1. Data on the dynamics of vapour bubbles obtained for water at atmospheric pressure.

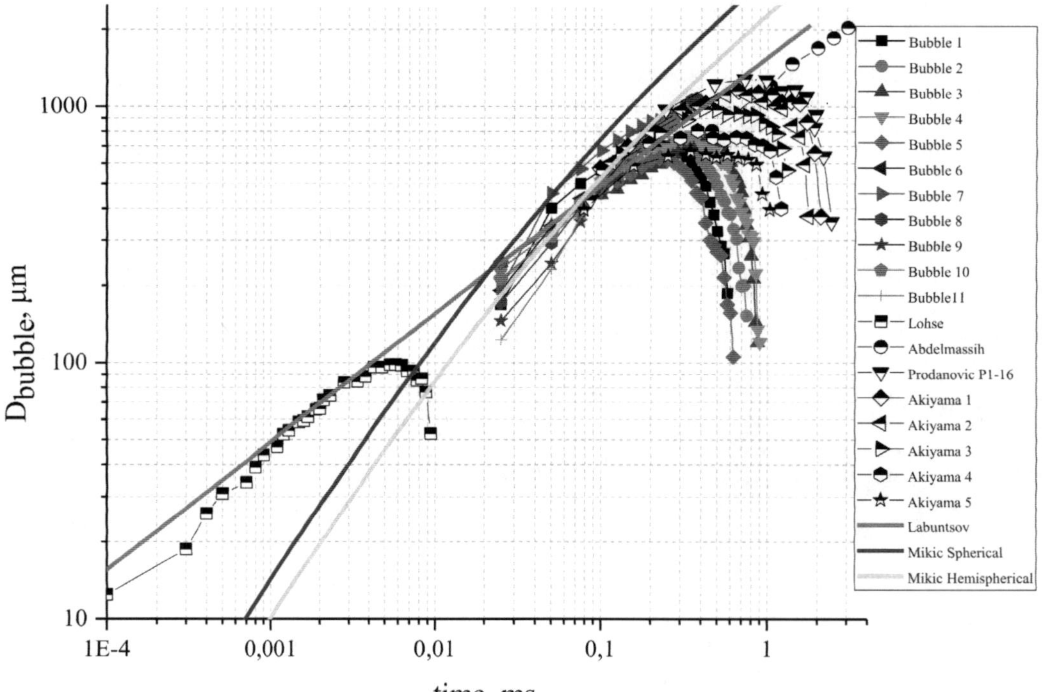


Fig. 3. Comparison of obtained data (for 11 bubbles) with correlations and data from the literature.

Since our bubbles is seen from top, we cannot know exact shape of our bubbles, perhaps, at the initial stage of intensive growth of the vapor bubble, due to the action of inertial forces of the liquid, the vapor bubble is hemispherical, as shown in [10]. The Jacob number used for both correlation formulas is 75, which corresponds to a heater wall temperature of 125 °C under the given experimental conditions. The good agreement between the experimental data of [10] (for semispherical bubbles) and the correlation formula of Labuntsov (for spherical bubbles) calculated for our experimental conditions is probably accidental, since the heating scheme in the experiment [10] is completely different and the Jacob number cannot be calculated, because the wall temperature is not measured in [10].

## 4. Conclusion

Boiling of subcooled water flow in a mini-channel under local heating was studied experimentally. Microbubble dynamics during subcooled flow boiling was studied with high-speed optical system with high spatial resolution. Obtained results on the bubble dynamics is in good agreement with correlations well-known from the literature. Obtained data show that in the case of 1 mm high channel the bubble can reach the upper wall of the channel in less than 1 ms, while in the case of 100 μm high channel it will reach the upper wall in about 10 μs, i.e. the use of cameras with speeds of at least 1 million fps is needed for investigation of bubble dynamics in microchannels.

The study was carried out under financial support of Russian Foundation for Basic Research (Grant No. 19-08-01235).

## References

[1] Bar-Cohen A and Wang P 2009 Second International Conference on Micro/Nanoscale Heat and Mass Transfer 553–567
[2] Zaitsev D, Tkachenko E, and Kabov O 2017 EPJ Web of Conferences, 159, 00054.
[3] Thome J 2004 International Journal of Heat and Fluid Flow, 25.
[4] Kim J 2007 International Journal of Heat and Fluid Flow, 28.
[5] Robinson J, Kempers R, Colenbrander J, Bushnell N, Chen R 2018 Applied Thermal Engineering, 136.
[6] Kabov O A, Lyulin Yu V, Marchuk I V, and Zaitsev D V 2007 International Journal of Heat and Fluid Flow, 28, 103–12
[7] Tkachenko E, Zaitsev D 2016 MATEC Web of Conferences, 72, 01114
[8] Zaitsev D, Tkachenko E, Belosludtsev V, Kreta A, Kabov O 2018 Journal of Physics: Conference Series, 1105 (1), 012142
[9] Prodanovic V, Fraser D and Salcudean M 2002 International Journal of Multiphase Flow, 28, 119334
[10] Wang Y, Zaytsev M., The H, Eijkel J, Zandvliet H, Zhang X, Lohse, D. 2017 ACS nano, 11(2), 2045-2051
[11] Labuntsov D 2000 Physical fundamentals of energy. Selected works on heat transfer, hydrodynamics, thermodynamics (in Russian) Moscow MEI
[12] Mikic B, Rohsenow W and Griffith P 1970 International Journal of Heat and Mass Transfer 13(4) 657-666
[13] Akiyama M, and Tachibana F 1974 Bulletin of JSME, 17(104), 241-247
[14] Abdelmessih A 1969 Spherical bubble growth in a highly superheated liquid pool, Plenum Press

Journal of Physics: Conference Series **1369** (2019) 012060 doi:10.1088/1742-6596/1369/1/012060

5th International Workshop on Heat/Mass Transfer Advances
for Energy Conservation and Pollution Control
August 13-16, 2019, Novosibirsk, Russia

# Numerical simulation of flow with evaporation in triangular grooves

**N Sibiryakov**[1,2] **and O Kabov**[1]

[1] Kutateladze Institute of Thermophysics, Russian Academy of Sciences, Prosp. Lavrentyeva 1, Novosibirsk, 630090, Russia
[2] Novosibirsk State Technical University, Prosp. K. Marksa 20, Novosibirsk, 630073, Russia

E-mail: kolyasibir@yandex.ru

**Abstract.** We study the flow of an evaporating liquid film in triangular open microchannel. Above the surface of the liquid flows a gas. The sear-stress on the gas-liquid surface drags the liquid and removes the vapor. It models either flow in a channel or a flow in a single groove. The novelty of this work is that we simulate gas-driven flow with evaporation. We use Boundary element method to solve Laplace equation with mixed boundary conditions. This method gives the normal derivative of vapor concentration at the surface of the liquid. The resulting evaporation flux is growing sharply near the contact line. Integral evaporation coefficient slightly depends on the depth of the liquid. The upper gas flow drives the liquid film, resisting a velocity drop and preventing the dry spot formation.

## 1. Introduction

The performance of modern computers highly depends on the processor frequency. Modern chips produce several billion elementary operations per second. Such a load on the processor causes serious heating of the electronics. Air fans can hardly cool personal computers, consume a lot of energy and cannot cope with the growing performance of processors. In addition, the control system on the space station or on the satellite is not profitable to cool the processor with a large and energy-intensive cooler. To divert large heat flows from a small area, it is necessary to create fundamentally new cooling systems.

Thin evaporating liquid films [1] can divert large heat fluxes from small arias. In the future, microchannels with a two-phase flow will become the basis for new heat removal systems. Even without evaporation [2], the flow in the triangular channel is more efficient than in rectangular due to the larger surface area of the channel. The heat flux reaches 50 W per cm square. Moreover, recent articles show that the flow in open triangular channels is very stable to external disturbances [3]. Cooling efficiency increases significantly if the fluid in the channel evaporates.

For the first time, the problem of a two-phase flow in a closed channel was solved by M. Soliman, J. R. Schuster, P. J. Berenson [4]. They considered a vertical channel of circular section, in which vapor condensation occurs on the walls. The fluid moved by gravity and the flow of steam, which flowed into the channel against the direction of liquid film. For channels of a triangular cross section, a similar problem was solved by V. Sartre et al. [5]. In their formulation, steam enters the channel and also condenses on the walls, but, unlike the cylindrical channel, vapor condenses in the corners and liquid films start there. A similar problem, but with a vertically arranged channel on which the vapor condenses, was solved by T.S. Zhao, Q. Liao [6]. They examined in more detail how the process of condensation occurs on the walls of the tube at different thicknesses of the liquid film. The problem with evaporation in a triangular channel was solved by Y. Peles and S. Haber [7]. In their formulation, the liquid evaporates in a closed triangular channel, as a result, the associated vapor flow moves at a high speed above the surface. The theoretical works of V.S. Ajaev and G.M. Homsy [8, 9] are devoted for the flow of liquid and saturated vapor in channels of rectangular cross section. They considered in detail how the liquid evaporates in the region of the contact line and found which heat flux can divert a channel of similar geometry.

There are not many experimental works on the flow of liquid with evaporation in the channels of a triangular cross section. T.S. Zhao, Q.C. Bi [10] experimentally studied vapor condensation on a pipe of a triangular cross section. G.P Peterson, H.B. Ma [11] studied the maximum heat flux that can divert the flow in an open triangular channel. The most detailed experimental study of the flow with evaporation was carried out by K. Helbig et al. [12]. They studied theoretically, numerically and experimentally the flow under the action of gravity in a channel with ribbed walls. In their work, they found that before forming a dry spot, the fluid in the grooves begins to behave unstable and is broken into drops. They did not give any explanations for this phenomenon in their article, because they focused on the calculation of the flow until it breaks into a flow in separate grooves. In this paper, since the flow is considered in a separate groove, we can predict the moment at which the flow of fluid in the corners or grooves becomes unstable.

## 2. Theoretical model

In my formulation of the problem, a constant temperature is set on the channel walls, the channel is open on top and evaporation occurs freely from the surface of the liquid (figure 1). This simulates both the flow in a triangular open channel and in each individual groove at the bottom of a structured channel. Let z-coordinate be directed along the channel, x and y in a cross-section.

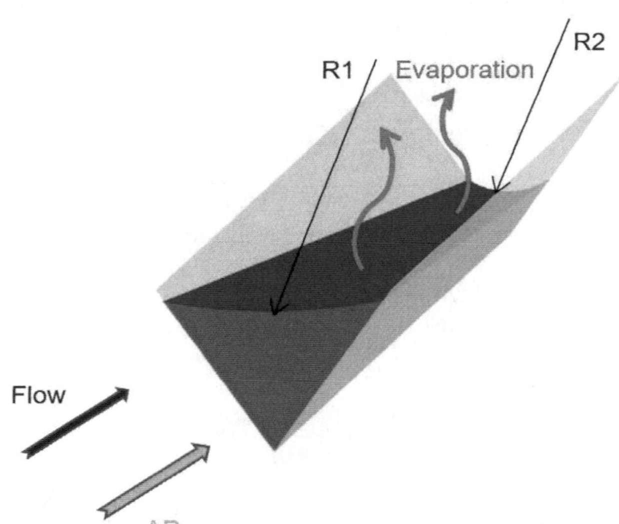


**Figure 1.** Channel layout. Depth of the liquid decreases due to evaporation.

Liquid flows into the channel with an initial velocity $v_0$ and with an initial thickness $h_0$. Evaporation from the surface of the liquid is proportional to the temperature of the liquid and the surface element:

$$dm = -C(T,h)dS,$$

where dm is evaporated mass of the liquid, $C(T,h)$ is the evaporation rate of the liquid, which depends on temperature ($T$) and on the depth ($h$) of the liquid, $dS$ is the surface element from which evaporation occurs.

Instead of finding the velocity at each point of the fluid, we search for the velocity $v$ averaged over the section. Since the thickness h of the fluid is small (less than a millimeter), the surface of the fluid in the cross section is an arc of a circle of radius $R$. Because of the curvature of the boundary and surface tension σ, the pressure in the fluid P is equal to the Laplace pressure:

$$P = P_0 - \frac{\sigma}{R}.$$

From geometric considerations:

$$R = h\frac{\sin\varphi}{\cos(\varphi+\theta)}.$$

The law of conservation of momentum in the integral form:

$$\int_V \rho g_z dV + \int_S \left(P - \zeta\frac{\rho v^2}{2} + \zeta_a\frac{\rho(v_a - v)^2}{2}\right)dS = \int_S \rho v(\boldsymbol{vn})ds.$$

Where $\rho$ is density of the liquid, $g_z$ is projection of free fall acceleration projection onto the z axis, $P$ is liquid pressure, $\zeta$ is the coefficient of wall-liquid friction force, $\zeta_a$ is the coefficient of gas-liquid friction force, $v_a$ is the velocity of gas above the liquid, $\boldsymbol{n}$ is normal vector. Averaging over the cross-section gives us following equation:

$$\frac{\rho}{S}(v^2 S)' = -P' + \rho g_z - \zeta\frac{\rho v^2}{2}\frac{L}{S} + \zeta_a\frac{\rho(v - v_a)^2}{2}\frac{2\alpha R}{S},$$

where S is derived from geometric considerations:

$$S = \frac{h^2}{2}\left[sin\,\varphi - \frac{\sin^2\varphi}{\cos^2(\varphi+\theta)}(\alpha - \sin\alpha)\right].$$

For a small sized system, the flow is laminar, so the shear stress is well described by poiseuille law:

$$\zeta = \frac{13.3}{\mathrm{Re}} = \frac{13.3\eta}{\rho v D_\Gamma}, \qquad D_\Gamma = \frac{4S}{L}, \qquad L = 2h,$$

where $\eta$ is liquid viscosity.

From the mass conservation law follows:

$$(vS)' = -\frac{C(T,h)}{\rho}h\frac{sin\,\varphi}{cos(\varphi+\theta)}\alpha.$$

Now we split variables and get system of two ordinary differential equations for velocity and depth distribution in the channel:

$$v' = \frac{g_z + \frac{\sigma A}{2\rho v h^2} - 2Ak\frac{v}{h} - B\frac{v}{h^2} + 2Bk\alpha\frac{v_a - v}{h^2}}{v - \frac{\sigma}{2\rho v R}},$$

$$h' = -\frac{A}{v}\frac{\sin\varphi}{\cos(\theta+\varphi)} - \frac{v'h}{2v},$$

where:

$$A = \frac{C(T,h)}{\rho}\left(\frac{\pi}{2} - \theta - \varphi\right),$$

$$B = \frac{26.6\nu}{\left(sin\,\varphi - \frac{sin^2\varphi}{cos^2(\varphi+\theta)}[\alpha - cos(\theta+\varphi)]\right)^2}.$$

Note that from the first equation for velocity follows that $v' \to \infty$ when

$$\rho v^2 = \frac{\sigma}{2R}.$$

After the velocity becomes critical, the theoretical model does not work. This criterion is a condition for the onset of the Rayleigh – Plateau instability. Apparently, soon after the model stops working, the liquid breaks up into drops and there is no longer a continuous flow, as it was already observed in the work of K. Helbig et al. [12].

## 3. Calculation of evaporation rate

Before solving a system of two ordinary differential equations, it is necessary to find the evaporation coefficient of the liquid. To do this, it is necessary to solve the diffusion equation, since the evaporation rate is determined by the vapor gradient on the surface of the liquid.

$$-\frac{dm}{dt} = D(\boldsymbol{n}, \nabla c_v)\rho_v dS = D\frac{\partial c_v}{\partial n}\rho_v dS$$

From here we obtain the connection between the desired function and the solution of the equation of vapor diffusion in the channel:

$$\mathrm{C}(T, h) = D\frac{\rho_v}{\rho_l}\frac{\partial c_v}{\partial n}.$$

Under stationary conditions and with negligible convective terms the diffusion equation is simplified to Laplace equation:

$$\mathrm{D}\Delta c = 0.$$

On the surface of the liquid, the vapor concentration must be equal to the saturation concentration $c = c_s(T)$. At the upper boundary almost all of the steam is removed and the concentration $c = 0$. Moreover, the solution is linear for these boundary conditions. Since we want to model the flow in one of the trenches of the finned channel, the symmetry condition must be satisfied on the side walls (it is also the periodic condition) $\frac{\partial c}{\partial n} = 0$.

We know the value of the function at some part of the boundary, and we know the value of the derivative on the other part. The best method to find the value of the normal derivative with the greatest accuracy is boundary integral equations method, or rather its modification, which is called the boundary elements method [13]. For the correct work of this method at the boundaries with a non-continuous normal (in the corners), it is necessary either to round them or to use smooth, finite kernels [14].

## 4. Results and discussion

We obtained the solution for normal derivative of vapor concentration at the liquid surface. The integral method automatically gives a large normal derivative in those places where the thickness of the liquid film is small. In real experiments, it is the microregion that accounts for the maximum evaporation rate, which corresponds to the obtained numerical solution. To obtain the evaporation rate of the fluid in the channel, it is necessary to integrate the normal derivative of the concentration over the surface of the fluid in this section. Having carried out calculations for different values of the depth of the fluid in the channel, we obtain the dependence of the integral evaporation coefficient on the depth of the fluid (figure 2).

This dependence is linear a large depth range. Because the main evaporation occurs in a microregion, evaporation rate integrated over the cross section weakly depends on liquid's depth in the channel. This result is in agree with experimental observations for droplets evaporation [15, 16].

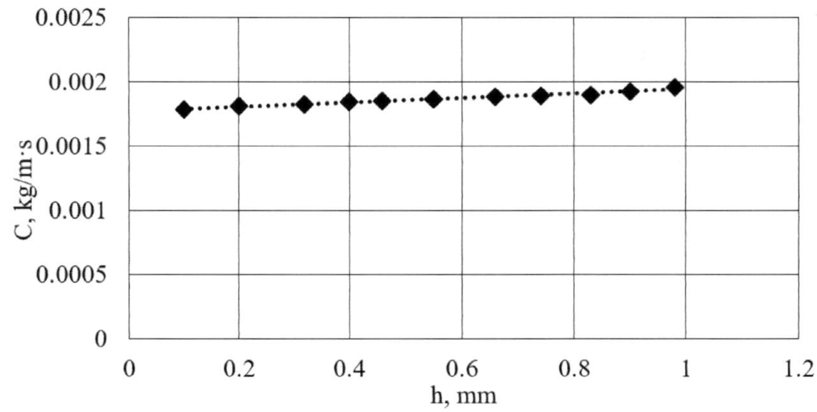


**Figure 2.** The dependence of the integral evaporation rate on the depth of the liquid in the channel.

Now, let us consider how the rate of flow of the associated gas flow affects the distribution of the thickness of the liquid in the channel. To do this, we take the flow rate of the same order with the initial velocity of the liquid (figure 3).

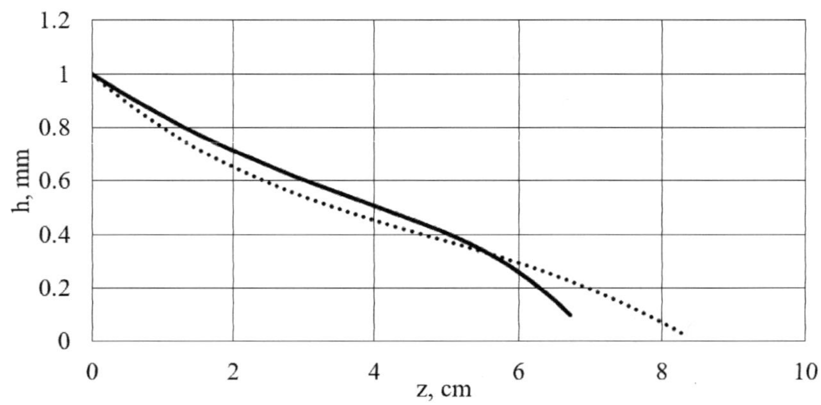


**Figure 3.** The liquid film thickness distribution in the channel. The initial velocity of the fluid is 2 m/s. For the dotted line the gas velocity is 4 m/s, for the solid line 2 m/s.

The velocity of the fluid grows to a limited extent, since the force of viscous friction against the walls gradually becomes greater than the capillary one. After reaching the maximum, the speed drops to a critical value, after which the simulation does not work (figure 4). Gas flow prevents a rapid drop in the velocity of the liquid film. This increases the effective cooling length.

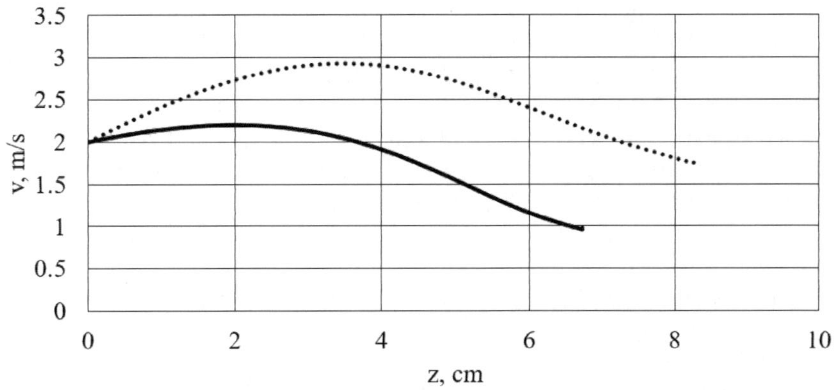


**Figure 4.** The velocity of the fluid in the channel. The initial depth of the fluid is 1 mm. For the dotted line the gas velocity is 4 m/s, for the solid line 2 m/s.

Let us estimate the effective heat flux along the canal:

$$q = \frac{\lambda\rho}{2h}\frac{d(vS)}{dx}.$$

It practically does not change, slightly decreasing and reaches 29 kW per m square. This means that reducing the size of the cooling device and increasing the initial velocity of the fluid and gas in the channel increases the efficiency.

## 5. Conclusion

In this work we model the flow of a liquid with evaporation in micro-grooves for the first time, taking into account the co-current gas and uneven evaporation from the surface. This is a big improvement of the previous work [17].

The obtained integral evaporation rate (Fig. 2) weakly depends on the depth of the fluid in the channel. This is due to the fact that evaporation is intensified in the microregion [15,16]. This effect was obtained numerically, since the method of boundary integral equations has an advantage over other methods for solving partial differential equations in problems with a rapidly changing normal vector (the presence of angles).

Numerical simulation shows that the viscosity of the fluid plays a significant role in this problem limiting the maximum velocity of the fluid and preventing the film from rapid thinning. Upper gas flow increases the velocity of the fluid in the channel, thereby increasing the effective cooling length.

Simulation works only while velocity is greater than critical (3). As soon as the speed drops to a critical, in a real liquid, the Rayleigh – Plateau instability develops and the film breaks into drops [12].

Increasing the initial velocity and reducing the size of the channel also improves the effective heat flux. This means that it is possible to make smaller channels if the co-current gas flow will blow with sufficient speed.

**Acknowledgment**

This work was supported by the Russian Scientific Foundation (project No. 18-79-10258).

**References**

[1] Gatapova E Ya and Kabov O A 2008 *Int. J. Heat and Mass Transfer* **51** 4797–10
[2] Tiselj I, Hetsroni G, Mavko B, Mosyak A, Pogrebnyak E and Segal Z 2004 *Int. J. Heat and Mass Transfer* **47** 2551–565
[3] White N C and Troian S M 2018 *Phys. Rev. Fluids*, arXiv:1812.09816
[4] Soliman M, Schuster J R and Berenson P J 1968 *ASME J. Heat Transfer* **90** 267–76

[5] Sartre V, Zaghdoudi M and Lallemand M 2000 *Int. J. Therm. Sci.* **39** 498–504
[6] Zhao T S and Liao Q 2002 *Int. J. Heat and Mass Transfer* **45** 2829–42
[7] Peles Y P and Haber S 2000 *Int. J. Multiphase Flow* **26** 1095–115
[8] Ajaev V S and Homsy G M 2001 *J. of Coll. and Interf. Sci.* **240** 259–71
[9] Ajaev V S and Homsy G M 2001 *J. of Coll. and Interf. Sci.* **244** 180–89
[10] Zhao T S and Bi Q C 2001 *Int. J. Heat and Mass Transfer* **44** 2523–34
[11] Peterson G P and Ma H B 1996 *ASME J. Heat Transfer* **118** 731–9
[12] Helbig K, Alexeev A, Gambaryan-Roisman T and Stephan P 2005 *Flow, Turbulence and Combustion* **75** 85–104
[13] Crouch S L and Starfield A M 1983 *Boundary Element Methods in Solid Mechanics* (George Allen & Unwill) p 322
[14] Sibiriakov E B 2017 *Physical Mesomechenics* **20** 98–104
[15] Gatapova E Ya, Semenov A A, Zaitsev D V and Kabov O A 2014 *Colloids and Surfaces A: Physicochem. Eng. Aspects* **441** 776–85
[16] Kabov O A, Gatapova E Ya, Semenov A A, Jutley M S, Ajaev V V, Kirichenko E O, Feoktistov D V, Kuznetsov G V and Zaitsev D V 2018 *Interfacial Phenomena and Heat Transfer* **6** 421–35
[17] Sibiryakov N, Kabov O and Belosludstev V 2019 *EPJ Web of Conferences* **196** 00051

5th International Workshop on Heat/Mass Transfer Advances
for Energy Conservation and Pollution Control,
August 13-16, 2019, Novosibirsk, Russia

# HEAT TRANSFER AT BOILING OF LIQUID FILM IRRIGATING A HORIZONTAL BUNDLE OF ROUGH TUBES

**I.I. Gogonin**

Kutateladze Institute of Thermophysics, Acad. Lavrentyev Ave. 1, Novosibirsk, Russia

**Abstract.** According to experiments with film irrigation of packing of tubes with artificial roughness, boiling on them begins at very low-temperature heads. Heat transfer at boiling on these tubes is noticeably more intense than that on a smooth tube.

## Introduction

Effect of wall roughness on boiling heat transfer under free convection is known long ago [1 -3]. In [4] a criterial dependence describing satisfactorily experimental results at boiling on walls with various roughness, is presented.

There are very few experimental data on heat transfer at boiling of liquid film irrigating walls with different roughness. It is known that boiling heat transfer is always accompanied by evaporation heat transfer. In the transition area between evaporation and boiling their contribution to the total heat removal is commensurate.

The below-described experiments were carried out in a closed circulation experimental test-bench. Copper and duralumin tubes both plain and rough in the bottom part of a horizontal tube bundle were used as the experimental tubes. The one-row tube bundle consisted of 20 tubes. The $2^{nd}$ from the bottom copper tube with the plain surface has the surface finish $R_z \cong 2.5$ microns, the wall thickness $\delta_w = 2$ mm, and $D = 10$ mm. The $3^{rd}$ from the bottom copper tubes has roughness $R_z \cong 20 - 30$ microns because of special treatment with corundum particles. The $4^{th}$ from the bottom duralumin tube has roughness $R_z \cong 20 - 30$ microns because of special treatment with corundum particles in the air stream.

The bank was irrigated with Freon R21 at $T \cong 40°C$. The tubes were heated by water flowing from a constant header tank inside the tubes. Measurements were carried out simultaneously for all experimental tubes at the same heating water flow rate.

## Measurement technique

The temperature of the water heating experimental section is measured by semiconductor temperature sensors. The sensors are calibrated with the accuracy of less than 0.02°C. The sensors calibration was carried out after their placement into the test volume by means of thermostats. For the plain tubes, the wall temperature was determined as an arithmetic mean value of measurement of 5 thermocouples sealed around perimeters of tubes from 0 to 180°. For the rough tubes, the wall temperature was calculated using the heat transfer coefficient.

A Freon flow rate irrigating the tube bundle was measured with a turbine flow meter.

The heat flux density on the experimental sections is calculated according to the dependence:

$$q=(m \cdot C_p \cdot \Delta T)/F \qquad (1)$$

The heating water flow rate is determined on the readings of rotameters as well as flowmeters with the accuracy rating 0.5. The rotameters installed in every section are calibrated individually.

The measured parameters were the following: temperature of heating water at the inlet and outlet of the experimental section, temperature drop between the inlet and outlet from the section, temperature of

Freon liquid and vapor in the experimental volume, pressure in the evaporator and a condenser, flow rate of liquid Freon irrigating the tube bundle, the power of electric heaters of water and Freon, geometry of the tube bundle.

Experimental results on heat transfer at boiling on the plain and rough tubes are shown in **Fig. 1**. It is seen that boiling began on the copper plain tube at $\Delta T = 4.5$ - 5°C ($q = 10^4$ W/m$^2$), on the copper rough tube at $\Delta T = 2.5 - 2.8$°C ($q = 6{\cdot}10^3$ W/m$^2$), and on the duralumin rough tube at $\Delta T = 3.5$°C ($q = 7{\cdot}10^3$ W/m$^2$). The effect of heat release of wall material is quite clear as well as at pool boiling [8, 9]. Heat transfer on the copper tube is considerably higher than on the duralumin tube with the same geometry characteristics. Reduction of the temperature drop at which the boiling begins because of roughness increase of a heat release wall is extremely important for heat exchangers with the limited value of available temperature drop. Boiling heat transfer on a copper rough tube at $q$ = idem increases one and a half times in comparison with a smooth wall. The use of rough tubes is a simple and reliable way to enhance heat transfer of the film irrigating a tube bundle.

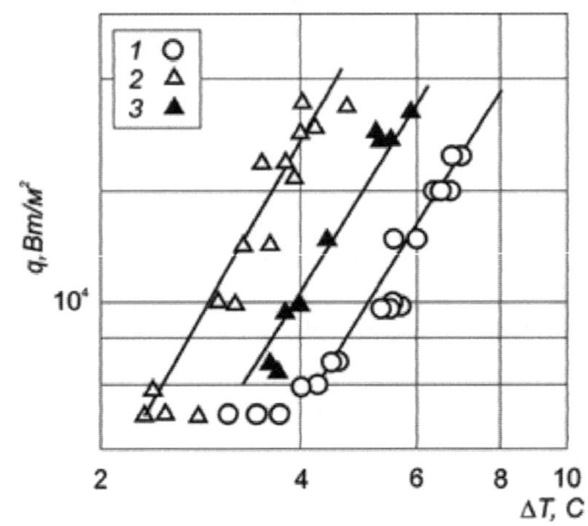


**Figure 1.** Heat transfer at film evaporation and boiling in the tube bundle.
R21, $Ts$ = 40°C, $D$ = 10 mm, $S/D$ = 1.2, $\delta_w$ = 2 mm;
1 – copper, $R_z$=2.5 microns, 2 – copper, $R_z$= 20 - 30 microns, 3 – duralumin D16T, $R_z$= 20 - 30 microns;
$Re$ = 300 – 600.

**Data representation in dimensionless coordinates**

As shown in [15], the total heat transfer at film evaporation and boiling can be calculated according to the dependence:

$$\bar{\alpha} = \alpha_0 \left(1 - \frac{L_d}{L}\right) + \alpha_d \frac{L_d}{L} - \alpha_c \left(1 - \frac{L_d}{L}\right) \qquad (2)$$

Here $L_d$ is a part of tube perimeter where convective heat transfer on the initial part of the thermal boundary layer, $L = \pi{\cdot}R$ is a tube semi-perimeter, $\alpha_d$ is convective heat transfer on the initial part of the thermal boundary layer, $\alpha_0$ – boiling heat transfer, $\alpha_e$ – evaporation heat transfer.

Temperature measurements along the perimeter of the plain tube wall showed that local $\Delta T = T_w$ - $T_s$ is independent of the thermocouple location on the tube perimeter. This dense tube bank should be considered as a vertical wall because there is no initial part of the thermal boundary layer on the bank

5th International Workshop on Heat/Mass Transfer Advances
for Energy Conservation and Pollution Control,
August 13-16, 2019, Novosibirsk, Russia

tubes. Here a drop dimension exceeds the distance between bank tubes, so the drop flows smoothly from an upper to the lower tube. The drop temperature is practically constant.

Without the initial part of the thermal boundary layer the dependence (2) is the following:

$$\overline{\alpha} = \alpha_0 + \alpha_c \quad (3)$$

A calculation algorithm for heat transfer at film evaporation and boiling at the wave and turbulent film flow is stated in [15]. The presented comparison between experimental data for a smooth tube bundle and calculated values is quite satisfactory.

As follows from Fig. 2 for rough tubes, heat transfer is changed only at film boiling. The boiling begins at considerably lower temperature difference between wall and liquid in comparison with smooth tubes.

The presented comparison between experimental data for a smooth tube bank and calculated values is quite satisfactory.

Heat transfer calculation at boiling must be performed according to the dependence:

$$Nu^* = 0.01 Re_*^{0.8} Pr^{1/3} b K_t^{0.4} \left(\frac{\lambda_L c_L \rho_L}{\lambda_{wall} c_{wall} \rho_{wall}}\right)^{-0.2} \bar{R}_z^{0.2} \quad (4)$$

Here $\left(\frac{\lambda_L c_L \rho_L}{\lambda_{wall} c_{wall} \rho_{wall}}\right)^{-0.2}$ is the ratio of physical properties of heat carrier to physical properties of the cooled

wall. This parameter takes into account the effect of the wall properties on boiling heat transfer rate. $\bar{R}_z = \frac{R_z}{(\nu_L^2/g)^{1/3}}$ is a parameter taking into account the effect of wall roughness on boiling heat transfer.

As shown in [4], at $\bar{R}_z < 0.9$ heat transfer is enhanced as wall roughness grows. At $\bar{R}_z > 0.9$ change of wall roughness has no effect on heat transfer.

It should be noted that film evaporation heat transfer is independent of tube roughness and agree satisfactorily with a calculation based on the three-layer model of turbulent flow stated in [16] for $Re > Re_{cr}$. $Re_{cr}$ is determined by [9].

A heat transfer calculation algorithm at film evaporation and boiling for wave and turbulent flow is stated in [15]. The presented comparison of experimental data with calculated dependencies is quite satisfactory [9 – 14].

At wave film, flow heat transfer is practically independent of irrigation density and determined by the thermal resistance of a "residual" film thickness [15].

At the beginning of the film boiling, it is turbulized by growing vapor bubbles, and it's surface area increases considerably. In this case, heat transfer determination for a single-phase flow should be considered as estimating and understated. As follows from Fig. 2, at maximum (in our experiments) heat fluxes the deviation of the experimental data from the calculated achieves 10 – 20%.

Journal of Physics: Conference Series **1369** (2019) 012061 doi:10.1088/1742-6596/1369/1/012061

5th International Workshop on Heat/Mass Transfer Advances
for Energy Conservation and Pollution Control,
August 13-16, 2019, Novosibirsk, Russia

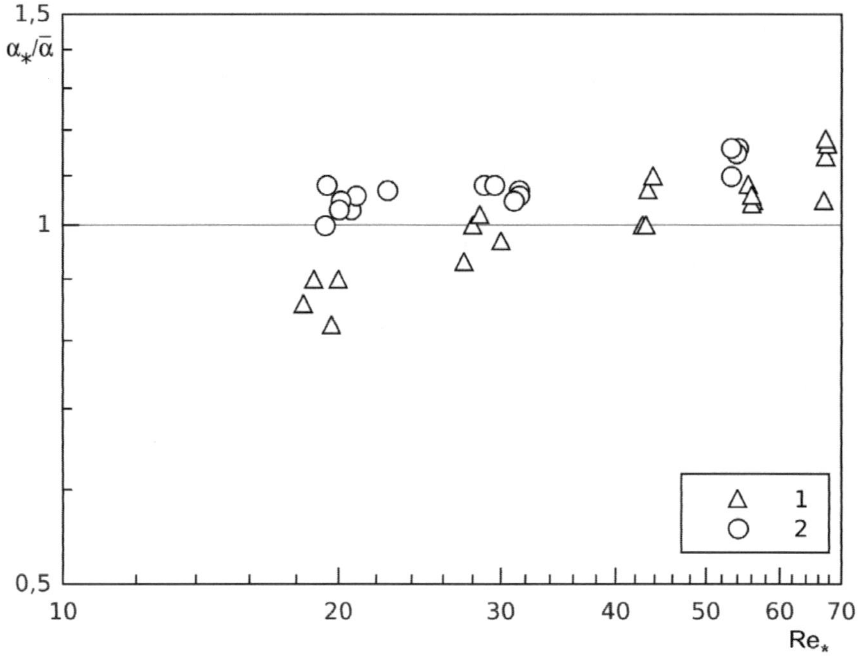


**Figure 2.** Deviation of the experimental data from the calculated values.

Currently, it is not possible to determine heat transfer during evaporation of the boiling film more accurately.

At $T_s$ = const, the dependence (4) takes the form of

$$Nu^* = C \cdot Re_*^{0.8} \tag{5}$$

Here $C = 0.01Pr^{1/3}bK_t^{0.4}\left(\frac{\lambda_L c_L \rho_L}{\lambda_{wall} c_{wall} \rho_{wall}}\right)^{-0.2} \bar{R}_z^{0.2}$ at $\bar{R}_z \leq 0.9$

or $C_1 = 0.01Pr^{1/3}bK_t^{0.4}\left(\frac{\lambda_L c_L \rho_L}{\lambda_{wall} c_{wall} \rho_{wall}}\right)^{-0.2}$ at $\bar{R}_z > 0.9$

In **Fig. 2** the comparison of experimental data at boiling on rough tubes is presented in the form of $\frac{\alpha_e}{\bar{\alpha}} = f(Re_*)$ according to the calculation on the dependence (3). Experimental data on tubes with increased roughness agree with the calculated data satisfactorily.

The maximum deviation of the experimental data from the calculated values of heat transfer coefficient is observed during comparison with experiments on the plain tube. In our opinion, it is due to the fact that at lower heat fluxes boiling did not take place around the whole tube perimeter. The similar phenomenon was observed in the experimental works [17, 18].

5th International Workshop on Heat/Mass Transfer Advances
for Energy Conservation and Pollution Control,
August 13-16, 2019, Novosibirsk, Russia

## Conclusions

1. The use of tubes with increased roughness is a simple and reliable way to enhance boiling heat transfer. Decrease of $\Delta T$, which corresponds boiling origin, is especially actual for devices with the limited temperature drops.
2. The proposed calculation of heat transfer at film evaporation and boiling for rough tubes agrees satisfactorily with measurement results.

**Acknowledgments**
The study was performed at the Institute of Thermophysics (SB RAS) with the support of the BSI SAS Program for 2017-2020 (project III.18.2.3, reg. no. AAAA-17-117030310025-3)

## Reference

[1] Kurihara H.M., Meyers J.K. *American Institute of Chemical Engineering Journal. 1960. V.6. N 1. P. 23.*
[2] Berenson P.J. *I.J. Heat Mass Transfer,* 1962, V. 2, P. 985-999.
[3] Danilova G.N., Belsky V.K. *Kholodilnaya tekhnika*, 1970, N 3, P. 24.
[4] Gogonin I.I. *TVT,* 2006, V. 44, N. 6, P. 918-925.
[5] Petukhov B.S., Kirillov V.V. *Teploenergetika,* 1958, N. 4, P. 63-68.
[6] Grigoryev V.A., Pavlov Yu.I., Ametistov E.V., Boiling of cryogenic liquids. Moscow, Energiya, 1977, 289 p.
[7] Galin N.M., Kirillov P.L., Heat mass transfer (in nuclear power engineering). Moscow, Energoatomizdat, 1987, 375 p.
[8] Brauer H. VDI Forschugsheft - 1956, H. 457, S. 1 – 40.
[9] Lorenz J.J., Yung D.A. *Trans. ASME, ser. E,* 1979, N. 1, P. 206.
[10] Chyu M.C., Bergles A.E. *Trans. ASME Ser. C*, 1988. N 3, p.175
[11] Rubanov E.A., Investigation of heat transfer to thin films under phase transformation. Dissertation to the degree of candidate of technical sciences, Novosibirsk, Institute of Thermophysics.
[12] Je Chin Fletcher L.S. *Int. Eng. Chem. Prog. Des.* 1985 v. 24, N 3, p. 570.
[13] Danilova G.N., Bukin V.G., Dyundin V.A. *Kholodilnaya tekhnika*, 1976, N. 6, P. 21.
[14] Fujita Y., Tsutsui M. *Proc. 10th Int. Heat Transfer Conf. Brighton* 1994 v. 6 p. 175.
[15] Gogonin I.I. Theoretical Fundamental Chemical Technology, 2014, V. 48, N. 1, P. 103.
[16] Kutateladze S.S., Nakoryakov V.E., Heat mass transfer and waves in gas-liquid systems. Novosibirsk, Nauka, 1984.
[17] Parken W.H., Fletcher F.L., Sernas S., Han J.C. *Trans ASME ser. A*, 1991, N. 2, P. 11.
[18] Danilova G.N., Dosov V.G. *Kholodilnaya tekhnika*, 1979, N. 8, P. 39.

5th International Workshop on Heat/Mass Transfer Advances
for Energy Conservation and Pollution Control,
August 13-16, 2019, Novosibirsk, Russia

# HEAT TRANSFER AND HYDRODYNAMIC CHARACTERISTICS AT EVAPORATION OF LIQUID FILM IRRIGATING A HORIZONTAL BUNDLE OF FINNED TUBES

**I.I. Gogonin.**

Kutateladze Institute of Thermophysics, Acad. Lavrentyev Ave. 1, Novosibirsk, Russia

**Abstract.** The paper presents experimental results on heat transfer during evaporation of an R21 freon film that irrigates a packing of horizontal finned tubes. It is shown that heat transfer on the finned tubes is noticeably more intense even with respect to the full surface of the finned tube.

## Introduction

The finning of tubes is a well-proven method of heat transfer enhancement at evaporation and boiling. Unfortunately, experimental data on heat transfer at evaporation and boiling of film irrigating a horizontal bundle of finned tubes are very few [1, 2]. Below there are heat transfer experimental results at the evaporation of Freon R21 irrigating a bundle of horizontal finned tubes. The experiments are carried out on an installation with forced circulation of the working substance.

## Measurement technique

The temperature of the water heating experimental section is measured by semiconductor temperature sensors as well as controlled by multi-junction differential thermocouples. The sensors are calibrated with the accuracy of not less than 0.02°C.

The wall temperature is calculated from the heat transfer coefficient.

The heat flux density on the experimental sections is calculated according to:

$$q=(m \cdot C_p \cdot \Delta T)/F \qquad (1)$$

The heating water flow rate is determined on the readings of rotameters as well as flowmeters with the accuracy rating 0.5. The rotameters installed in every section are calibrated individually.

The experiments are carried out at irrigation of the bundle with Freon R21 at $t_s = 40°C$. Reynolds number of the film irrigating the tube bundle is changed within the range of $380 \le Re \le 1500$. The bundle consists of 12 tubes arranged in a vertical row. Experimental tubes are placed in the bottom part of the bundle. The second from the bottom tube was a plain copper (M-1) tube with technical surface finish $R_z \approx 2.5$ microns, the wall thickness $\delta_w = 2$ mm, and D = 10 mm. The third from the bottom copper (M-1) tube is a finned tube with the pitch of helical finning 1.07 mm and the finning factor 1.46.

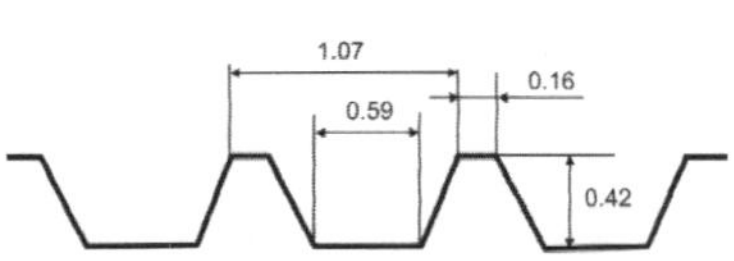


**Figure 1.** Dimensions (in mm) of the helical groove threaded on the outer surface of the 3rd copper experimental tube.

In **Fig. 1** the dimensions of the finned tube are given. According to the investigation results published in [3], the distance between ribs is sufficient to hold the minimal liquid amount within a groove between ribs.

**Measurement results**

In **Fig. 2** heat transfer experimental results are presented both for the plain and the finned tube in coordinates $q$ - $\Delta T$. It is seen that boiling on the plain copper tube took place at $\Delta T$ = 4.5 - 5°C ($q \cong 10^4$ W/m$^2$). The value of heat flux density on the finned tube is here conditional because it is calculated relative to the surface area of a plain tube with the outer diameter 10 mm. The experimental results on the finned tubes appeared to be non-traditional. Usually boiling on finned tubes takes place at considerably lower temperature drops than on plain tubes. In these experiments, there is no boiling within the whole investigated range of heat fluxes.

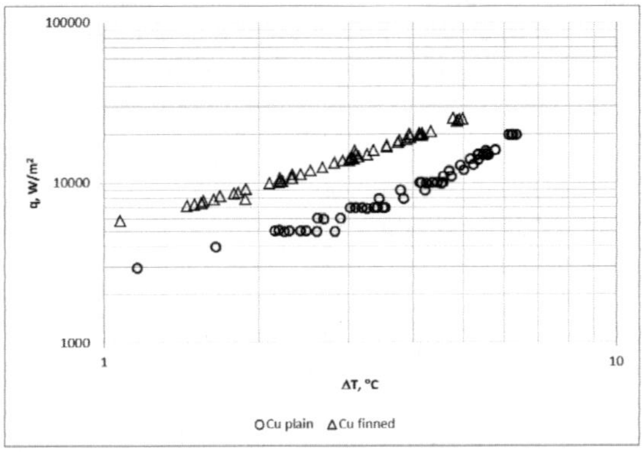


**Figure 2.** Experimental results for plain and finned copper tubes in coordinates $q$ - $\Delta T$.

It is especially distinctly clear in the **Fig. 3** showing that $\alpha \approx$ const within the whole range of parameters. In **Fig. 3** the heat flux density is calculated relative to the whole surface area of the finned tube. It follows from the **Fig. 3** that evaporation heat transfer is considerably higher on the finned tube than on the plain tube even if the heat flux density is calculated relative to the full surface of the finned tube.

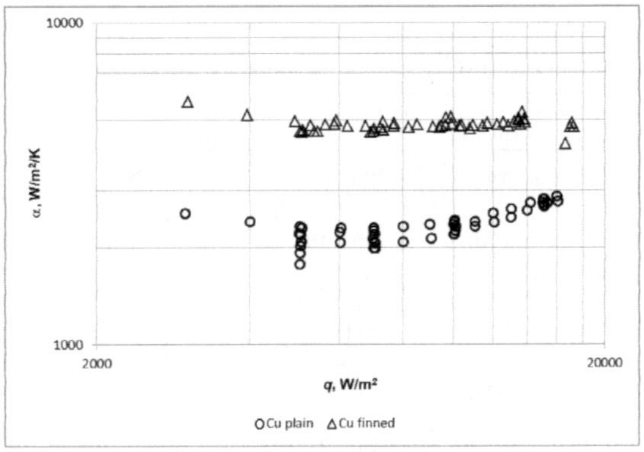


**Figure 3.** Experimental results for plain and finned copper tubes in coordinates $\alpha$ - $q$ where the heat flux density is calculated relative to the whole surface area of the finned tube.

5th International Workshop on Heat/Mass Transfer Advances
for Energy Conservation and Pollution Control,
August 13-16, 2019, Novosibirsk, Russia

In **Fig. 4** a physical model of film flow on a finned tube is presented. Supposed that surface tension force considerably exceeds gravitational force at film flow through the area between fins. Both visual observations and analysis of video films investigating hydrodynamics of film irrigating finned tube bundle justify this supposition. Micro-bubbles escaping at ethanol irrigation of finned tubes show the liquid motion trajectory along the lateral rib surface quite distinctly. A diagram of this movement is shown in **Fig. 4**.

Our visual observation, as well as video films, showed that under the force of surface tension liquid flows from a rib apex on the lateral rib surface to the groove bottom along the trajectory shown in **Fig. 4**. A thin film on the lateral area of the rib wets its surface and evaporates intensively.

It means that at liquid film irrigation of a finned tube the rib height becomes a characteristic linear dimension. Besides, the copper finned tube could be considered as an isothermal surface because the rib efficiency for this tube is determined according to the following dependency (4)

$$E = \frac{th(l\sqrt{2\alpha\cdot\delta/\lambda}}{l\sqrt{2\alpha\cdot\delta/\lambda}}$$

and close to 1. Here $l=h/\delta$; $\delta$ is the rib thickness near its base; $h$ is a height of the rib.

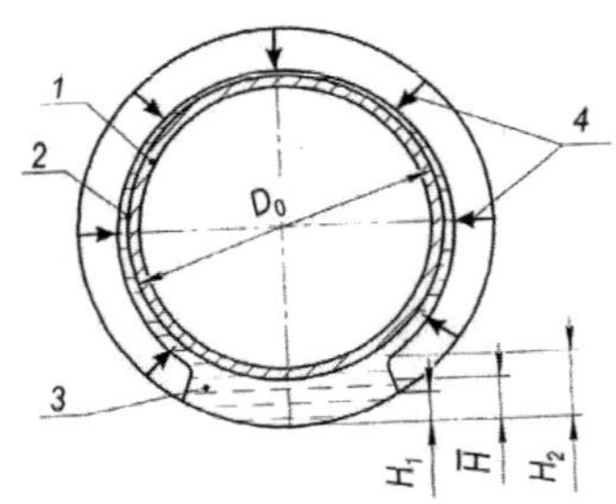


**Fig. 4.** Diagram of liquid movement along the rib.

1 – the base of the finned tube, 2 – liquid film, 3 –liquid layer held between ribs, 4 – the direction of liquid movement along the rib. H – the average height of the held liquid layer.

Calculation of heat transfer at the evaporation of film irrigating a bundle of finned tubes is difficult because of very complicated hydrodynamics at the irrigation of each rib.

The lateral rib surface is irrigated in its upper part with thin liquid laminar film flowing down from the rib edge.

The bottom part of the rib within the groove between ribs is irrigated with the main liquid flow that flows along the groove between ribs and is turbulent.

The boundary between laminar and turbulent flows cannot be determined correctly. It can be estimated at a first approximation.

Heat transfer coefficient at film evaporation from different rib surfaces can be calculated only approximately taking into account the following assumptions:

5th International Workshop on Heat/Mass Transfer Advances
for Energy Conservation and Pollution Control,
August 13-16, 2019, Novosibirsk, Russia

a) Surface tension force exceeds considerably the force of gravity. The upper part of the lateral rib surface is irrigated with liquid flowing from the edge rib surface to the groove along the trajectory shown in **Fig. 4**.
b) Heat transfer both on the edge part of a rib and within the groove between ribs takes place like on a smooth tube [5].
c) The mass flow rate irrigating the rib edge is determined from the balance dependencies.
d) He rib is within the initial section of the thermal boundary layer with the rib height as a characteristic linear dimension.
e) A copper finned tube can be considered as isothermal surface so the rib efficiency is close to 1.
f) The film thickness in the groove is calculated on the Dukler – Bergelin dependence [6]:

$$\delta = Re_{cr}^{-0.2}\left(\frac{3\nu^2}{g}\right)^{1/3} Re^{8/15} \qquad (2)$$

The critical Reynolds number is determined here according to the Brauer dependence [7]:

$$Re_{cr} = 35Ka^{\frac{1}{10}} \qquad (3)$$

The film thickness calculated on (2) is close to a film thickness determined according to the dependency from [8] and to the experimental results presented in [9].

The irrigation density of the upper rib part is calculated according to the dependence:

$$G_T = \frac{m_T}{2\pi R}\ [kg/m{\cdot}s],$$

where $m_T[\frac{kg}{s}]$ is a mass flow rate of the liquid flowing along the rib edge. The film Reynolds number $Re_2$ on the lateral rib surface is determined on the dependence $Re_2 = G_2/\mu$.

The heat transfer on the initial section of the thermal boundary layer (the upper part of the lateral rib surface) is determined on the dependence:

$$Nu_2^* = 0.6{\cdot}Pr^{\frac{1}{3}}{\cdot}(\frac{Re_2}{\bar{h}})^{1/9} \qquad (4)$$

Dependence (4) is an empirical one. It is recorded by analogy with the analytical formula from [1] obtained to describe heat transfer on the initial section of a vertical wall. In the dependence (4) the Galileo number is replaced by a dimensionless rib height:

$$\bar{h} = \frac{h}{\sqrt{\frac{\sigma}{g{\cdot}(\rho_L - \rho_V)}}}$$

To determine heat transfer coefficients both on the edge part of the rib and within the groove at turbulent film flow we can use the three-layer model described in [11], where turbulent viscosity is piecewise approximated by corresponding dependencies. In [11] a Table of numeral computation results of Nusselt number via film Reynolds number for some integer values of Prandtl number is presented.

Calculating film thickness in the groove according to (5) and considering that the groove has a trapezoidal shape, we can approximately determine areas of the lateral rib part streamlined by turbulent or laminar liquid flows. Using experimental data on $\Delta T = T_w - T_s$ we can calculate specific heat fluxes on each part of a finned tube. The total heat removal consists of the following components:

$$\Sigma Q = (Q_{\mathrm{edge}} + Q_{\mathrm{groove}} + 2Q_{\mathrm{lam}} + 2Q_{\mathrm{turb}})\cdot n \tag{5}$$

Here n is the number of fins on the experimental tube surface; $Q_{\mathrm{edge}}$ , $Q_{\mathrm{groove}}$, $2Q_{\mathrm{lam}}$, $2Q_{\mathrm{turb}}$ – heat removals from the tube edge, from the horizontal groove part between ribs, from the lateral rib part streamlined by laminar flow, by turbulent flow, n is a number of ribs on an experimental tube.

The calculated on (5) values of heat removals agree with experimental value with an error ±15% within the whole parameter range of the experimental investigation.

## Conclusion

The heat transfer at evaporation on finned tubes is so intense that excludes conditions which are necessary for vapor bubble origin. It should be noted that at the maximum heat flux value on the finned tube $\Delta T_w$ was always < 5°C (i.e. the temperature head of the boiling inception on the smooth tube).

The presented calculation algorithm describes the experimental results satisfactorily.

**Acknowledgements**
The study was performed at the Institute of Thermophysics (SB RAS) with the support of the BSI SAS Program for 2017-2020 (project III.18.2.3, reg. no. AAAA-17-117030310025-3)

## Reference

[1] Fujita J., Tsatsui T. 1994 *Proc. J. 10th Int. Heat Transfer conference* - Brighton UK, **6** 175-170.
[2] Pavlenko A.N., Pecherkin N.I., Volodin O.A. 2016 ed. SB RAS, Novosibirsk, 191
[3] Gogonin I.I., Kabov O.A. 1983 ed. SB RAS series of technical sciences **2**, 3-9.
[4] Isachenko V.P., Osipova V.A., Sukomel A.S. 1980 Energoizdat, 417.
[5] Gogonin I.I. 2014 Theoretical Fundamental Chemical Technology **48** 103.
[6] Dukler A.E., Bergelin O.F. 1952 *Chem Eng. Progress* **48** 557.
[7] Brauer H. 1956 *VDI Forschugsheft* 1 – 40.
[8] Geshev P.I. 2014 Thermophysics and Aeromechanics **21** 579-586.
[9] Asali J.C., Hanratty T.J. 1985 Andreussi P. AIChE J. **31** 895–902.

5th International Workshop on Heat/Mass Transfer Advances
for Energy Conservation and Pollution Control
August 13-16, 2019, Novosibirsk, Russia

# Energy conservation and the use of the heat of refrigeration of ice arenas to heat the school of figure skating of St. Petersburg

**A. Kondrashov[1*] and M. Egorov[1,2]**

[1]Peter the Great St.Petersburg Polytechnic University, St. Petersburg, 195251, Russia

* Corresponding author: kondrashov22@gmail.com

[2]Joint-Stock Company "I.I. Polzunov Scientific and Development Association on Research and Design of Power Equipment", St. Petersburg, 191167, Russia

**Abstract.** A variant has been proposed for using the system for utilizing the heat of condensation of heat engines, most of which has not been used and released into the environment, for air heating of ice arenas. An analysis of unused heat energy was made using the example of a refrigerating machine, which, in addition to cold production, also produces a significant amount of heat that is generated during the compression of the refrigerant in the compressor to the pressure and condensation temperature, and in most of the existing refrigeration units, thermal energy is not used and transferred to the environment using air condensers or cooling towers located on the street. The analysis of the economic efficiency of the use of heat recovery systems, for various technical needs. A study was carried out on the actual implementation of the system of partial utilization of the condensation heat of refrigerating machines, which is aimed at air heating of ice arenas at the St. Petersburg State Budgetary Institution of the specialized school of Olympic reserve on figure skating (hereinafter referred to as the "Institution").

## 1. Introduction

Energy saving is today considered one of the most relevant areas of Russia's development in connection with the entry into force of Federal Law No. 261 of November 23, 2009 on Energy Saving [1] and Energy Efficiency and on amendments to certain legislative acts of the Russian Federation [2].

The purpose of the current study is to analyze the economic efficiency of using heat recovery systems released into the environment from heat engines (Figure 1) for various technical needs [3-6].

The objectives are to conduct a study on the introduction of the system for utilization of heat energy from heat engines [7,8] released into the environment [9] for various technical needs, related to the need to reduce thermal pollution [10], as well as the continuous increase in energy prices

**Figure 1** Refrigeration room

Before the modernization of the heating system of ice arenas at the St. Petersburg State Budgetary Institution of the specialized school of Olympic reserve on figure skating (hereinafter referred to as the "Institution"), the heating was carried out at the expense of the air conditioning system, fan heaters and radiators (during the heating season).

Air conditioners of ice arenas (Figure 2) incl. consist of a section for the first heating of air (heat from GUP TEK), a section for drying (a cooling circuit from refrigerators) and a section for a second heating (a circuit of forcondensers of refrigerators).

**Figure 2** Ventilation room. Air conditioners ice arenas.

It is worth noting that in the Institution it is impossible to completely abandon the use of the primary circuit for heating air conditioners in ice arenas, for maximum loading of the second circuit, because of the possible freezing of the heat exchanger, as well as the operation of the automation system, therefore air gets to the secondary circuit already heated (above + 25°C). It is also not possible to load the second heating circuit as much as possible, since a significant increase in the temperature of the elements of the installation can cause damage to the engine, bearings, parts made of synthetic materials [10].

Including supply/exhaust pipelines to the ice arenas made at a height of more than 6 meters from the floor level, which does not allow to fulfill the regulatory requirements regarding the air temperature in the ice arena at a height of 1.5 m from the ice (+ 6 + 12 °C) [11], when using air conditioning ice arenas.

However, it is possible to heat the air in ice arenas at a height of 1.5 m from the ice, during the inter-heating period, using the VEABA W61 fan heaters (Figure 3) installed on the Institutional ice arenas.

**Figure 3** Main Arena. Fan Heaters

## 2. Materials and Methods.

*2.1. Equipment*

Institution in which the research is carried out:

- St. Petersburg State Budgetary Institution of the specialized school of Olympic reserve on figure skating, located at St. Petersburg ul. Tupolevskaya, Building 4

Equipment:

- three ice arenas (2 arenas – 30x60 sq.m. And 1 arena 20x24 sq.m., Total area 4100 sq.m.)
- three BE/SMED-BT2402 chillers manufactured by Climaveneta, each containing two Bitzer CSH 8571-140Y-40D semi-hermetic screw compressors with a total cold output of 900 kW.
- Three air condensers FRIGA-BOHN ECA08P9 P10C3 (Figure 4)
- Six Alfa Laval CPS/S 145 forcondenser
- Wilo IL50 / 170-7.2 / 2 Circulation Pump
- Twenty fan heaters VEABA W61

**Figure 4** Air Condenser. Roof

*2.2. Analysis of the economic feasibility of a design decision on the introduction of energy-saving technologies, which consists of two subsections.*

1. Analysis of the economic effect of the complete failure of the use of heat from the heat supplying organization GUP TEK SPb, for the hydraulic system of heat fans, air heating of ice arenas during the heating period. 7 months (from January to April and from October to December)
2. Analysis of the economic effect of the complete failure to use the heat pump (chiller), which was used only in the inter-heating period. 5 months (from May to September)

**Determination of the economic effect of the complete rejection of the use of heat from the heat supplying organization GUP TEC St. Petersburg, for the hydraulic system of fan heaters, air heating of ice arenas during the heating season. 7 months (from January to April and from October to December)**

$$Cyear.eff.of\ heat = Tyear.\text{work} \times Qrecycling \times Ctherm.energy \times Kuse\ of\ heat - Coperating\ costs\ of\ the\ system.$$

where:

$C_{year.eff.of\ heat}$ – total annual effect in case of heat rejection of GUP TEK, mln. rur.;

$T_{year.work}$– time of the heating season, hours/year;

$Q_{recycling}$– heat recovered on forcondensators, kW;

$K_{use\ of\ heat}$ – the coefficient of using the heat of condensation is set equal to "0.5", because of the inability to use the system to 100%;

$C_{therm.energy}$ – cost of heat energy, rur/kW;

$C_{operating\ costs\ of\ the\ system}$.. – the cost of the system, rur.;

$$Tyear.\text{work} = \frac{18hours}{day} \cdot \frac{7days}{a\ week} \cdot \frac{4{,}4weeks}{month} \cdot \frac{7months}{year} = \frac{3880{,}8hours}{year}$$

$Q_{recycling}$=6 forcondensators *75 kW =450 kW

$C_{therm.energy}$.= 2,087 rur./kW with VAT. (1 Gcal =1,163 MW =1163kW. At a cost of 1 Gcal, 2427 rubles. with VAT, we get the cost of 1kW of heat energy = 2427/1163 = 2,087 rubles/kW)

$C_{\text{operating costs of the system}}$ - for the operation of the system only a circulating pump with a capacity of 7.5kW is needed., that is, $N_{\text{consumption of the heat recovery system}}$=7,5 kW.

Due to the fact that our Institution pays for the consumed electric energy in the fourth price category of JSC «PSK», then:
• cost of 1 kW * hour of electrical energy Cel.en. = 1.84 rubles/kW*hour,
• the cost of 1 kW for the transmission of electrical power Cel.power transm.= 1583 rubles / kW
• the cost of 1 kW of consumed electric power Ccons.el.power = 775 rubles/kW
• the cost of the operation of the heat recovery system of condensation will be:
$C_{\text{operating costs of the system}}=T_{\text{year.work}} \times (N_{\text{consumption of the heat recovery system}} \times C_{\text{el.en.}}))+7_{\text{months}}\times(N_{\text{consumption of the heat recovery system}}\times(C_{\text{el.power transm.}}+C_{\text{cons.el.power}}))$
$C_{\text{operating costs of the system.}}$=(3880,8×(7,5kw×1,84))+7×(7,5kw×(1583+775))=177 350 rubles
The total annual economic effect of using the heat recovery system in case of failure during the heating period from the heat of the SUE TEC for fan heaters will be (Figure 5):

$$C\,year.eff.of\ heat = 3880{,}8\frac{hours}{year} \times 450\text{kW} \times 2{,}087\frac{\text{rur}}{\text{kW}} \times 0{,}5 - 177350\ \text{rur.} = \mathbf{1{,}65}\ \text{mln. rur.}/7months\ or\ 235714\ rur/months$$

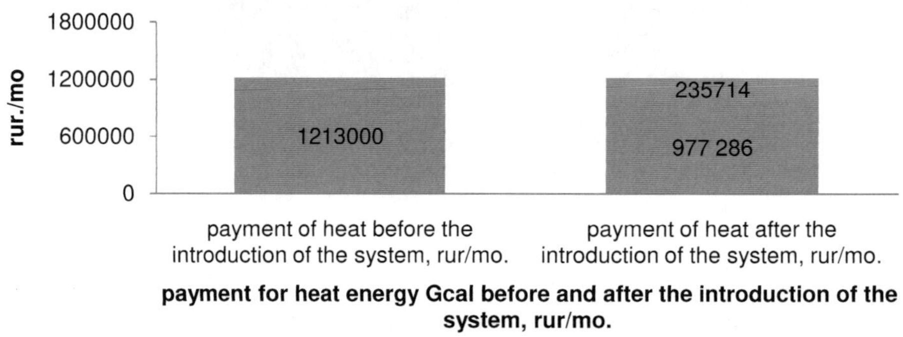


**Figure 5** Payment for heat energy Gcal, before and after the implementation of the system, rubles/month.

**Determination of the economic effect of the complete failure to use a heat pump (chiller), which was used only in the inter-heating period. 5 months (from May to September)**

$$Cheat\ pump\ savings = (Tyear.\text{work} \times ((Naver.pow.cons.heat\ pump + Naver.pow.cons.circ.pump) \times \text{Cel. en}) + 5months \times ((Naver.pow.cons.heat\ pump + Naver.pow.cons.circ.pump) \times (\text{Cel. power transm.} + \text{Ccons. el. power}))$$

where: $C_{\text{heat pump savings}}$ – savings from the use of a heat pump, rub.;
$T_{\text{year.work}}$– the time of operation of the heat pump in the inter-heating period, hours/year;
$N_{\text{aver.pow.cons.heat pump.}}$– average power consumption of the heat pump;
$N_{\text{aver.pow.cons.circ.pump}}$.– average power consumption of a single motor of a circulation pump, kW;
$C_{\text{el.en.}}$– cost of electric energy, rubles/kWh;
$C_{\text{el.power transm.}}$ – the cost of 1 kW for the transmission of electrical power;
$C_{\text{cons.el.power}}$ – cost of 1 kW of consumed electrical power;

$$Tyear.\text{work} = \frac{18hours}{day} \cdot \frac{7days}{a\ week} \cdot \frac{4{,}4weeks}{month} \cdot \frac{5months}{year} = 2772\frac{hours}{year}$$

$$Cheat\ pump\ savings = (2772 \times ((33kW + 7{,}5kW) \times 1{,}84) + 5 \times ((33kW + 7{,}5kW) \times (1583 + 775)) = \mathbf{684064}\ \text{rur.}/5months\ \text{or 136 812 rur/months}$$

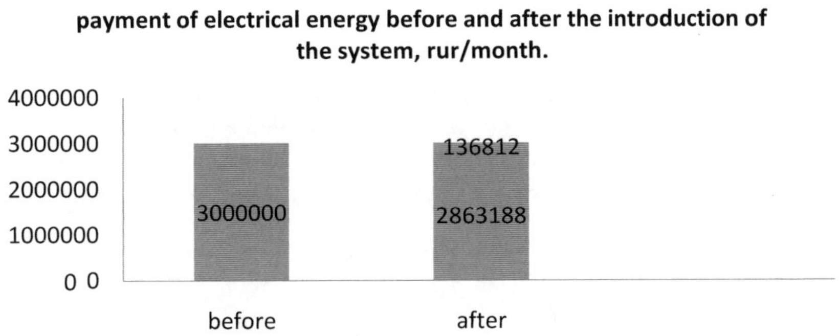


**Figure 6** Payment of electrical energy before and after the introduction of the system, rubles/month.

**Total annual savings from the operation of the heat recovery system.**

$C_{total\ savings} = C_{year.eff.of\ heat}+C_{heat\ pump\ savings}- C_{operating\ costs\ of\ the\ system}$

$C_{total\ savings}$=1,65mln.rur.+684064 rur. –177350 rur.=2,15 mln.rur./year

*2.3. Implementation costs:*

Costing for the implementation of the condensation heat recovery system for refrigerating machines (Figure 5)

The cost of equipment and installation work on the implementation of the system of partial utilization of the heat of condensation of refrigerating machines, for the air heating system of ice arenas - Cexpenses= 115,131.42 rubles.

The cost of operating the system per year (the work of a single electric motor 7.5 kW) – 307028 rubles. in year. (6652.8×(7,5kW×1,84))+12×(7,5 kW×(1583+775))=307028.64 rur/year

**Figure 7** The introduction of the system of utilization of the heat of condensation of refrigerators into the existing hydraulic system of air heating of the fan heaters of the ice arenas.

*2.4. The payback period for capital costs is:*

$C_{payback\ period}=C_{expenses}/C_{total\ savings}$= 115131rur/2.15mln.rur=1 month.

## 3. Introduction

The results of the study showed that to increase the energy efficiency of engineering systems, the high costs associated with equipment replacement and capital construction are not necessary.

In the future, it is planned to introduce a system for the complete utilization of the condensation heat of refrigerating machines, the almost complete absence of thermal pollution of the environment and the use of a system in addition to air heating of ice arenas for:

1. Thawing snow in the pit of snowmelt of ice-picker combines (currently in the Institution is carried out manually due to hot water from the GVS, and during the heating period due to the radiators in the pit of snowmelt)
2. Ground heating systems against freezing under ice arenas (at present, this system can only be used during the heating season)

## 4. Conclusions

The introduction of a system of partial utilization of the heat of condensation of refrigerators, for heating ice arenas at the expense of fan heaters, showed the following savings:

1. In case of refusal from the heat supplied by the SUE TEK to heat the coolant for fan heaters of ice arenas in favor of using the system of utilization of the heat of condensation of refrigerating machines, we obtain savings of 1 million 910 thousand rubles in year.
2. Reduction of thermal pollution [13] of the city of St. Petersburg from the Institution by not less than 1746360 kW per year.

## References

[1] Russian Federation. Laws. On energy saving and on increasing energy efficiency and on introducing amendments to certain legislative acts of the Russian Federation dated 11.23.2009 No. 261-FL. - Access mode: http://base.consultant.ru.
[2] Energy Saving: State Standard of the Russian Federation of November 30, 1999 No. 485-st (approved by a Resolution of the State Standard of Russia). [Electronic resource]. - Access mode: http://base.consultant.ru.
[3] V. Polyakova, T. Kalinina, K. Kichin. Analysis of the energy efficiency of heat pumps in heat supply systems according to the territorial-climatic basis. Young Scientist. 27, 142-146. 2016. URL https://moluch.ru/archive/131/36474/
[4] O. Averina, E. Moskaleva, T. Morozkina. Criteria for assessing energy efficiency. Young Scientist. 8, 427-429. 2014.URL https://moluch.ru/archive/67/11289/
[5] Energy saving technology: Textbook / Yu. D. Sibikin, M. Yu. Sibikin. - 3rd ed., Pererab. and add. - M .: Forum: SIC INFRA-M, 2013. - 352 p.Thermal pollution causes global warming, 2001, Bo Nordell, Lulea, Sweden
[6] Thermal pollution causes global warming, 2001, Bo Nordell, Lulea, Sweden
[7] Belokon NI, Thermodynamics, 1954, p. 117.
[8] Saveliev I.V. The course of general physics. T. 1. Mechanics. Molecular physics. - M., Science, 1987. - p. 341
[9] Kantalinsky V.P., Patrakeeva E.A. Research on heat recovery methods at a production plant. Improving the efficiency of power plants // International collection of scientific papers, Kaliningrad, 2002, p. 18 - 25.
[10] Thermal pollution causes global warming, 2001, Bo Nordell, Lulea, Sweden
[11] Technical and operational documentation. Units for ventilation and air conditioning of the type CV-A (internally) and CV-D (outdoor). VTS CLIMA. P.25
[12] A collection of rules CR31-112-2007 Part 3 table 6.2
[13] A. Soldatov, O. Skotnikova. Utilization of thermal emissions from NPP – creation of an energy-biological complex. Proceedings of International Student Scientific Conference "Polar Lights", 1999

5th International Workshop on Heat/Mass Transfer Advances
for Energy Conservation and Pollution Control
August 13-16, 2019, Novosibirsk, Russia

# Hydraulic resistance and heat transfer augmentation in pipes with inner helical micro-ribs: Review, data generalisation, and technical recommendations

**I. A. Popov[1], A. N. Skrypnik[1], A. V. Schelchkov[1,2]**

1. Kazan National Research Technical University named after A. N. Tupolev – KAI, Russia

2. All-Russian Research Institute of Flow Metering, Kazan, Republic of Tatarstan, Russian Federation

popov-igor-alex@yandex.ru

Abstract. The paper presents the data generalization and development of general correlations for heat transfer and pressure drop in pipes with inner helical micro-ribs. Heat transfer and friction coefficient data were gathered from related investigations on single-phase turbulent flow of Newtonian fluid in pipes with various shape inner helically-ribbed surfaces

## 1. Introduction

The turbulization and the periodical destruction of the boundary layer by inducing the swirling flow is known as an effective heat transfer augmentation technique. To date, a considerable number of scientific papers dedicated to investigations on heat transfer and friction of the flow in helically-ribbed pipes have been published.

There are various methods of producing the helically-ribbed pipe surfaces: by wire inserts, surface corrugation, machining on a lathe, etc. (fig. 1).

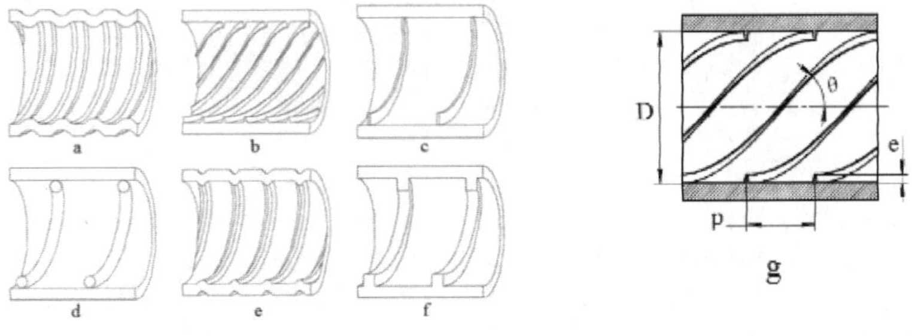


Figure 1. Various helically-ribbed pipe geometries (a-f) and characteristic dimensions (g): a – multistart corrugation; b – inner thread; c – tape insert, d – wire insert, e - corrugation, f - extrusion

Such pipes could be used for the purpose of heat transfer enhancement in single-phase flows, as well as for evaporating flow. Therefore, helically-ribbed pipes are widely used in boilers, air conditioning systems, chemical industry, and food processing. The literature indicates that pipes with the helical angle of ribs $\theta$=18-30° are widely used in the industry. However, research studies show greater efficiency for the pipes with the higher helical angle of ribs ($\theta$=70-86°). For the practical application, helically-ribbed pipe geometrical parameters must be justified in accordance with the demands.

## 2. Development of correlations for pressure drop and heat transfer for the single-phase flow in pipes with inner helical micro-ribs

The paper considers results on heat transfer and friction of the single-phase turbulent flow in pipes with inner helical micro-ribs. In order to obtain a database of friction and heat transfer coefficients, numerous investigations were examined: E. Sams (1956), D. Wilkie (1966), N. Sheriff (1966), P. Kumar (1970), R. Webb (1971, 1972, 1980, 2000), H. Yoshitomi (1976), R. Gupta (1979), T. Carnavos (1979, 1980), J. Withers (1980), В.Мигая (1980, 1981, 1989), H.M. Li (1982), Raja Rao (1982, 1983, 1985), W. Nakayama (1983), A.Bergles and T. Ravigururajan (1983, 1986, 1996), J.Chiou (1987), V. Zimparov (1991), Y. Nazmeev (1993), M. Jensen (1999), D. Yang (2001), S.Rainieri (2002), A Garcıa (2004, 2005, 2007, 2012, 2018), Г.Ефимова (2006) (30), G. Zdaniuk (2008), N.-H.Kim (2018), and authors [4] [5].

Heat transfer and friction factor data were correlated by the assumed model:

$$\xi(=2\cdot\Delta P/[(L/D)\cdot(\rho\cdot w^2)]), \mathrm{Nu}(=\alpha D/\lambda)=\mathrm{f}(\mathrm{Re}_D, e/D, p/D, p/e, N, \theta/90) \quad (1)$$

The choice of factors included in the model was based on the similarity theory, dimensional analysis, and literature review. The reviewed parameters and their physical meaning are explained in Table 1.

Table 1.

| Parameter | Description | Parameter meaning |
|---|---|---|
| $Re_D$ | Reynolds number | The ratio of inertial and viscous forces. Characterizes a flow regime. |
| p/e | Relative rib spacing | Characterizes a separation and reattachment of the flow after the ribs. |
| e/D | Relative rib height | The exposure of the rib height on the flow structure. The influence is subject to the boundary layer height (or roughness height e+). |
| p/D | Axial pitch | Characterizes a flow separation and reattachment zone. |
| N | Number of starts | Characterizes an integral roughness of the heat transfer surface. It unites geometrical parameters – *p* and $\theta$. |
| θ/90 | Normalized helix angle | Characterizes the flow rotation degree at the near-wall layers in helically-ribbed pipes. The helix angle value is normalized by the maximum angle of the helix (θ=90°). |

The analysis of the reciprocal influence between the factors mentioned above revealed the usage of the full range overdetermine the flow over the helically-ribbed pipe surface.

Geometrical parameters of the helically-ribbed pipe (besides pipes with transverse ribs $\theta$=90°) $N$, $p$ and $\theta$ are linked by the following relationship:

$$\xi, \mathrm{Nu}=f(\mathrm{Re}_D^{\,e/D,p/D,\theta/90}, (e/D)^{\mathrm{Re}_D,p/D,\theta/90}, (p/D)^{\mathrm{Re}_D,e/D,\theta/90}, (\theta/D)^{\mathrm{Re}_D,e/D,p/D}) \quad (2)$$

Thus, through an increase of the number of helical starts $N$ for the constant axial pitch $p$ the helix angle $\theta$ is changing.

Experimental results were obtained in the wide range of the helix angle values ($\theta$=*14-87°*). Thus, the number of starts N varied within the range $N$=1-54 for the fixed $\theta$ value.

According to the analysis of geometric simplices, which include *e, p* and $\theta$ were chosen. However, the axial pitch *p* could be included in two different dimensionless simplices: *p/e* и *p/D*. Therefore, *p/e* simplex was excluded since the rib height e is considered by the factor *e/D*.

In this regard, the generalization model (1) transforms to:

$$\xi, \text{Nu}=\text{f}(\text{Re}_D, e/D, p/D, \theta/90) \quad (3)$$

The model has been justified by the accuracy of the obtained correlations (described below).

The correlation equations were developed by polynomial regression method by means of the least square method.

The database for the friction factor and heat transfer coefficient contains more than 2200 points. Correlation equations for the friction factor $\xi$ and Nusselt number Nu were found to be:

$$\xi = 6.086 \cdot \text{Re}_\text{D}^{-0,226} \cdot (\text{e}/\text{D})^{0,657} \cdot (\text{p}/\text{D})^{-0,292} \cdot (\theta/90)^{0.543} \quad (4)$$

$$\overline{\text{Nu}} = 0.08 \cdot \text{Re}_\text{D}^{0,776} \cdot (\text{e}/\text{D})^{0,138} \cdot (\text{p}/\text{D})^{-0,186} \cdot (\text{q}/90)^{0.076} \cdot \text{Pr}^{0,4} \quad (5)$$

Equations (3) and (4) was developed in the following range of flow and dimensionless geometric parameters: $\text{Re}_\text{D}=4\cdot10^3$-$2\cdot10^5$; $p/D$ = 0.059 - 10; $e/D$ = 0.01 - 0.36; $\theta/90$ =0.16 - 0.99, $N$=1-54. A deviation of experimental data from the friction factor correlation (3) was ±50% with the confidence level of P=90% (coefficient of determination $R^2$=0.59).

For the Nusselt number correlation (4), experimental data deviation was found to be ±50% with the confidence level of P=95% and 35% for P=90% ($R^2 = 0.78$).

Similar results were obtained in T. Ravigururajan, A. Bergles [1] with a smaller number of experimental points (1650 and 1800 points for the friction factor and Nusselt number respectively).

In the view of high demands on the reliable prediction of thermal and hydraulic properties of heat exchange equipment (in the transport and energy sectors) mentioned experimental data deviation is not satisfactory.

The other model for the correlations was used in analogy to the model in A. Bergles and T. Ravigururajan (1986, 1996) [1]:

$$\xi, \text{Nu}=f(\text{Re}_D^{\,e/D,p/D,\theta/90}, (e/D)^{\text{Re}_D,p/D,\theta/90}, (p/D)^{\text{Re}_D,e/D,\theta/90}, (\theta/D)^{\text{Re}_D,e/D,p/D}) \quad (6)$$

This model considers all the mutual effects of the factors included with the maximal experimental data deviation of ±35% both to the Nusselt number and friction factor (*P=0.9*). However, analysis of developed correlations shows the absence of physical meaning despite the higher correlation accuracy.

In this regard, further research on the heat transfer and friction factor experimental data variation for the flow in helically-ribbed pipes were carried out. Based on the results obtained and analysis of the flow structure the experimental data were categorized by the flow parameters and ribs geometry.

One of the key parameters for the data characterization is the relative distance between the ribs *p/e*. As explained in Table 1, the parameter *p/e* characterizes the flow separation for the flow over the ribs. as shown in Webb (1971) [2].

As it is shown in Webb, for the pipes with repeated rib roughness ($\theta$=*90°*) the flow structure could be categorized by the flow patterns downstream of the rib as a function of the relative rib spacing *p/e* (fig. 2). Edwards and Sheriff [1961] have shown the heat transfer coefficient reaches the peak at the reattachment point downstream the rib at the relative distance between the rib equal p/e≈10. Thus, for the rib spacing *p/e<10* most of the rib spacing distance is filled by the recirculation zone and therefore lower values of the heat transfer coefficients obtained.

With an increase of the relative rib spacing *p/e>10* after the reattachment point downstream the rib the boundary layer growth begins. An increase of the relative rib spacing *p/e>10* after the reattachment point is followed by the boundary layer growth downstream the rib (if the rib height is comparable to the boundary layer thickness).

In the scope of this work, it was assumed that such patterns downstream the transverse ribs (*θ=90°*) are valid for the helical ribs (*θ<90°*). This assumption is supported by the analysis of the flow structure carried out in Bergles (1994), Garcia (2007).

Taking this into account, as the threshold value of *p/e,* at which the flow pattern changes, was chosen *p/e=10*. Accordingly, the database for the friction factor and heat transfer coefficient were divided into two groups- with *p/e<10* (predominance of the recirculation zone) and p/e>10 (with the declining impact of the rib on the flow structure).

However, it must be noted that the reattachment point varies *p/e=6-15* according to the mutual influence of geometrical and flow parameters.

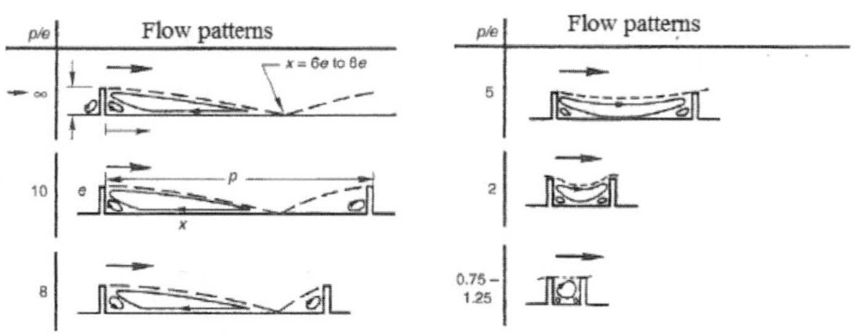


Figure 2. Flow patterns according to the relative distance between the ribs *p/e* (Webb [2])

Further, the experimental data were categorized by the value of the helix angle *θ* or by the normalized helix angle *θ/90*, relatively to the maximum angle of the helix (*θ=90°*-transverse ribs). This parameter is a characteristic of the flow rotation in helically-ribbed pipes near the wall region. The helix angle increase leads to a decrease in the flow rotation intensity. For the value of the helix angle of θ>50° the flow separation dominates over the flow rotation (Olimpiev 1992, Brodov 1987). The flow structure is similar to the flow over transverse ribs. Thus, the greatest contribution to heat transfer augmentation is made by flow separation.

A similar finding was observed in the work of Zimparov and Vulchanov [1990]. Results show that flow rotation prevails for the helix angle values of *θ<45°*. For the helix angle values over the *60* degrees the flow regime and heat transfer processes is governed by the cross-flow. In the range of *45°<θ<60°* the flow rotation and cross-flow with the flow separation make comparable contributions.

Consequently, according to the database analysis and the flow patterns the threshold value of B was found to be: for *p/e<10- θ=50°* and for *p/e>10- θ=45°.*

The friction factor *ξ* (5) and Nusselt number *Nu* (6) correlations for the turbulent flow of single phase Newtonian flow obtained by the authors are listed below:

$$\xi = A \cdot Re_D{}^n (e/D)^m \cdot (p/D)^k \cdot (\theta/90)^l \tag{7}$$

| p/e | θ | A | n | m | k | l | P | $R^2$ |
|---|---|---|---|---|---|---|---|---|
| ≤10 | θ<50° | 1,22 | -0,32 | 0,06 | -0,26 | 0,08 | ±22% | 0,74 |
| ≤10 | θ>50° | 2,28 | -0,01 | 1,05 | -0,41 | 0,61 | ±26% | 0,69 |
| >10 | θ<45° | 1,14 | -0,2 | 0,13 | -0,04 | 1,0 | ±18% | 0,91 |
| >10 | θ>45° | 8,16 | -0,18 | 0,87 | 0,34 | 0,63 | ±30% | 0,85 |

$$\overline{Nu} = B \cdot Re_D{}^n \cdot (e/D)^m \cdot (p/D)^k \cdot (\theta/90)^l \cdot Pr^{0.4} \tag{8}$$

| p/e | θ | B | n | m | k | l | P | $R^2$ |
|---|---|---|---|---|---|---|---|---|
| ≤10 | θ<50° | 0,0056 | 0,85 | -0,35 | -0,14 | 0,45 | ±12% | 0,97 |
| ≤10 | θ>50° | 0,054 | 0,82 | 0,017 | 0,15 | 1,0 | ±32% | 0,97 |

| | | | | | | | | |
|---|---|---|---|---|---|---|---|---|
| **>10** | θ<45° | 0,205 | 0,7 | 0,2 | -0,28 | -0,03 | ±26% | 0,96 |
| **>10** | θ>45° | 0,107 | 0,79 | 0,216 | -0,22 | -0,46 | ±34% | 0,94 |

where *A, B, n, m, k, l*- are coefficients of the model; *P*- the deviation of the experimental data from the correlation with the confidence interval of $P=0.95$., and $R^2$- the coefficient of determination.

Equation (5) statistics:

- for $p/e\leq 10$ and $\theta<50°$: the database contains 341 experimental points in the following range $Re_D=9{,}2\cdot 10^3$-$8{,}5\cdot 10^4$; $p/e$=2,27-9,88, $p/D$=0.059–0,816; $e/D$=0.02-0.36; θ=25-50°, $\theta/90$=0.277-0.533, $N$=1-18;
- for $p/e\leq 10$ и $\theta>50°$: the database contains 381 experimental points in the following range $Re_D=2{,}7\cdot 10^3$-$2{,}6\cdot 10^5$; $p/e$=1,95-10, $p/D$=0.156–2,17; $e/D$=0.02-0.236; $\theta$=50-90°, $\theta/90$=0.533-1.0, $N$=1-54;
- for $p/e>10$ и $\theta<45°$: the database contains 459 experimental points in the following range $Re_D=3{,}2\cdot 10^3$-$2{,}5\cdot 10^5$; $p/e$=13,11-167,72, $p/D$=0.15–6,9; $e/D$=0.01-0.2; $\theta$=24-46°, $\theta/90$=0.27-0.511, $N$=1-54;
- for $p/e>10$ и $\theta>45°$: the database contains experimental points in the following range $Re_D=3{,}1\cdot 10^3$-$4{,}9\cdot 10^5$; $p/e$=10,59-76,61, $p/D$=0.143–3,11; $e/D$=0.01-0.22; $\theta$=45-90°, $\theta/90$=0.5-1.0, $N$=1-54.

Equation (6) statistics:

- for $p/e\leq 10$ и $\theta<50°$: the database contains 110 experimental points in the following range $Re_D=1{,}2\cdot 10^3$-$7{,}6\cdot 10^4$; Pr=0,7-8, $p/e$=0,55-9,71, $p/D$=0.059–0,389; $e/D$=0.02-0.105; $\theta$=29-48°, $\theta/90$=0.33-0.53, $N$=1-18;
- for $p/e\leq 10$ и $\theta>50°$: the database contains 434 experimental points in the following range $Re_D=3{,}9\cdot 10^3$-$4\cdot 10^5$; Pr=0,7-8, $p/e$=1,13-10, $p/D$=0.1–2,17; $e/D$=0.01-0.298; $\theta$=55,8-90°, $\theta/90$=0.62-1.0, $N$=1-54;
- for $p/e>10$ и $\theta<45°$: the database contains experimental points in the following range $Re_D=3\cdot 10^3$-$1{,}1\cdot 10^5$; Pr=0,7-8, $p/e$=13,85-167,72, $p/D$=0.54–17,81; e/D=0.01-0.218; $\theta$=24-46°, $\theta/90$=0.11-0.51, $N$=1-54;
- for $p/e>10$ и $\theta>45°$: the database contains 1280 experimental points in the following range $Re_D=3\cdot 10^3$-$4{,}2\cdot 10^5$; Pr=0,7-8, $p/e$=10,72-81,71, $p/D$=0.14–4; $e/D$=0.01-0.21; $\theta$=46-90°, $\theta/90$=0.51-1.0, N=1-54.

Consequently, the experimental data deviation from correlations (5) and (6) considerably less than for correlations (3), (4) and similar, one obtained in A. Bergles and T. Ravigururajan [1]. The remained deviations, in the authors' view, might be associated with the fact that the rib shape influence on the friction factor augmentation had not been taken into consideration in the scope of this work. The rib shape impact on flow characteristics was stated in the studies of Hijikata [1987], Dreizer [1990], Tarasevich et al. [2015]. To date, however, there are no robust methods for accounting the rib shape influence on the flow structure. The other key factor affecting the accuracy of the correlations is the presence of the data related to the flow in pipes the wire coil inserts.

The wire coil inserts is a frequently used heat transfer augmentation technique. However, at low axial pitch values, the contact between the wall surface and the wire may be interrupted. This leads to the friction value change due to the flow underneath the wire inserts. Flow structure downstream the rib changes. This leads to the deviation of the friction factor and Nusselt number at the related pipe geometry in the database.

## 3. Conclusion

General correlations for friction factor and Nusselt number for the flow in helical-ribbed pipes are presented. Reported correlations are useful in the heat exchanger design within a wide range of rib geometry. The work was implemented as a part of the Russian Scientific Foundation grant number 19-19-00259.

**References**

[1] Ravigururajan TS, Bergles AE. 1996 Experimental Thermal and Fluid Science **13** 55–70
[2] Webb RL, Eckert ERG, Goldstein RJf 1971 International Journal of Heat and Mass Transfer **14** 601–617.
[3] Skrypnik A. N. et al. 2018 Journal of Engineering Physics and Thermophysics **91** 52-63.
[4] Popov I. A. et al. 2018 Journal of Physics: Conference Series. – IOP Publishing **980** 012004.
[5] Vicente P.G, Garcıa A., Viedma A. 2004 International Journal of Heat and Mass Transfer **47** 671–681.

5th International Workshop on Heat/Mass Transfer Advances
for Energy Conservation and Pollution Control
August 13-16, 2019, Novosibirsk, Russia

# Simulation of flows with phase transitions and heat transfer using mesoscopic methods

**A L Kupershtokh[1] and D A Medvedev[2]**

[1] National Research Novosibirsk State University, 2 Pirogova str., Novosibirsk, 630090, Russia
[2] Lavrentyev Institute of Hydrodynamics SB RAS, 15 Lavrentyev prosp., Novosibirsk, 630090, Russia

E-mail: skn@hydro.nsc.ru

**Abstract.** We use mesoscopic lattice Boltzmann and phase-field methods to simulate the growth of crystals from supercooled melt in the presence of melt convection and the behavior of a pinned droplet under the action of the electric field. In the first problem, the flow influences significantly the shape and the stability of growing patterns, leading to the enhanced development of fingers in the direction opposite to the flow. In the second problem, after the application of the electric field, the droplet begins to elongate in the field direction and the oscillations are produced. These oscillations decay in time due to a viscosity of a fluid. After several oscillations, the droplet acquires its equilibrium shape. The contact angle is essentially reduced compared to the case without an electric field.

## 1. Introduction
Flows of fluid with phase transitions (liquid-solid and liquid-vapor) are very important for both science and industry. Simulation of such flows is a complicated problem due to the evolution of phase boundaries. Mesoscopic methods that avoid boundary tracking are very promising for these applications.

In the present paper, we consider two problems. The first one is the growth of dendrites in the presence of fluid convection. This is a free boundary problem where boundary conditions need to be imposed at the surface with the shape which changes in time. In this case, the phase-field method is currently the choice for simulating the phase transformation [1]. Being mesoscopic by nature, this method is naturally combined with the lattice Boltzmann (LB) method for fluid flow simulation [2].

The second problem is the non-stationary three-dimensional dynamics of a pinned dielectric droplet in an electric field. Here, there is a complex interplay between the forces acting on the liquid (electric forces, gravity forces, surface tension, wetting of the substrate). These forces affect the shape of the droplet boundary. On the other hand, the distribution of the electric field, the electric and the capillary forces are determined by the droplet shape. Again, the mesoscopic LB method is very promising for simulating this process.

## 2. Lattice Boltzmann method
Lattice Boltzmann method considers the fluid flows as an ensemble of pseudoparticles that can move along the links of the 3D lattice with possible velocity vectors $\mathbf{c}_k$. We use the D3Q19 LB model ($k = 0,\ldots,18$). The distribution functions $N_k$ in each node are the main variables which are transferred

to the neighbor nodes of 3D lattice in accordance with the characteristics of differential equations $\mathbf{e}_k = \mathbf{c}_k \Delta t$. Then, the distribution functions are changed in a node due to the collision operator with the relaxation time $\tau$. The evolution equation for the distribution functions has the form

$$N_k(\mathbf{x} + \mathbf{c}_k \Delta t, t + \Delta t) = N_k(\mathbf{x}, t) + (N_k^{eq}(\rho, \mathbf{u}) - N_k(\mathbf{x}, t)) / \tau + \Delta N_k . \quad (1)$$

The hydrodynamic variables (the density $\rho$ and the velocity $\mathbf{u}$) can be calculated as the moments of the distribution functions $\rho = \sum_{k=0}^{b} N_k$ and $\rho\mathbf{u} = \sum_{k=1}^{b} \mathbf{c}_k N_k$. The process repeats at every time step. The actions of volume forces are taken into account as a change of distribution functions $\Delta N_k$ using the Exact Difference Method [3,4]. The pseudopotential method is used to simulate the liquid-vapor phase transition [5,6]. The total force acting on the fluid in a node from the matter in neighbor nodes has the form $\mathbf{F} = -\nabla(P(\rho,T) - \rho\theta)$, where $P(\rho,T)$ is the equation of fluid state.

## 3. Growth of dendrites

We simulate the thermally controlled growth of dendrites in the presence of fluid convection. The phase-field method [1] is used to calculate the phase transformation. The heat transport (convective and by heat conduction) is simulated using the additional LB component with zero mass (passive scalar) [7,8] taking into account the latent heat released during solidification. The symmetric model is used; the thermal diffusivity and the density of the liquid and the solid phases are the same. The solid-liquid interaction is described by dissipative forces acting in partially filled regions [9,10]. The two-dimensional case is considered with the four-fold anisotropy of the surface tension $\sigma = \sigma_0(1 + \varepsilon\cos 4\theta)$, ε is the degree of anisotropy. The growth of patterns is determined by the surface stiffness, which is defined as $\sigma + d^2\sigma / d\theta^2 = \sigma_0(1 - 15\varepsilon\cos 4\theta)$. The kinetic undercooling is neglected.

Figure 1 shows the simulation results for the growth of patterns in a shear flow (flow is directed from left to right). The left panel presents the results for different flow velocities for the non-dimensional initial undercooling Δ = 0.7 ($\Delta = c_p(T_m - T_0)/L$ where $c_p$ is the specific heat, $T_m$ is the melting temperature, and $L$ is the latent heat of melting), the anisotropy of surface stiffness 15ε = 0.15. The reduced velocity is calculated as $\tilde{U} = Ud_0 / D$, where $d_0 = \sigma_0 c_p T_m / L^2$ is the average capillary length with the average surface tension $\sigma_0$, $D$ is the thermal diffusivity. The Prandtl number is Pr = ν/$D$ = 1.78. Here, ν is the kinematic viscosity of the melt. Contours represent the boundary of the growing pattern at a different time (the difference between neighbor contours is 50-time steps). For these values of undercooling and anisotropy, the pattern shape is dendritic. With the increase of the flow velocity, the arm perpendicular to the flow direction is somewhat inclined to the inflow direction. Also, the growth of side branches at the upstream side is enhanced due to the supply of cold melt.

When the initial undercooling is increased to 0.8 keeping the same value of anisotropy, a seaweed pattern is obtained which consists of pair fingers (doublons), which is shown in the right panel of Fig. 1. Here, the increment between neighbor contours is 20-time steps. In this case, the influence of flow is much more pronounced. The perpendicular arm bends upstream and becomes parallel to the flow for high flow velocity. The additional side branches also appear.

Simulation of the crystal growth from the melt with natural thermal convection shows the enhanced propagation of arms pointing downwards and the hindering of upward arms. This effect is also due to the supply of cold melt in the first case and heated liquid in the second case.

In the case of a parallel flow, the growth of the finger is enhanced when the flow velocity increases, and the tip radius becomes smaller. For higher flow velocity, however, the sidebranching is observed and the instability of a needle crystal. For high undercooling, the increase of the flow velocity leads to the selection of doublons instead of dendrites.

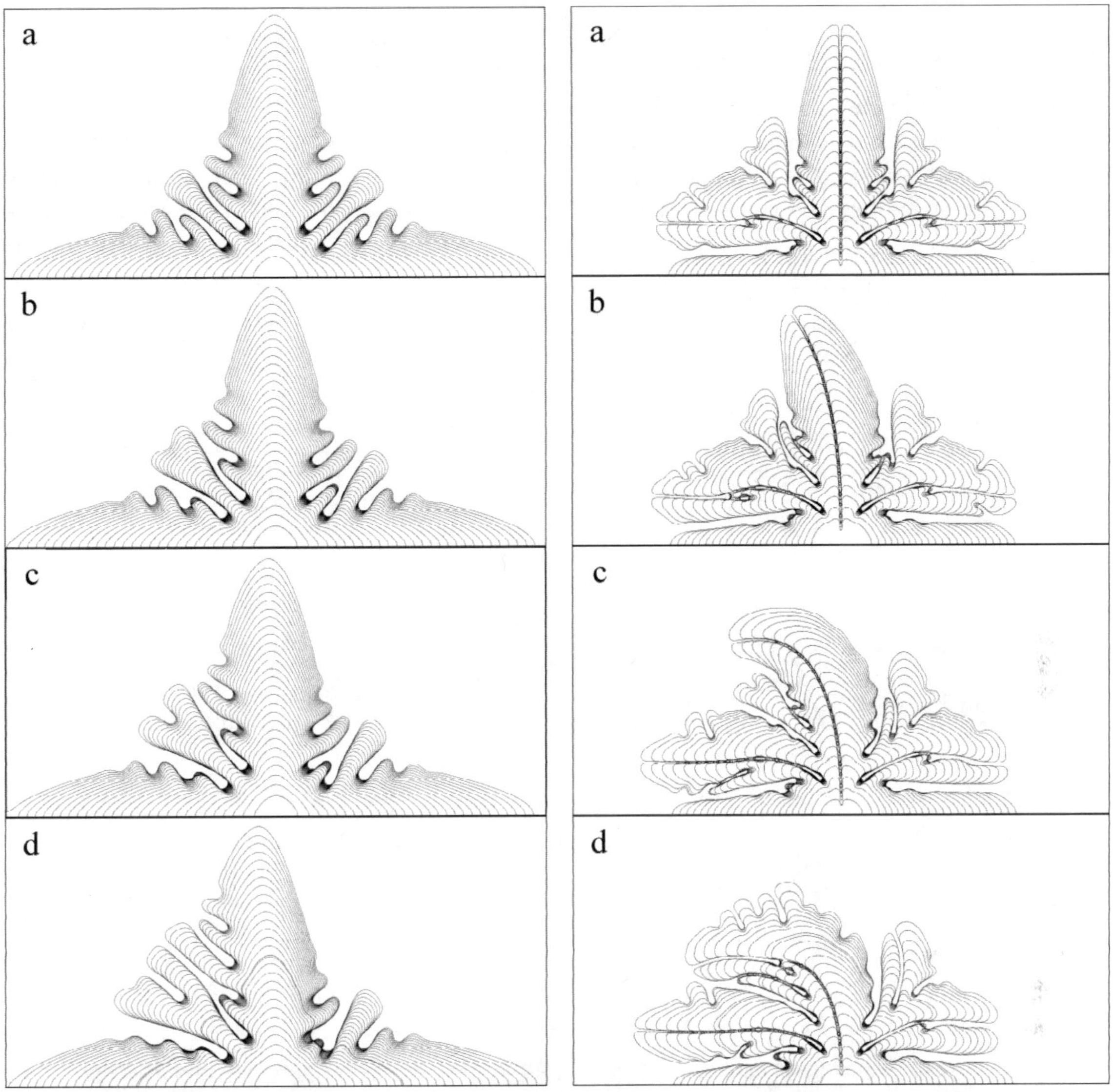


Fig. 1. Left: dendrite in a shear flow. Reduced velocity $\tilde{U}=0$ (a), 0.0123 (b), 0.0247 (c), 0.0493 (d). Right: seaweed pattern in a shear flow. Reduced velocity $\tilde{U}=0$ (a), 0.0247 (b), 0.0493 (c), 0.0987(d).

## 4. Dielectric droplet in an electric field

We also simulate the non-stationary three-dimensional dynamics of dielectric droplets in the electric field. The LB method is used for simulations of the pinned droplet. The position of a contact line for pinned droplet does not move. The van der Waals equation of state in the reduced variables is used for two-phase fluids in accordance with [6]. The dielectric permittivity of a liquid is $\varepsilon_l = 2.3$. The value of the Bond number $\mathrm{Bo} = \rho g R_0^2 / \sigma$ is equal to 0.67. The electrical Bond number $\mathrm{Bo}_e = \varepsilon_0(\varepsilon_l - 1)E^2 R_0 / \sigma$ is equal to 18.9. One of the simple models assumes the interaction forces between a fluid node $\mathbf{x}$ and the nearest five nodes on the solid surface in the form

$$\mathbf{F}(\mathbf{x}) = B\Phi(\mathbf{x})\sum_{k=1}^{5} w_k\,\Phi_{\mathrm{solid}}(\mathbf{x}+\mathbf{e}_k)\cdot\mathbf{e}_k \tag{2}$$

Here, $\Phi(\mathbf{x}) = \sqrt{\rho\theta - P(\rho,T)}$. For the pinned droplet, the value $\Phi_{\mathrm{solid}}(\mathbf{x}+\mathbf{e}_k)$ on the solid surface is equal to value $\Phi(\rho_l)$ in the circle of the predefined radius $R_0$. In general case, the parameter $B$ allows one to control the value of the wettability of the solid surfaces.

We solve Poisson's equation for the distribution of the electric field potential $\varphi$ between the flat electrodes at each time step using the well-known method of simple iterations. The periodic boundary conditions are used along the $x$ and $y$ coordinates. The value of the potential at the upper boundary of the calculation domain is $\varphi(x, y, L_z) = -V$. At the solid substrate, we have $\varphi(x, y, 0) = 0$. The electric field is calculated as $\mathbf{E} = -\nabla\varphi$.

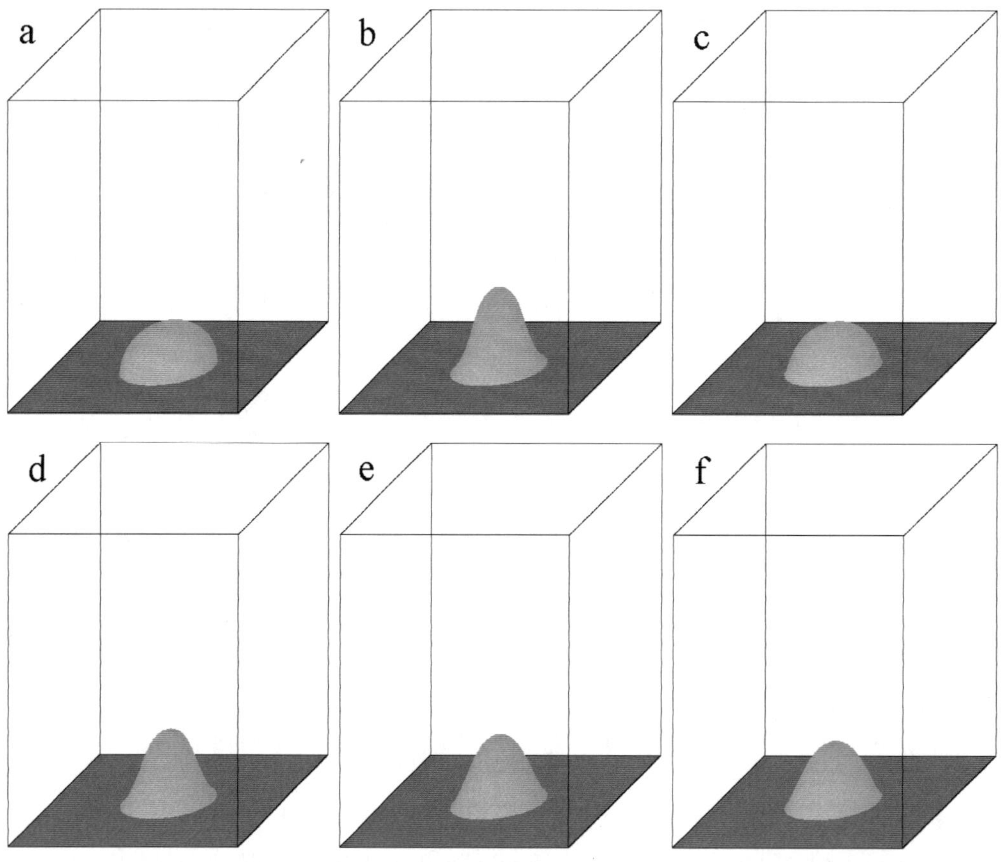


Fig. 2. The dynamics of the pinned droplet on the solid surface in the electric field. The initial hemispherical (a), the first maximum (b), the second maximum (d), the third maximum (e) and the final (f) shapes of the droplet. 3D lattice 400×400×544. $t = 0$ (a), 5400 (b), 11600 (c), 17200 (d), 27400 (e), 114000 (f). $R_0 = 80$, Bo = 0.67, $\varepsilon_l = 2.3$. $\mathrm{Bo}_e = 18.9$, Oh = 0.030.

Initially, the droplet is hemispherical (Fig. 2,a and Fig. 4, curve 1). After the electric field is applied, the droplet begins to elongate in the direction of the electric field. The maximum height of the droplet at the first oscillation is shown in Fig. 2,b. After several oscillations (Fig. 3), the droplet acquires its equilibrium shape. The final shape of the prolate droplet is shown in Fig. 2,f and in Fig. 4 (curve 2). Moreover, for the pinned droplet, the contact angle reduces. The similar experimental results were obtained in [11] for almost hemispherical conducting droplet and for the contact angle without electric field close to 90 degrees. However, the numerical results were obtained in [11] for the pinned droplet only as of the static solution.

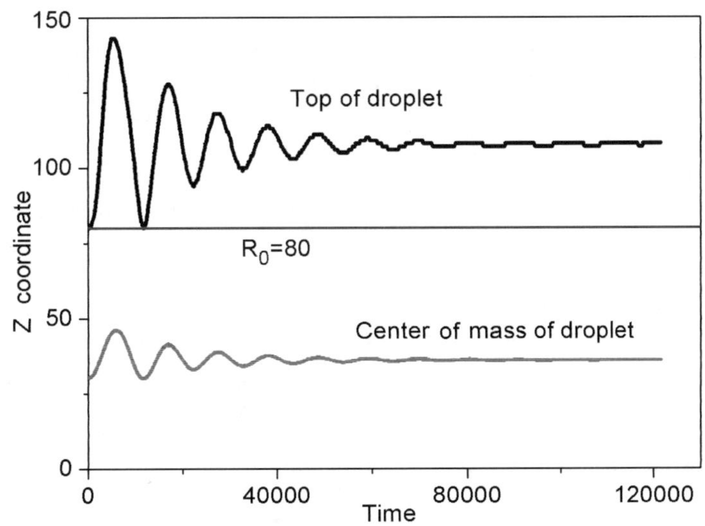


Fig. 3. Time dependences of the positions of the droplet top and the center of mass. $R_0 = 80$, Bo = 0.67, $\varepsilon_l = 2.3$. $\mathrm{Bo}_e = 18.9$, Oh = 0.030.

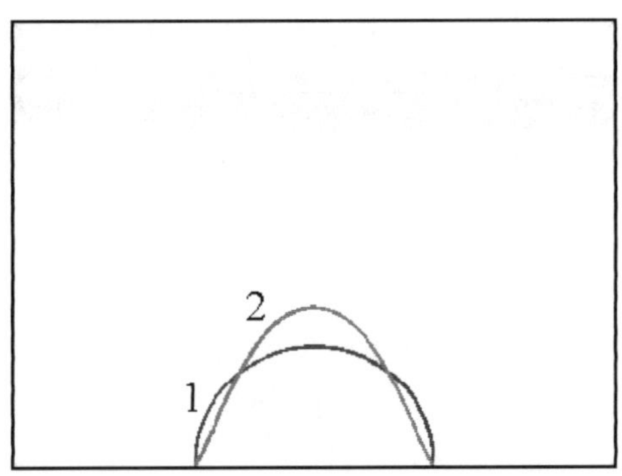


Fig. 4. The part of the central cross-section of the pinned droplet on the solid surface in the electric field. The initial hemispherical (curve 1) and the final (curve 2, $t = 114000$) shapes of the droplet. 3D lattice 400×400×544.

## 5. Conclusion

We use mesoscopic lattice Boltzmann and phase field methods to simulate flows with phase transitions, liquid-solid, and liquid-vapor ones. First, we investigate the growth of dendrites into a supercooled melt with different flows of liquid. The simulations show the strong influence of the shear flow on the shape and the stability of growing patterns, especially for low anisotropy, when seaweed patterns are present. The growth is enhanced in the direction parallel to the flow and hindered in the opposite direction due to the supply of the cold melt by the flow. For the natural convection and the parallel flow, the flow changes significantly the morphology of the patterns obtained.

Simulations of the behavior of a pinned droplet under the action of electric field reveal the initial oscillations of the droplet due to electric forces. Later, the droplet acquires the elongated shape, and the contact angle is decreased compared with the case without the electric field.

**Acknowledgements**
The study was supported by the Russian Science Foundation (grant No. 18-19-00538).

Journal of Physics: Conference Series **1369** (2019) 012065 doi:10.1088/1742-6596/1369/1/012065

**References**

[1] Karma A and Rappel W-J 1996 Phase-field method for computationally efficient modeling of solidification with arbitrary interface kinetics *Phys. Rev. E* **53** R3017–20

[2] Qian Y H, d'Humières D and Lallemand P 1992 Lattice BGK models for Navier–Stokes equation *Europhys. Lett.* **17**(6) 479–84

[3] Kupershtokh A L 2004 New method of incorporating a body force term into the lattice Boltzmann equation *Proc. 5th Int. EHD Workshop* (University of Poitiers, Poitiers, France) pp 241–6

[4] Kupershtokh A L 2010 Criterion of numerical instability of liquid state in LBE simulations *Computers and Mathematics with Applications* **59**(7) 2236–45

[5] Qian Y H and Chen S 1997 Finite size effect in lattice-BGK models *Int. J. Modern Physics C* **8**(4) 763–71

[6] Kupershtokh A L, Medvedev D A and Karpov D I 2009 On equations of state in a lattice Boltzmann method *Computers and Mathematics with Applications* **58**(5) 965–74

[7] Shan X 1997 Simulation of Rayleigh–Bénard convection using a lattice Boltzmann method *Phys. Rev. E* **55** 2780–8

[8] Kupershtokh A L, Medvedev D A and Gribanov I I 2018 Thermal lattice Boltzmann method for multiphase flows *Phys. Rev. E* **98**(2) 023308

[9] Beckermann C, Diepers H-J, Steinbach I, Karma A and Tong X 1999 Modeling melt convection in phase-field simulations of solidification *J. Comput. Phys.* **154** 468–96

[10] Medvedev D and Kassner K 2005 Lattice Boltzmann scheme for crystal growth in external flows *Phys. Rev. E* **72** 056703

[11] Corson L T, Tsakonas C, Duffy B R, Mottram N J, Sage I C, Brown C V and Wilson S K 2014 Deformation of a nearly hemispherical conducting drop due to an electric field: Theory and experiment *Phys. Fluids* **26**(12) 122106

# Calculations of shock-wave flow structure in axisymmetric channel with near-wall ethylene burning with throttle air jet

**V P Zamuraev**[1,2,a], **A P Kalinina**[1, 2,b], **DV Fedorova**[1,3]**andD.SPopova**[1,3]

[1]Khristianovich Institute of Theoretical and Applied Mechanics,
Siberian Division, Russian Academy of Science,
4/1, Institutskaya str. 630090, Novosibirsk, Russia
[2]Novosibirsk State University,
2, Pirogova St., 630090, Novosibirsk, Russia
[3]Novosibirsk State Technical University
630073, Novosibirsk, Russia
[a]Corresponding's author: zamuraev@itam.nsc.r
[b]Corresponding's author: kalinina@itam.nsc.ru

**Abstract**. The CFD modelling of the ramjet start with low losses of the stagnation pressure was carried out. The basic feature of this start technology was the jet of compressed air creating effect of a throttle which aids in preliminary decelerating of the flow up to transonic velocities. The comparison with experiment was performed for the hydrogen burning. We studied the axial and near-wall supply of gas fuel. The physical mechanisms stimulating stable presence of transonic mode in the part with constant section were investigated within the frameworks URANS approach. It was shown that fuel supply blockage is a way to control near-wall supplying of gaseous fuel at considerably low pressure falls. And blocking the ignition processes is a way to control the axial supply at quite high pressure falls.

## 1. Introduction

Gas jet efflux is widely used to control the process of burning, including the methods to intensify burning in the ramjet combustion chambers at high flight Mach numbers [1-8]. The quality of mixing and minimizing the losses of full pressure are of special importance, but the stability of the process is important too. The works [1, 2] suggest the method to initiate operation of the ramjet combustion chamber. This method is a way to form the transonic range in the parts with constant section. The transonic mode is naturally unstable. So, here is the risk to realize the process of locking the channel. For this reason, of great importance is analyzing the factors that eventually could make the burning stable are of great importance.

The process of burning intensification [1, 2] is based on the interaction of gas mixture (non-mixed previously) with shock-wave structure produced by the throttle air-jet. One of the problem associated with the burning initiation method are selection of injector location and the its shape to supply the gas fuel. We showed that the near-wall burning of hydrogen supplied at low pressure through the wall [3] results in considerably high amplitude of parameters pulsation. It is caused by practically full blockage of fuel inflow by the zone of high pressure near the fuel aperture. On one hand, this mechanism results in to higher pulsations, on the other hand it prevents penetrating ample amount of fuel into the channel, which could cause the lockage of the channel. The numerical study represented in [3] showed that in case of axisymmetric supersonic supply of ethylene throw the injector with big diameter, the average amplitude of the pulsations parameters is quite low. The present investigation aims at obtaining physical parameters stabilizing the burning process in the system analogic to [1, 2] for near-wall and axisymmetric fuel supply. In spite of observing axial fuel supply, the process of burning can be considered as the "near-wall", because the shock wave from the throttle jet supplied through the wall plays an important role. In case of considerably big diameters of the channel and fuel supply along the axis, burning initiation is not possible. The comparison was done for ethylene burning. The numerical modeling was done within URANS approach. The problem formulation for numerical modeling and its objectivation are represented in [3] (the Reynolds averaged Navier – Stokes equations, (k-ε and SST k-ω turbulence modules). The same work describes the geometrical dimension of the system and the values of physical parameters of the system modeled (the initial condition is supersonic flow with Mach number M=2 in the part with constant section). Three variants to initiate burning in the channel are considered for axisymmetric supply of ethylene. These examples show that the process of locking channel can be prevented by three physical mechanisms: the same one of pressure difference change due to burning, that in the case of supplying through the wall, weak mixing processes, and limiting the consumption of inflowing fuel by the injector shape.

## 1. Comparison with experiment data

Quite detailed description of the experiment being compared in order to verify the methods chosen for numerical modeling is represented in [1-3].

The pressure distribution obtained numerically along the length of the flat or axisymmetric channel was compared to the experimental distribution [2] for the channels with the same cross dimensions. The region between the jet supplying hydrogen and the throttling jet in numerical simulation was somewhat shorter, so we make a comparison only in this area. The frequency of 10 Hz corresponded to the time interval of 100 ms, the gas flies at least 30 m, or 60 channel lengths during this time. Therefore, in a numerical simulation for 10 ms after the start of the gas-dynamic pulse, the effect can be considered as at constant pressure in the gas tanks.

As mentioned above, the combustion mode was pulsating. The curve 1 in Figure 2(a,b) shows the dependence of the static pressure on the coordinate along the channel for the flat and axisymmetric channel respectively. The pressure and the Mach number were obtained along the line outstanding 10 mm from the lower wall of the channel. The results [2] corresponding to the prechamber pressure 7 atm were presented in this figure by the curves2. The Mach number obtained by numerical simulation presented by the curves 3, and the lines 4 corresponds to the level of Mach number M = 1. One can see that there are pulsating transonic mode with a Mach number value which is approximately equal to M =1 for both of cases (flat and axisymmetric channel). In the flat channel (Figure 2(a)) the minimum value of the Mach number is 0.8, the maximum is 1.5 and average value is approximately M = 1.2. In the axisymmetric channel (Figure 2(b)) the minimum value of the Mach number is 0.5, the maximum is 1.5 for the flat channel and average value is approximately M = 1.

The average value of the Mach number in the flat channel is more than in the axisymmetric one. The reason is the difference in the square area of the slot zone. A throttling jet acts on the subsonic flow similar to the Laval nozzle, which results in flow accelerating again to the supersonic speed. The pressure ranges corresponding to the experiment overlaps with pressure range obtained numerically, but the average pressure level in the experiment is lower than in the calculation. It can be seen from the comparison of numerical simulation and experimental data that when the pressure in the channel increases (curve 1) the Mach number decreases (curve 3) and vice versa. The range of experimental pressures intersects with the calculated range, but on average, the experimental values are approximately the same for the case of the flat channel and are smaller for the axisymmetric channel. One can conclude that the flow regime in [2] is also transonic, but the values of Mach numbers are greater than M=1 and approximately equal to the M=1.2. The pressure level and its decrease at the end of the channel part with constant section are similar in the calculation and experiment.

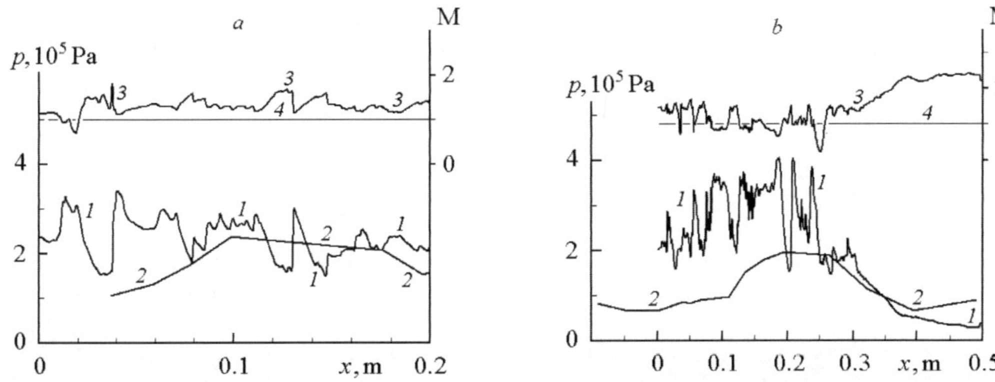

**Figure 2.** The distributions of pressure calculated (curve *1*) and experimental (curve *2*) pressure distributions. Mach number distribution (curve *3*), line *4* indicates level M = 1; a,b – hydrogen burning in the flat and axisymmetric channel respectively.

## 2. The axial and near-wall fuel supply comparison

For axisymmetric fuel supplying we studied three variants (the fuel temperature always was 300K). In the first case ethylene was supplied from the injector with inner diameter 10 mm at pressure 6 atm in the gas-generator. The ethylene consumption was so high, that if all ethylene supplied was burned, the channel would be blocked inevitably. No burning was observed in case of absence of throttle jet burning. When supplying the throttle jet a shock wave arose initiating burning in the circular zone close to the zone of oxygen contact with ethylene. No intensive gas motion along the radius was observed. Due to bad mixing, the biggest part of ethylene near the radius was not involved into burning process. The perturbation was moving upstream up to the injector hole. However, the channel was not locked, because the throttling jet formed Laval gas-dynamic nozzle. Near the throttling jet the pressure value was fluctuating close to the pressure level in the gas-generator. Here we could observe the manifestation of the same physical mechanism as in the case of near-wall fuel jet but in relation to the drossel jet. When much heat is generated, the pressure arises, and the throttling jet stops outflowing.

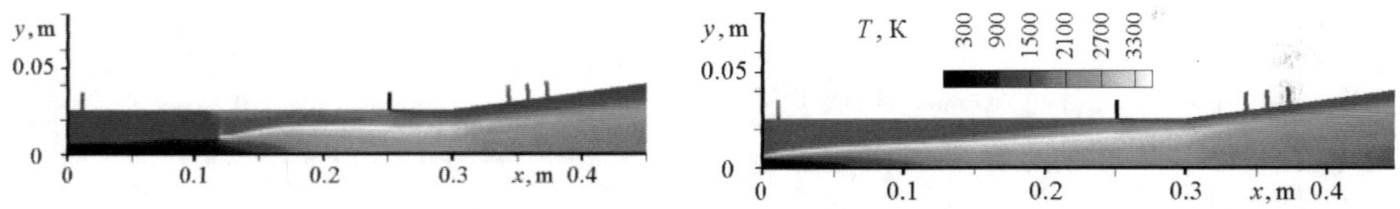

**Figure 3**. Temperature distribution at the beginning of process and in the quasi-stationary solution.

Absence of the throttling jet resulted in reducing burning intensity. The heat extraction was reduced; the pressure in the channel was lowered. But thank to this, the throttling jet could outflow again. And it initiated consequently the burning process. As a result, the transonic flow (M≈0.7-0.8) was formed in the circular zone. Weak oscillations were observed. The flow reached supersonic velocity due to gas-dynamic Laval nozzle created by the throttle-jet. The Fig. 3a and 3b show the distribution of temperature at the ignition and end of the process. Burning continued in the expanding part of the channel.

In the second studied case homogenous mixture of ethylene-oxygen was supplied to the channel at the beginning of the process. Then, after the ignition was initiated when the mixture met the throttling jet, pure ethylene was supplied as in the first variant. This process is pictured in Fig.4. Starting from $t$ = 3.34, pure ethylene was supplied from the output of the injector in order to prevent lockage of the channel. As in the first case, we didn't observe burning before the throttling jet contacted the shock wave. But in case of contact with throttling jet intensive burning arose with intensive gas turbulence, which is shown in Fig. 5-12.The, when proceeding to pure ethylene the process character became the same as in the first case studied.

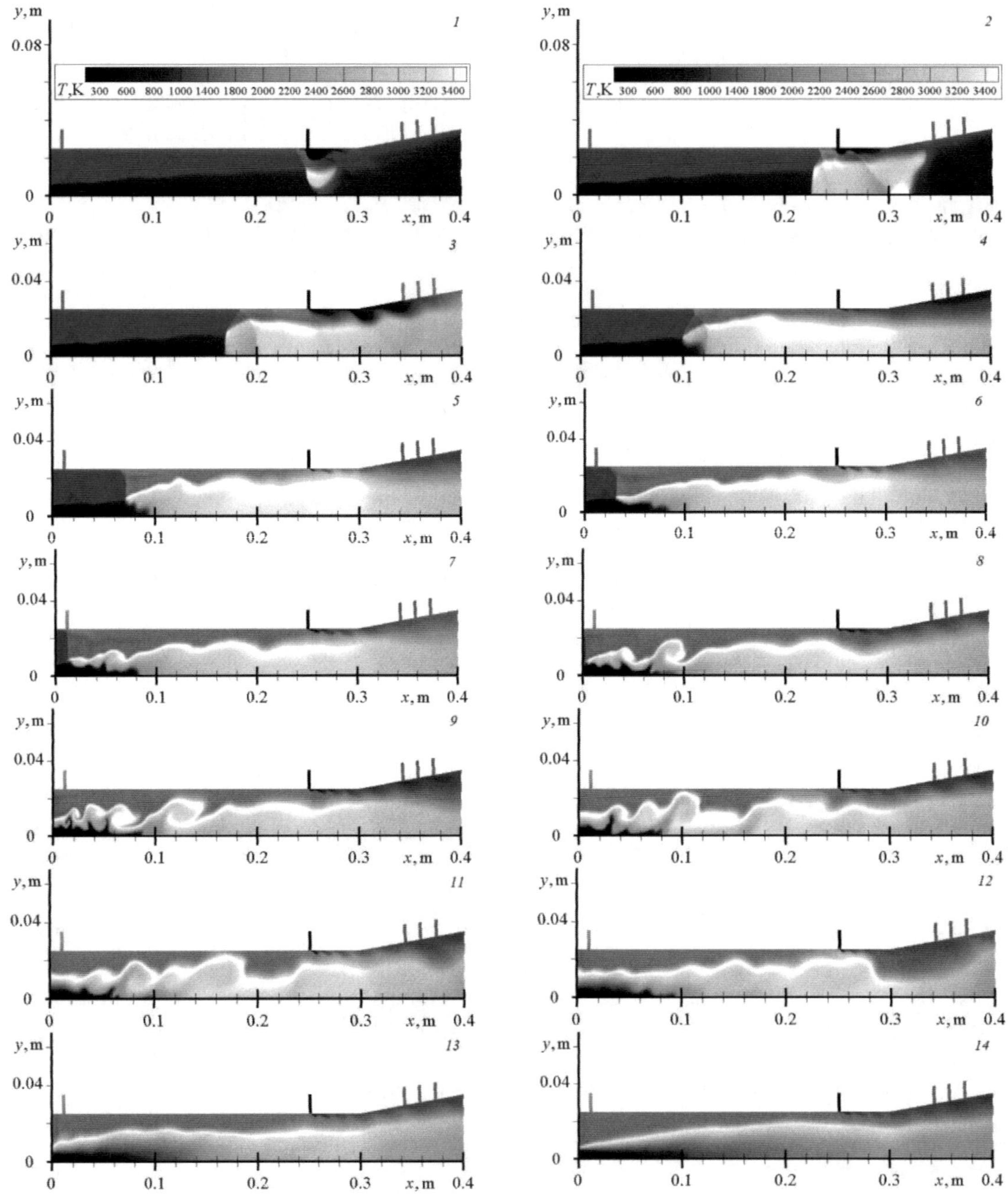

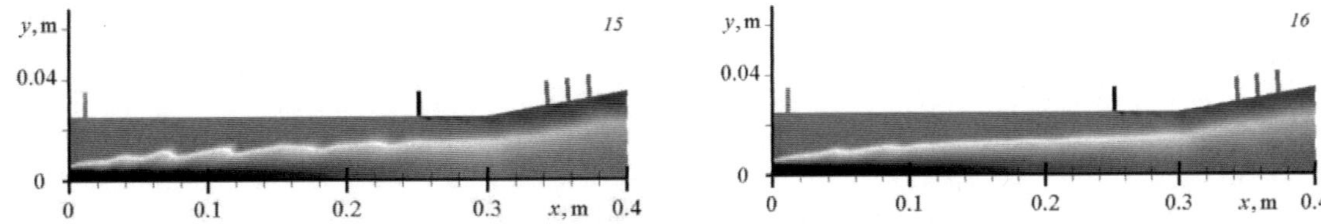


**Figure 4.** The time evolution of 2D temperature distribution for the axial ethylene supply: *1* – *t*(ms)= 2.14, *2* – 2.17, *3* – 2.25, *4* – 2.43, *5* – 2.58, *6* – 2.76, *7* – 2.88, *8* – 2.98, *9* – 3.07, *10* – 3.22, *11* – 3.34, *12* – 3.48, *13* – 3.74, *14* – 4.01, *15* – 4.78, *16* – 5.36

In the third case, we studied flowing of pure ethylene from the injector with small inner diameter 1.8 mm, and external diameter 10mm. The pressure in gas generator was about 15 atm, but thanks to small inner diameter the ethylene consumption was considerably low. Due to the big external diameter near the injector output the vortex zones were formed, which was resulted by ethylene going downstream mixed well with air. There was not burning before meeting the shock wave from the throttle jet. When the mixture contacted the shock wave, the process character was similar to the case with oxygen-ethylene mixture: considerable temperature non-uniformities were formed in transversal direction. These non-uniformities lasted during all the process. The Fig.5 represents two time points at the beginning of the process. The physical mechanism of process control (connected to the pressure difference between the gas generator of throttling jet and main stream in the channel) was acting here too. The inflow of ethylene into the channel was limited by the small inner injector diameter. Let us note, that in all three cases of axisymmetric supply the fuel consumption at the injectior output didn't change considerably.

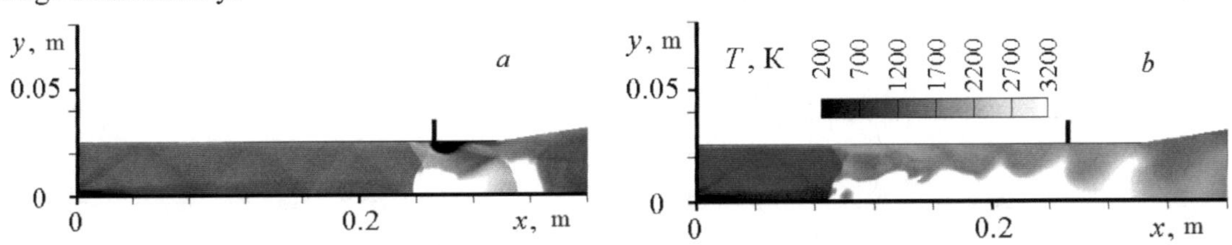


**Figure 5.** Temperature distribution for the case for injector with small inner diameter: *a* – *t*(ms) = 2.14, *b* – 2.17.

## 3. Conclusion

The comparison of the experiment and numerical data was performed for the case of the hydrogen burning for near-wall fuel supply for the flat and axisymmetric channel. The satisfactory coincidence of the numerical and experimental results was observed. It is shown that the ignition occurs due the interaction of shock wave from the throttle jet with fuel mixture, in spite of supplying method, disposition, and injector shape. In case of near-wall supply the heat extraction is limited due to blocking fuel inflow. In case of axisymmetric supply the heat extraction is limited by blockage of ignition processes. The scale of transversal temperature non-uniformities and other parameters rises sharply with increasing mixing degree of ethylene and air. The process of locking can be prevented by limiting fuel consumption or by low level of mixing. Nevertheless, in all cases the transonic mode is formed.

**Acknowledgments**

The work was performed under the program of fundamental scientific research of state academies of sciences for 2013–2020 (project AAAA-A17-117030610135-6) and with the financial support of the RFBR (grant No. 17-08-00183).

The authors gratefully acknowledge Christopher Hearsey and Valeria Hearsey for their assistance.

## References

[1] Tretyakov P, Zabaykin V and Prokhorov A 2015 *Proc. Russian Congress on fundamental problems of theoretical and applied mechanics* ed. Akhmetov D (Kazan: Kazan) p. 3778
[2] Abashev V, Korabelnikov A, Kuranov A and Tretyakov P 2017 *AIP Conference Proceedings* **1893** 020005
[3] Zamuraev V. et al 2019 *J. Phys.: Conf. Ser.* **1268** 012024
[4] Surzhikov S 2014 *Physical-chemical kinetics in gas dynamics* **16** 3
[5] Zamuraev V and Kalinina A *Vestnik NSU Series: Physics.* 2015, **10** 26
[6] Firsov A, Yarantsev D, Leonov S and Ivanov V *Computer research and modeling*, 2017, **9** 75
[7] Hudic G, Zhukov V, Manukovskiy K, Novikov N, Rykov Yu and Feodoritova O 2016 *Numerical simulation of high speed combustion using OpenFOAM package* 10 (Preprint IPM im. M. V. Keldysh) p 32
[8] Zamuraev V and Kalinina A 2018 *Thermophysics and Aeromechanics* **25**

# AUTHOR INDEX

# AUTHOR INDEX